Lecture Notes in Computer S

Edited by G. Goos, J. Hartmanis and J. van Leeuwen

Advisory Board: W. Brauer   D. Gries   J. Stoer

Laurent Fribourg   Franco Turini (Eds.)

Logic Program Synthesis and Transformation – Meta-Programming in Logic

4th International Workshops
LOPSTR '94 and META '94
Pisa, Italy, June 20-21, 1994
Proceedings

Springer-Verlag
Berlin Heidelberg New York
London Paris Tokyo
Hong Kong Barcelona
Budapest

Series Editors

Gerhard Goos
Universität Karlsruhe
Vincenz-Priessnitz-Straße 3, D-76128 Karlsruhe, Germany

Juris Hartmanis
Department of Computer Science, Cornell University
4130 Upson Hall, Ithaka, NY 14853, USA

Jan van Leeuwen
Department of Computer Science, Utrecht University
Padualaan 14, 3584 CH Utrecht, The Netherlands

Volume Editors

Laurent Fribourg
LIENS-CNRS
45 rue d'Ulm, F-75005 Paris, France

Franco Turini
Dipartimento di Informatica, Università di Pisa
Corso Italia, 40, I-56125 Pisa, Italy

CR Subject Classification (1991): F.4.1, F.3.2-3, I.2.2, I.2.4, D.3.1, D.3.4

ISBN 3-540-58792-6 Springer-Verlag Berlin Heidelberg New York

CIP data applied for

© Springer-Verlag Berlin Heidelberg 1994
Printed in Germany

Typesetting: Camera-ready by author
SPIN: 10479332   45/3140-543210 - Printed on acid-free paper

Preface

This volume contains the proceedings of the LOPSTR'94 and META'94 workshops, which were held in Pisa on June 20-21, 1994. The two workshops took place in parallel and were attended by 69 persons coming from 16 countries.

LOPSTR'94 is the Fourth International Workshop on Logic Program Synthesis and Transformation. Previous workshops had taken place in Manchester (1991, 1992) and in Louvain-la-Neuve (1993). This year, 19 submitted papers were selected for presentation in Pisa, among which 15 were accepted for inclusion in these proceedings, after some revisions had been made by the authors. The main topics addressed by the papers are: unfolding/folding, partial deduction, proofs as programs, inductive logic programming, automated program verification, specification and programming methodologies.

META'94 is the Fourth International Workshop on Metaprogramming in Logic. Previous workshops had taken place in Bristol (1988), in Leuven (1990), and in Uppsala (1992). This year, 21 papers were submitted and 11 were accepted for presentation in Pisa and inclusion in the proceedings. The main topics addressed by the papers are: language extensions in support of Meta-logic, semantics of Meta-logic, implementation of Meta-logic features, performance of Meta-logic, and applications of Meta-logic to non-monotonic reasoning, program control, and programming in the large,

In addition to LOPSTR-META regular contributions, two invited speakers gave very insightful and appreciated talks in Pisa: Giorgio Levi on "Abstract debugging of logic programs"(included in this book) and Burton Dreben on "Herbrand's contributions to logic ".

We would like to thank all the members of the Program Committees as well as the referees for their thorough job of reviewing.

We also express our gratitude to all the members of the local organizing committee for their help in organizing these successful events.

We gratefully acknowledge the financial support of the Compulog Network of Excellence.

September 1994   Laurent Fribourg, Franco Turini

Organizing Committee:
Antonio Brogi, Paolo Mancarella, Dino Pedreschi, Franco Turini (U. Pisa)

Program Committee of LOPSTR:

Dmitri Boulanger (K.U. Leuven)
Gerard Ferrand (U. Orleans)
Pierre Flener (Bilkent U., Ankara)
Laurent Fribourg (LIENS-CNRS, Paris)(Chair)
Jean-Marie Jacquet (U. Namur)
Maurizio Proietti (IASI-CNR, Roma)
Franco Turini (U. Pisa)
Geraint Wiggins (U. Edinburgh)

Reviewers for LOPSTR:

Michel Bidoit	Tadashi Kanamori
Paolo Ciancarini	Baudouin Le Charlier
Ilyas Çiçekli	Christophe Lecoutre
Koen De Bosschere	Vincent Lombart
Pierre Deransart	Anne Mulkers
Danny De Schreye	Guy Alain Narboni
Philippe Devienne	Chet Murthy
Yves Deville	Anne Parrain
Alexander Dikovsky	Sophie Renault
Maurizio Gabrielli	Christel Vrain

Program Committee of META:

L. Aiello (Rome, Italy)	J. Barklund (Uppsala, Sweden)
K. Bowen (Syracuse, USA)	D. De Schreye (Leuven, Belgium)
Y. Jiang (London, UK)	J. Komorowski (Trondheim, Norway)
G. Lanzarone (Milan, Italy)	A. Martelli (Turin, Italy)
L.M. Pereira (Lisbon, Portugal)	A. Pettorossi (Rome, Italy)
A. Porto (Lisbon, Portugal)	T. Sato (Tsukuba, Japan)
L. Sterling (Cleveland, USA)	S.A. Tarnlund (Uppsala, Sweden)
F. Turini (Pisa, Italy) (Chairman)	

Table of Contents

LOPSTR

META

Logic Frameworks for Logic Programs*

David A. Basin

Max-Planck-Institut für Informatik
Saarbrücken, Germany
Email:basin@mpi-sb.mpg.de

Abstract. We show how logical frameworks can provide a basis for logic program synthesis. With them, we may use first-order logic as a foundation to formalize and derive rules that constitute program development calculi. Derived rules may be in turn applied to synthesize logic programs using higher-order resolution during proof that programs meet their specifications. We illustrate this using Paulson's Isabelle system to derive and use a simple synthesis calculus based on equivalence preserving transformations.

1 Introduction

Background

In 1969 Dana Scott developed his Logic for Computable Functions and with it a model of functional program computation. Motivated by this model, Robin Milner developed the theorem prover LCF whose logic PPλ used Scott's theory to reason about program correctness. The LCF project [13] established a paradigm of formalizing a programming logic on a machine and using it to formalize different theories of functional programs (e.g., strict and lazy evaluation) and their correctness; although the programming logic was simple, within it complex theories could be developed and applied to functional program verification.

This paradigm can be characterized as *formal development from foundations*. Type theory and higher-order logic have been also used in this role. A recent example is the work of Paulson with ZF set theory. Although this theory appears primitive, Paulson used it to develop a theory of functions using progressively more powerful derived rules [24].

Most work in formalized program development starts at a higher level; foundations are part of an informal and usually unstated meta-theory. Consider, for example, transformation rules like Burstall and Darlington's well known fold-unfold rules [7]. Their rules are applied to manipulate formulas and derive new ones; afterwards

---

* This research was funded by the German Ministry for Research and Technology (BMFT) under grant ITS 9102. The responsibility for the contents lies with the author. I thanks Tobias Nipkow and Larry Paulson for advice on Isabelle. I also thank Alan Bundy, Ina Kraan, Sean Matthews, and members of the Edinburgh Mathematical Reasoning Group for helpful discussions and collaboration on related work.

some collection of the derived formulas defines the new program. The relationship of the new formulas to the old ones, and indeed which constitute the new program is part of their informal (not machine formalized) metatheory. So is the correctness of their rules (see [18, 8]). In logic programming the situation is similar; for example, [30, 29] and others have analyzed conditions required for fold-unfold style transformations to preserve equivalence of logic programs and indeed what "equivalence" means.

Development from Foundations in Logic Programming

We propose that, analogous to LCF, we may begin with a programming logic and derive within it a program development calculus. Derived rules can be applied to statements about program correctness formalized in the logic and thereby verify or synthesize logic programs. Logic programming is a particularly appropriate domain to formalize such development because under the declarative interpretation of logic programs as formulas, programs are formalizable within a fragment of first-order logic and are therefore amenable to manipulation in proof systems that contain this fragment. Indeed, there have been many investigations of using first-order logic to specify and derive correct logic programs [9, 10, 11, 17, 19]. But this work, like that of Burstall and Darlington, starts with the calculus rather than the foundations. For example in [17] formulas are manipulated using various simplification rules and at the end a collection of the resulting formulas constitutes the program. The validity of the rules and the relationship of the final set of formulas (which comprise the program) to the specification is again part of the informal meta-theory.

Our main contribution is to demonstrate that without significant extra work much of the informal metatheory can be formalized; we can build calculi from foundations and carry out proofs where our notion of correctness is more explicit. However, to do this, a problem must be overcome: first-order logic is too weak to directly formalize and derive proof rules. Consider for example, trying to state that in first-order logic we may replace any formula $\forall x.A$ by $\neg \exists x.\neg A$. We might wish to formulate this as $\forall x.A \rightarrow \neg \exists x.\neg A$. While this is provable for any instance A, such a generalized statement cannot be made in first-order logic itself; some kind of second-order quantification is required.[2] In particular, to formalize proof rules of a logic, one must express rules that (in the terminology of [15]) are *schematic* and *hypothetical*. The former means that rules may contain variables ranging over formula. The latter means that one may represent logical consequence; in the above example consequence has been essentially internalized by implication in the object language.

Rather than moving to a more powerful logic like higher-order logic, we show that one can formalize program development using weak logics embedded in logical frameworks such as Paulson's Isabelle system [25] or Pfenning's ELF [28]. In our

---

[2] First-order *logic* is too weak, but it is possible to formalize powerful enough first-order *theories* to express such rules by axiomatizing syntax, e.g., [32, 3, 23]. However, such approaches require some kind of reflection facility to establish a link between the formalized meta-theory and the desired theory where such rules are used. See [4] for a further discussion of this. Moreover, under such an approach unification cannot be used to identify program verification and synthesis.

work, a programming logic (also called the *object logic*) is encoded in the logic of the logical framework (the *meta-logic*). For example, the meta-logic of Isabelle, which we use, is fragment of higher-order logic containing implication (to formalize hypothetical rules) and universal quantification (to formalize schematic rules). Within this meta-logic we formalize a theory of relevant data-types like lists and use this to specify our notion of program correctness and derive rules for building correct programs. Moreover, Isabelle manipulates rules using higher-order unification and we use this to build programs during proof where meta-variables are incrementally instantiated with logic programs.

We illustrate this development paradigm by working through a particular example in detail. Within an Isabelle theory of first-order logic we formulate and derive a calculus for reasoning about equivalences between specifications and representations of logic programs in first-order logic. The derived calculus can be seen as a formal development of a logic for program development proposed in Wiggins (see Section 3.4). After derivation, we apply these rules using higher-order unification to verify that logic programs meet their specifications; the logic programs are given by meta-variables and each unification step during proof incrementally instantiates them.

Our experience indicates that this development is quite manageable. Isabelle comes with well-developed tactic support for rewriting and simplification. As a result, our derivation of rules was mostly trivial and involved no more than typing them in and invoking the appropriate first-order simplification tactics. Moreover, proof construction with these rules was partially automated by the use of Isabelle's standard normalization and simplification procedures. We illustrate this by developing a program for list subset.

2 Background to Isabelle

What follows is a brief overview of Isabelle [25, 26, 27] as is necessary for what follows. Isabelle is an interactive theorem prover developed by Larry Paulson. It is a logical framework: its logic serves as a meta-logic in which object logics (e.g., first-order logic, set theory, etc.) are encoded. Proofs are interactively constructed by applying proof rules using higher-order resolution. Proof construction may be automated using *tactics* which are ML programs in the tradition of LCF that construct proofs.

Isabelle provides a fine basis for our work. Since it is a logical framework, we may encode in it the appropriate object logic, first-order logic (although we indicate in Section 5 other possible choices). Isabelle's metalogic is based on the implicational fragment of higher-order logic where implication is represented by "==>" and universal quantification by "!!"; hence we can formalize and derive proof rules which are both hypothetical and schematic. Rules, primitive and derived, may be applied with higher-order unification during higher-order resolution; unification permits meta-variables to occur both in rules and proofs. We use this to build logic programs by theorem proving where the program is originally left unspecified as a higher-order meta-variable and is filled in incrementally during the resolution steps; the use of resolution is similar to the use of "middle out reasoning" to build logic programs as demonstrated in [20, 21, 6, 33].

Isabelle manipulates objects of the form[3] `[|F1; ...; Fn|] ==> F`. A proof proceeds by applying rules to such formulas which result in zero or more subgoals, possibly with different assumptions. When there are no subgoals, the proof is complete. Although Isabelle proof rules are formalized natural deduction style, the above implication can be read as an intuitionistic sequent where the `Fi` are the hypotheses. Isabelle has resolution tactics which apply rules in a way the maintains this illusion of working with sequents.

3 Encoding A Simple Equivalence Calculus

We give a simple calculus for reasoning about equivalence between logic programs and their specifications. Although simple, it illustrates the flavor of calculus and program development we propose.

3.1 Base Logic

We base our work on standard theories that come with the Isabelle distribution. We begin by selecting a theory of constructive first-order predicate calculus and augment this with a theory of lists to allow program development over this data-type (See *IFOL* and *List* in [27]). The list theory, for example, extends Isabelle's first-order logic with constants for the empty list "[ ]", cons ".", and standard axioms like structural induction over lists. In addition, we have extended this theory with two constants called `Wfp` (well-formed program) and `Def` with the property that `Wfp(P) = Def(P) = P` for all formulas `P`; their role will be clarified later.

The choice of lists was arbitrary; to develop programs over numbers, trees, etc. we would employ axiomatizations of these other data-types. Moreover, the choice of a constructive logic was also arbitrary. Classical logic suffices too as the proof rules we derive are clearly valid after addition of the law of excluded middle. This point is worth emphasizing: *higher-order unification, not any feature of constructivity, is responsible for building programs from proofs in our setting.*

In this theory, we reason about the equivalence between programs and specifications. "Equivalence" needs clarification since even for logic programs without impure features there are rival notions of equivalence. The differences though (see [22, 5]) are not so relevant in illustrating our suggested methodology (they manifest themselves through different formalized theories). The notion of equivalence we use is equivalence between the specification and a logic program represented as a *pure logic program* in the above theory. *Pure logic programs* themselves are equivalences between a universally quantified atom and a formula in a restricted subset of first-order logic (see [6] for details); they are similar to the *logic descriptions* of [12].

For example, the following is a pure logic program for list membership, where "." is *cons*.[4]

---

[3] We shall use `typewriter font` to display concrete Isabelle syntax.

[4] Unfortunately, "." is overloaded and also is used in the syntax of quantifiers; e.g., $\forall x\, y.\phi$ which abbreviates $\forall x.\forall y.\phi$.

$$\forall x\, y.p(x, y) \leftrightarrow (x = [] \land \mathit{False}) \lor (\exists v_0\, v_1.x = v_0.v_1 \land (y = v_0 \lor p(v_1, y))) \qquad (1)$$

Such programs can be translated to Horn clauses or run directly in a language like Gödel [16].

3.2 Problem Specification

As our notion of correctness is equivalence between programs and specifications, our proofs begin with formulas of the form $\forall \overline{x}.(spec(\overline{x}) \leftrightarrow E(\overline{x}))$. The variables in $\overline{x}$ represent parameters to both the specification $spec$ and the logic program E; we do not distinguish input from output. $spec$ is a first-order specification and E is either a concrete (particular) pure logic program or a schematic (meta) variable standing in for such a program. If E is a concrete formula then a proof of this equivalence constitutes a *verification* proof as we are showing that E is equivalent to its specification. If E is a second-order meta-variable then a proof of this equivalence that instantiates E serves as a *synthesis* proof as it builds a program that meets the $spec$. If $spec$ is already executable we might consider such a proof to be a *transformation* proof.

An example we develop in this report is synthesizing a program that given two lists l and m is true when l is a subset of m. This example has been proposed by others, e.g., [17, 33]. Slipping into Isabelle syntax we specify it as

```
ALL l m. (ALL z. In(z,l) --> In(z,m)) <-> ?E(l,m).
```

Note that ALL, --> and <-> represent first-order universal quantification, implication, and equivalence, and are declared in the definition of first-order logic. The "?" symbol indicates metavariables in Isabelle. Note that ?E is a function of the input lists l and m but z is only part of the specification. Higher-order unification, which we use to build an instance for ?E will ensure that it is only a function of l and m.

3.3 Rules

We give natural deduction rules where the conclusion explains how to construct ?E from proofs of the subgoals. These rules form a simple calculus for reasoning about equivalences and can be seen as a reconstruction of those of the Whelk system (see Section 3.4). Of course, since A <-> A is valid, the synthesis specification for subset can be immediately proven by instantiating ?E with the specification on the left hand side of the equivalence. While this would lead to a valid proof, it is uninteresting as the specification does not suggest an algorithm for computing the subset relation. To make our calculus interesting, we propose rules that manipulate equivalences with restricted right-hand sides where the right hand side can be directly executed.

Specifically, we propose rules that admit as programs (on the right hand sides of equivalences) formulas like the body of the membership predicate given above, but exclude formula like ALL z. In(z,l) --> In(z,m). To do this we define inductively the set of such admissible formulas. They are built from a collection of (computable) base-relations and operators for combining these that lead to computable algorithms provided their arguments are computable. In particular, our base relations are the relations True, False, equality and inequality. Our operators will be the propositional connectives and existential quantification restricted to a use like that in the membership example, i.e., of the form $\exists v_0 v_1. x = v_0.v_1 \wedge P$ where P is admissible. This limited use of existential quantification is necessary for constructing recursive programs in our setting; it can be trivially compiled out in the translation to Horn clauses.

Note that to be strictly true to our "foundations" paradigm, we would specify the syntax of such well-formed logic programs in our theory (which we could do by recursively defining a unary well-formedness predicate that captures the above restrictions). However, to simplify matters we capture it by only deriving rules for manipulating these equivalences where the right-hand sides meet these restrictions. To ensure that only these rules are used to prove equivalences we will resort to a simple trick. Namely, we wrap all the right hand sides of equivalences in our rule, and in the starting specification with the constructor Wfp. E.g., our starting goal for the subset proof would really be

```
ALL l m. (ALL z. In(z,l) --> In(z,m)) <-> Wfp(?E(l,m))).
```

As Wfp was defined to be the identity (i.e., Wfp(P) equals P) it does not effect the validity of any of the rules. It does, however, affect their applicability. That is, after rule derivation we remove the definition of Wfp from our theory so the only way we can prove the above is by using rules that manipulate equivalences whose right hand side is also labeled by Wfp. In particular, we won't be able to prove

```
ALL l m. (ALL z. In(z,l) --> In(z,m)) <-> Wfp(ALL z. In(z,l) --> In(z,m)).
```

Basic Rules Figure 1 contains a collection of typical derived rules about equivalences. Many of the rules serve simply to copy structure from the specification to the program. These are trivially derivable, for example

$$\frac{A \leftrightarrow \mathit{Wfp}(\mathit{ExtA}) \quad B \leftrightarrow \mathit{Wfp}(\mathit{ExtB})}{A \wedge B \leftrightarrow \mathit{Wfp}(\mathit{ExtA} \wedge \mathit{ExtB})}.$$

Translating this into Isabelle we have

```
[| A <-> Wfp(ExtA); B <-> Wfp(ExtB) |] ==> A & B <-> Wfp(ExtA & ExtB).
```

This rule is derivable (recall that Wfp(P) = P) in one step with Isabelle's simplification tactic for intuitionistic logic, so it is a valid rule. The rule allows us essentially to decompose synthesizing logic programs for a conjunction into synthesizing programs for the individual parts. Note that this rule is initially postulated with free variables like A and ExtA which are treated as constants during proof of the rule; this prevents their premature instantiation, which would lead to a proof of something more specialized. When the proof is completed, these variables are replaced by metavariables, so the rule may be later applied using unification.

There are two subtleties in the calculus we propose: parameter variables and induction rules. These are explained below.

```
RAllI: [| !!x. A(x) <-> Wfp(Ext) |] ==> (ALL x.A(x)) <-> Wfp(Ext)

RAndI: [| A <-> Wfp(ExtA); B <-> Wfp(ExtB) |] ==> A & B <-> Wfp(ExtA & ExtB)

ROrI: [| A <-> Wfp(ExtA); B <-> Wfp(ExtB) |] ==> A | B <-> Wfp(ExtA | ExtB)

RImpI: [| A <-> Wfp(ExtA); B <-> Wfp(ExtB) |] ==> (A --> B) <-> (Wfp(ExtA --> ExtB))

RTrue: [| A |] ==> A <-> Wfp(True)

RFalse: [| ~A |] ==> A <-> Wfp(False)

RAllE: [| ALL x.A(x) <-> Wfp(Ext(x)) |] ==> A(x) <-> Wfp(Ext(x))

ROrE: [| A ==> (C <-> Wfp(ExtA)); B ==> (C <-> Wfp(ExtB)); A | B |] ==>
          C <-> Wfp(ExtA | ExtB)

EqInstance:  A = B <-> Wfp(A = B)
```

Fig. 1. Examples of Basic Rules

Predicate Parameters Recall that problems are equivalences between specifications and higher-order meta-variables applied to parameters, e.g., 1 and m in the subset specification. We would like our derived rules to be applicable independent of the number of parameters involved. Fortunately, these do not need to be mentioned in the rules themselves (with one exception noted shortly) as Isabelle's higher-order unification properly propagates these parameters to subgoals. This is explained below.

Isabelle automatically *lifts* rules during higher-order resolution (see [25, 26]); this is a sound way of dynamically matching types of meta-variables during unification by applying them to new universally quantified parameters when necessary. This idea is best explained by an example. Consider applying the above conjunction rule to the following (made-up) goal.

```
ALL 1 m. ((ALL z. In(z,1)) & (Exists z. ~In(z,m))) <-> Wfp(?E(1,m)))
```

In our theory, we begin proving goals by "setting up a context" where initial universally quantified variables become eigenvariables.[5] Applying ∀-I (∀-intro of first-order logic) twice yields the following.

```
!! 1 m. ((ALL z. In(z,1)) & (Exists z. ~In(z,m))) <-> Wfp(?E(1,m)))
```

Now if we try to apply the above derived rule for conjunction, Isabelle will automatically lift this rule to

```
!! 1 m. [| ?A(1,m) <-> Extract(?ExtA(1,m)); ?B(1,m) <-> Extract(?ExtB(1,m)) |] ==>
    ?A(1,m) & ?B(1,m) <-> Extract(?ExtA(1,m) & ?ExtB(1,m)),
```

which now resolves with `?A(1,m) = ALL z. In(z,1)`, `?B(1,m) = Exists z. ~In(z,m)`, and the program is instantiated with `?E(1,m) = ?ExtA(1,m) & ?ExtB(1,m)`. As the proof proceeds `?ExtA` and `?ExtB` are further instantiated.

---

[5] By eigenvariables, we mean variables universally quantified outermost in the context. Recall universal quantification is the operator "!!" in Isabelle's meta-logic. See [26] for more details.

Recursive Definitions Our calculus so far is trivial; it copies structure from specifications into programs. One nontrivial way of transforming specifications is to admit proofs about equivalence by induction over the recursively defined data-types.
But this introduces a problem of how predicates recursively call themselves.

We solve this by proving theorems in a context and proof by induction can extend
this context with new predicate definitions.[6] In particular, the context will contain
not only axioms for defined function symbols (e.g., like In in the subset example)
but it also contains a meta-variable ("wrapped" by Def) that is instantiated during
induction with new predicate definitions.

Back to the subset example; our starting goal actually includes a context which
defines the axioms for In and includes a variable ?H which expands to a definition
or series of definitions. These will be called from the program that instantiates ?E.

```
[| ALL x. ~In(x,[]); ALL x h t. In(x,h.t) <-> x = h | In(x,t) |]
   ==> Def(?H) --> (ALL l m. (ALL z. In(z,l) --> In(z,m)) <-> Wfp(?E(l,m)))
```

The wrapper Def (recall this, like Wfp is the identity) also serves to restrict unification;
in particular, only the induction rule which creates a schematic pure logic program
can instantiate Def(?H).

Definitions are set up during induction. Consider the following rule corresponding to structural induction over lists. This rule builds a schematic program P(x,y)
contained in the first assumption. The second and third assumption correspond to
the base case and step case of a proof by induction showing that this definition is
equivalent to the specification formula A(x,y). This rule is derived in our theory by
list induction.

```
[| Def(ALL x y. P(x,y) <-> (x = [] & EA(y)) | (EX v0 v1. x = v0.v1 & EB(v0,v1,y)));
   ALL y. A([],y) <-> Wfp(EA(y));
   !!m. ALL y. A(m,y) <-> Wfp(P(m,y)) ==> ALL h y. A(h.m,y)  <-> Wfp(EB(h,m,y)) |]
 ==> A(x,y) <-> Wfp(P(x,y))
```

As in [2] we have written a tactic that applies induction rules. Resolution with
this rule yields three subgoals (corresponding to the three assumptions above) but
the first is discharged by unifying against a Def(?H) in the context which sets up a
recursive definition. This is precisely the role that Def(?H) serves. Actually, to allow
for multiple recursive definitions, the induction tactic first duplicates the Def(?H)[7]
before resolving with the induction rule. Also, it thins out (weakens) the instantiated
definition in the two remaining subgoals.

There is an additional subtlety in the induction rule which concerns parameter
arguments. Other rules did not take additional parameters but this is the exception;
P takes two arguments even though the induction is on only one of them. This is

---

[6] The ability to postulate new predicate definitions can, of course, lead to inconsistency.
We lack space here for details, but it is not hard to prove under our approach that
defined predicates are defined by well-founded recursion and may be consistently added
as assumptions.

[7] Def(?H) equals ?H and if we have an hypothesis ?H then we can instantiate it with ?H1 &
?H2. Instantiation is performed by unification with &-elimination and results in the new
assumptions ?H1 and ?H2 which are rewrapped with Def. This "engineering with logic"
is accomplished by resolving with a derived rule that performs these manipulations.

necessary as the rule must establish (in the first assumption) a definition for a predicate with a fixed number of universally quantified parameters and the number of these cannot be predicted at the time of the induction. Our solution to this problem is ad hoc; we derive in Isabelle a separate induction rule for each number of arguments needed in practice (two are needed for the predicates synthesized in the subset example). Less ad hoc, but more complex, solutions are also possible.

3.4 Relationship to Other Calculi

The derived calculus, although very simple, is motivated by and is similar to the Whelk Calculus developed by Wiggins in [33]. There Whelk is presented as a new kind of logic where specifications are manipulated in a special kind of "tagged" formal system. A tagged formula A is of the form $[\![A]\!]_{P(\bar{x})\leftrightarrow\phi}$. Both formulas and sequents are tagged and the tag subscript represents part of a pure logic program. The Whelk logic manipulates these tags so that the tagged (subscripted) program should satisfy two properties. First, the tagged program should be logically *equivalent* to formula it tags in the appropriate first-order theory. To achieve this the proof rules state how to build programs for a given goal from programs corresponding to the subgoals. Second, the tagged program should be *decidable*, which means as a program it terminates in all-ground mode. One other feature of Whelk is that a proof may begin with a subformula of the starting goal labeled by a special operator ∂. At the end of the proof the Whelk system *extracts* the tagged program labeling this goal; hence Whelk may be used to synthesize logic programs.

The rules I give can be seen as a reinterpretation of the rules of Whelk where tagged formulas are formulated directly as equivalences between specifications and program schemas (for full details see [1]); hence, seen in this light, the Whelk rules constitute a simple calculus for manipulating equivalences. For example, the Whelk rule for reasoning about conjunctions is

$$\partial\wedge\text{-}I\ \frac{[\![\Gamma\vdash\partial A]\!]_{P(\mathcal{E})\leftrightarrow\phi}\ \ [\![\Gamma\vdash\partial B]\!]_{P(\mathcal{E})\leftrightarrow\psi}}{[\![\Gamma\vdash\partial(A\wedge B)]\!]_{P(\mathcal{E})\leftrightarrow\phi\wedge\psi}}$$

and can be straightforwardly translated into the rule RAndI given in Section 3.3 (ϕ and ψ play the role of $ExtA$ and $ExtB$ and P and its parameters $\mathcal{E}$ are omitted.) Our derivation of many of these rules provides a formal verification that they are correctness preserving with respect to the above mentioned equivalence criteria. In [1] we describe the translation and derivation of the Whelk rules in considerable detail. Interestingly, not all of the Whelk rules given could be derived; the reinterpretation led to rules which were not derivable (counter models could be given) and hence helped uncover mistakes in the original Whelk calculus. This confirms that just as it is useful to have machine checked proofs of program correctness, it is also important to certify calculi formally.

4 Program Development

We now illustrate how the derived rules can be applied to program synthesis. Our example is synthesizing the subset predicate (over lists). We choose this as it is a standard example from the literature. In particular, our proof is almost identical to one given in [33].

Our proof requires 15 steps and is given in Figure 2 with comments. Here we replay Isabelle's response to these proof steps, i.e., the instantiated top-level goal and subgoals generated after each step. The output is taken directly from an Isabelle session except, to save space, we have combined a couple of steps, "pretty printed" formulas, and abbreviated variable names.

```
val [inbase,instep] = goal thy
" [| ALL x. ~In(x,[]); \
\    ALL x h t. In(x,h.t) <-> x = h | In(x,t) |] \
\ ==> Def(?H) --> (ALL l m. (ALL z. In(z,l) --> In(z,m)) <-> Wfp(?E(l,m)))";

(* After performing forall introductions, perform induction *)
by SetUpContext;
by (IndTac WListInduct2 [("x","l"),("y","m")] 1);

(* Base Case *)
br RAllI 1;
br RTrue 1;
by (cut_fast_tac [inbase] 1);

(* Step Case *)
by(SIMP_TAC (list_ss addrews [instep,AndImp,AllAnd,AllEqImp]) 1);
br RAndI 1;

(* Prove 2nd case with induction hypothesis! *)
by (etac allE 2 THEN assume_tac 2);

(* First Case --- Do an induction on y to synthesize member(h,y) *)
by (IndTac WListInduct2 [("x","y"),("y","h")] 1);
br RFalse 1;    (* Induction Base Case *)
by(SIMP_TAC (list_ss addrews [inbase]) 1);
by(SIMP_TAC (list_ss addrews [instep]) 1);  (* Induction Step Case *)
br ROrI 1;
br EqInstance 1;
by (etac allE 1 THEN assume_tac 1);  (* Apply induction hypothesis *)
```

Fig. 2. Isabelle Proof Script for Subset Proof

The proof begins by giving Isabelle the subset specification. Isabelle prints back the goal to be proved (at the top) and the subgoals necessary to establish it. As the proof proceeds, the theorem we are proving becomes specialized as ?H is incrementally instantiated with a program. We have also given the names `inbase` and `instep` to the context assumptions that define the membership predicate `In`.

```
Def(?H) --> (ALL l m. (ALL z. In(z, l) --> In(z, m)) <-> Wfp(?E(l, m)))
  1. Def(?H) --> (ALL l m. (ALL z. In(z, l) --> In(z, m)) <-> Wfp(?E(l, m)))
val inbase = "ALL x. ~ In(x, [])"
val instep = "ALL x h t. In(x, h . t) <-> x = h | In(x, t)"
```

The first proof step, invoked by the tactic `setUpContext`, moves the definition variable ?H into the assumption context and, as discussed in the previous section, promotes universally quantified variables to eigenvariables so our rules may be used via lifting.

```
Def(?H) --> (ALL l m. (ALL z. In(z, 1) --> In(z, m)) <-> Wfp(?E(1, m)))
  1. !!l m. Def(?H) ==> (ALL z. In(z, 1) --> In(z, m)) <-> Wfp(?E(1, m))
```

Next, we invoke our induction tactic that applies the derived list induction rule, specifying induction on 1. The execution of the tactic instantiates our schematic definition ?H with the first schematic definition ?P and a placeholder ?Q for further instantiation. Note too that ?E has been instantiated to this schematic program ?P.

```
Def((ALL x y. ?P(x, y) <-> x = [] & ?EA10(y) |
     (EX v0 v1. x = v0 . v1 & ?EB11(v0, v1, y))) &
     ?Q) -->
(ALL l m. (ALL z. In(z, 1) --> In(z, m)) <-> Wfp(?P(1, m)))
  1. !!l m y. Def(?Q) ==> (ALL z. In(z, [])) --> In(z, y)) <-> Wfp(?EA10(y))
  2. !!l m ma h y.
      [| Def(?Q); ALL y. (ALL z. In(z, ma) --> In(z, y)) <-> Wfp(?P(ma, y)) |] ==>
      (ALL z. In(z, h . ma) --> In(z, y)) <-> Wfp(?EB11(h, ma, y))
```

We now prove the first case, which is the base-case (and omit printing the step case in the next two steps — Isabelle maintains a goal stack). First we apply RA11I which promotes the ∀-quantified variable z to an eigenvariable. The new subgoal becomes (as this step does not instantiate the theorem we are proving, we omit redisplaying it) the following.

```
  1. !!l m y z. Def(?Q) ==> (In(z, []) --> In(z, y)) <-> Wfp(?EA10(y))
```

Next we apply RTrue which states if ?EA10(y) is True, the above is provable provided In(z, []) --> In(z, y) is provable. This reduces the goal to one of ordinary logic (without Wfp) as it instantiates the base case with the proposition True.

```
Def((ALL x y. ?P(x, y) <-> x = [] & True |
     (EX v0 v1. x = v0 . v1 & ?EB11(v0, v1, y))) &
     ?Q) -->
(ALL l m. (ALL z. In(z, 1) --> In(z, m)) <-> Wfp(?P(1, m)))
  1. !!l m y z. Def(?Q) ==> In(z, []) --> In(z, y)
```

Finally we complete this step by applying Isabelle's predicate-calculus simplification routines augmented with base case of the definition for In. Isabelle leaves us with the following step case (which is now the top goal on the stack and hence numbered 1).

```
  1. !!l m ma h y.
      [| Def(?Q); ALL y. (ALL z. In(z, ma) --> In(z, y)) <-> Wfp(?P(ma, y)) |] ==>
      (ALL z. In(z, h . ma) --> In(z, y)) <-> Wfp(?EB11(h, ma, y))
```

We now normalize this goal by applying the tactic

```
SIMP_TAC (list_ss addrews [instep,AndImp,AllAnd,AllEqImp]) 1
```

This calls Isabelle's simplification tactic which applies basic simplifications for the theory of lists, list_ss, augmented with the recursive case of the definition for In and the following lemmas AndImp, AllAnd and AllEqImp.

```
(A | B --> C) <-> (A --> C) & (B --> C)
ALL v. A(v) & B(v)) <-> (ALL v.A(v)) & (ALL v.B(v))
(ALL v. v = w --> A(v)) <-> A(w)
```

Each of these had been previously (automatically!) proven with Isabelle's predicate calculus simplifier. This normalization step simplifies our subgoal to the following.

```
1. !!l m ma h y.
      [| Def(?Q); ALL y. (ALL z. In(z, ma) --> In(z, y)) <-> Wfp(?P(ma, y)) |] ==>
      In(h, y) & (ALL v. In(v, ma) --> In(v, y)) <-> Wfp(?EB11(h, ma, y))
```

We decompose the conjunction with RAndI, which yields two subgoals.

```
Def((ALL x y. ?P(x, y) <-> x = [] & True |
      (EX v0 v1. x = v0 . v1 & ?EA21(v1, v0, y) & ?EB22(v1, v0, y))) &
   ?Q) -->
(ALL l m. (ALL z. In(z, l) --> In(z, m)) <-> Wfp(?P(l, m)))
 1. !!l m ma h y.
      [| Def(?Q); ALL y. (ALL z. In(z, ma) --> In(z, y)) <-> Wfp(?P(ma, y)) |] ==>
      In(h, y) <-> Wfp(?EA21(ma, h, y))
 2. !!l m ma h y.
      [| Def(?Q); ALL y. (ALL z. In(z, ma) --> In(z, y)) <-> Wfp(?P(ma, y)) |] ==>
      (ALL v. In(v, ma) --> In(v, y)) <-> Wfp(?EB22(ma, h, y))
```

We immediately solve the second subgoal by resolving with the induction hypothesis. I.e., after $\forall$-E we unify the conclusion with the induction hypothesis using Isabelle's assumption tactic. This instantiates the program we are building by replacing ?EB22 with a recursive call to ?P as follows.

```
Def((ALL x y. ?P(x, y) <-> x = [] & True |
      (EX v0 v1. x = v0 . v1 & ?EA21(v1, v0, y) & ?P(v1, y))) & ?Q) -->
(ALL l m. (ALL z. In(z, l) --> In(z, m)) <-> Wfp(?P(l, m)))
```

Returning to the first goal (to build a program for ?EA21), we perform another induction; the base case is proved as in the first induction except rather than introducing True with RTrue we introduce False with RFalse and solve the remaining goal by simplification. This leaves us with the step case.

```
Def((ALL x y. ?P(x, y) <-> x = [] & True |
      (EX v0 v1. x = v0 . v1 & ?Pa(v0, y) & ?P(v1, y))) &
   (ALL x y. ?Pa(y, x) <-> x = [] & False |
      (EX v0 v1. x = v0 . v1 & ?EB32(v0, v1, y))) &
   ?Q27) -->
(ALL l m. (ALL z. In(z, l) --> In(z, m)) <-> Wfp(?P(l, m)))
 1. !!l m ma h y mb ha ya.
      [| ALL y. (ALL z. In(z, ma) --> In(z, y)) <-> Wfp(?P(ma, y));
         Def(?Q27); ALL y. In(y, mb) <-> Wfp(?Pa(y, mb)) |] ==>
      In(ya, ha . mb) <-> Wfp(?EB32(ha, mb, ya))
```

As before, we normalize this subgoal with Isabelle's standard simplifier.

```
1. !!l m ma h y mb ha ya.
      [| ALL y. (ALL z. In(z, ma) --> In(z, y)) <-> Wfp(?P(ma, y));
         Def(?Q27); ALL y. In(y, mb) <-> Wfp(?Pa(y, mb)) |] ==>
      ya = ha | In(ya, mb) <-> Wfp(?EB32(ha, mb, ya))
```

Applying ROrI unifies ?EB32(v0, v1, y) with ?EA40(v1, v0, y) | ?EB41(v1,v0, y) and yields a subgoal for each disjunct.

```
1. !!l m ma h y mb ha ya.
     [| ALL y. (ALL z. In(z, ma) --> In(z, y)) <-> Wfp(?P(ma, y));
        Def(?Q27); ALL y. In(y, mb) <-> Wfp(?Pa(y, mb)) |] ==>
     ya = ha <-> Wfp(?EA40(mb, ha, ya))
2. !!l m ma h y mb ha ya.
     [| ALL y. (ALL z. In(z, ma) --> In(z, y)) <-> Wfp(?P(ma, y));
        Def(?Q27); ALL y. In(y, mb) <-> Wfp(?Pa(y, mb)) |] ==>
     In(ya, mb) <-> Wfp(?EB41(mb, ha, ya))
```

In the first we apply EqInstance which instantiates ?EA40(v1, v0, y) with y = v0.
This completes the first goal leaving only the following.

```
Def((ALL x y. ?P(x, y) <-> x = [] & True |
       (EX v0 v1. x = v0 . v1 & ?Pa(v0, y) & ?P(v1, y))) &
    (ALL x y. ?Pa(y, x) <-> x = [] & False |
       (EX v0 v1. x = v0 . v1 & (y = v0 | ?EB41(v1, v0, y)))) &
    ?Q27) -->
(ALL l m. (ALL z. In(z, l) --> In(z, m)) <-> Wfp(?P(l, m)))
 1. !!l m ma h y mb ha ya.
     [| ALL y. (ALL z. In(z, ma) --> In(z, y)) <-> Wfp(?P(ma, y));
        Def(?Q27); ALL y. In(y, mb) <-> Wfp(?Pa(y, mb)) |] ==>
     In(ya, mb) <-> Wfp(?EB41(mb, ha, ya))
```

We complete the proof by resolving with the induction hypothesis. Isabelle prints
back the following proven formula with no remaining subgoals.

```
[| ALL x. ~ ?In(x, []);
   ALL x h t. ?In(x, h . t) <-> x = h | ?In(x, t) |] ==>
Def((ALL x y. ?P(x, y) <-> x = [] & True |
       (EX v0 v1. x = v0 . v1 & ?Pa(y, v0) & ?P(v1, y))) &
    (ALL x y. ?Pa(x, y) <-> x = [] & False |
       (EX v0 v1. x = v0 . v1 & (y = v0 | ?Pa(v1, y)))) &
    ?Q) -->
(ALL l m. (ALL z. ?In(z, l) --> ?In(z, m)) <-> Wfp(?P(l, m)))
```

Note that the context remains open (?Q) as we might have needed to derive additional predicates. Also observe that Isabelle never forced us to give the predicates built (?P and ?Pa) concrete names; these were picked automatically during resolution when variables were renamed apart by the system.

The constructed program can be simplified and translated into a Gödel program similar to the one in [33]. Alternatively it can be directly translated into the following Prolog program.

```
p([],Y).                         pa([],Y)        :- false.
p([V0|V1],Y) :- pa(Y,V0), p(V1,Y).   pa([V0|V1],V0).
                                 pa([V0|V1],Y) :- pa(V1,Y).
```

5   Conclusion, Comparisons, and Future Work

The ideas presented here have applicability, of course, outside logic programming and Isabelle can be used to derive other calculi for verification and synthesis. [2, 4] describes other applications of this methodology. But logic programming seems an especially apt domain for such development due to the close relationship between the specification and programming language.

Other authors have argued that first-order logic is the proper foundation for reasoning about and transforming logic programs (e.g., [11, 9]). But there are benefits to using even richer logics to manipulate first-order, and possibly higher-order, specifications. For example, in this paper we used a recursion schema corresponding to structural induction over lists. But synthesizing logic programs with more complicated kinds of recursion (e.g., quick sort) requires general well-founded induction. But providing a theory where the user can provide his own well-founded relations necessitates formalizing well-foundedness which in turn requires quantifying over sets or predicates and, outside of set-theory, this is generally second-order. We are currently exploring synthesis based on well-founded induction in higher-order logic.

Another research direction is exploring other notions of equivalence. Our calculus has employed a very simple notion based on provability in a theory with induction principles over recursive data-types. There are other notions of equivalence and ways of proving equivalence that could be formalized of course. Of particular interest is exploring schematic calculi like that proposed by Waldau [31]. Waldau presents a calculus for proving the correctness of transformation schemata using intuitionistic first-order logic. In particular he showed how one can prove the correctness of fold-unfold transformations and schemata like those which replace recursion by tail recursion. The spirit of this work is similar to our own: transformation schemata should be proven correct using formal proofs. It would be interesting to carry out the kinds of derivations he suggests in Isabelle and use Isabelle's unification to apply his transformation schemata.

We conclude with a brief comparison of related approaches to program synthesis based on unification. This idea can be traced back to [14] who proposed the use of resolution not only for checking answers to queries, but also for synthesizing programs and the use of second-order matching by Huet and Lang to apply schematic transformations. Work in unification based program synthesis that is closest in spirit to what we described here is the work of [20, 21], which used higher-order (pattern) unification to synthesize logic programs in a "middle-out" fashion. Indeed, synthesis with higher-order resolution in Isabelle is very similar as in our work, the meta-variable standing in for a program is a second-order pattern and it is only unified against second-order patterns during proof. [20, 21] emphasizes, however, the automation of such proofs via rippling while we concentrate more on the use of logical frameworks to give formal guarantees to the programming logic itself. Of course, these approaches are compatible and can be combined.

References

1. David Basin. Isawhelk: Whelk interpreted in Isabelle. Abstract accepted at the 11th International Conference on Logic Programming (ICLP94). Full version available via anonymous ftp to mpi-sb.mpg.de in pub/papers/conferences/Basin-ICLP94.dvi.Z.
2. David Basin, Alan Bundy, Ina Kraan, and Sean Matthews. A framework for program development based on schematic proof. In *7th International Workshop on Software Specification and Design*, Los Angeles, December 1993. IEEE Computer Society Press.
3. David Basin and Robert Constable. Metalogical frameworks. In Gérard Huet and Gordon Plotkin, editors, *Logical Environments*, pages 1–29. Cambridge University Press, 1993.

4. David Basin and Sean Matthews. A conservative extension of first order logic and its applications to theorem proving. In *13th Conference of the Foundations of Software Technology and Theoretical Computer Science*, pages 151–160, December 1993.

5. A. Bundy. Tutorial notes; reasoning about logic programs. In G. Comyn, N.E. Fuchs, and M.J. Ratcliffe, editors, *Logic programming in action*, pages 252–277. Springer Verlag, 1992.

6. A. Bundy, A. Smaill, and G. A. Wiggins. The synthesis of logic programs from inductive proofs. In J. Lloyd, editor, *Computational Logic*, pages 135–149. Springer-Verlag, 1990. Esprit Basic Research Series. Also available from Edinburgh as DAI Re search Paper 501.

7. R.M. Burstall and J. Darlington. A transformation system for developing recursive programs. *Journal of the Association for Computing Machinery*, 24(1):44–67, 1977.

8. Wei Ngan Chin. *Automatic Methods for Program Transformation*. Ph. D. thesis, Imperial College Department of Computer Science, March 1990.

9. K. L. Clark and S-Å. Tärnlund. A first order theory of data and programs. In B. Gilchrist, editor, *Information Processing*, pages 939–944. IFIP, 1977.

10. K.L. Clark. Predicate logic as a computational formalism. Technical Report TOC 79/59, Imperial College, 1979.

11. K.L. Clark and S. Sickel. Predicate Logic: a calculus for deriving programs. In R. Reddy, editor, *Proceedings of IJCAI-77*, pages 419–420. IJCAI, 1977.

12. Pierre Flener and Yves Deville. Towards stepwise, schema-guided synthesis of logic programs. In T. Clement and K.-K. Lau, editors, *Logic Program Synthesis and Transformation*, pages 46–64. Springer-Verlag, 1991.

13. Michael J. Gordon, Robin Milner, and Christopher P. Wadsworth. *Edinburgh LCF: A Mechanized Logic of Computation*, volume 78 of *Lecture Notes in Computer Science*. Springer-Verlag, 1979.

14. Cordell Green. Application of theorem proving to problem solving. In *Proceedings of the IJCAI-69*, pages 219–239, 1969.

15. Robert Harper, Furio Honsell, and Gordon Plotkin. A framework for defining logics. *Journal of the Association for Computing Machinery*, 40(1):143–184, January 1993.

16. P. Hill and J. Lloyd. The Gödel Report. Technical Report TR-91-02, Department of Computer Science, University of Bristol, March 1991. Revised in September 1991.

17. C.J. Hogger. Derivation of logic programs. *JACM*, 28(2):372–392, April 1981.

18. L. Kott. About a transformation system: A theoretical study. In *Proceedings of the 3rd International Symposium on Programming*, pages 232–247, Paris, 1978.

19. Robert A. Kowalski. Predicate logic as a programming language. In *IFIP-74*. North-Holland, 1974.

20. Ina Kraan, David Basin, and Alan Bundy. Logic program synthesis via proof planning. In *Proceedings of LoPSTr-92*. Springer Verlag, 1992.

21. Ina Kraan, David Basin, and Alan Bundy. Middle-out reasoning for logic program synthesis. In *10th International Conference on Logic Programming (ICLP93)*, pages 441–455, Budapest Hungary, 1993.

22. M.J. Maher. Equivalences of logic programs. In J. Minker, editor, *Foundations of Deductive Databases and Logic Programming*. Morgan Kaufmann, 1987.

23. Sean Matthews, Alan Smaill, and David Basin. Experience with FS_0 as a framework theory. In Gérard Huet and Gordon Plotkin, editors, *Logical Environments*, pages 61–82. Cambridge University Press, 1993.

24. Lawrence C. Paulson. Set theory for verification: I. From foundations to functions. *Journal of Automated Reasoning*. In press; draft available as Report 271, University of Cambridge Computer Laboratory.

25. Lawrence C. Paulson. The foundation of a generic theorem prover. *Journal of Automated Reasoning*, 5:363–397, 1989.

26. Lawrence C. Paulson. Introduction to Isabelle. Technical Report 280, Cambridge University Computer Laboratory, Cambridge, January 1993.
27. Lawrence C. Paulson. Isabelle's object-logics. Technical Report 286, Cambridge University Computer Laboratory, Cambridge, February 1993.
28. Frank Pfenning. Logic programming in the LF logical framework. In *Logical Frameworks*, pages 149 – 181. Cambridge University Press, 1991.
29. Taisuke Sato. Equivalence-preserving first-order unfold/fold transformation systems. *Theoretical Computer Science*, 105:57–84, 1992.
30. H. Tamaki and T. Sato. Unfold/fold transformations of logic programs. In *Proceedings of 2nd ICLP*, 1984.
31. Mattias Waldau. Formal validation of transformation schemata. In T. Clement and K.-K. Lau, editors, *Logic Program Synthesis and Transformation*, pages 97–110. Springer-Verlag, 1991.
32. Richard W. Weyhrauch. Prolegomena to a theory of formal reasoning. *Artificial Intelligence*, 13:133–170, 1980.
33. Geraint A. Wiggins. Synthesis and transformation of logic programs in the Whelk proof development system. In K. R. Apt, editor, *Proceedings of JICSLP-92*, 1992.

AN EXTENDED TRANSFORMATION SYSTEM FOR CLP PROGRAMS

N. Bensaou - I. Guessarian
LITP - Université Paris 6 - email: ig@litp.ibp.fr, bensaou@litp.ibp.fr

Abstract. We extend the Tamaki–Sato transformation system for logic programs into a transformation system for constraint logic programs including fold, unfold, replacement, thinning and fattening, and constraint simplifications; we give a direct proof of its correctness.

1 INTRODUCTION

Logic programming has been extended to constraint logic programming to integrate the resolution of numerical, boolean or set-theoretical constraints together with symbolic evaluation methods. Program transformations have been introduced to improve efficiency of programs; the usual strategy is to first write a simple, but may be non efficient program, which can easily be proved correct, and to then improve its efficiency by applying to it program transformations which preserve correctness. For logic programs, a well-known such system is the one introduced by Tamaki–Sato [24], following a methodology first defined by Burstall–Darlington [6]; this system has then been improved and generalized in e.g. [1, 12, 20]. Other transformation systems have been proposed in different frameworks and properties of these transformations, e.g. the associativity of the unfolding rule, have been proved in [8].

From an operational viewpoint, it was recently proved in [2] that unfolding preserves universal termination and cannot introduce infinite derivations, both when it is applied alone or in a "restricted" Tamaki-Sato transformation sequence, while in [3] it was proved that the Tamaki-Sato system preserves the acyclity of the initial program.

In the framework of Prolog, transformations preserving the solutions of programs together with their computation times are given [21], while in [23] syntactical conditions have been defined to ensure that the transformations preserve the sequence of answer substitutions semantics for Prolog. In [22], an abstract strategy for automatic transformation of logic programs has been presented. In [14] transformation methods involving partial evaluations and designed for a program-query pair are studied. In [20] a transformation system for CLP programs with a stronger form of correctness is proposed, and finally, a fold/unfold transformation system for CLP programs has recently been introduced in [7].

Semantics of logic programs and constraint logic programs have been studied in [10, 13, 18]. Finally, a survey of the state of the art about constraint logic programs can be found in [15].

A sequence of program transformations is correct if the final program is equivalent to the original one, namely the two programs have the same semantics, cf. Section 4.1. Usually, we will be given a basic program P, together with new predicates defined in terms of the predicates of P, and the goal of the program transformations will be to find direct definitions, more efficiently computable, of these new predicates. We will then say that a constraint logic program (CLP program in short) has been successfully transformed when the sequence of transformations ends with a final program which is equivalent to the original one (and is of course hopefully simpler or better). We here extend the Tamaki–Sato fold/unfold program transformations [24] and the transformations proposed in [1], to CLP programs.

In [5] we presented a fold/unfold transformation system for CLP programs and proved its correctness with respect to a denotational and an operational semantics. The main contributions of the present paper are the following

1) we improve the fold transformation by making it more general,

2) we define new transformation rules for CLP programs, namely replacement, deletion and addition of redundant clauses and atoms, constraint simplifications, and give the conditions under which these transformations are correct.

The paper is organized as follows: in Section 2, we recall the basic definitions about CLP programs and we define the semantics of our programs; we define the transformation rules in Section 3, and we finally prove the correctness of the proposed transformation system with respect to the defined semantics in Section 4; technical proofs are given in the appendix. The present paper considerably improves [5] where only a fold/unfold transformation was considered, with the more restrictive fold of [24].

2   CLP PROGRAMS AND THEIR SEMANTICS

2.1   CLP programs

A CLP program is a set of definite clauses, of the following form:

$$p_0(\vec{t_0}) \longleftarrow p_1(\vec{t_1}), \ldots, p_n(\vec{t_n}) : c_1(\vec{u_1}), \ldots, c_m(\vec{u_m}) \tag{1}$$

where for $0 \leq i \leq n$, p_i is a predicate symbol, for $j = 1, \ldots, m$, c_j is a constraint, and for $k = 1, \ldots, m$, $i = 0, \ldots, n$, $\vec{t_i}, \vec{u_k}$ are vectors of terms, n and m can be equal to 0.

In the sequel, clauses will be written in the following form:

$$p_0(\vec{t_0}) \longleftarrow p_1(\vec{t_1}) : d_1, \ldots, p_n(\vec{t_n}) : d_n \tag{2}$$

where for $k = 1, \ldots, n$, $i = 0, \ldots, n$, $\vec{t_i}$ are terms, and d_k are constraints. Both forms 1 and 2 are equivalent (cf. [5]).

2.2   Semantics

Our semantics is based on the C-models [10, 5]. Along similar lines, a very general and elegant semantics has been recently introduced for CLP programs in [13]: it uses cylindric algebras as interpretations for CLP programs. We however will use here a less general but simpler and more intuitive semantics, based on a generalization of the well-known Herbrand interpretations.

We will consider a language $CLP(X)$, where X is an appropriate constraint domain; the semantic interpretation of the domain X is a structure R; R has underlying domains, in which variables are assigned values and on which predicates and functions operate; R will be fixed throughout and will be implicitly understood in the sequel.

The languages $CLP(X)$ have been given both operational semantics, computed by SLD refutation, and fixpoint semantics, based on immediate consequence operators T_P; these semantics have been proved equivalent [15, 11, 5]. We will use a fixpoint semantics to prove the correctness of our transformations. We only define here the notions of constrained atoms, solvable constraints, and Herbrand basis characterizing the semantics.

Definition 2.1 • *solvable constraint: A constraint c is said to be R-solvable (in short solvable), if and only if there exists a valuation* $\theta: Var(c) \mapsto D_R$ *such that* $R \models \theta(c)$. θ *is called a solution of c.*

θ *is called a solution of c restricted to the set X, and we write* $R \models_X \theta(c)$ *if and only if there exists a solution* θ' *such that* $R \models \theta'(c)$ *and* $\theta'_{|X} = \theta$.

• *preorder on constraints:* c_1 *is said to be stronger than* c_2 *if and only if any solution of* c_1 *is also a solution of* c_2; *this is denoted by* $c_1 \Longrightarrow c_2$. *The induced equivalence on constraints is denoted by* $c_1 \Longleftrightarrow c_2$ *and is called logical equivalence.*

Let $Var(c_1) \subseteq X \cup Y$ *and* $Var(c_2) \subseteq X \cup Y$; c_1 *is said to be X-stronger than* c_2 *if and only if* $\forall \theta$, $R \models_X \theta(c_1) \Longrightarrow R \models_X \theta(c_2)$, *this is denoted by* $c_1 \Longrightarrow_X c_2$. $\Longrightarrow_X$ *extends* $\models_X$ *of [9].*

• *constrained atom: A constrained atom is a pair* $p(\vec{X}) : c$ *where c is a constraint and p an atom. Two constrained atoms* $p(\vec{X}) : c$ *and* $p(\vec{X'}) : c'$ *are called variants [10] iff they are identical up to variable renamings. "to be variants" is an equivalence relation denoted by* $\approx$. *A constrained atom* $p(\vec{X}) : c$ *is said to be solvable if and only if c is solvable.*

Intuitively, a solvable constrained atom represents all possible instanciations of p with a valuation $\theta: \vec{X} \mapsto D_R$ which is a solution of c. The "Herbrand Basis" $\mathcal{A}$ will be the set of equivalence classes of constrained atoms modulo $\approx$; the $\approx$ equivalence class of $p(\vec{X}) : c$ will also be denoted by $p(\vec{X}) : c$.

Definition 2.2 *clause and atom instance*

A constrained clause (ρ) $A \longleftarrow A_1 : c_1, \ldots, A_n : c_n$ *is an instance of a constrained clause* (ρ') $B \longleftarrow B_1 : e_1, \ldots, B_n : e_n$ *if there exists a substitution* θ *such that:*

$$\theta(B) = A , \quad \text{and for } i = 1, \ldots, n, \quad \theta(B_i) = A_i , \quad c_i \Longrightarrow \theta(e_i) \qquad (3)$$

i.e. the atoms of (ρ) *are (possibly non ground) instanciations of the atoms of* (ρ') *and the constraints occurring in* (ρ) *are stronger than (or equal to) the instanciations of the corresponding constraints of* (ρ'), *namely any solution of* c_i *is also a solution of* $\theta(e_i)$; *thus the application of* (ρ') *is valid.*

Similarly, a constrained atom $A : c$ *is an instance of a constrained atom* $B : e$ *if there exists a substitution* θ *such that:* $\theta(B) = A$ *and c is stronger than (or equal to)* $\theta(e)$.

Definition 2.3 • *preorder: let* $\mathcal{A}$ *be the set of* $\approx$ *equivalence classes of constrained atoms of program P. A preorder* $\sqsubseteq$ *is defined on* $\mathcal{A}$ *by:* $(p(\vec{X}) : c_1) \sqsubseteq (p(\vec{X}) : c_2)$ *if and only if* c_1 *is stronger than* c_2 . *The induced equivalence on the set of constrained atoms is denoted by* $\equiv$.

• *basis: The basis* $\mathcal{B}$ *is the factor set of* $\mathcal{A}$ *by the equivalence relation* $\equiv$. *A subset I of* $\mathcal{B}$ *is said to be upwards closed if whenever* $A' : c' \in I$, *and* $A : c$ *is an instance of* $A' : c'$ *then* $A : c \in I$.

• *interpretation: an interpretation is an upwards closed subset of* $\mathcal{B}$. *A constrained atom* $A : c$ *is true in interpretation I if and only if* $A : c \in I$. *A constrained clause* (ρ') $B \longleftarrow B_1 : e_1, \ldots, B_n : e_n$ *is true in interpretation I if and only if for each instance (see Definition 2.2)* (ρ) $A \longleftarrow A_1 : c_1, \ldots, A_n : c_n$ *of* (ρ') *such that* $A_i : c_i \in I$ *for* $i = 1, \ldots, n$, $(A : c_1 \wedge \cdots \wedge c_n) \in I$.

Note that: 1) if $p(\vec{X'}) : c'$ is a variant of $p(\vec{X}) : c$, then it is also an instance of $p(\vec{X}) : c$; if $(p(\vec{X}) : c_1) \sqsubseteq (p(\vec{X}) : c_2)$, then $p(\vec{X}) : c_1$ is an instance of $p(\vec{X}) : c_2$.

2) if I is upwards closed and $A : d \in I$, then I contains all constrained atoms $A : c \sqsubseteq A : d$ with constraints c stronger than d, since they are instances of $A : d$ (see Definition 2.2). It will be fundamental in our semantics that the interpretations be upwards closed.

The above defined interpretations enable us to give a declarative fixpoint semantics based on an immediate consequence operator T_P [5]. We will only illustrate this semantics by an example here.

Example 2.4 Let P be defined on $\mathbb{Z}$ by

$$A(z,y) \longleftarrow A(x,z) : z \geq x, 1 \leq y \leq 5 \qquad (4)$$
$$A(x,x) \longleftarrow x \leq 0 \qquad (5)$$
$$A(x,x) \longleftarrow x \geq 100 \qquad (6)$$
$$A(1,1) \longleftarrow \qquad (7)$$

the semantics of P is defined as follows.

$$
\begin{aligned}
M(P) = &\{A(1,1), A(1,2), A(1,3), A(1,4), A(1,5)\} \cup \{A(x,x) : x \leq 0\} \\
&\cup \{A(x,x) : x \geq 100\} \cup \{A(2,y) : 1 \leq y \leq 5\} \cup \{A(3,y) : 1 \leq y \leq 5\} \\
&\cup \{A(4,y) : 1 \leq y \leq 5\} \cup \{A(5,y) : 1 \leq y \leq 5\} \\
&\cup \{A(x,y) : x \leq 0, 1 \leq y \leq 5\} \cup \{A(x,y) : x \geq 100, 1 \leq y \leq 5\}
\end{aligned}
$$

$M(P)$ contains closed and non closed atoms; for each non closed atom, $M(P)$ also contains all instances of that atom: e.g. the atoms $A(x,x) : x \geq 100 \wedge x \leq 200$, $A(x,x) : x \geq 100 \wedge x \leq 1000, \ldots, A(x,x) : d(x)$, with $d(x)$ a constraint stronger than $x \geq 100$ all are instances of $A(x,x) : x \geq 100 \in M(P)$, hence belong to $M(P)$. □

Definition 2.5 *1) A conjunction of constrained atoms $A_1 : c_1 \wedge \cdots \wedge A_n : c_n$ is said to be satisfiable by P, I if there exists a (not necessarily ground) substitution θ such that $\theta(A_i : c_i) \in T_P(I)$ for $i = 1, \ldots, n$, and we will write:*

$$P, I \vdash \theta(A_1 : c_1 \wedge \cdots \wedge A_n : c_n)$$

Similarly, $P \vdash A_1 : c_1 \wedge \cdots \wedge A_n : c_n$ if and only if for every interpretation I, $I \vdash P$ implies $I \vdash A_1 : c_1 \wedge \cdots \wedge A_n : c_n$.

2) Let P be a constrained logic program. A conjunction of constrained atoms $A_1 : c_1 \wedge \cdots \wedge A_k : c_k$ is said to be equivalent to a conjunction $B_1 : e_1 \wedge \cdots \wedge B_m : e_m$ if for every interpretation I and every substitution θ, $P, I \vdash \theta(A_1 : c_1 \wedge \cdots \wedge A_k : c_k)$ if and only if $P, I \vdash \theta(B_1 : e_1 \wedge \cdots \wedge B_m : e_m)$.

The notation is: $P \Vdash [A_1 : c_1 \wedge \cdots \wedge A_k : c_k \Longleftrightarrow B_1 : e_1 \wedge \cdots \wedge B_m : e_m]$

Example 2.6 Condition 2) will allow us to move constraints around since for any P, $P \Vdash [A_1 : c_1 \wedge \cdots \wedge A_k : c_k \Longleftrightarrow A_1 : e_1 \wedge \cdots \wedge A_k : e_k]$, as soon as $c_1 \wedge \cdots \wedge c_k \Longleftrightarrow e_1 \wedge \cdots \wedge e_k$. See Sections 3.4, 3.10. □

3   THE TRANSFORMATION RULES

The transformation process consists in applying an arbitrary number of times the transformation rules which are: definition, and unfolding as in the Tamaki-Sato system [24], folding defined as in Gardner–Sheperdson [12], together with replacement, thinning, fattening and pruning. The basic difference between the Tamaki–Sato system and the systems of [6, 12, 20] is in folding, which is quite special in the Tamaki–Sato system. As folding should preserve total correctness, it can never be arbitrary, but must obey some restrictions. Unfolding is always totally correct (cf. Corollary 4.7) but arbitrary folding is well known to be only partially correct in the absence of constraints, and it is not even partially correct for constraint logic programs (cf. Example 3.4). The folding of [12] is however both simpler and more general than the one in [24], this is the reason why we adopted it.

Definition clauses introduce new predicates, and the program transformation will find a simpler form, e.g. a synthetic form, or a direct recursion, for these new predicates; this will decrease the number of recursive calls, hence increase program efficiency.

In the rest of this section, we will define the syntax of the transformation rules. Their correctness with respect to the semantics given formally in Section 2.2 will be shown in Section 4.

3.1   Definition rule

Introduce a clause of the form

$$(\delta) \quad p(x_1, x_2, \ldots, x_n) \longleftarrow A_1 : c_1, \ldots, A_m : c_m$$

where p is a new predicate name not appearing in P, $x_1, x_2, \ldots, x_n$ are pairwise distinct variables, $A_1, A_2, \ldots, A_m$ are predicates appearing in the initial program P, $c_1, c_2, \ldots, c_m$ are (possibly empty) constraints.

The clause (δ) is added to P, so that $P' = P \cup \{\delta\}$. This implies that, for each given new predicate p, there can be at most one definition clause with head p.

Example 3.1 Let P be the following program, defining the length of a list and the matching of two lists:

$$
\begin{align}
long([], X) &\longleftarrow \qquad\qquad\quad : X = 0 \tag{8}\\
long([N|Y], X) &\longleftarrow long(Y, X') : X = X' + 1 \tag{9}\\
match([], Y_2) &\longleftarrow \tag{10}\\
match([N_1|Y_1], [N_2|Y_2]) &\longleftarrow match(Y_1, Y_2), N_1 = N_2 \tag{11}
\end{align}
$$

Let the predicate $mtlg$ define the matching of two lists, such that the length of the first list is shorter than the length of the second list; $mtlg$ is defined by:

$$
\begin{align}
mtlg(Y_1, Y_2, X_1, X_2) \longleftarrow\ &match(Y_1, Y_2), long(Y_1, X_1) : X_1 < X_2,\\
&long(Y_2, X_2) : X_1 < X_2 \tag{12}
\end{align}
$$

3.2 Unfolding rule

Let (ρ) be a clause in the program P.

$$(\rho) \quad A \longleftarrow A_1 : c_1, \ldots, A_q : c_q, \ldots, A_n : c_n$$

Let (π_k), for $k = 1, \ldots, m$, be *all* the clauses in P whose head can be unified with A_q:

$$(\pi_k) \quad T_k \longleftarrow T_{k1} : c'_1, \ldots, T_{kj} : c'_j$$

The unfolding of (ρ) by these m clauses at the constraint atom $(A_q : c_q)$ is obtained by unifying A_q with each of the T_k's.

Let the substitution μ_k $(k = 1, \ldots, m)$ be an m.g.u. of A_q and T_k, hence: $\mu_k(A_q) = \mu_k(T_k)$ and substitute for $\mu_k(A_q : c_q)$

$$\mu_k(T_{k1} : c'_1 \wedge c_q, \ldots, T_{kj} : c'_j \wedge c_q),$$

for $k = 1, \ldots, m$. Let now clauses (τ_k) be defined by, for $k = 1, \ldots, m$:

$$(\tau_k) \quad \mu_k(A) \longleftarrow \mu_k(A_1 : c_1, \ldots, A_{q-1} : c_{q-1}, T_{k1} : (c'_1 \wedge c_q), \ldots,$$
$$T_{kj} : (c'_j \wedge c_q), A_{q+1} : c_{q+1}, \ldots, A_n : c_n)$$

clause (ρ) is replaced by the set of clauses (τ_k), for $m \geq k \geq 1$, and $P' = (P - \{\rho\}) \cup \{\tau_1, \ldots, \tau_m\}$.

Note that after an unfolding, the constraints can fail to be distributed properly (i.e. constraints involving a variable X should be affected to atoms involving the variable X, cf. [5]) among the atoms in the body of the unfolded rule, but this does not affect the correctness of our transformation.

Example 3.2 (3.1 continued) Unfolding clause 12 at its first atom results in two new clauses
a) unfolding with clause 10 we obtain:

$$mtlg([], Y_2, X_1, X_2) \longleftarrow long([], X_1) : X_1 < X_2,$$
$$long(Y_2, X_2) : X_1 < X_2 \tag{13}$$

b) unfolding with clause 11 we obtain:

$$mtlg([N_1|Y_1], [N_2|Y_2], X_1, X_2) \longleftarrow match(Y_1, Y_2), N_1 = N_2,$$
$$long([N_1|Y_1], X_1) : X_1 < X_2, long([N_2|Y_2], X_2) : X_1 < X_2 \tag{14}$$

Unfolding 13 at its first predicate once and simplifying we obtain:

$$mtlg([], Y_2, X_1, X_2) \longleftarrow long(Y_2, X_2) : 0 = X_1 < X_2 \tag{15}$$

unfolding 14 twice we obtain:

$$mtlg([N_1|Y_1], [N_2|Y_2], X_1, X_2) \longleftarrow N_1 = N_2, match(Y_1, Y_2),$$
$$long(Y_1, X'_1) : (X_1 = X'_1 + 1, X'_1 < X'_2),$$
$$long(Y_2, X'_2) : (X_2 = X'_2 + 1, X'_1 < X'_2) \tag{16}$$

Finally clause 12 is replaced by clause 15 and clause 16.

3.3 Folding rule

Let (ρ) and (δ) be clauses in the program P:

$$(\rho) \quad A \longleftarrow A_1 : c_1, \ldots, A_q : c_q, A_{q+1} : c_{q+1}, \ldots, A_{q+r} : c_{q+r}, \ldots, A_n : c_n$$

$$(\delta) \quad B \longleftarrow B_1 : c'_1, \ldots, B_r : c'_r$$

and μ a substitution such that the following conditions hold:

1. $\forall j = 1, \ldots, r, \; A_{q+j} = \mu(B_j)$ and moreover (cf. Definition 2.1)

$$\forall j = 1, \ldots, r \quad (c_{q+j} \Longrightarrow \mu(c'_j)) \tag{17}$$

2. (δ) is the only clause in P whose head unifies with $\mu(B)$.
3. μ substitutes *distinct variables* for the internal variables of (δ), and these variables do not occur in the head predicate A of (ρ), nor in the $\{A_1 : c_1, \ldots, A_q : c_q\} \cup \{A_{q+r+1} : c_{q+r+1}, \ldots, A_n : c_n\}$, nor in $\mu(B) : c_{q+1} \wedge \cdots \wedge c_{q+r}$.

Recall that a variable Y is said to be *internal* to a clause (δ) if Y occurs only in the body of (δ) and Y does not occur in its head.

The folding of (ρ) by (δ) at the subset $\{A_{q+1} : c_{q+1}, \ldots, A_{q+r} : c_{q+r}\}$ is the clause (τ) which will replace (ρ):

$$(\tau) \quad A \longleftarrow A_1 : c_1, \ldots, A_q : c_q, \mu(B) : c_{q+1} \wedge \cdots \wedge c_{q+r},$$
$$A_{q+r+1} : c_{q+r+1}, \ldots, A_n : c_n$$

(ρ) is called the *folded* rule and (δ) is called the *folder*. Note that conditions 1., 2. and 3. are needed in order to insure the correctness of the transformation (see [5, 12]).

Finally $P' = (P - \{\rho\}) \cup \{\tau\}$, and P' is equivalent to P. The folding considered here is the inverse of the unfolding rule as in [12]; it is more general than the folding of [24, 5], and its correctness proof is much easier.

Example 3.3 (3.2 continued) Folding 16 by 12 gives:

$$(\tau) \quad mtlg([N_1|Y_1], [N_2|Y_2], X_1, X_2) \longleftarrow N_1 = N_2,$$
$$mtlg(Y_1, Y_2, X'_1, X'_2) \{ X_1 = X'_1 + 1, X_2 = X'_2 + 1, X'_1 < X'_2 \}$$

Example 3.4 This example shows that arbitrary folding not taking into proper account the constraints is not even partially correct. Let P

$$(\rho) \quad s(X, Y) \longleftarrow q(X, Z) , \; q(Z, Y) : Y > 0$$
$$(\delta) \quad p(X, Y) \longleftarrow q(X, Z) , \; q(Z, Y) : Y > 100$$
$$(\sigma) \quad p(X, Y) \longleftarrow \quad\quad r(X, Y) : Y > 0$$

Folding (ρ) by (δ) gives P_1:

$$(\tau) \quad s(X, Y) \longleftarrow \quad\quad p(X, Y) : Y > 0$$
$$(\delta) \quad p(X, Y) \longleftarrow q(X, Z) , \; q(Z, Y) : Y > 100$$
$$(\sigma) \quad p(X, Y) \longleftarrow \quad\quad r(X, Y) : Y > 0$$

which is no longer equivalent to P, and is not even partially correct since according to P, s is defined as soon as $q(X, Z) , \; q(Z, Y)$ and $Y > 0$, while according to P_1, s is defined as soon as $q(X, Z) , \; q(Z, Y)$ and $Y > 100$ or $r(X, Y)$ and $Y > 0$.

3.4 Substitution rule

Let P be a logic program and (ρ) a clause in P:

$$(\rho) \quad A \longleftarrow A_1 : c_1, \ldots, A_k : c_k, \ldots, A_n : c_n$$

such that the following equivalence holds (cf. Definition 2.5, 3.):

$$P \parallel\!\!-A_1 : c_1 \wedge \cdots \wedge A_k : c_k \Longleftrightarrow B_1 : e_1 \wedge \cdots \wedge B_m : e_m.$$

Using this equivalence we can replace the conjunction $A_1 : c_1 \wedge \cdots \wedge A_k : c_k$ by the conjunction $B_1 : e_1 \wedge \cdots \wedge B_m : e_m$ and we will obtain the clause

$$(\rho') \quad A \longleftarrow B_1 : e_1, \ldots, B_m : e_m, A_{k+1} : c_{k+1}, \ldots, A_n : c_n$$

and the program $P' = (P - \{\rho\}) \cup \{\rho'\}$ will be equivalent to P.

The substitution rule seems quite restrictive, it is however useful in optimizing the number of transformation steps, and also for proving the correctness of folding and of moving constraints around.

3.5 Replacement rule

Let P be a logic program and (ρ) a clause in P:

$$(\rho) \quad A \longleftarrow A_1 : c_1, \ldots, A_k : c_k, \ldots, A_n : c_n$$

assume that the following conditons 1) and 2) hold (cf. Definition 2.5):

1) $P \vdash \theta(A_1 : c_1 \wedge \cdots \wedge A_k : c_k)$ implies $P - (\rho) \vdash \theta'(B_1 : e_1 \wedge \cdots \wedge B_m : e_m)$ for some θ' such that θ and θ' agree on all the variables of $A, A_{k+1} : c_{k+1}, \ldots, A_n : c_n$, and $\theta(c_1 \wedge \cdots \wedge c_k) \Longrightarrow \theta'(e_1 \wedge \cdots \wedge e_m)$,

2) $P \vdash \theta(B_1 : e_1 \wedge \cdots \wedge B_m : e_m)$ implies $P - (\rho) \vdash \theta'(A_1 : c_1 \wedge \cdots \wedge A_k : c_k)$ for some θ' such that θ and θ' agree on all the variables of $A, A_{k+1} : c_{k+1}, \ldots, A_n : c_n$, and $\theta(e_1 \wedge \cdots \wedge e_m) \Longrightarrow \theta'(c_1 \wedge \cdots \wedge c_k)$.

Using these conditions we can replace the conjunction $A_1 : c_1 \wedge \cdots \wedge A_k : c_k$ by the conjunction $B_1 : e_1 \wedge \cdots \wedge B_m : e_m$, we will obtain the clause

$$(\rho') \quad A \longleftarrow B_1 : e_1, \ldots, B_m : e_m, A_{k+1} : c_{k+1}, \ldots, A_n : c_n$$

and the program $P' = (P - \{\rho\}) \cup \{\rho'\}$ will be equivalent to P.

This corresponds to the notions of sub-molecule, rule-subsumption and replacement of [20]: there is however a difference with the replacement rule of [20], namely: our notion of replacement which is taken from [12], where it is called permissible goal replacement, is more liberal (Maher's rule requires that none of the predicates of $A_1, \ldots, A_k, B_1, \ldots, B_m$ depend on the predicate of A, hence implies an implicit stratification of the program). There are cases when our replacement rule applies and not Maher's (cf. Example 3.5). Thinning and fattening can be viewed as particular cases of this rule [24, 1].

Example 3.5 Let P be the program

$$\begin{array}{l} p(X, X) \longleftarrow q(Y) \, , \, p(X, Y) : Y > 100 \\ p(X, Y) \longleftarrow r(Y) : Y > 100 \\ q(Y) \longleftarrow r(Y) : Y > 100 \end{array}$$

Then, according to our rule

$$P \vdash [(q(Y) \, , \, p(X, Y) : Y > 100) \Longleftrightarrow (q(Y) : Y > 100)]$$

Hence the first rule can be simplified into $p(X, X) \longleftarrow q(Y) : Y > 100$. Maher's replacement rule cannot be applied here because the predicate p occurs in the replaced conjunction.

3.6 Elimination of redundant atoms: thinning

This transformation enables us to suppress some useless constrained atoms in clause bodies (cf. thinning of [1]). Let (ρ) be the clause:

$$(\rho) \quad A \longleftarrow A_1 : c_1, \ldots, A_{i-1} : c_{i-1}, A_i : c_i, A_{i+1} : c_{i+1}, \ldots, A_n : c_n$$

If there exists in P a clause (π) such that, for instance,

$$(\pi) \quad B \longleftarrow A'_1 : c'_1, \ldots, A'_{i-1} : c'_{i-1}$$

and a substitution θ verifying:

$$A_i = \theta(B), \quad \text{and for } j = 1, \ldots, i-1, \quad A_j = \theta(A'_j),$$
$$\text{for } j = 1, \ldots, i-1, \quad c_j \Longrightarrow \theta(c'_j), \tag{18}$$
$$c_i = \theta(c'_1) \wedge \cdots \wedge \theta(c'_{i-1}) \tag{19}$$

this implies that $A_i : c_i \longleftarrow A_1 : c_1, \ldots, A_{i-1} : c_{i-1}$ is an instance of (π). Let the internal variables of (π) be those variables which occur in the body of (π) and do not occur in the head of (π). If

1. θ substitutes *distinct variables* for the internal variables of (π), and these variables do not occur in the head predicate A of (ρ), nor in the $\{A_1 : c_1, \ldots, A_{i-1} : c_{i-1}\} \cup \{A_{i+1} : c_{i+1}, \ldots, A_n : c_n\}$.

Then we can substitute for clause (ρ) the clause (ρ'):

$$(\rho') \quad A \longleftarrow A_1 : c_1, \ldots, A_{i-1} : c_{i-1}, A_{i+1} : c_{i+1}, \ldots, A_n : c_n$$

henceforth $P' = (P - \{\rho\}) \cup \{\rho'\}$ and P' is equivalent to P.

Thinning can be viewed as a particular case of replacement:

Example 3.6 Let (ρ) be clause 20 and (π) be clause 21 in the following program P:

$$r(X) \longleftarrow q(X) : X > 100 \, , \, p(X) : X > 100 \, , \, t(X) \tag{20}$$
$$p(X) \longleftarrow q(X) : X > 100 \tag{21}$$
$$p(X) \longleftarrow s(X) \tag{22}$$

The atom $p(X)$ can be deleted from the first clause 20. We obtain the equivalent program P':

$$P' : \begin{array}{l} r(X) \longleftarrow q(X) : X > 100 \, , \, t(X) \\ p(X) \longleftarrow q(X) : X > 100 \\ p(X) \longleftarrow s(X) \end{array}$$

Maher's replacement rule could also have been applied here.

Example 3.7 This example shows the necessity for condition 19, i.e. the conjunction of constraints of the clause π must be equivalent to the constraint of the redundant atom, namely c_i:

$$r(X) \longleftarrow q(X) : X > 100 \, , \, p(X) : X > 0 \, , \, t(X)$$
$$p(X) \longleftarrow q(X) : X > 200$$
$$p(X) \longleftarrow s(X)$$
$$q(X) \longleftarrow$$

In this case, we cannot suppress $p(X)$ from the first clause.

Example 3.8 This counter example shows the need for condition 1. concerning internal variables of the clause (π):

$$r(X) \longleftarrow q(Y) : X > 100 \,,\, p(X) : X > 100$$
$$p(X) \longleftarrow q(Y)$$
$$q(X) \longleftarrow$$

Deleting $p(X) : X > 100$ from the first clause would result in the program:

$$r(X) \longleftarrow q(Y) : X > 100$$
$$p(X) \longleftarrow q(Y)$$
$$q(X) \longleftarrow$$

which is different from the original program and equivalent to

$$r(X) \longleftarrow q(Y) : X > 100 \,,\, p(Z)$$
$$p(X) \longleftarrow q(Y)$$
$$q(X) \longleftarrow$$

Remark 3.9 If the applicability conditions of fold are satisfied, i.e. if (π) is a definition clause, then a better simplification will be achieved by first folding (ρ) by (π), then merging the two remaining atoms in the body of the folded clause into a single atom; i.e. the folded clause being:

$$A \longleftarrow A_i : c_i, \theta(B) : \theta(c'_1 \wedge \cdots \wedge c'_i), A_{i+1} : c_{i+1}, \ldots, A_n : c_n$$

with $A_i = \theta(B)$, and $c_i = \theta(c'_1) \wedge \cdots \wedge \theta(c'_i)$, the simplified clause is

$$A \longleftarrow A_i : c_i, A_{i+1} : c_{i+1}, \ldots, A_n : c_n.$$

In an application strategy of these rules, we will consider that in such a case folding has precedence over redundance elimination.

3.7 Introduction of a redundant atom in a clause body: fattening

It is the inverse of the previous operation (cf. [1]). Let (ρ) $A \longleftarrow A_1 : c_1, \ldots, A_i : c_i, A_{i+1} : c_{i+1}, \ldots, A_n : c_n$ assume there exists in P a clause (π) $B \longleftarrow B_1 : e_1, \ldots, B_i : e_i$ and a substitution θ such that:

$$A_j = \theta(B_j) \text{ for } j = 1, \ldots, i$$
$$c_j \Longrightarrow \theta(e_j) \text{ for } j = 1, \ldots, i$$

and $\bigwedge_{j=1,\ldots,i} c_j \Longrightarrow c = \bigwedge_{j=1,\ldots,i} \theta(e_j)$, then we can substitute (ρ') for (ρ):

$$(\rho') \quad A \longleftarrow A_1 : c_1, \ldots, A_n : c_n, \theta(B) : c$$

$P' = (P - \{\rho\}) \cup \{\rho'\}$ is equivalent to P.
For instance, in Example 3.6, P can be obtained by fattening from P'.

3.8 Suppressing useless (redundant) clauses: generalized pruning

A program can be simplified by suppressing its useless clauses, namely clauses which are subsumed by other clauses (cf. generalized pruning of [1]).

Let P be a program containing a clause (ρ). The clause (ρ) is said to be redundant if the set of atoms that it can produce is contained in the set of atoms produced by $P - \{\rho\}$. Such a clause cannot add any atom to the model of the program. A sufficient applicability condition is to check whether all the atoms produced by clause (ρ) are already in $T_{P-\{\rho\}} \uparrow \omega$.

Namely, letting P be a program containing clauses:

$$(\rho) \quad A \longleftarrow A_1 : c_1, \ldots, A_n : c_n$$

and

$$(\pi) \quad B \longleftarrow B_1 : e_1, \ldots, B_m : e_m \text{ with } m \leq n$$

and such that there exists a substitution θ verifying: $A = \theta(B)$, $A_i = \theta(B_i)$, and $c_i \Longrightarrow \theta(e_i)$, for $i = 1, \ldots, m$, i.e.: (ρ) contains as a subset in its body an instance of (π), then (ρ) is a redundant clause and can be deleted, and $P' = P - \{\rho\}$ is equivalent to P.

Example 3.10 Let P be the program:

$$(\rho) \quad p(x,y) \longleftarrow r(y) : x = y+1, z = y-1, y = 5;$$
$$q(x) : x = y+1, z = y-1, x > 0 \qquad (23)$$
$$(\pi) \quad p(x,y) \longleftarrow q(x) : x = y+1, z = y-1 \qquad (24)$$
$$q(x) \longleftarrow$$
$$r(y) \longleftarrow$$

Clause 23 contains an instance of clause 24, hence can be deleted.

Remark 3.11 A particular case of this simplification takes place when $m = n$. Then clause (ρ) is redundant, and it is an instance of clause (π) (cf. Definition 2.2). Hence every clause which is an instance of another clause can be deleted.

Example 3.12 Let P be the program:

$$p(X) \longleftarrow X = Y$$
$$p(X) \longleftarrow X = a \qquad (25)$$

Clause (25) is redundant and can be deleted. P is equivalent to P':

$$p(X) \longleftarrow X = Y$$

3.9 Suppression of clauses which can never be used

We can also simplify a program by deleting clauses which can never be applied because they contain an inconsistance in their body or are inconsistant with the rest of the program. Let $(\rho) \quad A \longleftarrow A_1 : c_1, \ldots, A_n : c_n$ be a clause of P. If the conjunction of constrained atoms in the body of (ρ) is unsatisfiable, the clause (ρ) can then be deleted, and $P' = P - \{\rho\}$ is equivalent to P.

3.10 Constraint simplifications

Equivalent constraints :
We can replace a clause by another clause whose constraints are equivalent to the constraints of the first clause.
Let (ρ) be the clause: (ρ) $A \longleftarrow A_1 : c_1, \cdots, A_n : c_n$ and assume that $\bigwedge_{i=1,\ldots,n} c_i \Longleftrightarrow \bigwedge_{i=1,\ldots,n} e_i$. Then we can replace (ρ) by (ρ')

$$(\rho') \quad A \longleftarrow A_1 : e_1, \cdots, A_n : e_n$$

It is a particular case of the substitution rule since clearly

$$P \Vdash (A_1 : c_1, \cdots, A_k : c_k) \Longleftrightarrow (B_1 : e_1, \cdots, B_m : e_m)$$

is satisfied with $k = m = n$, and, for $i = 1, \ldots, m$, $A_i = B_i$; the set of constrained atoms $A_{k+1} : c_{k+1}, \ldots, A_n : c_n$ in (ρ) is empty.
Hence $P' = (P - \{\rho\}) \cup \{\rho'\}$ is equivalent to P.
A particular case of the previous simplification is when for $i = 1, \ldots, m$, $c_i \Longleftrightarrow e_i$ and e_i is a simplified form of c_i.

Eliminating intermediate variables in constraints :
An intermediate variable in a clause, is a variable occurring in the body but not the head of the clause. Such variables can be deleted if they occur only in the constraints and not in the atoms of the body of the rule, we will say in such a case that the intermediate variable is a local variable. However, we will no longer have correctness, i.e. the simplified program can possibly have more solutions than the original program, so we can only hope for a notion of weak correctness. The notion of weak satisfaction of constraints (cf. [9]), see Definition 2.1, enables us to express solvability of constraints while deleting intermediate local variables. Let $X \cup Y$ be the variables of the clause, X being the set of variables occurring either in the head or in the atoms of the body, hereafter called *global variables*, and Y the set of variables occurring only in constraints called *intermediate local variables*; θ is said to satisfy the constraint c on the global variables X if and only if there exists a solution θ' of c on $(X \cup Y)$ such that $\theta'_{|X} = \theta$. We can simplify the original constraint by deleting the constraints on local intermediate variables and will obtain as simplified constraint the projection of the original constraint on the global variables.

Proposition 3.13 *Let P contain clause (ρ)* $A \longleftarrow A_1 : c_1, \ldots, A_n : c_n$. *Let* (ρ') $A \longleftarrow A_1 : c'_1, \ldots, A_n : c'_n$ *where* $c'_i = c_{i|X}$ *for* $i = 1, \ldots, n$. *Let* $P' = (P - \{\rho\}) \cup \{\rho'\}$. *Then* $M(P) \subseteq M(P')$.

However this result is not very interesting, in that it does not even imply partial correctness; henceforth, in order to obtain correctness, we must take a weaker semantics and assume the stronger hypotheses which are given below. If in addition the following holds (cf. Definition 2.1)

$$c'_i \Longrightarrow_X c_i \text{ for } i = 1, \ldots, n$$

then, since $c'_i = c_{i|X}$ implies that $c_i \Longrightarrow c'_i$, clause (ρ') will be weakly equivalent to clause (ρ) and $P' = (P - \{\rho\}) \cup \{\rho'\}$ will be weakly equivalent to P.

Example 3.14 (ρ) $\quad Q(u, v) \longleftarrow T(u) : v = u + 3, x > u, y = u + v$.
The constraints are $\{v = u + 3, x > u, y = u + v\}$; since e.g. on the reals or the integers $\{v = u + 3\} \Longrightarrow_{u,v} \{v = u + 3, x > u, y = u + v\}$, x and y are local intermediate variables which can be deleted; take for instance $\{x = u + 1\}$ and the constraint $\{x > u\}$ will be satisfied. We can substitute for (ρ) the clause (ρ'):
(ρ') $\quad Q(u, v) \longleftarrow T(u) : v = u + 3$.

4   CORRECTNESS OF THE TRANSFORMATION SYSTEM

The program transformation consists in applying to a program P together with a set of definitions D, the transformation rules. The result is another program $Tr(P)$. If a transformation preserves program equivalence with respect to a given semantics Sem, i.e. $Sem(Tr(P)) = Sem(P \cup D)$ then the transformation is said to be *totally correct*. It is said to be *partially correct* if $Sem(Tr(P)) \subseteq Sem(P \cup D)$. The system proposed by Burstall and Darlington [6] preserves only partial correctness of programs. The system of Tamaki-Sato was proved totally correct with respect to the least Herbrand model semantics [24] and with respect to the computed answer substitution semantics [16]. In [1], this system was also proved correct with respect to the S-semantics of [10].

All the rules of the system proposed in the previous section except for the elimination of intermediate variables, are correct with respect to a fixpoint semantics inspired from the usual semantics of [15] and [17], from the C-semantics of [10], and based on a notion of model in a Herbrand Universe with variables. Our proof method is inspired by the proof of Tamaki and Sato [24] but is much simpler.

4.1   Correctness of the transformation system

Proposition 4.1 *If P is transformed into P' via one unfold transformation, then $M(P) = M(P')$.*

Proposition 4.1 is proved in [5]. See also Lemma 6.4 of [20].

Proposition 4.1 showed that unfold transformations preserve total correctness. Conversely, the following lemma shows that the fold transformation can be viewed as the inverse of the unfold transformation, hence is also correct.

Lemma 4.2 *If P is transformed into P' via a fold transformation, then, P' can be transformed into P via an unfold transformation followed by a substitution rule.*

Correctness of the fold transformation then follows from the correctness of unfold and substitution (Propositions 4.1 and 4.4).

Proposition 4.3 *If P is transformed into P' by applying a fold transformation rule, then $M(P) = M(P')$.*

Proposition 4.4 *If P results in P' by applying the substitution rule, then $M(P) = M(P')$.*

Proposition 4.5 *1) If P results in P' by applying the replacement rule, then*
    (i) if only condition 1) of the replacement rule is satisfied, then $M(P) \subset M(P')$;
    (ii) if only condition 2) of the replacement rule is satisfied, then $M(P') \subset M(P)$, i.e. the replacement rule with only condition 2) is partially correct.
    2) If P results in P' by applying the replacement rule, with both conditions 1) and 2) satisfied, then $M(P) = M(P')$ and P' is equivalent to P.

Proposition 4.6 *If P results in P' by applying thinning (suppressing a redundant atom), fattening (adding a redundant atom), or generalized pruning, then $M(P) = M(P')$.*

Corollary 4.7 *If P is transformed into P' via a transformation $\mathcal{R}$, with $\mathcal{R}$ among unfold, replacement, thinning, fattening or generalized pruning, then $M(P) = M(P')$.*

Corollary 4.7, even though it may seem trivial, gives a very useful tool for checking the correctness of potential transformation rules.

Corollary 4.8 *Let $\mathcal{R}'$ be a transformation rule, let $P' = \mathcal{R}'(P)$ be the result of applying transformation $\mathcal{R}'$ to P, then $\mathcal{R}'$ is correct if and only if $M(P) = M(\mathcal{R}(P'))$, where $\mathcal{R}(P')$ is a suitable transformation of $P' = \mathcal{R}(P)$, chosen among unfold, replacement, thinning, fattening or generalized pruning.*

Proof: Since $\mathcal{R}$ is correct, $M(P') = M(\mathcal{R}(P'))$; $\mathcal{R}'$ is correct if and only if $M(P) = M(P')$, and this is equivalent to $M(P) = M(\mathcal{R}(P'))$. □

4.2 Weak correctness of the elimination of intermediate variables in constraints

We define a weaker semantics of a program P by taking into account the solvability of constraints.

Definition 4.9 *Let X be the global variables of P and Y its intermediate local variables and let $I \subset \mathcal{B}$ be an interpretation; the immediate consequence operator $T_{P,S}$ is defined as follows $A : c$ is an immediate consequence of P, I if and only if there exists an instance (cf. Definition 2.2) (ρ) $A \longleftarrow A_1 : c_1, \ldots, A_n : c_n$ of a clause (ρ') $B \longleftarrow B_1 : e_1, \ldots, B_n : e_n$ of P, there exist $A_i : d_i \in I$ for $i = 1, \ldots, n$, such that $c_i \Longrightarrow_X d_i$ for $i = 1, \ldots, n$, and $c \Longleftrightarrow_X c_1 \wedge \cdots \wedge c_n$ is solvable. Let $M_S(P)$ be defined as the least fixpoint of $T_{P,S}$ $M_S(P) = lfp(T_{P,S}) = T_{P,S} \uparrow \omega$.*

Proposition 4.10 *Assume that*

$$
\begin{array}{ll}
(\rho) & A \longleftarrow A_1 : c_1, \ldots, A_n : c_n \\
(\rho') & A \longleftarrow A_1 : c_1', \ldots, A_n : c_n'
\end{array}
$$

If P is tranformed in $P' = (P - \{\rho\}) \cup \{\rho'\}$ by deleting some intermediate local variables in (ρ), and moreover $c_i' \Longrightarrow_X c_i$ for $i = 1, \ldots, n$, then $M_S(P) = M_S(P')$.

5 CONCLUSION

We gave in this paper a transformation system for CLP programs and proved its correctness. This system is an extension of the Tamaki-Sato system to CLP programs; it includes fold, unfold, replacement, thinning and fattening, pruning and constraint simplifications; its correctness proof, which is based on the T_P operator, is simpler than the Tamaki-Sato proof; this simplicity stems from the naturalness of the T_P operator. Our formalism is a very natural extension of the Logic Programming methods to the CLP case, and it immediately gives back the results known in Logic Programming when the constraints are empty.

A general transformation system for CLP programs with a more restrictive and more syntactic fold has been studied in [20], and a fold/unfold transformation system for CLP programs with again a more restrictive fold has been recently introduced by [7]: this last system was proven correct with respect to the Ω-semantics, which has the liability of being very strong (few programs will be equivalent, e.g. programs P and P' of Example 3.12 are not equivalent with respect to the Ω-semantics) but the asset of being modular.

We prove for all transformations except constraint simplifications the strongest form of correctness from which one can deduce simply other equivalences with respect to weaker semantics; we thus conserve the advantage of proposing a simple and practical tool for automatic transformation systems. For constraint simplification however, we can only prove a weaker form of correctness taking into account the solvability of constraints.

Acknowledgments: We thank the referees and A. Naït Abdallah for insightful comments.

References

1. A.Bossi; N.Cocco. Basic transformation operations which preserve answer substitutions of logic programs, *Journal of logic programming*, Vol. 16, 1993, 47-87.
2. A.Bossi; N.Cocco. Preserving universal termination through unfold/fold, *Proceedings ALP'94*, to appear.
3. A.Bossi; S.Etalle. Transforming Acyclic Programs, *ACM Transactions on Programming Languages and Systems*, to appear 1994.
4. A. Bossi; N. Cocco; S. Dulli. A method for specializing logic programs, *ACM Trans. on programming langages and systems*, Vol.12, 2, April 1990, 253-302.
5. N. Bensaou; I. Guessarian. Transforming constraint logic programs, 11th *Symp. on Theoretical Aspects of Computer Science*, LNCS 775, 1994, 33-46.
6. R.M. Burstall; J. Darlington. A transformation system for deriving recursive programs, *J. ACM*, Vol. 24, 1, 1977, 44-67.
7. S. Etalle; M. Gabrielli. Modular transformations of CLP programs. *Proc. GULP-PRODE 1994*, to appear.
8. F. Denis; J.P. Delahaye. Unfolding, procedural and fixpoint semantics of logic programs, *Proc. STACS'1991*, LNCS 480, 1991, 511-522.
9. S.V. Denneheuvel; K.L. Kwast. Weak equivalence for constraint sets, *IJCAI*, 1991, 851-856.
10. M. Falaschi; G. Levi; M. Martelli; C. Palamidessi. Declarative Modeling of the Operational Behavior of Logic Languages, *Theoretical Computer Science* 69 , 1989, 289-318.
11. M Gabbrielli ; G. Levi. Modeling answer constraints in Constraint Logic Programs, *Proc. eight int. conf. on Logic Programming*, eds. Koichi & Furukawa, 1991, 238-252.
12. P.A. Gardner; J.C. Sheperdson. Unfold/fold transformations of logic programs, *Computational logic, essays in honor of Alan Robinson*, MIT Press, London, 1991, 565-583.
13. R. Giacobazzi; S.K. Debray; G. Levi. A generalized semantics for constraint logic programs, *Proc. Int. Conf. on Fifth Gen. Computer Systems*, Tokyo, 1992, 581-591.
14. T.J. Hickey; D.A. Smith. Toward the partial evaluation of CLP languages, *Proc. PEPM'91*, ACM-SIGPLAN Notices Vol. 26, 9, 1991, 43-51.
15. J. Jaffar; M.J. Maher. Constraint logic programming: a survey, to appear in J. Logic Programming.
16. T.Kawamura; T.Kanamori. Presrvation of stronger equivalence in unfold/fold logic program transformation, *Proc. Intern. Conf. on FGCS*, Tokyo (1988), 413-421.
17. P. Kanellakis; G. Kuper; P. Revesz. Constraint Query Languages, Tech. report, Department of Computer Science, Brown university, November 1990.
18. G. Levi. Models, unfolding rules and fixpoint semantics, *Proc. of the fifth international conf. on Logic programming*, 1988, 1649-1665.
19. M.J. Maher. Correctness of a logic program transformation system, IBM Research Report RC 13496, T.J. Watson Research center, 1987.
20. M.J. Maher. A transformation system for deductive database modules with perfect model semantics, *Theoretical Computer Science* 110, 1993, 377-403.
21. A. Parrain; P. Devienne; P. Lebegue. Techniques de transformations de programmes généraux et validation de meta-interpréteurs, *BIGRE* 1991.
22. M. Proietti; A. Pettorossi. An abstract strategy for transforming logic programs, *Fundamenta Informaticae*, Vol. 18, 1993, 267-286.
23. M. Proietti; A. Pettorossi. Semantics preserving transformation rules for Prolog, *Proc. PEPM'91*, ACM-SIGPLAN Notices Vol. 26, 9, 1991, 274-284.

24. H. Tamaki; T. Sato. Unfold/Fold transformation of logic programs, *Proc. 2nd logic programming conference*, Uppsala, Sweden, 1984.

6   APPENDIX: TECHNICAL PROOFS

Definition 6.1 *Immediate consequence operator T_P – Let P be a CLP program and $I \subset B$ be an interpretation; a constrained atom $A : c$ is an immediate consequence of P, I if and only if there exists an instance (cf. Definition 2.2) (ρ) $A \longleftarrow A_1 : c_1, \ldots, A_n : c_n$ of a clause (ρ') $B \longleftarrow B_1 : e_1, \ldots, B_n : e_n$ of P such that $A_i : c_i \in I$ for $i = 1, \ldots, n$, and $c = c_1 \wedge \cdots \wedge c_n$. Let $T_P(I)$ denote the set of immediate consequences of P, I, and $\hat{T}_P(I) = \{A : c \mid A : c \text{ is an instance of } A' : c' \in T_P(I)\}$ denote the upwards closure of the set of immediate consequences of P, I.*

Proof of Proposition 3.13 Recall that (ρ) $A \longleftarrow A_1 : c_1, \ldots, A_n : c_n$ (ρ') $A \longleftarrow A_1 : c'_1, \ldots, A_n : c'_n$ and $P' = P - \{\rho\} \cup \{\rho'\}$. Note that $c'_i = c_{i|X}$ implies that $c_i \Longrightarrow c'_i$, hence (ρ) is an instance of (ρ').

By induction on k we check that $T_P^k(\emptyset) \subseteq T_{P'}^k(\emptyset)$ for all $k \in \mathbb{N}$.

It is clear for $k = 0$. Assume it holds for k, i.e. $I_k = T_P^k(\emptyset) \subseteq T_{P'}^k(\emptyset) = I'_k$, and let $A' : c' \in T_P^{k+1}(\emptyset)$: we check that $A' : c' \in T_{P'}^{k+1}(\emptyset)$.

a) If $A' : c'$ is obtained by instanciating a clause in $P - \{\rho\}$, then $A' : c' \in T_{P'}(I'_k)$.

b) If $A' : c'$ is obtained by an instance θ' of (ρ), then $\exists \theta'(A_1 : c_1), \ldots, \theta'(A_n : c_n) \in I_k$ such that $\theta'(A : c) = A' : c' \in T_P(I_k)$. Hence $\theta'(A_j : c_j) \in I_k \subseteq I'_k$, for $j = 1, \ldots, n$. But $\theta'(A_j : c_j)$, for $j = 1, \ldots, n$, is an instance of $\theta'(A_j : c'_j)$, hence by 6.1, rule (ρ') can be applied and $\theta'(A : c) \in T_{P'}(I'_k)$. $\quad\square$

Proof of Proposition 4.1 See [5]. $\quad\square$

Proof of Lemma 4.2 Recall that

(ρ) $\quad A \longleftarrow A_1 : c_1, \ldots, A_q : c_q, A_{q+1} : c_{q+1}, \ldots, A_{q+r} : c_{q+r}, A_{q+r+1} : c_{q+r+1}, \ldots, A_n : c_n$
(δ) $\quad B \longleftarrow \qquad\qquad\qquad B_1 : c'_1, \ldots, B_r : c'_r$
(τ) $\quad A \longleftarrow \quad A_1 : c_1, \ldots, A_q : c_q, \mu(B) : c_{q+1} \wedge \cdots \wedge c_{q+r}, A_{q+r+1} : c_{q+r+1}, \ldots, A_n : c_n$

and $P' = (P - \{\rho\}) \cup \{\tau\}$. We will show that P can be retrieved from P' by unfolding and substitution; the correctness of folding will follow from the correctness of unfolding and substitution.

Let us try to unfold (τ) at the atom $\mu(B)$ in P'. Indeed, since (δ) is the only clause in P whose head is unifiable with $\mu(B)$ and since the heads of the clauses in P' are the same as the heads of the clauses in P, (δ) is also the only clause in P' whose head is unifiable with $\mu(B)$, and (τ) can only be unfolded by (δ). Assuming $B = B(X_1, \ldots, X_n)$, θ defined by $\theta(X_i) = \mu(X_i)$ for $i = 1, \ldots, n$ is clearly a most general unifier of B and $\mu(B)$. Because of the condition 2. of folding, (μ substitutes *distinct variables* for the internal variables of (δ), and these variables do not occur in (τ)), θ can be extended to all the variables of $(\tau) \cup (\delta)$ by letting

$$\theta(y) = \begin{cases} \mu(y), & \text{if } y \text{ is an internal variable of } (\delta); \\ y, & \text{if } y \text{ is a variable of } (\tau). \end{cases}$$

Hence, letting $c = c_{q+1} \wedge \cdots \wedge c_{q+r}$ and unfolding (τ) at $\mu(B) : c$ with most general unifier θ results in (ρ')

(ρ') $\quad \theta(A \longleftarrow A_1 : c_1, \ldots, A_q : c_q, B_1 : c \wedge c'_1, \ldots, B_r : c \wedge c'_r,$
$\qquad\qquad A_{q+r+1} : c_{q+r+1}, \ldots, A_n : c_n)$
(ρ') $\quad A \longleftarrow A_1 : c_1, \ldots, A_q : c_q, \mu(B_1) : c \wedge \mu(c'_1), \ldots, \mu(B_r) : c \wedge \mu(c'_r),$
$\qquad\qquad A_{q+r+1} : c_{q+r+1}, \ldots, A_n : c_n)$
(ρ') $\quad A \longleftarrow A_1 : c_1, \ldots, A_q : c_q, A_{q+1} : c \wedge \mu(c'_1), \ldots, A_{q+r} : c \wedge \mu(c'_r),$
$\qquad\qquad A_{q+r+1} : c_{q+r+1}, \ldots, A_n : c_n)$

since by the condition 1. of the folding rule (see implication 17), for $j = 1, \ldots, r$, $c_{q+j} \implies \mu(c'_j)$, we also have $c \implies \mu(c'_j)$ for $j = 1, \ldots, r$ and (ρ') can be replaced by

$$(\rho'') \quad A \longleftarrow A_1 : c_1, \ldots, A_q : c_q, A_{q+1} : c, \ldots, A_{q+r} : c, A_{q+r+1} : c_{q+r+1}, \ldots, A_n : c_n$$

it is now immediate by the substitution rule that (ρ'') can be replaced by (ρ) (cf. Section 3.10). □

Proof of Proposition 4.4 Let P, (ρ) and (ρ') be such that

$$(\rho) \quad A \longleftarrow A_1 : c_1, \ldots, A_k : c_k, \ldots, A_n : c_n$$
$$P \Vdash [A_1 : c_1 \wedge \cdots \wedge A_k : c_k \iff B_1 : e_1 \wedge \cdots \wedge B_m : e_m] \quad (26)$$
$$(\rho') \quad A \longleftarrow B_1 : e_1, \ldots, B_m : e_m, A_{k+1} : c_{k+1}, \ldots, A_n : c_n$$

and let $P' = P - \{\rho\} \cup \{\rho'\}$. P' is equivalent to P, i.e. $M(P) = M(P')$.

Let $I_n = T_P^n(\emptyset)$ and $T_{P'}^n(\emptyset) = I'_n$: we prove by induction on n that $I_n = I'_n$ for all n. The base case is clear. Let us check the inductive step: assume $I_n = I'_n$ and let $\theta(A' : c') \in T_P(I_n) = I_{n+1}$;
– if $\theta(A' : c')$ is obtained by applying a rule in $P - \{\rho\}$, the inductive step is clear;
– if $\theta(A' : c')$ is obtained by applying the rule (ρ), then $\theta(A_1 : c_1 \wedge \cdots \wedge A_k : c_k) \in I_n = T_P(I_{n-1})$; by 26, (cf. Definition 2.5), $\theta(B_1 : e_1 \wedge \cdots \wedge B_m : e_m) \in I_n$ and by the induction hypothesis $\theta(B_1 : e_1 \wedge \cdots \wedge B_m : e_m) \in I'_n$, hence $\theta(A' : c') \in T_{P'}(I'_n) = I'_{n+1}$. □

Proof of Proposition 4.5 2) clearly follows from 1) (i) and (ii). So, let us prove 1). Let P, (ρ) and (ρ') be such that

$$(\rho) \quad A \longleftarrow A_1 : c_1, \ldots, A_k : c_k, \ldots, A_n : c_n$$
$$(\rho') \quad A \longleftarrow B_1 : e_1, \ldots, B_m : e_m, A_{k+1} : c_{k+1}, \ldots, A_n : c_n$$

and let $P' = P - \{\rho\} \cup \{\rho'\}$.

Let us first prove 1) (i); assume that condition 1) of the replacement rule, which we recall below, is satisfied.

1) $P \vdash \theta(A_1 : c_1 \wedge \cdots \wedge A_k : c_k)$ implies $P - (\rho) \vdash \theta'(B_1 : e_1 \wedge \cdots \wedge B_m : e_m)$ for some θ' such that θ and θ' agree on all the variables of $A, A_{k+1} : c_{k+1}, \ldots, A_n : c_n$, and $\theta(c_1 \wedge \cdots \wedge c_k) \implies \theta'(e_1 \wedge \cdots \wedge e_m)$,

Let us then show that $M(P) \subset M(P')$. Let $I_n = T_P^n(\emptyset)$ and $T_{P'}^n(\emptyset) = I'_n$: we prove by induction on n that

$$\text{for all } n \text{ there exists } m \text{ such that } I_n \subset I'_m \quad (27)$$

The base case is clear. Let us check the inductive step: assume $I_l \subset I'_j$ and let $\theta(A' : c') \in T_P(I_l) = I_{l+1}$;
– if $\theta(A' : c')$ is obtained by applying a rule in $P - \{\rho\}$, the inductive step is clear;
– if $\theta(A' : c') = \theta(A : c')$ is obtained by applying the rule (ρ), then

$$\theta(A_1 : c_1 \wedge \cdots \wedge A_k : c_k) \in I_l = T_P(I_{l-1}) \quad (28)$$
$$\text{and}$$
$$\theta(A_{k+1} : c_{k+1} \wedge \cdots \wedge A_n : c_n) \in I_l = T_P(I_{l-1}) \quad (29)$$

with $c' = c_1 \wedge \cdots \wedge c_k \wedge c_{k+1} \wedge \cdots \wedge c_n$; 29 together with the induction hypothesis implies that $\theta(A_{k+1} : c_{k+1} \wedge \cdots \wedge A_n : c_n) \in I'_j$; 28 implies that $P \vdash \theta(A_1 : c_1 \wedge \cdots \wedge A_k : c_k)$, hence by condition 1) $P - (\rho) \vdash \theta'(B_1 : e_1 \wedge \cdots \wedge B_m : e_m)$, hence there exists a j' such that $\theta'(B_1 : e_1 \wedge \cdots \wedge B_m : e_m) \in T_{P-(\rho)}^{j'}(\emptyset) \subset I'_{j'}$. Let $j'' = \sup(j, j')$, then $\theta'(B_1 : e_1 \wedge \cdots \wedge B_m : e_m) \in I'_{j''}$, and $\theta'(A_{k+1} : c_{k+1} \wedge \cdots \wedge A_n : c_n) = \theta(A_{k+1} : c_{k+1} \wedge \cdots \wedge A_n : c_n) \in I'_{j''}$, hence, applying rule (ρ') we obtain $\theta'(A : e_1 \wedge \cdots \wedge e_m \wedge c_{k+1} \wedge \cdots \wedge c_n) \in T_{P'}(I'_{j''}) = I'_{j''+1}$; as $\theta'(e_1 \wedge \cdots \wedge e_m \wedge c_{k+1} \wedge \cdots \wedge c_n) = \theta'(e_1 \wedge \cdots \wedge e_m) \wedge \theta(c_{k+1} \wedge \cdots \wedge c_n)$ and since $\theta(c_1 \wedge \cdots \wedge c_k) \implies \theta'(e_1 \wedge \cdots \wedge e_m)$ and $\theta'(A) = \theta(A)$, $\theta(A : c') = \theta(A : c_1 \wedge \cdots \wedge c_k \wedge$

$c_{k+1} \wedge \cdots \wedge c_n)$ is an instance of $\theta'(A : e_1 \wedge \cdots \wedge e_m \wedge c_{k+1} \wedge \cdots \wedge c_n)$, hence $\theta'(A : c')$ is in $I'_{j''+1}$.

To prove 1) (ii), we assume that the condition 2) of the replacement rule, is satisfied, namely

  2) $P \vdash \theta(B_1 : e_1 \wedge \cdots \wedge B_m : e_m)$ implies $P - (\rho) \vdash \theta'(A_1 : c_1 \wedge \cdots \wedge A_k : c_k)$ for some θ' such that θ and θ' agree on all the variables of $A, A_{k+1} : c_{k+1}, \ldots, A_n : c_n$, and $\theta(e_1 \wedge \cdots \wedge e_m) \Longrightarrow \theta'(c_1 \wedge \cdots \wedge c_k)$.
Mutatis mutandis, a proof similar to the previous one will show that

$$\text{for all } n \text{ there exists } m \text{ such that } I'_n \subset I_m \tag{30}$$

Whence it will follow that $M(P') \subset M(P)$. □

Proof of Proposition 4.6 1) *Correctness of thinning*
Let
  (ρ) $A \longleftarrow A_1 : c_1, \ldots, A_{i-1} : c_{i-1}, A_i : c_i, A_{i+1} : c_{i+1}, \ldots, A_n : c_n$
  (π) $B \longleftarrow A'_1 : c'_1, \ldots, A'_{i-1} : c'_{i-1}$
  (ρ') $A \longleftarrow A_1 : c_1, \ldots, A_{i-1} : c_{i-1}, A_{i+1} : c_{i+1}, \ldots, A_n : c_n$

be as in Section 3.6, with a substitution θ such that

$$A_i = \theta(B) , \quad \text{and for } j = 1, \ldots, i-1 , \quad A_j = \theta(A'_j),$$
$$\text{for } j = 1, \ldots, i-1 , \quad c_j \Longrightarrow \theta(c'_j), \tag{31}$$
$$c_i = \theta(c'_1) \wedge \cdots \wedge \theta(c'_{i-1}) \tag{32}$$

(i) $M(P) \subset M(P')$ since (ρ') subsumes (ρ), i.e. whenever (ρ) applies, (ρ') also applies and gives the same consequence;
(ii) moreover, since condition 2) of the replacement rule is satisfied by taking

$$B_1 : e_1, \ldots, B_m : e_m = A_1 : c_1, \ldots, A_{i-1} : c_{i-1}$$
$$A_1 : c_1, \ldots, A_k : c_k = A_1 : c_1, \ldots, A_{i-1} : c_{i-1}, A_i : c_i$$

$M(P') \subset M(P)$ (cf. Proposition 4.5 1) (ii)).

2) *Correctness of fattening*
  Let (ρ) $A \longleftarrow A_1 : c_1, \ldots, A_i : c_i, A_{i+1} : c_{i+1}, \ldots, A_n : c_n$ and (π) $B \longleftarrow B_1 : e_1, \ldots, B_i : e_i$ be in P, and θ be a substitution such that: $C \longleftarrow A_1 : c_1, \ldots, A_i : c_i$ is an instance of (π), with $\theta(B) = C$ and moreover $\bigwedge_{j=1,\ldots,i} c_j \Longrightarrow c = \bigwedge_{j=1,\ldots,i} \theta(e_j)$, $c_j \Longrightarrow \theta(e_j)$ for $j = 1, \ldots, i$. Let (ρ'):

$$(\rho') \quad A \longleftarrow A_1 : c_1, \ldots, A_n : c_n, C : c$$

and $P' = P - \{\rho\} \cup \{\rho'\}$. Apply now the thinning transformation to P' and (ρ'); (ρ') contains in its body an instance $C \longleftarrow A_1 : c_1, \ldots, A_i : c_i$ of (π), the applicability conditions for the thinning transformation are satisfied, and applying this transformation we obtain $P'' = P' - \{\rho'\} \cup \{\rho''\}$. Moreover, it is immediate to check that $(\rho'') = (\rho)$, hence $P'' = P$ and $M(P) = M(P'') = M(P')$ by the correctness of thinning, and P' is equivalent to P.

3) *Correctness of pruning*
  Let (ρ) $A \longleftarrow A_1 : c_1, \ldots, A_n : c_n$ and (π) $B \longleftarrow B_1 : e_1, \ldots, B_m : e_m$ with $m \leq n$ and such that there exists a substitution θ verifying: $A = \theta(B)$, $A_i = \theta(B_i)$, $c_i \Longrightarrow \theta(e_i)$, for $i = 1, \ldots, m$.
  Let $P' = P - \{\rho\}$, then $T_P \uparrow \omega = T_{P'} \uparrow \omega$.
  1) Clearly, $T_{P'} \uparrow \omega \subseteq T_P \uparrow \omega$ since the set of clauses of P' is contained in P.
  2) $T_P \uparrow \omega \subseteq T_{P'} \uparrow \omega$. By induction on n we check that $T_P^n(\emptyset) \subseteq T_{P'}^n(\emptyset)$ for all $n \in \mathbb{N}$.
  It is clear for $n = 0$. Assume it holds for n, i.e. $I_n = T_P^n(\emptyset) \subseteq T_{P'}^n(\emptyset) = I'_n$, and let $A' : c' \in T_P^{n+1}(\emptyset)$: we check that $A' : c' \in T_{P'}^{n+1}(\emptyset)$.

a) If $A' : c'$ is obtained by instanciating a clause in $P - \{\rho\}$, then $A' : c' \in T_{P'}(I_n) \subseteq T_{P'}(I'_n)$.

b) If $A' : c'$ is obtained by an instance θ' of (ρ), then $\exists \theta'(A_1 : c_1), \ldots, \theta'(A_n : c_n) \in I_n$ such that $\theta'(A : c) = A' : c' \in T_P(I_n)$. Hence $\theta'(A_j : c_j) \in I_n \subseteq I'_n$, for $j = 1, \ldots, m, \ldots, n$. There exists a substitution θ such that $A = \theta(B)$ $A_i = \theta(B_i)$ $c_i \implies \theta(e_i)$, for $i = 1, \ldots, m$. Hence $\theta'(\theta(B_j) : \theta(e_j)) \in I'_n$, for $j = 1, \ldots, m$. By clause (π) we obtain $\theta'(\theta(B) : \bigwedge_{j=1,\ldots,m} \theta(e_j)) \in T_{P'}(I'_n)$. Since all interpretations are upwards-closed and $c = \bigwedge_{j=1,\ldots,n} c_j \implies \bigwedge_{j=1,\ldots,m} \theta(e_j)$, we have $\theta'(A : c) \in T_{P'}(I'_n)$. $\square$

Proof of Proposition 4.10 Let P contain clause (ρ) $A \longleftarrow A_1 : c_1, \ldots, A_n : c_n$. We can substitute for (ρ) the clause ρ' (ρ') $A \longleftarrow A_1 : c'_1, \ldots, A_n : c'_n$, where $c'_i = c_{i|X}$ for $i = 1, \ldots, n$. Let $P' = (P - \{\rho\}) \cup \{\rho'\}$.

Assume that the local intermediate variables have been renamed so that the local variables occurring in the c_i's for $i = 1, \ldots, n$ are pairwise distinct.

a) $M_S(P) \subseteq M_S(P')$ can be proved as in Proposition 3.13 above: i.e. $c'_i = c_{i|X}$ implies that $c_i \implies c'_i$, hence (ρ) is an instance of (ρ').

b) Let us check conversely that $M_S(P') \subseteq M_S(P)$.

Let $I_{S,k} = T_{P,S}^k(\emptyset)$ and $I'_{S,k} = T_{P',S}^k(\emptyset)$. By induction on k we prove that $I'_{S,k} \subseteq I_{S,k}$, $\forall k \in \mathbb{N}$.

This is clear for $k = 0$; assume it holds for k and check that $I'_{S,k+1} \subseteq I_{S,k+1}$. Let $A' : c' \in I'_{S,k+1}$ then

1. If $A' : c'$ is obtained by a clause of $P' - \{\rho'\}$, then $A' : c' \in I_{S,k+1}$.
2. If $A' : c'$ is obtained by an instance of clause (ρ') then $\exists \theta'$ instance of (ρ') such that $\theta'(A : \bigwedge_{j=1,\ldots,n} c'_j) = A' : d' \in I'_{S,k+1}$ and $\theta'(\bigwedge_{j=1,\ldots,n} c'_j) \Longleftrightarrow_X c'$ is solvable; whence $\theta'(A_j) : \theta'(d'_j) \in I'_{S,k}$ with $\theta'(c'_j) \implies_X \theta'(d'_j)$ and $\theta'(c'_j)$ is solvable for $j = 1, \ldots, n$. By the induction hypothesis, we deduce that $\theta'(A_j) : \theta'(d'_j) \in I_{S,k}$, hence also $\theta'(A_j) : \theta'(c'_j) \in I_{S,k}$ since $\theta'(c'_j) \implies_X \theta'(d'_j)$. Since $c'_j \implies_X c_j$ for $j = 1, \ldots, n$, $\theta'(c'_j)$ solvable implies that $\theta_j(c_j)$ solvable for some θ_j such that $\theta_{j|X} = \theta'$.

Since the local variables occurring in the c_i's for $i = 1, \ldots, n$ are pairwise distinct, μ can be well-defined by

$$\mu(y) = \begin{cases} \theta'(y) = \theta(y) & \text{if } y \in X \text{ is a global variable} \\ \theta_j(y) & \text{if } y \text{ is a variable local to constraint } c_j \end{cases}$$

and μ is such that for $j = 1, \ldots, n$, $\mu(c_j) = \theta_j(c_j)$, $\mu(A_j) = \theta'(A_j) = \theta_j(A_j)$, $\mu(A) = \theta'(A) = A'$, $(\mu(A_j) : \mu(c_j)) = (\theta'(A_j) : \theta_j(c_j)) \in I_{S,k}$ and $\mu(c_j)$ solvable. We can thus apply rule (ρ) and $\mu(A : \bigwedge_{j=1,\ldots,n} c_j) = A' : d \in I_{S,k+1}$. Hence also, by the definition of $I_{S,k+1}$, since c' is solvable, and $c' = \bigwedge_{j=1,\ldots,n} c'_j \Longleftrightarrow_X d = \bigwedge_{j=1,\ldots,n} c_j$ (because on the one hand c_j is an instance of c'_i, and on the other hand $c'_j \implies_X c_j$), $A' : c' = \mu(A) : c' \in I_{S,k+1}$. $\square$

Using Call/Exit Analysis
for Logic Program Transformation

Dmitri Boulanger*   Maurice Bruynooghe**

Department of Computer Science, Katholieke Universiteit Leuven
Celestijnenlaan 200 A, B-3001, Heverlee, Belgium
email: {dmitri, maurice}@cs.kuleuven.ac.be

Abstract. A technique for transformation of definite logic programs is presented. A first phase performs an analysis of the extended call/exit patterns of the source program. It is shown that a particular form of correct abstract call/exit patterns can be used as a guide to control the transformation itself and can help to generate the target program having desired properties. The technique provides a framework which, combined with problem specific information concerning the source program, can lead to nontrivial transformations.

1 Introduction

The most popular approaches for definite logic program transformations are based on unfold/fold and goal replacement operations applied to the clauses of a source program. The unfold/fold transformations have been introduced in [5] and later were applied to logic programming in [25, 15, 24]. The recent paper [13][3] revisits the framework developed by Tamaki and Sato.

In this paper we are developing another approach to logic program transformation. Namely, we consider transformations which are performed into two phases: a first phase performs a complete static data flow analysis to derive a transformation guide of the source program, the second phase follows the guide to transform the source logic program (here we follow some basic ideas of [14, 12]). In [1, 2] we have introduced a rather general and powerful approach to derive unfold/fold transformations of definite logic programs by abstract interpretation. The unfold/fold transformations, which can be obtained in this way, have been shown to be more general than those of Tamaki and Sato. Indeed, as it was pointed out in [18, 13], SLD-like tree analysis can produce transformations which cannot always be expressed in terms of unfold/fold transformations as described in [25, 15, 24].

---

* Supported by the K.U.Leuven. Permanent address: Keldysh Institute for Applied Mathematics, Russian Academy of Science, Miusskaya sq., 4, 125047 Moscow, Russia.
** Supported by the Belgian National Fund for Scientific Research.
[3] The examples presented in this paper show that the approach of Tamaki and Sato is not sufficiently sound.

The paper aims at introducing a logic program transformation framework, which is capable of performing complex transformations by applying unfold/fold and goal replacement transformations which are specialised by problem specific information which are specialised . It is clear, that a toolkit having such a rich set of elementary operations has to be controlled by high level specifications, which are to be sufficiently expressive to describe the target of the transformation of the source program. In this respect the central idea of our approach is to specify the behaviour of the target program in the form of its possible call/exit patterns. The source program can have a lot of different sets of call/exit patterns (by choosing different computation rules), so it is possible to choose the most suitable ones and to obtain by transformation the program, having the specified call/exit patterns, i.e. to fix by transformation the desired behaviour of the program.

The paper is organised as follows: after some preliminaries in section 3 we introduce a special variant of SLD resolution, which is used as basis for our approach. Afterwards we present a technique to represent call/exit patterns and use them as a guide during execution of the definite logic program. The results of the execution are used to transform the program. The new program reflects the specified call/exit structure. In section 4 we give an extensive example, which shows an advanced technique usable for deriving non trivial transformations.

2 Preliminaries

In what follows the standard terminology and the basic knowledge of the theory of logic programming, as can be found in [20], is assumed. We will use the standard notions of SLD derivation and refutation, unification and idempotent substitution. The properties of unification and substitutions presented in [9, 19] are used indirectly throughout the paper.

The capital letters B, Q and R denote conjunctions of atoms. In the sequel they are considered as *collections* and are called *blocks*. The letters A and H denote atoms. Where convenient, we write $E(\overline{X})$ to denote a syntactical object E (block, atom, goal or other expression over the set of terms) with variables $\overline{X}$. By $var(E)$ we denote the set of variables, occurring in the syntactical object E (so $var(E(\overline{X})) = \overline{X}$).

The greek letters θ, σ, φ and ψ will be used to denote idempotent substitutions. Given a substitution θ, $\theta \mid_{var(E)}$ will denote the restriction of the substitution θ to the variables occurring in E. Given an expression E over the set of terms, $E\theta$, the instance of E by θ, is defined as usual.

The greek letter χ is used to denote set of constraints or equations. The most general unifier of a solvable set of constraints is an idempotent substitution σ which we denote by $solv(\chi)$. Given the well known correspondence between solved form and idempotent substitutions [19], we sometimes write $eq(\sigma)$ to denote the solved form corresponding to a substitution σ. With χ a set of constraints, by $\chi\sigma$ we mean $\chi \cup eq(\sigma)$. An empty set of constraints $\emptyset$ corresponds to the identity substitution $\{\}$. We consider a definite logic program P to be a set of definite clauses $\{c_1, \ldots, c_n\}$ equipped with an initial goal G_0. The successive goals in a SLD derivation for the initial goal wrt P will be denoted by G_i, $i = 0, 1, \ldots, n$. In the context of this paper we will always deal with programs having exactly the same initial goal $G_0 =\leftarrow \epsilon$, where the predicate symbol ϵ never appears in the bodies of the clauses $\{c_1, \ldots, c_n\}$. The predicate ϵ is used to declare the "entry points" of a program. Moreover, given two logic programs P_1 and P_2, the intersection of the corresponding languages includes all function symbols but only one predicate symbol, namely ϵ. It is clear, that all these restrictions cannot influence the generality of the presentation. Finally, we have a simple notion of equivalence of programs: two programs P_1 and P_2 are equivalent iff they are logically equivalent wrt the common language (an extensive discussion of the topic can be found in [21]). This notion of logic program equivalence is the most appropriate one for a wide class of applications.

3 Using Extended Call/Exit Patterns for Controlling Logic Program Transformation

In this section we present an algorithm for logic program transformation based upon call/exit analysis of an SLD-like tree of the source program P. So firstly we describe a special variant of SLD-resolution to be used as the main engine for the call/exit analysis.

3.1 Extended OLD Resolution

Extended OLD resolution was introduced by us in [1, 2] and deviates from the standard SLD resolution only by imposing certain restrictions on the computation rule. Namely, the SLD derivation

$$G_0 \xrightarrow{c_1 \; \theta_1} G_1, \ldots, \xrightarrow{c_n \; \theta_n} G_n$$

is an EOLD derivation for the initial goal G_0 wrt a definite logic program P provided that any goal G_i in the derivation is represented as an ordered sequences $< B_1, B_2, \ldots B_m >$, $m \geq 0$ of conjunctions of atoms (blocks) and the computation rule always selects an atom from the first block. Given a goal G represented as an *ordered* sequence of blocks, $\|G\|$ will denote the collection of all atoms occurring in G (In OLD resolution [26, 16] blocks have only one atom).

The following definition gives a more precise description of an EOLD resolution step.

Definition 1. *EOLD Resolvent*
Let G_i be a goal represented as $< B_1, B_2, \ldots, B_m >$, $m \geq 1$, A the atom selected from B_1, c_{i+1} a standardised apart input clause from P, θ_{i+1} the *mgu* of A and the head of the clause c_{i+1}. Then G_{i+1} is derived from G_i and c_{i+1} using *mgu* θ_{i+1} via the EOLD computation rule if G_{i+1} is obtained by applying the following two steps:

- construct the auxiliary goal $G' =< B'_1, B_2, \ldots, B_m >$, where $B'_1 = B_1$ with A replaced by the possibly empty body of c_{i+1}
- if B'_1 is empty, then $G_{i+1} =< B_2, \ldots, B_m > \theta_{i+1}$; else partition the first block B'_1 of G' into a number of non empty parts (blocks) $B'_{11}, \ldots, B'_{1k}$, $k \geq 1$, and let $G_{i+1} =< B'_{11}, \ldots, B'_{1k}, B_2, \ldots, B_m > \theta_{i+1}$ blocks

$\square$

In what follows the definition of a subrefutation in an EOLD derivation will play an important role.

Definition 2. *EOLD Subrefutation*

An EOLD subderivation of the form $G_i \xrightarrow{c_{i+1} \, \theta_{i+1}} G_{i+1}, \ldots, \xrightarrow{c_n \, \theta_n} G_n$ is an EOLD subrefutation for the first block B_1 of the goal $G_i =< B_1, B_2, \ldots, B_k >$, $k \geq 1$ if either $G_n =<>$ and $k = 1$ (i.e. the subderivation is a refutation for B_1) or there exists a substitution σ^4, such that $|| < B_2, \ldots B_k > \sigma || = ||G_n||$ and there is no goal G_j, $i \leq j < n$ having this property in the subderivation. The substitution $\theta_{i+1}\theta_{i+2} \ldots \theta_n$ is called an *answer substitution* of the subrefutation. $\square$

Notice that EOLD resolution is sound and complete because soundness and completeness of SLD resolution is independent from the computation rule [20].

Any expression identical, up to variable renaming, with the first block B of the goal G is said to be an *extended call* of the goal (it will be denoted by $call(G)$). The corresponding *extended answer* (it will be denoted by $answ(G)$) is any expression identical, up to variable renaming, with $B\theta$, where θ is the answer substitution of some refutation of block B. Extended call/exit patterns of EOLD trees describe important properties of programs. So below we present a technique for safely approximating the complete set of call/exit patterns.

3.2 A Safe Approximation of Call/Exit Patterns

Consider a complete (and, thus, possibly infinite) EOLD tree for a logic program P. The *complete Call/Exit cover* of the tree is the set of all call/exit pairs of the form $[K \Rightarrow Answ(K)]$, where the key K is an extended call and the set $Answ(K)$ is the set of all corresponding extended answers. In general the complete CE-cover is infinite. So we need a finite abstraction which is a safe approximation. A finite correct abstract CE-cover for an EOLD tree can be defined as follows. Suppose there is some equivalence relation over the set of all blocks such that the set of distinct equivalence classes is finite. Then the set of keys occurring in the complete CE-cover can be represented by a finite set $\{K_1^{\sim}, K_2^{\sim}, \ldots, K_n^{\sim}\}$ of equivalence classes. Let $Answ^{\sim}(K^{\sim})$ be the finite set of equivalence classes of blocks occurring in a set

$$\bigcup_{K \in K^{\sim}} Answ(K).$$

---

[4] In the context of EOLD resolution the substitution σ is exactly the substitution $\theta_{i+1} \ldots \theta_n$. Below we will use the modified EOLD resolution, where this property is not always the case.

Then the *complete abstract CE-cover* is a finite set of abstract call/exit pairs

$$\{[K_i^{\sim} \Rightarrow Answ^{\sim}(K_i^{\sim})] \mid i = 1, \ldots, n\}.$$

Let G be a goal occurring in an EOLD tree. Given a complete abstract CE-cover, the extended call $call(G)$ is said to be *covered* by the key $K^{\sim}$ iff $call(G) \in K^{\sim}$. Similarly, the extended answer $answ(G)$ is said to be *covered* iff the corresponding extended call $call(G)$ is covered by the key $K^{\sim}$ and $answ(G)$ belongs to some equivalence class occurring in the set $Answ^{\sim}(K^{\sim})$.

The complete abstract cover requires all call/exit pairs to be covered. In the sequel we will need a more flexible description of program properties: we will allow that not all call/exit pairs are covered. This will be used to control the transformation of the source program P (see below). Consider the following more flexible description of call/exit patterns.

Definition 3. *Safe Abstract Call/Exit Cover*
Given an EOLD tree, a safe abstract CE-cover of the tree is a subset of the complete abstract finite CE-cover such that the number of not covered extended calls and answers occurring in the EOLD tree is finite.   □

In what follows we will always assume that the EOLD tree and the corresponding safe abstract CE-cover satisfy the following conditions:[5]

- Every equivalence class can be represented by a block of atoms. All elements of an equivalence class are instances of its corresponding block. Thus we will not distinguish between equivalence classes and their corresponding blocks.
- The block size of an EOLD tree is bounded by some constant.

The conditions above ensure that any EOLD tree has a (safe) abstract finite CE-cover. For example, a trivial CE-cover can be obtained using only recursive predicate symbols of the program P with distinct variables as arguments to construct the blocks of the equivalence classes.

Example 1. The empty set is a safe abstract CE-cover for any EOLD tree, which can be constructed for any logic program P having a finite extended minimal model EMM_P (see [10]) provided that the block size is bounded. Also the set

$$\{[q(\overline{X}) \Rightarrow EMM_P(q)] \mid q \in P\}$$

is always a safe (possibly infinite) CE-cover for any program P, where $q \in P$ are the predicates of P and $EMM_P(q) \subseteq EMM_P$ are the elements of the extended minimal model of P having a predicate symbol q. Moreover it is complete CE-cover.

On the other hand, the empty set is not a safe cover of any OLD tree, which can be constructed for the program

$$\varepsilon(X) \leftarrow even(s(X)).$$
$$even(0).$$
$$even(s(s(X))) \leftarrow even(X).$$

---
[5] These conditions are not strictly necessary. We use them to simplify the presentation.

with the standard goal $\leftarrow \varepsilon(X)$, because the number of not covered answers is infinite. Notice, that the number of not covered calls is finite. The abstract CE-cover

$$\{[even(X) \Rightarrow \{even(Y)\}]\}$$

is safe for any OLD tree, which can be constructed for the above program. $\quad\square$

A safe abstract CE-cover can be used to control EOLD resolution and to construct a tree having a special shape.

3.3 Modified EOLD Resolution

Our framework uses a special variant of EOLD resolution. In order to introduce it we need some auxiliary definitions. Firstly, we will use *extended goals*, which are sequences of blocks and *exit markers*. An exit marker is a block which is syntactically different from any block of atoms and has the form $\triangle_G(\sigma, \chi)$, where G is a goal in the current EOLD derivation, σ is a substitution and χ is a set of *constraints*. The goal G is the goal, where the exit marker has been inserted. The initial goal always has the form $G_0 =< \varepsilon, \triangle_{G_0}(\{\}, \emptyset) >$. Given a substitution θ, we will assume that

$$\triangle_G(\sigma, \chi)\theta = \triangle_G(\sigma, \chi \cup eq(\theta)).$$

Secondly, we will use two special operations: a key factorisation and an exit marker elimination. The former inserts an exit marker, while the latter deletes it. *Key factorisation* is a transformation of a goal $G =< B_1, B_2, \cdots, B_n >$, where B_1 is a block of atoms. If there exists a standardised apart key $K = K(\overline{W})$ in the CE-cover, i.e. $var(K) = \overline{W}$, $\overline{W} \cap var(G) = \emptyset$, and a substitution σ, $dom(\sigma) = \overline{W}$ such that $B_1 = K\sigma$ (i.e. the block B_1 is covered by the key K), then the key factorisation *replaces* the goal G by $< K, \triangle_G(\sigma, \emptyset), B_2, \cdots, B_n >$. It is important to notice that after key factorisation the first block of the goal has "fresh" variables, which differ from the variables in the other blocks of atoms.

The *exit marker elimination* is applicable to a non singleton goal if the goal has at least two exit markers and if its first element is an exit marker. Let

$$G =< \triangle_{G_1}(\sigma_1, \chi_1), B_1, \ldots, B_m, \triangle_{G_2}(\sigma_2, \chi_2), \cdots >, \ m \geq 0$$

be a goal, where $\triangle_{G_2}(\sigma_2, \chi_2)$ is the *second* exit marker in the goal. There exists always at least one exit maker because any goal contains the exit marker of the initial goal, i.e. any goal has the form $< \cdots, \triangle_{G_0}(\{\}, \chi) >$. Suppose that $\theta = solv(\chi_1) \neq fail$ and $G_1 =< K(\overline{W}), \triangle_{G_1}(\sigma_1, \emptyset), \cdots >$ with $dom(\sigma) = \overline{W}$ (cf. key factorisation above). Then the safety of the CE-cover ensures (see def. 4 below, which describes the structure of a derivation), that there exist a key $K = K(\overline{W})$ and a corresponding answer $K\sigma_{answ}$ with $dom(\sigma_{answ}) = \overline{W}$ such that $K\theta$ is an instance of $K\sigma_{answ}$, i.e.

$$K\theta = (K\sigma_{answ})\varphi, \text{ and } solv(eq(\sigma_{answ}) \cup eq(\varphi)) = solv(\chi_1)$$

Then the exit marker elimination *creates* the new goal $G' = exit(G)$,

$$G' =< B_1, \ldots, B_m, \triangle_{G_2}(\sigma_2, \chi_2 \cup eq(\varphi)) \cdots > \psi,$$

where $\psi = solv(eq(\sigma_1) \cup eq(\sigma_{answ}))$. The substitution ψ is exactly the part of the answer which is "prescribed" by the CE-cover to instantiate the remaining part of the goal. Therefore, ψ is applied to the whole goal, while the not allowed part φ is added to the constraints of the next exit marker. In this way φ is isolated (but not lost!) in the next exit marker. This explains why it is always necessary to have at least one exit marker – it accumulates the parts of the answers which have to be delayed.

The extensions above allow to modify standard EOLD resolution as follows. Let us denote by $resolve_{EOLD}(G, c)$ the goal, which can be derived from the goal G and input clause $c \in P$ by applying a standard EOLD resolution step (cf. def. 1).

Definition 4. *(Modified) EOLD* Resolvent*
Given a CE-cover and an extended goal $G =< B_1, B_2, \cdots, B_n >$, the (modified) EOLD* resolution step

$$G_i \xrightarrow{\theta_{i+1}} G_{i+1}, \quad G_{i+1} = resolve^*_{EOLD}(G_i, c_{i+1})$$

consists of *one* of the following operations:

1. If B_1 is a block of *atoms* (not an exit marker), derive the intermediate goal

$$G' = resolve_{EOLD}(G_i, c_{i+1})$$

and construct the corresponding substitution θ_{i+1} and:
If the goal G' has the form $< B'_1, B'_2, \cdots >$, where B'_1 is a block of atoms such that the key factorisation on B'_1 is applicable,
then $G_{i+1} =< K, \triangle_{G_{i+1}}(\sigma, \emptyset), B'_2, \cdots >$
else (the key factorisation is not applicable) $G_{i+1} = G'$.
2. If the goal G has the form $< \triangle_G(\sigma, \chi), \cdots >$ (the first block is an exit marker) do:
if the exit marker elimination is applicable on $\triangle_G(\sigma, \chi)$
then $G_{i+1} = exit(G_i)$, $\theta_{i+1} = \sigma$
else (exit marker elimination is not applicable) the goal is non-extendable

$\square$

In the sequel we will assume that *all* failed branches of the modified EOLD* tree are dropped. A branch of the tree is considered to be a failed branch iff it contains a non extendable goal G such that it is either a failed goal in standard sense (cf. def. 1) or the set of all constraints occurring in G is not solvable. Then definition 4 above ensures the following simple properties of the modified EOLD* trees:

– A goal G is a non extendable goal iff it is a success goal of the form $G =< \triangle_{G_0}(\{\}, \chi) >$, $solv(\chi) \neq fail$.

- The definition of subrefutation (cf. def. 2) and the notions of extended calls and answers, which have been introduced above in the context of standard EOLD resolution (cf. section 3.1), are also applicable to the goals of the modified EOLD* trees (including the goals having an exit marker as a first element), but the answer substitutions have another meaning (see prop. 5 and 6 below).
- Each covered extended call occurring in the tree occurs as a key in the CE-cover

Consider a standard EOLD tree and the corresponding modified EOLD* tree constructed wrt some safe finite abstract CE-cover. The relation between them is given in the following proposition:

Proposition 5. *The EOLD* tree constructed using some safe finite abstract CE-cover contains a derivation*

$$G_0 \xrightarrow{\theta_1} G_1, \ldots, \xrightarrow{\theta_n} G_n$$

having the final success goal G_n of the form $< \triangle_{G_0}(\{\}, \chi) >$, $\sigma = solv(\chi) \neq fail$ (i.e. the refutation of the initial goal) iff the standard EOLD tree contains a refutation

$$G_0 \xrightarrow{\theta'_1} G'_1, \ldots, \xrightarrow{\theta'_m} G'_m, \; n \geq m$$

such that $G'_n = <>$ and $solv(eq(\theta_1) \cup \ldots \cup eq(\theta_n) \cup \chi) = solv(\chi) = \theta'_1 \ldots \theta'_m$. $\square$

The following proposition describes the most important feature of EOLD* resolution.

Proposition 6. *Given an EOLD* tree constructed using some safe finite abstract CE-cover, any extended call of an extendable goal occurring in the tree has a finite number of corresponding extended answers, while the length of the corresponding subrefutation and their total number can be unbounded.* $\square$

The propositions 5 and 6 are very important for the transformation algorithm introduced below.

3.4 Transformation Algorithm

Given an EOLD* tree constructed wrt some *safe* abstract CE-cover, the algorithm for generating a new logic program Π is the following:

1. *Introducing New Predicate Symbols*: for each key K different from ϵ in the CE-cover generate a fresh predicate symbol π/n, where n is the number of distinct variables occurring in the key K [6]. The predicate symbol ϵ is also included in the set of predicate symbols of the program Π. In this way each *covered* extended call in the EOLD* tree is *associated* with a new predicate symbol of Π.

---

[6] In [3] we give powerful conditions, which allow for dropping some arguments of the new predicates.

2. *Synthesising New Program Clauses*: for each subrefutation of the first block of a goal G having the form

$$G \xrightarrow{\theta_1} G_1, \ldots, \xrightarrow{\theta_n} G_n$$

where $G = < K, \triangle_G(\sigma, \emptyset), \cdots >$ and $G_n = < \triangle_G(\sigma, \chi), \cdots >$ (i.e. the key factorisation has been successful for the goal G), construct a new clause of the program Π of the form

$$(\ \pi_0(\overline{X}_0) \leftarrow \pi_{i_1}(\overline{Y}_{i_1}), \ldots, \pi_{i_m}(\overline{Y}_{i_m}) \ )solv(\Sigma)$$

where:

- π_0 is a new predicate symbol associated with the block of atoms B_0 (covered extended call of G) and $\overline{X}_0 = var(B_0)$
- $G_{i_j} \neq G, \ j = 1, \ldots, m$ are all goals occurring in the subrefutation B_0 of the form

$$G_{i_j} = < B_{i_j}, \triangle_{G_{i_j}}(\sigma_{i_j}, \chi_{i_j}), \cdots, \triangle_G(\cdots) \cdots >,$$

such that the exit marker $\triangle_G(\sigma_{i_j}, \chi_{i_j})$ is exactly the *second* element in the goal (the latter means that we consider only the top level covered extended calls occurring in the subrefutation of B_0)
- π_{i_j} is the predicate associated with the covered extended call of G_{i_j} (which corresponds to the block of atoms B_{i_j}) and $\overline{Y}_{i_j} = var(B_{i_j})$, $j = 1, \ldots, m$
- $\Sigma = \bigcup_k eq(\theta_{i_k})$, where θ_{i_k} is the substitution labelling the EOLD* step

$$G_{i_k - 1} \xrightarrow{\theta_{i_k}} G_{i_k}$$

and the goals $G_{i_k} \neq G, \ k = 1, \ldots, l$ are all the goals in the subrefutation of B_0 having the form

$$G_{i_k} = < Bs_{i_k}, \triangle_G(\sigma, \chi_{i_k}), \cdots >$$

or

$$G_{i_k} = < B_{i_k}, \triangle_{G_{i_k}}(\sigma_{i_k}, \chi_{i_k}), \cdots, \triangle_G(\cdots) \cdots >$$

where Bs_{i_k} is a (possibly empty) sequence of blocks of atoms (here we consider all top level calls occurring in the subrefutation of B_0)

For the clauses which are synthesised from the refutations of the initial goal $G_0 = < \varepsilon, \triangle_{G_0}(\{\}, \emptyset) >$ the predicate ε is used to construct the clause head. Notice that $\overline{X}_0, \overline{Y}_{i_j}, \ j = 1, \ldots, m$ are disjoint sets of variables.

3. *The Final Program*: consider the set of all clauses, which can be obtained following the algorithm above. If clauses which are renaming of each other are considered identical, then the set of clauses is finite. Moreover, the new program Π is logically equivalent to the source program P wrt the common language, which consists of only one predicate symbol ε and all function symbols.

To be convinced that the algorithm always generates a finite program, it is sufficient to notice the following:

- All goals of the form $G = < K, \Delta_G(\cdots) \cdots >$ have exactly the same (up to the variable renaming) set of subrefutations for the first block, because the key factorisation always creates the first block with fresh variables, and thus the set of subrefutations is independent from the particular goal in the tree. In other words, the influence of the other blocks on the goal is isolated (delayed) in the exit marker in the form of a set of constraints. This means, that the final program has a finite number of predicate definitions (one for each key in the CE-cover and one for ϵ).
- The number of clauses in each definition is finite. Suppose that there is a definition having an infinite number of clauses. This implies, that the extended call, which corresponds to the predicate symbol of the definition, has an infinite number of extended answers. This contradicts proposition 6. For the same reason the number of atoms in the bodies of the clauses is also bounded.

The equivalence of the source program P and the program Π can be derived by using proposition 5 and by showing that any EOLD* refutation of P, which is constructed wrt a chosen safe finite CE-cover, can be simulated by some OLD* refutation (EOLD* with singleton blocks!) of Π^7, which is constructed wrt the trivial CE-cover (constructed from the new predicates having distinct variables as arguments) and vice versa.

Let us provide some important remarks concerning the applicability of the logic program transformation framework, which has been introduced above. The presentation above was given under assumption that there exists an EOLD tree having a safe CE-cover. Finding the EOLD tree and the CE-cover is intended to be solved separately during a first phase. In the next section we will give several remarks concerning this problem.

It should be clear from the discussion above, that we really need to construct only some upper portion of the EOLD* tree. Namely, all keys occurring in the CE-cover and some portion of the corresponding subrefutations have to be discovered in the tree[8]. This problem is also addressed in the next section.

The presentation itself of the transformation algorithm was given in a form, which was as simple as possible (the price of the simplicity is that the algorithm uses too many intermediate variables, which are not always necessary). For the same reason a tabulation mechanism [26]) was not included too. Here we were mainly interested in the careful investigation of the class of logic program transformations (see sections 4 and 5 below), which can be derived in our framework. So we have omitted technical details related to an efficient implementation of the

---

[7] This means, that the algorithm is capable to derive *FOLD* transformations.

[8] Using the specially constructed CE-cover we significantly extend the "upper portion" algorithms of [22, 23]. For example, in this way we avoid *Eureka*-steps when generating complex folding transformations (see also [1, 2]).

algorithm (a variant of the algorithm. which is rather efficient from an implementation point of view, but has a more narrow class of derivable transformations can be found in [1, 2]).

Finally, notice, the great importance of the particular structure of the chosen safe abstract CE-cover. It can be used to control the structure of the new program: the keys specify the granularity of the definitions of Π, while the answers can be used to control the internal structure of the definitions and the structure of the clauses of Π. By allowing some extended calls and answers not to be covered (cf. example 1), we can specify what should be factored out by partial evaluation: all procedure calls not corresponding to the keys in the CE-cover are partially evaluated (cf. [18]). Thus, the particular form of the equivalence relation over the set of blocks is crucial. It seems that, by choosing non trivial equivalence relations, one can obtain rather complex and deep transformations[9].

4 Deriving Logic Programs Using Call/Exit Patterns of the Source Program

As could be seen from above the EOLD* interpretation and subsequent transformation algorithm are strictly controlled by the safe abstract CE-cover of some EOLD tree. So the first problem is to prove the existence of the desired EOLD tree and to find a safe approximation of its call/exit patterns. This problem is well known and several frameworks based on abstract top-down computation equipped with the tabulation mechanism have been suggested [26, 16, 7]. It was shown, that abstract tabled computations can be done correctly for a very wide class of abstract domains [16, 7]. These algorithms allow to construct complete finite abstract CE-covers of some OLD tree using some equivalence relation over the set of atoms. Definition 1 has been elaborated keeping in mind the applicability of these algorithms to the extended call/exit patterns, which are used in our framework. Thus, we can assume, that we are given some equivalence relation $\mathcal{R}$ and the corresponding complete abstract CE-cover $\mathcal{CE}(\mathcal{R})$. Notice, that $\mathcal{CE}(\mathcal{R})$ can imply a particular blocking strategy of some EOLD tree, i.e. the rule for partitioning atoms of the current goal into blocks (cf. def. 1).

Below we provide an example of a program transformation, which uses a more general form of $\mathcal{R}$ than was described in section 3.1. This example is intended to demonstrate the capabilities of the framework and the usefulness of nontrivial equivalence relations over the set of atoms.

Consider the logic program P given in example 1. The abstract CE-cover

$$\{[even(s(X)) \Rightarrow \{even(s(s(0))), < even(s(s(Y))), Y \neq 0 >\}]\}$$

is *not* a safe abstract CE-cover for P, because the not covered call $even(X)$ has an infinite number of exit patterns. Some modification of the framework presented in section 3 allows to derive a finite new program. The OLD* tree

---

[9] In [3] we use an equivalence relation over the set of blocks, which considers two blocks to be equivalent if they are equivalent goals wrt the source program P.

contains the following simple derivation

- $G_0 = <\epsilon(X), \Delta_0(\{\}, \emptyset)>$

  $\downarrow$ $\qquad\qquad\qquad\qquad\qquad\qquad\qquad\qquad\qquad \theta_1 = \{\}$

- $G_1 = <even(s(W1)), \Delta_1(\{W1 \leftarrow X\}, \emptyset), \Delta_0(\{\}, \emptyset)>$

  $\downarrow$ $\qquad\qquad\qquad\qquad\qquad\qquad\qquad\qquad \theta_2 = \{W1 \leftarrow s(W2)\}$

- $G_2 = <even(W2), \Delta_1(\{W1 \leftarrow X\}, \{W1 = s(W2)\}), \Delta_0(\{\}, \{W1 = s(W2)\})>$

  $\downarrow$ $\qquad\qquad\qquad\qquad\qquad\qquad\qquad\qquad\qquad \theta_3 = \{W2 \leftarrow 0\}$

- $G3 = \Delta_1(\{W1 \leftarrow X\}, \{W1 = s(0), W2 = 0\}), \Delta_0(\{\}, \{W1 = s(0), W2 = 0\}))>$

  $\downarrow$ $\qquad\qquad\qquad\qquad\qquad\qquad\qquad\qquad\qquad \theta_4 = \{W1 \leftarrow X\}$

- $G4 = \Delta_0(\{\}, \{W1 = s(0), W2 = 0, X = s(0)\})>$

which produces the clauses:

- the clause $\epsilon(X) \leftarrow \pi_1(X)$, which was obtained from the subrefutation of $call(G_0)$ by

$$(\epsilon(X) \leftarrow \pi_1(W1))solv(eq(\theta_1) \cup eq(\theta_4))$$

- the clause $\pi_1(s(0))$, which was obtained from the subrefutation of $call(G_1)$ by

$$(\pi_1(W1))solv(eq(\theta_2) \cup eq(\theta_3))$$

The subrefutations for $call(G_1)$ of the form $G_1 \xrightarrow{\theta_2} G_2 \xrightarrow{\theta_5} G_5 \longrightarrow \cdots$, where $\theta_5 = \{W2 \leftarrow s(s(W3))\}$ and
$G_5 = <even(W3), \Delta_1(\{W1 \leftarrow X\}, \{W1 = s(s(s(W3))), W2 = s(s(W3))\}), \Delta_0(\{\}, \{\cdots\})>$
will produce an infinite number of clauses $\pi_1(s(s(s(0)))), \pi_1(s(s(s(s(s(0)))))), \cdots$, if the algorithm from section 3.4 is used.

Thus we have to control the uncovered exit patterns of the call $even(X)$ "manually". We can apply apply some "massaging" on G_5 such that we avoid the the infinite number of distinct subrefutations for $even(W3)$. It can be done as follows. The substitution θ_5 is the first one, which creates an "incomplete" but already not allowed answer for $call(G_1)$. Thus θ_5 has to be made more flat. In general it is a complex operation, which is not always possible (because we should preserve correctness of all refutations in the tree), but in this particular case the following reconstruction is correct: $\theta_5' = \{W2 \leftarrow s(W4)\}$ and
$G_5' = <even(s(W4)), \Delta_1(\{W1 \leftarrow X\}, \{W1 = s(s(W4)), W2 = s(W4)\}), \Delta_0(\{\}, \{\cdots\})>$,
Now key factorisation on G_5' is applicable! One obtains: $G_5'' =$
$< even(s(W5)), \Delta_5(\{W5 \leftarrow W4\}, \emptyset), \Delta_1(\{W1 \leftarrow X\}, \{W1 = s(s(W4)), W2 = s(W4)\}),$
$\Delta_0(\{\}, \{\cdots\})>$.
The reconstruction above is ensured by the available abstract CE-cover. Indeed, the variable $W3$ occurs in only one atom of the goal G_5 and it can possibly be instantiated by resolving $even(W3)$. On the other hand, the CE-cover ensures, that the new introduced variable $W4$ occurring in G_5' can be instantiated only by $s(U)$ when resolving $even(s(W4))$ in G_5' and thus, the variable $W2$ will have the same value as in the original derivation, provided that all constraints in G_5' are modified accordingly. This reconstruction can be considered as applying a special equivalence relation over the set of atoms, which can be defined wrt the source program P. On the other hand, this reconstruction is something more

powerful than the well known goal replacement (see [13, 17, 3]). In [2] it is shown how the formal proof of the equivalence above can be obtained.

The reconstructed subrefutations of the form $G_1 \xrightarrow{\theta_2} G_2 \xrightarrow{\theta'_5} G''_5 \xrightarrow{\theta_6} G_6 \cdots$, where $call(G_6) = even(W6)$ and $\theta_6 = \{W5 \leftarrow s(W6)\}$, will produce only one clause $\pi_1(s(s(W4))) \leftarrow \pi_1(W4)$ by

$$(\pi_1(W1) \leftarrow \pi_1(W5))solv(eq(\theta_2) \cup eq(\theta'_5) \cup \{W5 = W4\}),$$

i.e. the following new program is generated

$$\Pi = \{\varepsilon(X) \leftarrow \pi_1(X), \quad \pi_1(s(0)), \quad \pi_1(s(s(X))) \leftarrow \pi_1(X)\}.$$

Notice, that it was sufficient to consider only some upper portion of the tree. In general, if a covered call has s answers in the CE-cover, then only s derivations, which are passing it should be considered (this can be implemented by applying the corresponding tabulation mechanism). In the case of OLD* resolution by allowing to consider several answers we extend the algorithms of [11, 12], where only one answer can be considered. In complex applications similar to meta-programs specialisation multiple answers can help to increase the granularity of residual programs. By using OLD* resolution with CE-covers having only singleton sets of answers, our algorithm reduces to that [12]. On the other hand in the example above the specification of the pair of answers was important and was used in another way (otherwise the goal reconstruction would not be possible).

The new program Π is the *odd numbers* program. This kind of transformations cannot be produced using the frameworks of [11, 12, 17, 22, 23] (in our case it has been possible due to incorporating of special information about the source program). This example shows that our approach can be used in a problem specific way by introducing specialised equivalence relations. We expect that very interesting transformations can be obtained by using equivalence relations over blocks of unbounded size (the number of equivalence classes should be finite). In this way folding of an infinite number of atoms can be derived (the recent paper [4] gives an example). Equivalence relations of the latter kind can produce transformations similar to [6, 8]. Extensive examples of standard "finite" folding transformations, which can be obtained by applying a weaker version of the presented algorithm are given in [1, 2].

5 Conclusion and Related Work

Our framework allows to consider the source logic program as a "runnable" specification, which should be used to derive a "real" program, which satisfies some requirements. Indeed, one can perform the following step:

1. Choose the most suitable abstract domain for the program at hand. The domain should satisfy the necessary conditions for the correct application of an abstract tabulation mechanism (see for example [16, 7]). This step is

not formal and should take into account all available information about the program and its application (sometimes it can be provided by the users of the program).

2. In terms of the abstract domain specify the desired call/exit patterns of the target program. The following items can be specified:
   - What should be factored out by partial evaluation
   - The macro structure of the program, i.e. the number of the definitions and their granularity
   - The internal structure of the clauses

   Prove by applying an abstract tabulation technique, that the specified call/exit patterns imply the existence of an EOLD tree.
3. Apply the algorithms described in sections 3.3 and 3.4.

The basic ideas of the approach allow to integrate in one toolkit lots of existing transformation techniques for definite logic programs [18, 13, 11, 12, 22, 23] combined with high level concepts to specify the properties of the target program and possibilities to incorporate problem specific information. It seems, that the development of a general form for the requirements on the equivalence relations over the set of blocks is a promising direction of research.

References

1. Boulanger,D., Bruynooghe,M., *Deriving Transformations of Logic Programs using Abstract Interpretation*, Logic Program Synthesis and Transformation (LOPSTR'92), eds. K.K.Lau and T.Clement, Workshops in Computing, Springer Verlag, 1993, 99-117.

2. Boulanger,D., Bruynooghe,M., *Deriving Fold/Unfold Transformations of Logic Programs Using Extended OLDT-based Abstract Interpretation*, J. Symbolic Computation, 1993, Vol.15, 495-521.

3. Boulanger,D., Bruynooghe,M., *Using Abstract Interpretation for Goal Replacement*, Logic Program Synthesis and Transformation LOPSTR'93, Workshops in Computing, Springer-Verlag, 1993.

4. Boulanger,D., De Schreye,D., *Compiling Control Revisited: A New Approach based upon Abstract Interpretation*, Proc. 11-th Int. Conf. Logic Programming, 1994, 699-713.

5. Burstall,R., Darlington,J., *A Transformation System for Developing Recursive Programs*, JACM, Jan.1977, Vol.24, No.1, 44-67.

6. Bruynooghe,M., De Schreye,D., Krekels,B., *Compiling Control*, J. Logic Programming, 1989, Vol.6, Nos.2-3, 135-162.

7. Codognet,P., File, G., *Computations, Abstractions and Constraints in Logic Programs*, Proc. 4th Int. Conf. on Programming Languages, Oakland, USA, April 1992.

8. De Schreye,D., Martens,B., Sablon,G., Bruynooghe,M., *Compiling Bottom-Up and Mixed Derivations into Top-Down Executable Logic Programs*, J. Automated Reasoning, 1991, 337-358.

9. Eder,E., *Properties of Substitutions and Unifications*, J. Symbolic Computation, 1985, Vol.1, No.1, 31-46.

10. Falaschi,M., Levi,G., Martelli,M., Palamidessi,C., *Declarative Modelling of the Operational Behaviour of Logic Languages*, Theoretical Computer Science, 1989, Vol.69, No.3, 289-318.

11. Gallagher,J., Bruynooghe,M., *Some Low Level Transformations of Logic Programs*, Proc. 2nd Workshop in Meta-Programming in Logic, Leuven, 1990, 229-244.

12. Gallagher,J., Bruynooghe,M., *The Derivation of an Algorithm for Program Specialisation*, New Generation Computing, 1991, Vol.9, 305-333.

13. Gardner, P., Shepherdson, J., *Unfold/Fold Transformations of Logic Programs*, Computational Logic, Essays in Honour of Alan Robinson, eds. J.L.Lassez and G.Plotkin, MIT Press, 1991, 565-583.

14. Gallagher, J., Codish M., Shapiro E., *Specialisation of Prolog and FCP Programs Using Abstract Interpretation*, New Generation Computing, 1988, Vol.6, Nos.2-3, 159-186.

15. Kawamura, T., Kanamori, T., *Preservation of Stronger Equivalence in Unfold/Fold Logic Program Transformation*, Proc. 4th Int. Conf. on FGCS, Tokyo,1988.

16. Kanamori, T., Kawamura, T., *Abstract Interpretation Based on OLDT Resolution*, J. Logic Programming, 1993, Vol.15, Nos.1-2, 1-30 .

17. Kawamura, T., *Derivation of Efficient Logic Programs by Synthesising New Predicates*, Proc. 1991 Int. Symp. on Logic Programming, San Diego, 1991, 611-625.

18. Lloyd,L., Shepherdson,J., *Partial Evaluation in Logic Programming*, J. Logic Programming, 1991, Vol.11, Nos.3-4, 217-242.

19. Lassez, J.-L., Maher, M., Mariott, K., *Unification Revisited*, Foundations of Deductive Databases and Logic Programming, ed. J.Minker, Morgan-Kaufmann, 1988, 587-625.

20. Lloyd,L., *Foundations of Logic Programming*, Springer-Verlag, Berlin, 1987.

21. Maher,M., *Equivalences of logic Programs*, Foundations of Deductive Databases and Logic Programming, ed. J.Minker, Morgan-Kaufmann, 1988, 627-658.

22. Proietti,M., Pettorossi,A., *Construction of Efficient Logic Programs by Loop Absorption and Generalisation*, Proc. 2nd Workshop in Meta-Programming in Logic, Leuven, 1990, 57-81.

23. Proietti,M., Pettorossi,A., *Unfolding - Definition - Folding, In this Order, For Avoiding Unnecessary Variables in Logic Programs*, Proc. 3rd Int. Symp. on Programming Languages Implementation and Logic Programming, Aug. 1991, LNCS No.528, Springer-Verlag, 1991, 347-358.

24. Seki,H., *Unfold/Fold Transformation of stratified programs*, J. Theoretical Computer Science, 1991, Vol.86, 107-139.

25. Tamaki,H., Sato,T., *Unfold/Fold Transformation of Logic Programs*, Proc. 2nd International Conference on Logic Programming, Uppsala, 1984, 127-138.

26. Tamaki,H., Sato,T., *OLD Resolution with Tabulation*, Proc. 3rd Int. Conf. on Logic Programming, London, July 1986, 84-98.

A Transformation System for Definite Programs Based on Termination Analysis

J. Cook and J.P. Gallagher*

Department of Computer Science, University of Bristol, Bristol BS8 1TR, U.K.

Abstract. We present a goal replacement rule whose main applicability condition is based on termination properties of the resulting transformed program. The goal replacement rule together with a multi-step unfolding rule forms a powerful and elegant transformation system for definite programs. It also sheds new light on the relationship between folding and goal replacement, and between different folding rules. Our explicit termination condition contrasts with other transformation systems in the literature, which contain conditions on folding and goal replacement, often rather complex, in order to avoid "introducing a loop" into a program. We prove that the goal replacement rule preserves the success set of a definite program. We define an extended version of goal replacement that also preserves the finite failure set. A powerful folding rule can be constructed as a special case of goal replacement, allowing folding with recursive rules, with no distinction between *old* and *new* predicates. A proof that Seki's transformation system preserves recurrence, an important termination property, is outlined.

1 Introduction

In this paper we define a goal replacement rule (for definite programs) based on termination analysis. Under the conditions of the rule, a replacement can only be made if the resulting program terminates for all ground calls to the head of the transformed clause. We show that application of this rule to a program will preserve its success set. The finite failure set can also be preserved by adding a similar termination condition on the program being transformed. We also give a definition of a more general unfolding rule which is based on a partial evaluation rule discussed in [LS91] and [GS91].

On reviewing the literature regarding folding and goal replacement, it was apparent that the underlying issue (when considering the correctness of such rules) was the termination of the resulting program. This was particularly evident in [Sek89] and [BCE92].

In the first of these papers, Seki, discussing the unfold/fold transformations of Tamaki and Sato [TS84], noted that a goal may not terminate in a program resulting from a Tamaki-Sato folding rule, even though it finitely failed in the original program. To combat this, Seki proposed a reformulated folding rule.

---

* Correspondence to jpg@compsci.bristol.ac.uk

Under the conditions of [BCE92], a goal replacement can be made provided that 'a loop is not introduced' into the program. This is clearly a reference to some termination properties of the transformed program.

It should be noted that the original unfold/fold transformations for functional programs [BD77] preserves only partial correctness. Total correctness, that is, termination, has to be established separately. [2] In logic programming, starting with the work of Tamaki and Sato [TS84], the tradition has been to devise conditions on unfolding and folding that implicitly guarantee total correctness results. Recently, conditions on goal replacement requiring 'reversibility' of goal replacement have been proposed by Proietti and Pettorossi [PP93], stressing the preservation of partial and total correctness as separate concerns. This work is in the same spirit as ours though the approach differs. Our goal replacement is not reversible, but still preserves total correctness (in the sense of the program's success set).

The link between transformation rules and termination analysis provided us with the motivation for the system we present. The main difference of approach between our work and the works just cited (and also [GS91], [Mah87], [Sek89] and others) is that in our goal replacement rule termination of the transformed program is a condition of application of the rule, and we leave open the means by which termination is checked. Although the problem of checking goal replacement conditions is thereby shifted to checking termination properties, the advantage gained is that the link between folding and goal replacement is clarified. More flexible folding rules can be invented, as we show below. Also, research on termination analysis can be brought to bear on the problem.

We note that Boulanger and Bruynooghe [BB93] also developed an approach to goal replacement that is based on generation of replacement lemmata during an abstract interpretation of the program. This approach also seems capable of performing fold-like operations that are less restrictive than usual, allowing folding with recursive rules, for instance. [3]

There are two alternative strategies for implementing our transformation rules. One possibility is to use termination analysis techniques, to check termination of a transformed program directly. Secondly, special cases of the goal replacement can be derived, in which syntactic conditions ensuring termination are checked. This suggest a reconstruction of 'fold' transformations. In general the first approach seem the more promising, since it seems very difficult to find syntactic conditions that guarantee useful termination properties (such as acceptability [AP90]).

Section 2 contains a review of logic program transformation systems. In Section 3 we introduce our transformation rules and prove the correctness results. In Section 4 termination analysis is reviewed. An example of our system is given in Section 5. We show that our replacement rule allows 'foldings' to be based on

---

[2] We are grateful to A. Pettorossi for this pertinent comment

[3] We have been informed that folding using recursive rules was also considered in unpublished technical reports by Tamaki and Sato (1986), and by Kanamori and Fujita (1986).

equivalences that depend on recursive definitions of predicates. This is not allowed in other unfold/fold transformation systems. An informal proof that Seki's transformation system preserves recurrence can be found in Section 6. Finally, Section 7 is a short concluding section and some thoughts on future research directions.

The terminology used throughout this paper is consistent with that of [Llo87].

2 Background

2.1 Unfold/Fold of Definite Programs

In [TS84] the notion of a transformation sequence for definite programs was introduced.

Definition 1. An initial program, P_0, is a definite program satisfying the following conditions:

- P_0 is divided into two sets of clauses, P_{new} and P_{old}. The predicates defined in P_{new} are called new predicates, those defined in P_{old} are called old predicates.
- Predicates defined in P_{new} do not appear in P_{old} nor in the body of any clause in P_{new}.

The clauses defining new predicates are considered as *new definitions*; we note that they cannot be recursive.

Definition 2. Let P_0 be an initial program, and P_{i+1} a program obtained from P_i by applying either unfolding or folding for $i \geq 0$. The sequence of programs $P_0, \ldots, P_n$ is called a transformation sequence starting from P_0.

We do not repeat the definition of the unfold and fold rules from [TS84] here, but note that folding depends on the distinction between *old* and *new* predicates in the initial program. In a folding transformation applied in P_i in a transformation sequence, the clause used for folding is one of the clauses defining a new predicate. Thus folding in [TS84], and other systems derived from it, is not defined as a transformation on a program, but rather as a transformation on a sequence of programs.

Denoting the least Herbrand model of a program P by $M[P]$, the main result of [TS84] is the following.

Proposition 3. *Let $P_0, \ldots, P_n$ be a transformation sequence starting from P_0 which uses Tamaki-Sato unfolding and folding. Then $M[P_0] = M[P_n]$.*

Although the unfolding and folding rules of [TS84] preserve the success set of a program, they do not preserve its finite failure set. Some queries that finitely failed in an original program may not terminate after the transformation has been applied. [Sek89] proposes a modified folding rule based on the notion of an inherited atom. (Although Seki's transformations are defined for normal programs, we restrict our attention to definite programs here).

The fold-unfold system using Seki's fold rule preserves the finite failure set of a program as well as its minimal model. Let $FF[P]$ denote the finite failure set of P.

Proposition 4. *Let $P_0, \ldots, P_n$ be a transformation sequence starting from P_0 which uses unfolding as in [TS84], and the folding rule in [Sek89]. Then $FF[P_0] = FF[P_n]$.*

A property of Seki's transformation system will be considered in Section 6. For the rest of the paper we refer to the success set $SS[P]$ rather than $M[P]$ for a program P, for symmetry with $FF[P]$.

Transformation rules (for normal programs) are also considered by Gardner and Shepherdson in [GS91]. The unfolding rule (restricted to definite programs) is the same as that of [TS84]. The unfolding rule is used in conjunction with a reversible folding rule, similar to that of [Mah87], in which the folding clause is taken from the current program rather than from the initial program. Folding can be reversed by unfolding using these rules, and thus the correctness of folding follows from the correctness of unfolding. This elegant framework is independent of any transformation sequence. However, as a consequence of this, the framework appears to be less useful for practical purposes than the transformation system proposed in [TS84], in which the equivalences on which the foldings are based come from a set of clauses in the fixed initial program, and so are not altered during a transformation sequence.

By proving that their folding rule is the inverse of unfolding, Gardner and Shepherdson are able to show that their system preserves procedural equivalence based on SLDNF-resolution. For definite programs, therefore, both the minimal model and the finite failure set are preserved.

2.2 Goal Replacement

The goal replacement operation was defined in [TS84] to increase the power of the unfold/fold transformation system. In general, the rule is the replacement of a clause in P of the form

$$C : A \leftarrow A_1, \ldots, A_k, A_{k+1}, \ldots, A_n$$

by the clause

$$C' : A \leftarrow B_1, \ldots, B_m, A_{k+1}, \ldots, A_n$$

to form the program $P' = P - \{C\} \cup \{C'\}$, where $A_1, \ldots, A_k$ and $B_1, \ldots, B_m$ are in some sense equivalent in P and P'. (We consider replacing only some left conjunct of the body, without loss of generality).

Tamaki and Sato claimed that their goal replacement rule preserved the least Herbrand model of P. However, [GS91] gives a counter-example to this claim and defined a goal replacement rule with stronger applicability conditions. Gardner and Shepherdson show that the least Herbrand model is preserved by their goal replacement transformation.

One motivation of our work is to provide a goal replacement rule so that folding *is* a special case, and more flexible folding rules can be derived.

This was also one of the aims of Bossi *et al.* [BCE92]. A replacement operation (for normal programs) is defined in [BCE92]. Applicability conditions are given in order to ensure the correctness of the operation with respect to Fitting's semantics for normal programs. These semantics are based on Kleene's three-valued logic, the truth values being true, false, and undefined.

The conditions for goal replacement there are based on the rather complex notions of 'semantic delay' and 'dependency degree'. It is explained in [BCE92] that the applicability conditions are to ensure that there is 'no room for introducing a loop'. When the program P is transformed to P', under the applicability conditions:

- the dependency degree is an indication of how big a loop would be.
- the semantic delay is an indication of the space in which the loop would be introduced.

The transformation rules in [BCE92], though complex, provided us with the inspiration to search for more general conditions for goal replacement by investigating the problem of termination, since avoiding the 'introduction of a loop' seems to be the main intuition behind the work outlined above.

3   A Transformation System Based on Termination Analysis

In this section, we apply some of the ideas from the survey in Section 2 to develop a new transformation system. The main idea is that a goal replacement operation should depend on the notion of termination. Furthermore, we claim that this approach sheds light on other transformation systems.

We present two transformation rules for definite programs and prove that if the program P' results from the transformation of P under these rules then P and P' have the same least Herbrand model. The rules are, the unfold rule (Definition 5) based on partial evaluation, and the goal replacement rule (Definition 8) based on termination analysis. We do not consider a rule to introduce new definitions since we may consider all such definitions to have been introduced in the initial program of a transformation sequence.

3.1   Unfolding

Definition 5. Let P be a definite program and C a clause in P of the form

$$A \leftarrow Q$$

where Q is a conjunction of atoms (possibly empty).

Build an (incomplete) SLD-tree for $P \cup \{\leftarrow Q\}$ via some computation rule.

Choose goals $\leftarrow G_1, \ldots, \leftarrow G_r$ such that every non-failing branch of the SLD-tree contains precisely one of them. Let θ_i be the computed answer for the derivation from $\leftarrow Q$ to $\leftarrow G_i$, $(i = 1, \ldots, r)$.

The result of unfolding C on B in P is the definite program P' formed from P by replacing C by the set of clauses $\{C_1, \ldots, C_r\}$, where $C_i = A\theta_i \leftarrow G_i$.

Theorem 6. *Let the definite program P' be obtained by applying an unfold transformation to a clause in P. Then $SS[P] = SS[P']$ and $FF[P] = FF[P']$.*

The proof is drawn from a special case of partial evaluation which is discussed in [LS91] and [GS91]. Note that the extension of the usual 'one-step' unfold rule to the 'multi-step' rule above is more than just convenience; there are unfoldings using Definition 5 that cannot be reproduced by a series of one-step unfoldings, since a clause being used for unfolding may itself be modified as the result of a single unfolding step.

3.2 Goal Replacement

Let $Q_1 \equiv_V Q_2$ in P denote a notion of computational equivalence of formulas, in the following sense.

Definition 7. Let Q_1 and Q_2 be conjunctions of atoms and V a set of variables. We write $Q_1 \equiv_V Q_2$ in P if

- $\exists \sigma$ such that $P \cup \{\leftarrow Q_1\}$ has an SLD-refutation with computed answer σ, where $\sigma|_V = \theta$ $\Leftrightarrow$
  $\exists \rho$ such that $P \cup \{\leftarrow Q_2\}$ has an SLD-refutation with computed answer ρ, where $\rho|_V = \theta$

In other words the computed answers for goals $\leftarrow Q_1$ and $\leftarrow Q_2$ agree on the variables in V.

Definition 8. goal replacement
  Let P be a definite program and C a (non-unit) clause in P of the form

$$A \leftarrow Q_1, Q$$

where A is an atom and Q_1, Q conjunctions of atoms. Let R be a computation rule, and let vars(A, Q) denote the set of variables occurring in either A or Q. Form P' from P by replacing C by the clause

$$A \leftarrow Q_2, Q$$

where both of the following conditions are satisfied:

- $Q_1 \equiv_{\text{vars}(A,Q)} Q_2$ in P.
- For all ground instances A' of A, the SLD-tree for $P' \cup \{\leftarrow A'\}$ via R is finite.

Note that the definition refers to termination (finiteness of an SLD-tree) via a particular computation rule. One can use whatever computation rule one chooses in order to check termination. In practice one is probably interested in 'standard' computation rules such as the Prolog leftmost selection rule. The order of atoms in clause bodies and the computation rule are interdependent. For simplicity in the definition we have assumed that goal replacement takes place on the left part of the body, but with a fixed computation rule such as Prolog's this restriction needs to be relaxed. The definition is also asymmetric, in that we check the termination property in P' only. This reflects the directed nature of the transformation sequence, and the concern not to introduce loops during transformations. The second condition could also be weakened by checking termination of derivations starting with the transformed clause only (rather than all derivations of atomic goals unifying with its head).

The main correctness result for the goal replacement transformation is the preservation of the success set (Theorem 10). We shall see later that a symmetric version of the goal replacement rule, with an extended notion of goal equivalence, preserves the finite failure set as well (Theorem 15).

We start by proving a lemma stating that goals which succeed in P do not finitely fail in P'. Note that this lemma does not require the termination condition in the goal replacement rule, but simply uses the properties of goal equivalence. In a sense this lemma establishes a "partial correctness" result, and the termination condition will be used to establish total correctness.

Lemma 9. *Let the definite program P' be obtained by applying a goal replacement transformation to a clause in P. Then for all definite goals $\leftarrow B$,*

$P' \cup \{\leftarrow B\}$ has a finitely failed SLD-tree implies that $P \cup \{\leftarrow B\}$ has no SLD-refutation.

The proof of this lemma is in the appendix.

Theorem 10. *Let the definite program P' be obtained by applying a goal replacement transformation to a clause in P. Then $SS[P] = SS[P']$.*

Proof. The proof is in two stages. We show that:

1. $SS[P'] \subseteq SS[P]$
2. $SS[P] \subseteq SS[P']$

Stage 1: let $A' \in SS[P']$; show by induction on the length of the shortest SLD-refutation of $P' \cup \{\leftarrow A'\}$, that there is an SLD-refutation of $P \cup \{\leftarrow A'\}$. Let $SS^k[P']$ be the set of atoms A such that $P' \cup \{\leftarrow A\}$ has an SLD-refutation of length at most k.

For the base case, suppose $A' \in SS^1[P']$. This means that A' is a ground instance of the head of a unit clause D in P'. Goal replacement does not affect unit clauses, so D is also in P. Therefore $P \cup \{\leftarrow A'\}$ has an SLD-refutation of depth 1.

For the induction step, assume that $P \cup \{\leftarrow A'\}$ has an SLD-refutation for all $A \in SS^k[P']$. Suppose that $A' \in SS^{k+1}[P']$ and let T' be an SLD-refutation of $P' \cup \{\leftarrow A'\}$ of length $k+1$. Suppose the first clause (variant) used in T' is $C_r : H \leftarrow S$.

If C_r is not the result of the goal replacement, then C_r is also in P. Let $mgu(A', H) = \rho$; then $P' \cup \{\leftarrow S\rho\}$ has an SLD-refutation of length k with computed answer σ, say. Let $S\rho\sigma\gamma$ be any ground instance of $S\rho\sigma$. Then for each ground atom $\bar{A}$ in $S\rho\sigma\gamma$, $P' \cup \{\leftarrow \bar{A}\}$ has an SLD-refutation of length at most k and so by the induction hypothesis, $P \cup \{\leftarrow \bar{A}\}$ has an SLD-refutation. It follows that $P \cup \{\leftarrow S\rho\sigma\gamma\}$ has an SLD-refutation and hence so has $P \cup \{\leftarrow S\rho\}$. Therefore $P \cup \{\leftarrow A'\}$ has an SLD-refutation using C_r on the first step.

If C_r is the clause produced by goal replacement we have

$$C_r : A \leftarrow Q_2, Q \text{ in } P'$$

which replaced

$$C : A \leftarrow Q_1, Q \text{ in } P$$

where A' is unifiable with A with mgu θ. Then $P' \cup \{\leftarrow (Q_2, Q)\theta\}$ has an SLD-refutation and hence $P' \cup \{\leftarrow Q_2\theta\}$ has an SLD-refutation with computed answer α, say, and $P' \cup \{\leftarrow Q\theta\alpha\}$ has a refutation with computed answer σ, say.

Using the inductive hypothesis applied to the atoms in any ground instance of $Q_2\theta\alpha$ we can show that $P \cup \{\leftarrow Q_2\theta\}$ also has an SLD-refutation with computed answer α. Since $Q_1 \equiv_{\text{vars}(A,Q)} Q_2$ in P, and since θ acts on variables in A, $P \cup \{\leftarrow Q_1\theta\}$ has an SLD-refutation with computed answer $\hat{\alpha}$, where α agrees with $\hat{\alpha}$ in variables of $A\theta$ and $Q\theta$. Also by the inductive hypothesis applied to any ground instance of $Q\theta\alpha\sigma$ it follows that $P \cup \{\leftarrow Q\theta\alpha\}$ has a refutation with computed answer σ. Hence $P \cup \{\leftarrow Q\theta\hat{\alpha}\}$ has a refutation with computed answer σ. Putting the goals together it follows that $P \cup \{\leftarrow (Q_1, Q)\theta\}$ has an SLD-refutation. Therefore, $P \cup \{\leftarrow A'\}$ has an SLD-refutation using clause C on the first step.

Stage 2: let $A' \in SS[P]$ and show that $A \in SS[P']$. Proof is by induction on the length of an SLD-refutation of $P \cup \{\leftarrow A'\}$.

The base case is similar to the base case in Stage 1. For the inductive step, consider an SLD-refutation of $P \cup \{\leftarrow A'\}$ of length $k+1$, using C_r on the first step. If C_r is not the clause used in goal replacement the argument is identical to the corresponding step in Stage 1.

If C_r is the clause used in goal replacement, then A' is a ground instance of the head of C_r, and thus also of the head of the clause that replaces C_r. By the termination condition on goal replacement, $P' \cup \{\leftarrow A'\}$ has a finite SLD-tree, so it is either a finitely-failed tree, or contains a refutation. Suppose $P' \cup \{\leftarrow A'\}$ has a finitely-failed SLD-tree. Then by Lemma 9 $P \cup \{\leftarrow A'\}$ has no SLD-refutation. But this contradicts the inductive hypothesis. Hence $P' \cup \{\leftarrow A'\}$ has an SLD-refutation. $\square$

We conjecture that the set of computed answers is also preserved, since goal equivalence is based on equivalence of computed answers.

There is an obvious corollary to Theorems 6 and 10, which is the basis for reconstructing folding as a special case of goal replacement.

Corollary 11. *Let* $P_0, \ldots, P_k$ *be a transformation sequence using unfold and goal replacement rules. Let* $Q_1 \equiv_V Q_2$ *in* P_i *for some* $0 \le i < k$. *Then* $Q_1 \equiv_V Q_2$ *in* P_j, $0 \le j \le k$.

In particular, if there is a clause $A \leftarrow Q$ in some P_i, where A has no common instance with any other clause head in P_i, then $A \equiv_{\mathrm{vars}(A)} Q$ holds in P_i. Therefore it holds in subsequent programs in the sequence and the clause can be used for goal replacement in subsequent programs. In other words, if a clause of form $H \leftarrow Q\theta, R$ occurs in some P_j, where $j > i$. it may be replaced by $H \leftarrow A\theta, R$ in P_{j+1} provided that

1. Variables in $\mathrm{vars}(Q)/\mathrm{vars}(A)$ are mapped by θ into distinct variables not occurring in H, $A\theta$ or R. (This is sufficient to establish the first goal replacement condition that $Q\theta \equiv_{\mathrm{vars}(H,R)} A\theta$).
2. $P_{j+1} \cup \{\leftarrow H'\}$ has a finite SLD-tree (via some computation rule) for all ground instances H' of H.

This avoids the restrictions on folding present in [TS84], [Sek89], [GS91] and so on, that folding clauses define "new" predicates and are non-recursive. In Section 5 there is an example of a "fold" transformation using a recursive definition, which would not be possible with other fold rule based on [TS84]. Our goal replacement condition also avoids the rather complex syntactic conditions that are typical of folding rules.

In the literature there are two distinct flavours of folding. In [TS84] and related systems, folding is defined on a transformation sequence, and the clause used for folding might not be in the program to which folding is applied. By contrast, [GS91] and [Mah87] give a more elegant folding rule in which the folding clause comes from the program being transformed. These two folding rules are quite different in power. Both of them can be viewed as special cases of our goal replacement rule.

3.3 Preserving the Finite Failure Set

The goal replacement rule does not preserve the finite failure set. Preserving finite failure is desirable in some contexts, and so we strengthen the definition of goal equivalence in Definition 7, by insisting that the equivalent goals have the same failing behaviour as well as computed answers.

Definition 12. Let Q_1 and Q_2 be conjunctions of atoms and V a set of variables. We write $Q_1 \equiv_V Q_2$ in P if

- $\exists \sigma$ such that $P \cup \{\leftarrow Q_1\}$ has an SLD-refutation with computed answer σ, where $\sigma|_V = \theta$ $\Leftrightarrow$
  $\exists \rho$ such that $P \cup \{\leftarrow Q_2\}$ has an SLD-refutation with computed answer ρ, where $\rho|_V = \theta$, and

− for all substitutions θ acting only on variables in V,
$P \cup \{\leftarrow Q_1\theta\}$ has a finitely failed SLD-tree $\quad \Leftrightarrow$
$P \cup \{\leftarrow Q_2\theta\}$ has a finitely failed SLD-tree.

Assume now that the goal replacement transformation (Definition 8) refers to equivalence in the sense of Definition 12. We can then prove a stronger version of Lemma 9.

Lemma 13. *Let the definite program P' be obtained by applying a goal replacement transformation to a clause in P (using Definition 12 and Definition 8). Then for all definite goals $\leftarrow B$,*

$P' \cup \{\leftarrow B\}$ *has a finitely failed SLD-tree*
$\Rightarrow P \cup \{\leftarrow B\}$ *has a finitely failed SLD-tree.*

Proof. (outline)
The proof is similar in structure to the proof of Lemma 9, and does not use the termination condition on goal replacement. The key step is the construction of a finitely-failed SLD tree for a computation of form $P \cup \{\leftarrow (Q_1, R)\theta\}$ given a finitely-failed SLD tree for $P \cup \{\leftarrow (Q_2, R)\theta\}$, where $Q_1 \equiv_V Q_2$, θ acts only on V, and R is some conjunction. □

Note that the converse of Lemma 13 does not hold. A goal may fail finitely in P but loop in P', as the following example shows.

Example 1. Let $P = \{p \leftarrow q(X), \quad q(f(X)) \leftarrow r(X)\}$. Then $q(X) \equiv_{\{X\}} r(X)$ and we can replace $r(X)$ by $q(X)$ in the second clause, since the termination condition is satisfied. We obtain $P' = \{p \leftarrow q(X), \quad q(f(X)) \leftarrow q(X)\}$. $P \cup \{\leftarrow p\}$ has a finitely failed SLD tree but $P' \cup \{\leftarrow p\}$ loops.

As noted above, the goal replacement rule is asymmetric in that termination is checked only in the program after transformation. Clearly by virtue of Lemma 13 a symmetric version preserves the finite failure set as well as the success set.

Definition 14. (symmetric goal replacement) Let P be transformed to P' by goal replacement as defined in Definition 8 with Definition 12. If P can also be obtained from P' by goal replacement, then we say P' is obtained from P by symmetric goal replacement.

Theorem 15. *Let P' be obtained from P by symmetric goal replacement. Then $SS[P] = SS[P']$ and $FF[P] = FF[P']$.*

Proof. Using Theorem 10 and Lemma 13. □

Corollary 11 holds with respect to the symmetric goal replacement transformation and the extended definition of goal equivalence. Thus folding transformations that preserve the finite failure set can be constructed as goal replacements.

4 Termination of Definite Programs

We now turn our attention to the topic of termination analysis, which is needed in order to verify the conditions for our goal replacement rule. There has been a great deal of research effort addressing many aspects of termination analysis. This research has branched into many different directions because of the variety of definitions that exist for termination. The question of what is meant by a terminating program depends principally on such matters as the procedural semantics employed, the different modes of use of a program, and the fact that logic programs can be nondeterministic. This summary relies to a great extent on the survey in [DV92].

We may wish to consider either existential or universal termination, the former being termination because of finite failure, or because at least one solution is found (even though the program could have entered an infinite computation after finding such a solution), the latter being termination of the entire computation (all solutions having been found).

There are two different approaches in the research. The approach taken in, for example, [UV88] and [Plü92] looks to the provision of sufficient conditions for the termination of a logic program with respect to certain classes of queries that can be automatically verified. An alternative approach is taken in such papers as [VP86], [Bez89], [AP90] and [DVB92], where theoretical frameworks to solve questions about termination are offered.

Both [UV88] and [Plü92] use reasoning with linear predicate inequalities to generate termination proofs. The termination dealt with in [Plü92] is *left-termination* (i.e. termination under the left-to-right computation rule) of pure Prolog programs with respect to goals with input terms of known mode.

The concept of a level mapping is introduced by Bezem in [Bez89]. Denoting the Herbrand base of a program P by B_P:

Definition 16. A level mapping for a definite program P is a mapping $|.| : B_P \to N$.

The necessary decrease in measure for establishing termination through recursive clauses is ensured by the restriction of Bezem's work to a particular class of programs - those which are recurrent.

Definition 17. A definite program P is recurrent if there is a level mapping $|.|$ such that for each ground instance $A \leftarrow B_1, \ldots, B_n$ of a clause in P, we have $|A| > |B_i|$ for each $i = 1, \ldots, n$. A program is recurrent if it is recurrent with respect to some level mapping.

Note that, in the above definition, there is a decrease in measure between the head and body of *every* clause of the program. The main result of [Bez89] is that a logic program is terminating (with respect to ground goals and an arbitrary computation rule) if and only if it is recurrent.

Recurrency with respect to $|.|$ can be used to infer termination of non-ground goals whose ground instances have a maximum value with respect to $|.|$.

The ideas of [Bez89] are addressed with respect to termination in the left-to-right computation rule by Apt and Pedreschi in [AP90]. Many practical programs terminate under the Prolog selection rule (when given the correct class of inputs) even though they are not recurrent. As a result, in [AP90] the notion of a recurrent program is replaced by that of an *acceptable* one. Apt and Pedreschi prove that the notions of left-termination and acceptability coincide.

Some limitations of these approaches, with respect to automating the analysis of termination, are discussed by De Schreye *et al.* in [DVB92]. where extended notions of level mappings and recurrency are introduced. They introduce the notion of recurrency with respect to a set of atoms S. The key results in [DVB92] are that: P is recurrent with respect to S if and only if it is terminating with respect to S, and that P is recurrent if and only if it is recurrent with respect to B_P.

5 Example

This section contains an example of a program transformation under our rules in which the goal replacement stages cannot be made under the folding rule of [TS84], [Sek89] or [GS91]. This suggests that our system could be used to achieve more flexible folding rules, using, say, recursive definitions of new predicates. The example also illustrates the use of both recurrence and acceptability to establish applicability of goal replacements.

Example 2. Let $P^0 = \{C_1, \ldots, C_4\}$, where

$C_1 : \text{append}([], xs, xs).$
$C_2 : \text{append}([x|xs], ys, [x|zs]) \leftarrow \text{append}(xs, ys, zs).$
$C_3 : \text{leaves}(\text{leaf}(x), [x]).$
$C_4 : \text{leaves}(\text{tree}(x, y), zs) \leftarrow$
$\qquad \text{leaves}(x, xs), \text{leaves}(y, ys), \text{append}(xs, ys, zs).$

Unfold C_4 on the first atom to obtain $P^1 = (P^0/\{C_4\}) \cup \{C_5, C_6\}$, where

$C_5 : \text{leaves}(\text{tree}(\text{leaf}(x'), y), [x'|zs]) \leftarrow \text{leaves}(y, zs),$
$C_6 : \text{leaves}(\text{tree}(\text{tree}(x', y'), y), zs) \leftarrow$
$\qquad \text{leaves}(x', xs'), \text{leaves}(y', ys'), \text{append}(xs', ys', zs'),$
$\qquad \text{leaves}(y, ys), \text{append}(zs', ys, zs).$

Goal Replacement 1
Replacement of $\text{append}(xs', ys', zs'), \text{append}(zs', ys, zs)$ by
$\text{append}(ys', ys, vs), \text{append}(xs', vs, zs)$ in C_6 to give P^2 :

$C_1 : \text{append}([], xs, xs).$
$C_2 : \text{append}([x|xs], ys, [x|zs]) \leftarrow \text{append}(xs, ys, zs).$
$C_3 : \text{leaves}(\text{leaf}(x), [x]).$
$C_5 : \text{leaves}(\text{tree}(\text{leaf}(x'), y), [x'|zs]) \leftarrow \text{leaves}(y, zs),$
$C_7 : \text{leaves}(\text{tree}(\text{tree}(x', y'), y), zs) \leftarrow$
$\qquad \text{leaves}(x', xs'), \text{leaves}(y', ys'), \text{leaves}(y, ys),$
$\qquad \text{append}(ys', ys, vs), \text{append}(xs', vs, zs).$

Goal Replacement 2

Goal replace $\texttt{leaves}(y', ys'), \texttt{leaves}(y, ys), \texttt{append}(ys', ys, vs)$ by
$\texttt{leaves}(\texttt{tree}(y', y), vs)$ in C_7 to obtain P^3 :

$C_1 : \texttt{append}([], xs, xs).$
$C_2 : \texttt{append}([x|xs], ys, [x|zs]) \leftarrow \texttt{append}(xs, ys, zs).$
$C_3 : \texttt{leaves}(\texttt{leaf}(x), [x]).$
$C_5 : \texttt{leaves}(\texttt{tree}(\texttt{leaf}(x'), y), [x'|zs]) \leftarrow \texttt{leaves}(y, zs),$
$C_8 : \texttt{leaves}(\texttt{tree}(\texttt{tree}(x', y'), y), zs) \leftarrow$
$\qquad\qquad \texttt{leaves}(x', xs'), \texttt{leaves}(\texttt{tree}(y', y), vs), \texttt{append}(xs', vs, zs).$

Goal Replacement 3

Replace $\texttt{leaves}(x', xs'), \texttt{leaves}(\texttt{tree}(y', y), vs), \texttt{append}(xs', vs, zs)$ by
$\texttt{leaves}(\texttt{tree}(x', \texttt{tree}(y', y)), zs)$ in C_8 to obtain the final program P^4 :

$C_1 : \texttt{append}([], xs, xs).$
$C_2 : \texttt{append}([x|xs], ys, [x|zs]) \leftarrow \texttt{append}(xs, ys, zs).$
$C_3 : \texttt{leaves}(\texttt{leaf}(x), [x]).$
$C_5 : \texttt{leaves}(\texttt{tree}(\texttt{leaf}(x'), y), [x'|zs]) \leftarrow \texttt{leaves}(y, zs),$
$C_9 : \texttt{leaves}(\texttt{tree}(\texttt{tree}(x', y'), y), zs) \leftarrow \texttt{leaves}(\texttt{tree}(x', \texttt{tree}(y', y)), zs).$

We must show that the conditions of goal replacement (Definition 8) are satisfied for the three replacements steps in this transformation sequence. The recurrence of P^4 can easily be shown, which establishes the condition for the third goal replacement. For the first and second goal replacements, P^2 and P^3 are not recurrent, so we need a more refined notion of termination than recurrency. The notion of acceptability from [DVB92] is appropriate for these cases, since then we can prove termination for ground goals via a left-to-right computation rule. In [Coo92] the detailed demonstrations of the acceptability properties are given.

The applicability of the transformations does not actually require the notions of recurrence or applicability, which concern termination of all ground atoms, not just instances of the head of the transformed clause. There is clearly scope for using still more limited notions of termination (such as those in [DV92]) in cases where neither applicability nor recurrence can be shown.

6   Termination Preservation

It is interesting to examine current logic program transformation systems with regard to termination properties, since our work is based on the premise that termination properties are an essential ingredient in correct transformation systems. In this section we consider termination properties of the Seki transformation system [Sek89] (restricted to definite programs). We show that Seki's transformation system preserves recurrence. This shows that, given a recurrent initial program, a Seki transformation sequence is a special case of our transformations, since the Seki folding conditions are sufficient to establish recurrence and hence termination of the folded program.

Recently, Bossi and Etalle [BE94] have proved the stronger result that the Tamaki-Sato transformation system preserves another termination property, namely acyclicity. Although many typical terminating programs are neither recurrent nor acyclic these results are further evidence of the importance of termination properties in program transformation.

6.1 Preservation of Recurrence in Seki's System

We first recall that a transformation sequence is a sequence $P_0, \ldots, P_n$ where $P_0 = P_{new} \cup P_{old}$. The predicates in the heads of clauses in P_{new} are *new* predicates and the other predicates are *old* predicates.

The unfolding rule is standard and we do not repeat it here. The folding rule depends on the notion of an *inherited* atom. We do not give the full definition of an inherited atom here. Intuitively, an atom A in the body of a clause in P_i is inherited from P_0 if

- A was in the body of a clause in P_{new}, and
- for each transformation in the sequence up to P_i, A was neither the unfolded atom nor one of the folded literals.

Seki's folding rule can now be defined.

Definition 18. Let P_i be a program in a transformation sequence and C a clause in P_i of form $A \leftarrow Q_1\theta, Q$, where A is an atom and Q_1, Q conjunctions of atoms. Let D be a clause in P_{new} of form $B \leftarrow Q_1$. Then the result of folding C using D is

$$C' : A \leftarrow B\theta, Q$$

provided that all the following conditions hold.

- For each variable occurring in Q_1 but not in B, θ substitutes a distinct variable not appearing in C'.
- D is the only clause in P_{new} whose head is unifiable with $B\theta$.
- Either the predicate in the head of C is an old predicate, or there is no atom in $Q_1\theta$ which is inherited from P_0.

The resulting program is $P_{i+1} = (P_i/\{C\}) \cup \{C'\}$.

We first state a lemma (whose proof is in the appendix), and then the final result that recurrence is preserved by Seki's transformations.

Lemma 19. *Let $P_0, \ldots, P_n$ be a Seki transformation sequence and let P_0 be recurrent. Then for all clauses $A \leftarrow B_1, \ldots, B_m$ in P_i:*

(i) If A has an old predicate, none of the B_j is inherited from P_0.

(ii) If B_j has an old predicate and is not inherited from P_0, then for all ground instances $(A \leftarrow B_1, \ldots, B_m)\theta$ we have $|A\theta| - |B_j\theta| \geq 2$ for some level mapping $|.|$.

An outline proof of the lemma is in the appendix.

Theorem 20. *Suppose $P_0, \ldots, P_n$ is a Seki transformation sequence and P_0 is recurrent. Then P_n is recurrent.*

Proof. (Outline:) Suppose P_0 is recurrent by some level mapping$|.|'$. Construct a level mapping $|.|$ in the same way as in Lemma 19. The proof is by induction on the length of the transformation sequence. The case for P_0 is established by the construction of $|.|$. The inductive case for unfolding is a straightforward case analysis of the unfolding rule. For the folding case suppose P_{i+1} is obtained from P_i by folding $C : A \leftarrow Q_1\theta, Q$ using the folding clause $C_0 : B \leftarrow Q_1$ in P_{new} to obtain $C' : A \leftarrow B\theta, Q$ in P_{i+1}. By Lemma 19 and using the definition of folding, the 'distance' between A and each atom in $Q_1\theta$ is at least 2 (with respect to level mapping $|.|$). Also, since $C_0 \in P_{new}$ the distance between B and Q_1 is at least 1. Therefore the 'distance' between the head of C' and each atom in its body is at least 1. Therefore P_{i+1} is recurrent with respect to $|.|$. $\square$

7 Conclusion

The results of this work can be summarised as follows: we have introduced a goal replacement rule based on termination analysis. By making the requirement that all ground calls to the head of the transformed clause terminate, we have shown that the replacement rule preserves the success set of a definite program. Finite failure is preserved when a symmetric goal replacement rule is used, with an extended notion of goal equivalence. We have shown that a transformation system based on our rules can perform 'foldings' based on equivalences that depend on recursive definitions of predicates in the original program. Tamaki and Sato's folding rule and other similar versions do not allow this. Finally we have shown that Seki's unfold/fold rules (restricted to definite programs) preserve recurrence.

The identification of termination as the key property could be said merely to shift the problem of checking goal replacement conditions to that of establishing termination properties. In practice this is so, but the advantages of doing this are that promising research on termination analysis can be brought to bear, and links between folding and goal replacement are clarified and strengthened.

Two approaches to implementation can be envisaged. Firstly, automated termination analysis tools could provide direct checking of the goal replacement conditions (e.g. checking the transformed program for recurrence or acceptability). Secondly, we could search for some syntactic conditions that guaranteed applicability of the replacement rule, as in forms of 'folding' that are special cases of our replacement rule.

Future research possibilities include the following. The replacement rule could be extended to normal programs. This is obviously dependent on advances in the termination analysis of such programs, since finite failure is closely bound up with termination analysis. The relationship between the goal replacement rule we have introduced and other existing goal replacement/folding rules, such as [GS91] could be further investigated. The termination properties preserved by folding/goal replacement conditions in systems other than Seki's can be investigated. Application of the goal replacement rule can be extended by using weaker notions of termination (such as in [DVB92]).

Acknowledgements

We thank M. Proietti for pointing out a serious error in a previous version. We also thank A. Pettorossi and the LOPSTR'94 referees for comments and corrections.

References

[AP90]   K.R. Apt and D. Pedreschi. Studies in Pure Prolog: Termination. In J.W. Lloyd, editor, *Proceedings of the Esprit symposium on computational logic*, pages 150–176, 1990.

[BB93]   D. Boulanger and M. Bruynooghe. Using abstract interpretation for goal replacement. In Y. Deville, editor, *Logic Program Synthesis and Tranformation (LOPSTR'93)*, Louvain-la-Neuve, 1993.

[BCE92] A. Bossi, N. Cocco, and S. Etalle. Transforming Normal Programs by Replacement. In *Third Workshop on Metaprogramming in Logic*, Uppsala, 1992. META92.

[BD77]   R.M. Burstall and J. Darlington. A transformation system for developing recursive programs. *Journal of the ACM*, 24:44–67, 1977.

[BE94]   A. Bossi and S. Etalle. Transforming Acyclic Programs. *ACM Transactions on Programming Languages and Systems*, (to appear); also available as CWI Technical Report CS-R9369 December 1993, CWI, Amsterdam, 1994.

[Bez89] M. Bezem. Characterising Termination of Logic Programs with Level Mappings. In E.L. Lusk and R.A. Overbeek, editors, *Proceedings NACLP89*, pages 69–80, 1989.

[Coo92] J. Cook. A transformation system for definite logic programs based on termination analysis. Master's thesis, School of Mathematics, University of Bristol, 1992.

[DV92]   D. De Schreye and K. Verschaetse. Termination of Logic Programs:Tutorial Notes. In *Third Workshop on Metaprogramming in Logic*, Uppsala, 1992. META92.

[DVB92] D. De Schreye, K. Verschaetse, and M. Bruynooghe. A Framework for Analysing the Termination of Definite Logic Programs with respect to call patterns. In ICOT, editor, *Proceedings of the International Conference on Fifth Generation Computer Systems*, 1992.

[GS91]   P.A. Gardner and J.C. Shepherdson. Unfold/fold transformations of logic programs. In J.L Lassez and G. Plotkin, editors, *Computational Logic: Essays in Honour of Alan Robinson*. MIT Press, 1991.

[Llo87] J.W. Lloyd. *Foundations of Logic Programming, 2nd Edition*. Springer-Verlag, 1987.

[LS91]   J.W. Lloyd and J.C. Shepherdson. Partial Evaluation in Logic Programming. *Journal of Logic Programming*, 11(3 & 4):217–242, 1991.

[Mah87] M.J. Maher. Correctness of a Logic Program Transformation System. Research Report RC13496, IBM, T.J. Watson Research Center, 1987.

[Plü92] L. Plümer. Automatic Termination Proofs for Prolog Programs Operating on Nonground Terms. In *Proceedings ILPS'91,San Diego*, pages 503–517. MIT Press, 1992.

[PP93] M. Proietti and A. Pettorossi. Synthesis of programs from unfold/fold proofs. In Y. Deville, editor, *Logic Program Synthesis and Tranformation (LOP-STR'93)*, Louvain-la-Neuve, 1993.

[Sek89] H. Seki. Unfold/Fold Transformation of Stratified Programs. In G. Levi and M. Martelli, editors, *Sixth Conference on Logic Programming,Lisboa,Portugal*. The MIT Press, 1989.

[TS84] H. Tamaki and T. Sato. Unfold/Fold Transformation of Logic Programs. In *Proceedings of the Second international Logic Programming Conference*, pages 127–138, Uppsala, 1984.

[UV88] J.D. Ullman and A. Van Gelder. Efficient tests for top-down termination of logical rules. *Journal of the ACM, 35(2)*, pages 345–373, 1988.

[VP86] T. Vasak and J. Potter. Characterisation of Terminating Logic Programs. In *Proceedings 1986 Symposium on Logic Programming, Salt Lake City*, pages 140–147, 1986.

Appendix: Proofs of Lemma 9 and Lemma 19

Lemma 9: Let the definite program P' be obtained by applying a goal replacement transformation to a clause in P. Then for all definite goals $\leftarrow B$,

- $P' \cup \{\leftarrow B\}$ has a finitely failed SLD-tree implies that $P \cup \{\leftarrow B\}$ has no SLD-refutation.

Proof. Proof is by induction on the depth of the smallest finitely failed SLD-tree for $P' \cup \{\leftarrow B\}$. Let $B = B_1, \ldots, B_m$.

For the base case, suppose that $P' \cup \{\leftarrow B\}$ has a finitely-failed SLD-tree of depth 0. This means that some atom B_j ($1 \leq j \leq m$) fails to unify with the head of any clause in P'. Goal replacement has no effect on clause heads, so B_j also fails to unify with the head of any clause in P. Hence $P \cup \{\leftarrow B\}$ has no SLD-refutation.

For the induction step, assume that the property holds for goals G such that $P' \cup \{G\}$ has a finitely-failed SLD-tree of depth at most k, and that $P' \cup \{\leftarrow B\}$ has a finitely-failed SLD-tree F' of depth $k + 1$. Let $\leftarrow B$ have n immediate successor nodes, $\leftarrow S_1, \ldots, \leftarrow S_n$, in F', and let $D_1, \ldots, D_n$ be the corresponding clauses used. Each $\leftarrow S_i$ ($i = 1 \ldots n$) is at the root of a finitely failed SLD-tree of depth at most k. By the induction hypothesis $P \cup \{\leftarrow S_i\}$ (for $i = 1 \ldots n$) has no SLD-refutation.

If none of the clauses $D_1, \ldots, D_n$ is the clause produced by the goal replacement, then clearly $P \cup \{\leftarrow B\}$ has no SLD-refutation, since otherwise at least one $P \cup \{\leftarrow S_i\}$ would have a refutation, violating the induction hypothesis.

If D_i, say, is the clause produced by the goal replacement, then we have

$$D_i : A \leftarrow Q_2, Q \text{ in } P'$$

which replaced

$$C : A \leftarrow Q_1, Q \text{ in } P$$

where for some j $(1 \leq j \leq m)$ B_j is unifiable with A with mgu θ. Assume without loss of generality that $j = 1$.

Then $S_i = (Q_2, Q, B_2, \ldots, B_m)\theta$ and $P \cup \{\leftarrow S_i\}$ has no SLD-refutation, by the induction hypothesis. Consider the computation of $P \cup \{\leftarrow S_i\}$. There are two cases to consider.

(1) Suppose $P \cup \{\leftarrow Q_2\theta\}$ succeeds with computed answer α. Then $P \cup \{\leftarrow (Q, B_2, \ldots, B_m)\theta\alpha\}$ has no SLD-refutation. By the definition of goal replacement, $Q_1 \equiv_{\text{vars}(A,Q)} Q_2$ in P. The mgu θ acts on the variables of A so $P \cup \{\leftarrow Q_1\theta\}$ succeeds with computed answer $\hat{\alpha}$ where α and $\hat{\alpha}$ agree on variables in $A\theta$ and $Q\theta$. It follows that $P \cup \{\leftarrow (Q, B_2, \ldots, B_m)\theta\hat{\alpha}\}$ has no SLD-refutation, and therefore $P \cup \{\leftarrow (Q_1, Q, B_2, \ldots, B_m)\theta\}$ has no SLD-refutation.

(2) Suppose $P \cup \{\leftarrow Q_2\theta\}$ has no SLD-refutation. Again, since $Q_1 \equiv_{\text{vars}(A,Q)} Q_2$ in P and θ acts on the variables of A, it follows that $P \cup \{\leftarrow Q_1\theta\}$ has no SLD-refutation. Therefore $P \cup \{\leftarrow (Q_1, Q, B_2, \ldots, B_m)\theta\}$ has no SLD-refutation.

It now follows that $P \cup \{\leftarrow B\}$ has no SLD-refutation, since otherwise at least one of the goals $\leftarrow S_1, \ldots, \leftarrow S_{i-1}, \leftarrow S_{i+1}, \ldots, \leftarrow S_n$ or $\leftarrow (Q_1, Q, B_2, \ldots, B_m)\theta$ would have an SLD-refutation, violating the induction hypothesis.□

Lemma 19: Let $P_0, \ldots, P_n$ be a Seki transformation sequence and let P_0 be recurrent. Then for all clauses $A \leftarrow B_1, \ldots, B_m$ in P_i:

(i) If A has an old predicate, none of the B_j is inherited from P_0.
(ii) If B_j has an old predicate and is not inherited from P_0, then for all ground instances $(A \leftarrow B_1, \ldots, B_m)\theta$ we have $|A\theta| - |B_j\theta| \geq 2$ for some level mapping $|.|$.

Proof. (Outline): Suppose P_0 is recurrent by level mapping $|.|'$. Construct a level mapping $|.|$ as follows:

- $|A\alpha| = 2 * |A\alpha|'$ if A has an old predicate and $A\alpha$ is a ground atom.
- $|A\alpha| = \max\{|A_1\alpha|, \ldots, |A_p\alpha|\} + 1$ if $(A \leftarrow A_1, \ldots, A_p)\alpha$ is a ground instance of a clause in P_{new}.

Note that (i) holds trivially from the definition of inherited atom. We prove (ii) by induction on the length of the transformation sequence. The base case (that is, for P_0) is established by the construction of the level mapping $|.|$. The inductive case is a straightforward case analysis using the definitions of unfold and fold.□

On the Use of Inductive Reasoning in Program Synthesis: Prejudice and Prospects

Pierre Flener
Department of Computer Engineering
and Information Science
Faculty of Engineering
Bilkent University
TR-06533 Bilkent, Ankara, Turkey
pf@bilkent.edu.tr

Luboš Popelínský
Department of Computing Science
Faculty of Informatics
Masaryk University
Burešova 20
CZ-60200 Brno, Czech Republic
popel@fi.muni.cz

Abstract – In this position paper, we give a critical analysis of the deductive and inductive approaches to program synthesis, and of the current research in these fields. From the shortcomings of these approaches and works, we identify future research directions for these fields, as well as a need for cooperation and cross-fertilization between them.

1 Introduction

Many Software Engineering tasks—such as algorithm design, algorithm transformation, algorithm implementation into programs, and program transformation—are usually perceived as requiring sound reasoning—such as deduction—in order to achieve useful results. We believe that this perception is based on some faulty assumptions, and that there *is* a place for not-guaranteed-sound reasoning—such as induction, abduction, and analogy—in these tasks.

In this position paper, we analyze this case within the research of the Logic Programming community, and more specifically within logic program synthesis research. Moreover, we concentrate on the use of inductive inference as a complement to sound reasoning. This paper is then organized as follows. In Section 2, we take a critical look at the use of deductive reasoning in synthesis, so as to identify the shortcomings of that approach. In Section 3, we do the same thing with the use of inductive reasoning in synthesis. This allows us, in Section 4, to plead for a cooperation and cross-fertilization between the two approaches. In Section 5, we conclude this apology of the use of not-guaranteed-sound reasoning in Software Engineering.

2 On the Use of Deductive Reasoning in Program Synthesis

The Software Engineering tasks of algorithm design, algorithm transformation, algorithm implementation into programs, and program transformation are often believed to be the exclusive realm of sound reasoning, such as deduction. Note that the last two tasks are often merged into the first two, provided the focus is on some form of pure logic programs. However, it is clearly advantageous to separate algorithms from programs, that is to distinguish between the declarative aspects (such as logical correctness) and the procedural aspects (such as control, operational correctness, and efficiency) of software. This distinction is often blurred by the promises of Logic Programming, promises that have however not been fulfilled by languages such as Prolog. It is but a recent trend to dissociate algorithms and programs in Logic Programming

[16] [24] [32] [56] [86]. This corresponds to recasting Kowalski's equation "Algorithm = Logic + Control" [55] as "Program = Algorithm + Control". Since there is no consensus yet on these issues, and in order to keep our terminology simple, the word "algorithm" stands in the sequel for "algorithm or program", hence encompassing all viewpoints.

In the Logic Programming community, this trend of deduction-based approaches is clearly dominant in the proceedings of dedicated workshops such as the LOPSTR series (LOgic Program Synthesis and TRansformation, where "synthesis" stands for some form of (semi-)automated algorithm design) [19] [25] [57] [this volume], and of dedicated sessions at most Logic Programming conferences (ICLP [50], SLP, META, ...).

Let's focus our attention now on the most challenging of the four tasks enumerated above, namely algorithm design, and, more precisely, on algorithm synthesis. It should be noted that the line between synthesis and transformation is a very subjective one, and that the synthesizers of some researchers would be transformers for other researchers. For the sake of this paper, we assume the following purely syntactic criterion for distinguishing between synthesis and transformation: if the input and output languages of a tool are the same, then it is a transformer, otherwise it is a synthesizer. Usually, synthesis amounts to the translation from a rich, high-level input language (the specification language) into a less rich, lower-level language (the algorithm language). It is in this sense that we consider synthesis more challenging than transformation. Our restriction of the focus on algorithm synthesis can now be motivated as follows: synthesis generates the objects that can be transformed, so synthesis could just as well generate the transformed version right away. In this sense, it suffices to discuss synthesis here.

The rest of this section is now organized as follows. Section 2.1 briefly relates the various approaches to using deductive reasoning in synthesis. This allows us, in Section 2.2, to enumerate the problems of such deduction-based synthesis. Finally, Section 2.3 contains a partial conclusion.

2.1 Approaches to Deduction-based Program Synthesis

In Logic Programming, synthesis is usually the process of (semi-)automatically translating a specification (usually in a language quite close to full first-order logic plus equality) into a logic algorithm (usually in a language that is a proper subset of the specification language). Such a first-order axiomatization is often just assumed to be a faithful formalization of the intentions[1], though the synthesis process may provide feedback to the specification elaboration process [60]. From such a formal specification, a natural thing to do is to proceed with sound reasoning, so as to obtain an algorithm that is logically correct with respect to the specification. Research from this mind-set can be divided into two main categories [27] [31]:

- *Deductive Synthesis* (also known as *Transformational Synthesis*): meaning-preserving transformations are applied to the specification, until an algorithm is obtained. Sample works are those of Clark [18], Hogger [48], Bibel *et al.* [6], Sato and Tamaki [73] [74], Lau and Prestwich [62], Kraan *et al.* [56], and so on. The theoretical foundations to deductive synthesis are being laid out by Lau and Ornaghi [58]–[61].

---

1. Let's ignore here the problem that the intentions tend to be unknown or to change over time.

- *Constructive Synthesis* (also known as *Proofs-as-Programs Synthesis*): an algorithm is extracted from a (sufficiently) constructive proof of the satisfiability of the specification. Sample works are those of Tärnlund *et al.* [29] [46], Takayama [83], Bundy *et al.* [16], Wiggins [86], Fribourg [35], and so on.

These two categories are not as clear-cut as one might think, as some synthesis mechanisms can be successfully classified in the two of them [56]. The two approaches are probably only facets of the same technique.

A third category, namely *Schema-guided Synthesis*, should be added: an algorithm is obtained by successively instantiating the place-holders of some algorithm schema, using the specification, the algorithm synthesized so far, and the integrity constraints of the chosen schema. This usually involves a lot of deductive inference sub-tasks, hence the justification for viewing this as deduction-based synthesis. Curiously, this category seems almost absent from the Logic Programming world, although spectacular results are being obtained with this approach in Functional Programming by D.R. Smith [78] [79]. The work of Fuchs [40] and Kraan [56] is definitely schema-based, but their logic program schemas are not finegrained enough to more effectively *guide* the synthesis.

2.2 The Problems with Deduction-based Program Synthesis

Now, what are the problems with deduction-based synthesis, and the current approaches to it? We see basically two problems.

The first problem is related to the current synthesis approaches. A lot of wonderful theoretical effort is being directed at building *theories of algorithms* that may underlie synthesis. Almost any algorithm can be synthesized using these approaches. Another crux is that although these approaches show the derivability of many algorithms (which is in itself a very positive result), the search spaces are exponentially large, making these synthesizers impossible to use in automatic mode. So interactive usage is recommended. But a close look at such syntheses shows that one almost needs to know the final algorithm if any serious speedup is to be expected, which is not what one would expect to have to do with a synthesizer. What is needed *now* is also a *theory of algorithm design* in order to allow efficient traversal of these huge search spaces. The investigations on proof planning [17] and lemma generation [36] are first steps into that direction. But we believe that even more can be done, maybe along the lines of the schema-guided synthesis paradigm mentioned above. Indeed, one of the conclusions of the early efforts at automatic programming was that algorithm design knowledge and domain knowledge are essential if synthesis wants to scale up to realistic tasks [42]. Curiously, few people in the Logic Programming community seem to pay heed to that recommendation by Green and Barstow.

The second problem with deduction-based synthesis is related to the formality of the specifications. Where do the formal specifications come from? If deduction-based synthesizers guarantee total correctness of the resulting algorithms with respect to their specifications, what guarantee do we have that these specifications correctly capture our intentions? These questions are often either dismissed as irrelevant or considered as intractable issues that are left for future research. But what use are these synthesizers if we don't even know whether they solve our problems or not?

Before continuing, we must define the concept of specification, as there is little consensus on such a definition. For the purpose of this paper, a *specification* is a description of *what* a program does, and of *how to use* that program. Such specifications are totally declarative in that they don't express how the program actually works, and they thus don't bias the programmer in her/his task. Specifications have been alternately required to be not (necessarily) executable [47], (preferably) executable [37] [38], and so on.

So what are the problems with formal specifications? As seen above, there is no way to construct formal specifications so that we have a formal proof that they capture our intentions. So an informal proof is needed *somewhere* (at worst by testing a prototype, although testing never amounts to proving), as the purpose of Software Engineering is after all to obtain programs that implement our informal intentions. Writing formal specifications just shifts the obligation of performing an informal proof from the program *vs.* specification verification to the specification *vs.* intentions verification, but it doesn't rid us of that obligation. So specifications could just as well be informal, to prevent a delaying of an informal proof that has to be done anyway.

So we are actually hoping for somebody to write a new paper stating that specifications ought to be informal! Informal specifications should not be mixed up with natural language specifications: they are just specifications written in a non-formal language (without a predefined syntax and semantics). So natural language statements augmented with *ad hoc* notations, or statements in a subset of natural language with a "clear" semantics, would constitute an informal specification [63].

Another indicator why specifications ought to be informal can be obtained by observing the history of program synthesis research [72]: the first assemblers and compilers were seen as automatic programmers, as they relieved the programmers from many of the burdens of binary programming. Some programmers felt like they were only writing some form of specifications, and that the compilers took care of the rest. But after a while, the new "specification languages" were perceived as programming languages! The same story is happening over and over again with each new programming paradigm. One of the latest examples is Prolog: initially perceived by many as a specification language (and still being perceived as such by some people), the consensus now seems that it is a programming language. So how come that one-time specification languages are sooner or later perceived as programming languages, or that there never is a consensus about the difference between formal specifications and programs? The formal specifications of some people are indeed often very close to what other people would call programs. Moreover, formal specifications are often as difficult to elaborate and to maintain as the corresponding programs.

Some researchers don't require formal specifications to be totally declarative, but only as declarative as possible. Indeed, if a specification language allows the procedural expression of knowledge, practice shows that specifiers will use these features. But what does it mean for a formal specification to be declarative? As *writing* recursive statements seems to reflect a very procedural way of thinking[2], a possible criterion for declarativeness could be the absence of explicit recursion in the specification itself as well as in the transitive closure of the union of the specifications of the predicate-symbols used in that specification. But such recursion-free specifications are (possibly?) im-

---

2. However, recursive statements can be *understood* declaratively: the meaning is "... and so on".

possible to write, as sooner or later one gets down to such fundamental concepts as integer-addition or list-concatenation, which don't have non-recursive finite-length descriptions that completely capture them. So if the most evident symptom of "procedurality", namely recursion, seems impossible to avoid in formal specifications, this would imply that declarative formal specifications don't exist. And since specifications ought to be declarative, this would in turn imply that formal specifications cannot be written.

In our opinion, the solution to all these problems with formality is that formal specifications and programs are intrinsically the same thing! The inevitable intertwining between the formal specification elaboration process and the algorithm design process has already been pointed out by Swartout and Balzer [82].

As algorithm synthesis research aims at raising the level of language in which we can interact with the computer, synthesizers and compilers perform intrinsically the same process. In other words, the "real" programming is being done during the formalization process while going from an informal specification to a formal specification/program (which is then submitted to a synthesizer/compiler). In Logic Programming, there is little research about this formalization process, a laudable exception being Deville's work on hand-constructing logic algorithms from informal specifications [24], a process for which some mechanization opportunities have been pointed out [26].

Note that we are not saying that formal specifications are useless: of course it is important to be able to check whether a formal specification is internally consistent, and to generate prototypes from executable specifications, because this allows early error detection and hence significant cost-cutting. At any given time, formal specifications, even if written in a programming language, will have different purposes (validation, prototyping, contracts, ...) and attributes (readability, efficiency, ...) than programs [39]. We here just say that formal/executable specifications are already programs, though not in a conventional sense. But conventions change in time, and the specifications of today may well be perceived tomorrow as programs. To understand this, it helps to define programming from a process-theoretic viewpoint (that is, as an activity of carefully crafting, debugging, and maintaining a formal text) rather than from a product-theoretic viewpoint (programming yields a formal text).

2.3 Partial Conclusion about Deduction-based Program Synthesis

Let's summarize in a very synoptic way the situation about deduction-based synthesis:
- Deduction-based synthesis translates a formal specification (with assumed-to-be-complete information about the intentions) into an algorithm.
- Deduction-based synthesis research, in Logic Programming, should incorporate (more) explicit algorithm design knowledge, such as algorithm schemas.
- The deduction-based synthesis approach suffers from the following problems:
  - where do the formal specifications come from?
  - it's impossible to have a formal guarantee that a formal specification correctly captures the intentions;
  - formal specifications are often as difficult to write as programs;
  - there is even no consensus on the difference between formal specifications and programs; in fact, the expression "formal specification" is a contradiction in

terms, and formal specifications and programs are intrinsically the same thing; so synthesis and compilation also are intrinsically the same process.

Let's now move on to the use of inductive reasoning in algorithm synthesis.

3 On the Use of Inductive Reasoning in Program Synthesis

Human beings often understand a new concept after just seeing a few positive (and negative) examples thereof. Machine Learning is the branch of Artificial Intelligence that explores the mechanization of concept learning (from examples). Important sub-branches are Empirical Learning (from a lot of examples, but only a little background knowledge) and Analytical Learning (from a few examples, but a lot of background knowledge), the latter being also known as Explanation-Based Learning (EBL) or Explanation-Based Generalization (EBG) [85]. Machine Learning was long cast in the framework of propositional logic, but since the early 1980s, the results are being upgraded to first-order logic. These efforts are nowadays collectively referred to as ILP (Inductive Logic Programming, a term coined by Muggleton [65]), because concept descriptions are there written as logic programs. ILP is somehow a cross-fertilization between Logic Programming and Machine Learning, and between Empirical Learning and Analytical Learning, and is divided into Empirical ILP (heuristic-based learning of a single concept from many examples) and Interactive ILP (algorithmic and oracle-based learning of many concepts from a few examples). The base cycle of every learner is that it reads in examples from a teacher and periodically turns out hypotheses (conjectures at concept descriptions).

An important distinction needs to be made here. Algorithm synthesis from examples is but a niche (albeit a significant one) of ILP. Indeed, algorithm synthesis in general is only useful if the algorithm actually performs some "computations", via some looping mechanism such as recursion or iteration. Straight-line code is always very close to its full specification, and its synthesis is thus a mere rewriting process. So, in particular, algorithm synthesis from examples is only useful if the algorithm actually performs some "computations". But recursive algorithms are only a subclass of all possible concept descriptions, so algorithm synthesis from examples effectively is a niche of ILP. In the following, by "algorithms" we mean recursive concept descriptions, and by "identification procedures" we mean non-recursive concept descriptions. Other differences between ILP in general and induction-based algorithm synthesis in particular are summarized in Table 1 [31].

The central column of Table 1 shows the spectrum of situations covered by ILP research, but it doesn't mean to imply that all learners do cover, or should cover, this full spectrum. The right-hand column however shows the most realistic situation for induction-based algorithm synthesis, that is a situation that *should* be covered by every synthesizer. Let's have a look now at these two columns.

In ILP, the agent who provides the examples can be either a human being or some automated device (such as a robot, a satellite, a catheter, ...). It is possible for this agent not to know the intended concept, which means that it may give examples that are not consistent with the intended concept, or that it may give wrong answers to queries from the learner. Examples can be given in any amounts: Empirical ILP systems expect numerous examples, while Interactive ILP systems expect only a few examples, and often

	Inductive Logic Programming	Induction-based Algorithm Synthesis
Class of hypotheses	any concept descriptions	algorithms
Specifying agent	human or machine	human
Intended concept	sometimes unknown	always known
Consistency of examples	any attitude	assumed consistent
# examples	any	a few
# predicates in examples	at least 1	exactly 1
Rules of inductive infer.	selective & constructive	necessarily constructive
Correctness of hyp.s	any attitude	total correct. is crucial
Existence of hyp. schemas	hardly any	yes, many
# correct hypotheses	usually only a few	always many

Table 1: Induction-based Algorithm Synthesis as a Niche of Inductive Logic Programming

construct their own examples so as to submit them to the teacher. Examples may involve more than one predicate-symbol: the instance "Tweety" of the concept "canary" could yield the example:

```
mouth(tweety,beak) ∧ legs(tweety,2) ∧ skin(tweety,feather) ∧
        utterance(tweety,sings) ∧ color(tweety,yellow),
```

which involves many predicate-symbols, but not a *canary*/1 predicate-symbol. The used rules of inductive inference can be either selective (only the predicate-symbols of the premise may appear in the conclusion) or constructive (the conclusion "invents" new predicate-symbols). Selective rules are often sufficient to learn concepts, such as "canary", from multi-predicate examples. There are many learning situations where an approximately correct concept description is sufficient, whereas in other situations a totally correct description is hoped for. For general concept descriptions, there are hardly any useful schemas (template concept descriptions): indeed, such schemas tend to spell out the entire search space, and thus don't decrease its size. For general concepts, there are usually only a few correct hypotheses: for instance, there is probably only one correct definition of the "canary" concept, in any given context.

But in induction-based algorithm synthesis, the most realistic setting is where the specifier is a human being who knows the intended concept and who is assumed to provide only examples that are consistent with that intended concept.[3] "Knowing a concept" means that one can act as a decision procedure for answering membership queries

---

3. There are of course other settings for induction-based synthesis, such as the intelligent system that re-programs itself in the face of new problems [11]. We think that in such cases a general Machine Learning approach is more adequate as the system can't know in advance whether the new concept has an algorithm or an identification procedure.

for that concept [2], but it doesn't necessarily imply the ability to actually write that decision procedure.[4] Such a specifier cannot be expected to be willing to give more than just a few examples. Examples only involve one predicate-symbol, namely the one for which an algorithm is to be synthesized: for instance, an example of a sorting algorithm could be *sort*([2,1,3],[1,2,3]). The used rules of inductive inference thus necessarily include constructive rules, as algorithms usually use other algorithms than just themselves. Total correctness of the synthesized algorithm with respect to the intended concept is crucial in induction-based synthesis. Algorithms are highly structured, complex entities that are usually designed according to some strategy, such as divide-and-conquer, generate-and-test, global search [79]: algorithm synthesis can thus be effectively guided by an algorithm schema that reflects some design strategy. The existence of many such schemas, and the existence of many choice-points within these strategies entail the existence of many correct algorithms for a given "computational" concept. For instance, sorting can be implemented by Insertion-sort, Merge-sort, Quicksort algorithms, and many more.

So there is a dream of actually synthesizing algorithms from specifications by examples. Since many intentions are covered by an infinity of examples, finite specifications by examples cannot faithfully formalize such intentions, and the synthesizer needs to extrapolate the full intentions from the examples. This is necessarily done by not-guaranteed-sound reasoning, such as induction, abduction, or analogy.

The rest of this section is now organized as follows. Section 3.1 briefly relates the various approaches to using inductive reasoning in synthesis. This allows us, in Section 3.2, to enumerate the problems of such induction-based synthesis. In Section 3.3, we tackle the most commonly encountered prejudice about induction-based synthesis. Finally, Section 3.4 contains a partial conclusion.

3.1 Approaches to Induction-based Program Synthesis

In the early 1970s, some researchers investigated how to synthesize algorithms from traces of sample executions thereof. However, traces are very procedural specifications, and constructing a trace means knowing the algorithm, which rather defeats the purpose of synthesis. Sample work is related in [7] and, more recently, [52]. Regarding induction-based synthesis from examples, there are basically two approaches [31] [27]:

- *Trace-based Synthesis*: positive examples are first "explained" by means of traces (that fit some predefined algorithm schema), and an algorithm is then obtained by generalizing these traces, using the above-mentioned techniques of induction-based synthesis from traces. Sample works are those of Biermann *et al.* [8] [12], Summers [81], and so on, and they are surveyed by D.R. Smith [77]. This research was a precursor to the EBL/EBG research of Machine Learning.

- *Model-based Synthesis*: a logic program is "debugged" with respect to positive and negative examples until its least Herbrand model coincides with the intentions. This is the ILP approach. Sample works are those of E.Y. Shapiro [76], and many others are compiled by Muggleton [66] and surveyed in [67].

---

4. It would be interesting to examine specifiers (oracles) that are capable of answering other kinds of queries (subset, superset, ... [2]) and to investigate other meanings of the phrase "knowing a concept".

Historically speaking, the two approaches barely overlap in time: trace-based synthesis research took place in the mid and late 1970s, whereas model-based synthesis research is ongoing ever since the early 1980s. Indeed, in the late 1970s, trace-based synthesis research hit a wall and partly declared defeat considering that the techniques found didn't seem to scale up to realistic problems. But then, E.Y. Shapiro [76] and others published their first experiments with model-based approaches, and model-based synthesis took over, not only for induction-based algorithm synthesis, but for inductive concept learning in general.

"Linguistically" speaking, the two approaches also barely overlap: trace-based synthesis was pursued by the Functional Programming community, whereas model-based learning is being investigated by the Logic Programming community. Revivals of trace-based synthesis in the Logic Programming community have been suggested by Flener [31] [32] and Hagiya [45].

3.2 The Problems with Induction-based Program Synthesis

Now, what are the problems with induction-based synthesis, and the current approaches to it? We see basically two problems.

The first problem is related to the current synthesis approaches: there seems to be little dedicated induction-based synthesis research any more in the Logic Programming community, as most research seems directed at model-based learning in general. However, as conveyed by Table 1, induction-based synthesis is a sufficiently restricted sub-area of induction-based learning to justify very dedicated techniques. It is illusory to hope that very general learning techniques carry over without major efficiency problems to particular tasks such as induction-based synthesis: since synthesis is akin to compilation (see Section 2.2), this illusion amounts to looking for a universal programming language. The phrase ILP is ambiguous in that it can be understood in two different ways: ILP could mean "writing Logic Programs using Inductive reasoning" (I-LP), or it could mean "Programming (in the traditional sense of the word) in Logic using Inductive reasoning" (I-L-P). The bulk of ILP research accepts the first interpretation.

Some good ideas of trace-based synthesis (such as schema-guidance) haven't received much attention by model-based learning research. Indeed, as seen above, for general concepts there are hardly any schemas that wouldn't spell out the entire search space. Of course, one can use application-specific schemas, but then the question arises as to the acquisition of these schemas. Now, for the particular task of model-based synthesis, there *is* room for schemas [80] [84]: algorithm schemas significantly reduce the search space, they bring "discipline" into an otherwise possibly anarchic debugging process, and they convey part of the algorithm design knowledge called for by Green and Barstow [42]. The other entries of Table 1 provide an agenda for future, dedicated research in model-based synthesis.

Also, there is a fundamental difference between a teacher/learner relationship and a specifier/synthesizer relationship. A teacher usually is expected to know *how* to compute/identify the concept s/he is teaching to the learner, whereas a specifier usually only knows *what* the concept is about, the determination of *how* to compute it being precisely the task of the synthesizer. So a teacher can guide a learner who is "on the wrong track", but a specifier usually can't. A teacher can, right before the learning session, set the learner "on the right track" by providing carefully chosen examples and/or background knowledge, but a specifier often can't. For instance, most ILP systems can learn the Quicksort algorithm from examples of *sort*/2 plus logic procedures for *partition*/3 and *append*/3 as background knowledge. But this amounts to a "specification of *quick-sort*/2", which is a valid objective for a teacher, but not for a specifier: one specifies *sort*/2, a problem, not *quicksort*/2, a solution! We really wonder about the efficiency of ILP-style learners in a true specifier/synthesizer setting, where *a lot of* relevant *and* irrelevant background knowledge is provided. A solution to the ensuing inefficiency would be structured background knowledge, such as classifying the *partition*/3 procedure as a useful instance of the induction-parameter-decomposition placeholder in a divide-and-conquer algorithm schema.

The second problem with induction-based synthesis is that examples alone are too weak a specification approach. Incompleteness results are indeed abundant [3] [10] [41] [51] [68] [69]. It is true that in Machine Learning in general, examples are often all one can hope for. But, as conveyed by Table 1, in synthesis, we usually have the setting of a human specifier who *knows* the intended relation. So s/he probably knows quite a bit more about that relation, but can't express it by examples alone. For instance, it is unrealistic that somebody would want a sorting program and not know the reason why [2,1] is sorted into [1,2] rather than into [2,1]. The reason of course is that $1 \leq 2$, but the problem here is that the $\leq$/2 predicate-symbol cannot appear in the examples.

More generally, the problem is about the lack of provision of domain knowledge to the synthesizer (another recommendation by Green and Barstow [42]), and has been perceived a while ago. Various proposed solutions are type declarations for the parameters [76], type assertions about the intended relation [28], properties of the intended relation [31] [32], integrity constraints about a set of intended relations [20] [21], and bias (all knowledge potentially useful for narrowing the search space [71]), as generally used in ILP. Note that special care needs to be taken not to *require* complete knowledge about the intentions in the assertions/properties/constraints/bias, because otherwise a deduction-based synthesizer would be more appropriate. This is a problem with some of the proposed solutions [23]. Of course, if someone wants to give complete knowledge about the intentions, then the synthesizer should be able to handle it.

Some other often mentioned "problems" with induction-based synthesis are, in our opinion, no problems at all, and we discuss them in the next sub-section.

3.3 Induction-based Program Synthesis: Prejudice and Reality

When faced with research about synthesizing algorithms from examples, some deduction-based synthesis researchers react somewhere in between the paternalistic smile of a father at his child who just completed her/his first Lego house and aggressive attacks about the uselessness of such research. Let's have a look at the most frequently encountered prejudices, and debunk them in the face of reality.

Prejudice: Induction-based synthesis researchers think that they will provide *the* solution to synthesis.

Reality: Induction-based synthesis researchers are fully aware of the limitations of their research. They view it as just the provision of components and tools for software engineering environments. In the synthesizer-as-a-workbench-of-powerful-mini-synthesizers approach advocated by A.W. Biermann [9] and schema-guided synthesis researchers such as D.R. Smith [78] [79], there *is* a place for induction-based synthesizers, because certain classes of algorithms can be reliably synthesized with little effort from a few examples. As an illustration, the first author estimates that about 50% of the code of his induction-based SYNAPSE synthesizer [31] falls into such categories of algorithms (divide-and-conquer algorithms in this case), and could thus have been written by SYNAPSE itself.

Prejudice: Induction-based synthesis can at most pretend to aim at programming-in the-small. So it is useless as such algorithms are trivial and can often be written faster than the specifications by examples.

Reality: Even though deduction-based synthesis can effectively hope to scale up to programming-in-the-medium (though probably not to programming-in-the-large?), this doesn't mean that strictly less powerful approaches are not useful. One should not forget that synthesis aims at helping all sorts of programmers, not only the skilled ones. Moreover, synthesis aims at the design of any algorithms, not only the complex ones. Finally, synthesis aims at raising the level of language in which the programmer can communicate with the computer, and thinking in terms of examples seems to us of a higher level than thinking in terms of recursion. During the implementation of his SYNAPSE system, the first author felt many times that he would rather use SYNAPSE (if only it existed already!) than work out himself the recursive calls and other more low-level details. In ILP research on algorithm synthesis, a lot of denotation is now being paid to minimizing the number of examples, and the first results are promising [1] [5] [43] [70].

Prejudice: Induction-based synthesis research is useless because it offers no guarantee that the synthesized algorithms are correct with respect to our intentions.

Reality: In *both* the deduction-based and the induction-based synthesis approaches, it takes specification debugging and maintenance to achieve correctness with respect to the intentions. In the *two* approaches, completeness of the algorithm with respect to the specification is guaranteed, but only in the deduction-based approach does partial correctness with respect to the specification make sense. The problem does not lie in the use of not-guaranteed-sound *vs.* sound reasoning, but in the fact that synthesis starts from formal specifications. Whereas with example-based specifications one knows that the specification is but a fragmentary description of the intentions, such is usually not the case with axiomatic specifications, where one only knows that there is a problem when something goes wrong during the synthesis or during the execution of an implementation of the synthesized algorithm. The line of reasoning for the prejudice above could thus also be used to claim that deduction-based synthesis research is useless because it also doesn't offer a guarantee that the synthesized algorithms are correct with respect to our intentions, this because we have no guarantee that our formal specifications are correct with respect to our intentions.

There certainly are other prejudices, but let's leave it at these for now.

3.4 Partial Conclusion about Induction-based Program Synthesis

Let's summarize in a very synoptic way the situation about induction-based synthesis:

- Induction-based synthesis generalizes a formal specification (with known-to-be-fragmentary information about the intentions) into an algorithm.
- Induction-based synthesis research, in Logic Programming, is a niche of ILP, but its specifics are not being catered for. The results are usually inefficiency and inadequateness. For instance, synthesizers should incorporate (more) explicit algorithm design knowledge, such as algorithm schemas.
- The induction-based synthesis approach suffers from the problem that examples alone are too weak a specification approach: additional domain and problem knowledge must be provided.
- There *is* a place for the induction-based synthesis approach.

This finishes our discussion of the deduction-based and induction-based approaches to algorithm synthesis. Let's now plead for a cooperation and cross-fertilization between the two approaches, and actually also with abduction- and analogy-based approaches.

4 Towards a Cooperation and Cross-Fertilization between Deduction-based Synthesis and Other Approaches to Synthesis

From the beginning, ILP has sought cross-fertilization with other fields, be it by definition (ILP is an attempt at cross-fertilizing Machine Learning and Logic Programming) or by "charter" (ILP aims at the cross-fertilization of Empirical Machine Learning and Analytical Machine Learning). Other opportunities for cross-fertilization have been discovered and added to the ILP "charter". Some successful attempts have been with:

- *Data/Knowledge-Base Updating*: cross-fertilization resulted in that learning can now be done from (clausal) integrity constraints, a generalization of examples, and that non-unit clauses can now be asserted [20] – [22]. The extended field is known as Belief Updating or Theory Revision. The discovery of data dependencies in relational and deductive databases has also been examined [30] [53] [75].
- *Theorem Proving*: a procedure may be constructed by an analysis of a failed proof of a formal specification of its predicate-symbol [34] [49] [64]. However, in some cases, an inductive theorem prover is not able to process a formula and thus fails to finish a proof. Induction-based learning methods are then used for inventing new predicates and a new formula is built. It is shown that even in the case where the new formula is not equivalent to the original one, the prover is able to make the next step and to finish the proof [33].
- *Logic Program Transformation*: cross-fertilization with Analytical Learning (EBL/EBG) resulted in Explanation-Based Program Transformation (EBPT) [14], where sample concrete transformations guide the overall abstract transformation process.

So the question now arises as to whether cooperation and cross-fertilization are possible with (other) branches of deduction-based Software Engineering? Some attempts at solving Software Engineering tasks with induction-based techniques have been made, such as logic program synthesis and debugging [5] [23] [28] [43] [76], test case generation [4], and program verification [13], but there was no sign of actual cross-fertilization. So what is the potential of such cooperation and cross-fertilization?

One of the major problems we pointed out with the deduction-based synthesis paradigm is due to the formality of the needed specifications: where do the formal specifications (logical axiomatizations, that is) come from? Following Muggleton's summary of the importance of inductive reasoning in scientific discovery [65], the popular answer is that axioms, representing generalized beliefs, can be constructed from particular facts, which are in turn derived from the senses. Turing is reported to have believed that the problems due to Gödel's incompleteness theorem could be overcome by learning from examples. So a possible cooperation would be mixed-inference specification acquisition followed by deduction-based synthesis/transformation. Deduction-based synthesis can provide feedback to the specification elaboration process [60], but this doesn't assist in the *initial* formalization process and only detects inconsistencies *within* the specification, but not inconsistencies with respect to the intentions.

Another opportunity for cooperation lies in the synthesizer-as-a-workbench-of-powerful-mini-synthesizers approach advocated by A.W. Biermann [9] and schema-guided synthesis researchers such as D.R. Smith [78] [79]: such a workbench should include induction-based synthesizers that are known to reliably converge very quickly to correct algorithms from just a few examples. This would be handy synthesis tools, and we believe they would be used very often. Similarly for tools based on other kinds of inference.

A little-explored avenue for cross-fertilization is the use of deductive reasoning *within* induction-based synthesis, and vice-versa. Indeed, synthesis (especially if schema-guided) can often be broken down into very different sub-tasks (such as instantiating some place-holder of an algorithm schema). So it is likely that some sub-tasks are easier to solve by deductive reasoning, whereas others are more amenable to other kinds of reasoning.

For instance, the SYNAPSE system [31] [32] is schema-guided, starts from specifications by examples and some strong form of axioms (called properties), and features deductive *and* inductive reasoning, according to whichever is preferable for each placeholder of a divide-and-conquer schema. The given properties are used in a constructive way: formulas are extracted from an explanation of the failure of a deductive proof that the current algorithm satisfies the properties, and these formulas are then added to that algorithm. This is a different approach from the usage of assertions [28] or integrity constraints [21], which are used in a destructive way (to reject parts of the current hypothesis) and without any actual deductive reasoning.

This technique is related to abduction [54], which plays an important role when the incorporation of incoming information to an existing theory is impossible or inconvenient: for example in deductive databases and knowledge bases, where adding a new piece of knowledge may cause an inconsistency [15] [44], or in a fault diagnosis, where we are interested in a cause of failure rather than in the failure itself. By abduction, we look for an explanation of the new knowledge, consistent with the existing theory, and then add it to the theory.

Abductive reasoning is non-monotonic, as many explanations may exist for a given piece of knowledge. Another evidence for non-monotonicity is that explanations of two different pieces of incoming knowledge may contradict each other.

In the field of algorithm synthesis, De Raedt [20] pointed out that Shapiro's MIS [76] actually also performs abduction: in the context of multiple predicate learning, abduced facts are not added to the theory, but rather used as the starting point for a synthesis phase. In [20], an abductive technique for the inductive learner CLINT is described, making it capable of learning multiple predicates. The use of abduction in interactive algorithm synthesis from examples is also explored in [70].

We believe that deductive/abductive inference from, and constructive usage of, oracle-answers and extensions of example-based specifications will play an important role in induction-based synthesis (and learning). Conversely, we also believe that deduction-based synthesis will greatly benefit from the inclusion of other kinds of reasoning.

5 Conclusion

In this essay, we have given a critical analysis of the deduction-based and induction-based approaches to algorithm synthesis, and of the current research in these fields within the Logic Programming community. We have identified some future research directions for these approaches, as well as a clear need for cooperation and cross-fertilization between them.

The two approaches and their associated current research efforts have their shortcomings, and, upon close inspection, they even share the most fundamental shortcomings:

- the two current efforts suffer from the fact that (more) algorithm design knowledge (such as algorithm schemas) ought to be injected into the synthesizers;
- the two approaches suffer from the fact that there is no formal guarantee that the synthesized algorithms correctly cover our intentions; so in the two approaches an informal proof of correctness is needed somewhere (usually via specification debugging and maintenance).

We hope to have convinced initially suspicious readers that the intuitive argument of the superiority of the deduction-based approach is based on some faulty and prejudiced assumptions.

Acknowledgments

The first author gratefully acknowledges stimulating discussions with Baudouin Le Charlier (University of Namur, Belgium), Yves Deville (Catholic University of Louvain, Belgium), and Norbert Fuchs (University of Zürich, Switzerland). Thanks of the second author are due to his supervisor Olga Štěpánková (Czech Technical University, Prague, Czech Republic) and to Pavel Brazdil (University of Porto, Portugal) for useful discussions and suggestions. The first author also benefitted from the feedback of the students of his Automatic Program Synthesis course at Bilkent University.

References

ICLP = International Conference on Logic Programming
ILP = International Workshop on Inductive Logic Programming
LOPSTR = International Workshop on LOgic Program Synthesis and TRansformation
LPAR = International Conference Logic Programming and Automated Reasoning
SLP = Symposium on Logic Programming

[1] David W. Aha, Stéphane Lapointe, Charles X. Ling, and Stan Matwin. Inverting implication with small training sets. In F. Bergadano and L. De Raedt (editors), *Proceedings of the 1994 European Conference on Machine Learning*, pages 31–48. LNCS 784, Springer-Verlag, 1994.

[2] Dana Angluin. Queries and concept learning. *Machine Learning* 2(4):319–342, April 1988.

[3] Dana Angluin and Carl H. Smith. Inductive inference: Theory and methods. *Computing Surveys* 15(3):237–269, September 1983.

[4] Francesco Bergadano *et al.* Inductive test case generation. In *Proceedings of ILP'93*, pages 11–24.

[5] Francesco Bergadano and Daniele Gunetti. Inductive synthesis of logic programs and inductive logic programming. In [25], pages 45–56.

[6] Wolfgang Bibel. Syntax-directed, semantics-supported program synthesis. *Artificial Intelligence* 14(3):243–261, October 1980.

[7] Alan W. Biermann. On the inference of Turing machines from sample computations. *Artificial Intelligence* 3(3):181–198, Fall 1972.

[8] Alan W. Biermann. The inference of regular LISP programs from examples. *IEEE Transactions on Systems, Man, and Cybernetics* 8(8):585–600, 1978.

[9] Alan W. Biermann. Dealing with search. In Alan W. Biermann, Gérard Guiho, and Yves Kodratoff (editors). *Automatic Program Construction Techniques*, pages 375–392. Macmillan, 1984.

[10] Alan W. Biermann. Fundamental mechanisms in machine learning and inductive inference. In W. Bibel and Ph. Jorrand (editors), *Fundamentals of Artificial Intelligence*, pages 133–169. LNCS 232, Springer-Verlag, 1986.

[11] Alan W. Biermann. Automatic programming. In S. C. Shapiro (editor), *Encyclopedia of Artificial Intelligence*, pages 59–83. John Wiley, 1992. Second, extended edition.

[12] Alan W. Biermann and Douglas R. Smith. A production rule mechanism for generating LISP code. *IEEE Transactions on Systems, Man, and Cybernetics* 9(5):260–276, May 1979.

[13] Ivan Bratko and Marko Grobelnik. Inductive learning applied to program construction and verification. In *Proceedings of ILP'93*, pages 279–292.

[14] Maurice Bruynooghe and Danny De Schreye. Some thoughts on the role of examples in program transformation and its relevance for explanation-based learning. In K. P. Jantke (editor), *Proceedings of the 1989 International Workshop on Analogical and Inductive Inference*, pages 60–77. LNCS 397, Springer-Verlag, 1989.

[15] François Bry. Intensional updates: Abduction via deduction. In D. H. D. Warren and P. Szeredi (editors), *Proceedings of ICLP'90*, pages 561–578. The MIT Press, 1990.

[16] Alan Bundy, Alan Smaill, and Geraint Wiggins. The synthesis of logic programs from inductive proofs. In J. W. Lloyd (editor), *Proceedings of the ESPRIT Symposium on Computational Logic*, pages 135–149. Springer-Verlag, 1990.

[17] Alan Bundy, Andrew Stevens, Frank van Harmelen, Andrew Ireland, and Alan Smaill. Rippling: A heuristic for guiding inductive proofs. *Artificial Intelligence* 62(2):185–253, 1993.

[18] Keith L. Clark. *The synthesis and verification of logic programs*. Technical Report DOC-81/36, Imperial College, London (UK), September 1981.

[19] Tim Clement and Kung-Kiu Lau (editors), *Proceedings of LOPSTR'91*. Workshops in Computing Series, Springer-Verlag, 1992.

[20] Luc De Raedt. *Interactive Theory Revision: An Inductive Logic Programming Approach*. Academic Press, 1992.

[21] Luc De Raedt and Maurice Bruynooghe. Belief updating from integrity constraints and queries. *Artificial Intelligence* 53(2–3):291–307, February 1992.

[22] Luc De Raedt and Maurice Bruynooghe. A theory of clausal discovery. In *Proceedings of ILP'93*, pages 25–40.

[23] Nachum Dershowitz and Yuh-Jeng Lee. Logical debugging. *Journal of Symbolic Computation* 15(5–6):745–773, May/June 1993. Early version, entitled "Deductive debugging", in *Proceedings of SLP'87*, pages 298–306.

[24] Yves Deville. *Logic Programming: Systematic Program Development*. International Series in Logic Programming, Addison Wesley, 1990.

[25] Yves Deville (editor), *Proceedings of LOPSTR'93*. Workshops in Computing Series, Springer-Verlag, 1994.

[26] Yves Deville and Jean Burnay. Generalization and program schemata: A step towards computer-aided construction of logic programs. In E. L. Lusk and R. A. Overbeek (editors), *Proceedings of the North American Conference on Logic Programming'89*, pages 409–425. The MIT Press.

[27] Yves Deville and Kung-Kiu Lau. Logic program synthesis: A survey. *Journal of Logic Programming, Special Issue on 10 Years of Logic Programming*, 1994.

[28] Wlodek Drabent, Simin Nadjm-Tehrani, and Jan Maluszynski. Algorithmic debugging with assertions. In H. Abramson and M. H. Rogers (editors), *Meta-Programming in Logic Programming: Proceedings of META'88*, pages 501–521. The MIT Press.

[29] Agneta Eriksson and Anna-Lena Johansson. Computer-based synthesis of logic programs. In M. Dezani-Ciancaglini and U. Montanari (editors), *Proceedings of an International Symposium on Programming*, pages 105–115. LNCS 137, Springer-Verlag, 1982.

[30] Peter Flach. Predicate invention in inductive data engineering. In *Proceedings of the 1993 European Conference on Machine Learning*, pages 83–94. LNAI 667, Springer-Verlag, 1993.

[31] Pierre Flener. *Logic Algorithm Synthesis from Examples and Properties*. Ph.D. Thesis, Université Catholique de Louvain, Louvain-la-Neuve (Belgium), June 1993.

[32] Pierre Flener and Yves Deville. Logic program synthesis from incomplete specifications. *Journal of Symbolic Computation* 15(5-6):775–805, May/June 1993.

[33] Marta Fraňová. Fundamentals of a new method for inductive theorem proving: CM—construction of atomic formulae. In *Proceedings of the 1988 European Conference on Artificial Intelligence*.

[34] Marta Fraňová and Yves Kodratoff. *Predicate synthesis from formal specifications or using mathematical induction for finding the preconditions of theorems*. Technical Report 646, LRI, Université Paris-Sud, 1991.

[35] Laurent Fribourg. Extracting logic programs from proofs that use extended Prolog execution and induction. In D. H. D. Warren and P. Szeredi (editors), *Proceedings of ICLP'90*, pages 685–699. The MIT Press, 1990. Updated and revised version in [50], pages 39–66.

[36] Laurent Fribourg. Automatic generation of simplification lemmas for inductive proofs. In V. Saraswat and K. Ueda (editors), *Proceedings of SLP'91*, pages 103–116. The MIT Press, 1991.

[37] Norbert E. Fuchs. Hoare logic, executable specifications, and logic programs. *Structured Programming* 13:129–135, 1992.

[38] Norbert E. Fuchs. Specifications are (preferably) executable. *Software Engineering Journal* 7:323–334, September 1992.

[39] Norbert E. Fuchs. Private communications, March–June 1994.

[40] Norbert E. Fuchs and Markus P. J. Fromherz. Schema-based transformations of logic programs. In [19], pages 111–125.

[41] E. Mark Gold. Language identification in the limit. *Information and Control* 10(5):447–474, 1967.

[42] Cordell Green and David R. Barstow. On program synthesis knowledge. *Artificial Intelligence* 10(3):241–270, November 1978.

[43] Marko Grobelnik. Induction of Prolog programs with Markus. In [25], pages 57–63.

[44] A. Guessoum and John W. Lloyd. Updating knowledge bases. *New Generation Computing* 8:71–88, 1990.

[45] Masami Hagiya. Programming by example and proving by example using higher-order unification. In M. E. Stickel (editor), *Proceedings of the 1990 Conference on Automated Deduction*, pages 588–602. LNCS 449, Springer-Verlag, 1990.

[46] Åke Hansson. *A Formal Development of Programs*. Ph.D. Thesis, University of Stockholm (Sweden), 1980.

[47] I.J. Hayes and C.B. Jones. Specifications are not (necessarily) executable. *Software Engineering Journal* 4(6):330–338, November 1989.

[48] Christopher J. Hogger. Derivation of logic programs. *Journal of the ACM* 28(2):372–392, April 1981.

[49] Andrew Ireland. The use of planning critics in mechanizing inductive proofs. In A. Voronkov (editor), *Proceedings of LPAR'92*, pages 178–189. LNCS 624, Springer-Verlag, 1992.

[50] Jean-Marie Jacquet (editor). *Constructing Logic Programs*. John Wiley, 1993.

[51] Klaus P. Jantke. Algorithmic learning from incomplete information: Principles and problems. In J. Dassow and J. Kelemen (editors), *Machines, Languages, and Complexity*, pages 188–207. LNCS 381, Springer-Verlag, 1989.

[52] Alípio M. Jorge and Pavel Brazdil. Learning by refining algorithm sketches. *Proceedings of ECAI'94*. 1994.

[53] Jyrki Kivinen and Heikki Mannila. Approximate dependency inference from relations. In *Proceedings of the 1992 International Conference on Database Theory*.

[54] Andonakis C. Kakas, Robert A. Kowalski, and F. Toni. Abductive logic programming. *Journal of Logic and Computation* 2:719–770, 1992.

[55] Robert A. Kowalski. *Logic for Problem Solving*. North Holland, 1979.

[56] Ina Kraan, David Basin, and Alan Bundy. Middle-out reasoning for logic program synthesis. In D.S. Warren (editor), *Proceedings of ICLP'93*, pages 441–455. The MIT Press, 1993.

[57] Kung-Kiu Lau and Tim Clement (editors), *Proceedings of LOPSTR'92*. Workshops in Computing Series, Springer-Verlag, 1993.

[58] Kung-Kiu Lau and Mario Ornaghi. An incompleteness result for deductive synthesis of logic programs. In D.S. Warren (editor), *Proceedings of ICLP'93*, pages 456–477. The MIT Press, 1993.

[59] Kung-Kiu Lau and Mario Ornaghi. A formal view of specification, deductive synthesis and transformation of logic programs. In [25], pages 10–31.

[60] Kung-Kiu Lau and Mario Ornaghi. On specification frameworks and deductive synthesis of logic programs. In this volume.

[61] Kung-Kiu Lau, Mario Ornaghi, and Sten-Åke Tärnlund. The halting problem for deductive synthesis of logic programs. In P. van Hentenryck (editor), *Proceedings of ICLP'94*, pages 665–683. The MIT Press, 1994.

[62] Kung-Kiu Lau and S. D. Prestwich. Top-down synthesis of recursive logic procedures from first-order logic specifications. In D. H. D. Warren and P. Szeredi (editors), *Proceedings of ICLP'90*, pages 667–684. The MIT Press, 1990.

[63] Baudouin Le Charlier. *Réflexions sur le problème de la correction des programmes*. Ph.D. Thesis (in French), Facultés Universitaires Notre-Dame de la Paix, Namur (Belgium), 1985.

[64] Raul Monroy, Alan Bundy, and Andrew Ireland. Proof plans for the correction of false conjectures. In F. Pfenning (editor), *Proceedings of LPAR'94*. LNCS, Springer-Verlag, 1994.

[65] Stephen Muggleton. Inductive logic programming. *New Generation Computing* 8(4):295–317, 1991.

[66] Stephen Muggleton (editor). *Inductive Logic Programming*. Volume APIC-38, Academic Press, 1992.

[67] Stephen Muggleton and Luc De Raedt. Inductive logic programming: Theory and methods. *Journal of Logic Programming, Special Issue on 10 Years of Logic Programming*, 1994.

[68] Leonard Pitt and Leslie G. Valiant. Computational limits on learning from examples. *Journal of the ACM* 35(4):965–984, October 1988.

[69] Gordon D. Plotkin. A note on inductive generalization. In B. Meltzer and D. Michie (editors). *Machine Intelligence* 5:153–163, 1970. Edinburgh University Press, Edinburgh (UK).

[70] Luboš Popelínský, Pierre Flener, and Olga Štěpánková. ILP and automatic programming: Towards three approaches. Submitted to *ILP'94*, Bonn (Germany).

[71] *Proceedings of the International Workshop on Machine Learning'92 Workshop on Biases in Inductive Learning*. Aberdeen (Scotland, UK), 1992.

[72] Charles Rich and Richard C. Waters. Automatic programming: Myths and prospects. *IEEE Computer* 21(8):40–51, August 1988.

[73] Taisuke Sato and Hisao Tamaki. Transformational logic program synthesis. In *Proceedings of the International Conference on Fifth-Generation Computer Systems*, pages 195–201, 1984.

[74] Taisuke Sato and Hisao Tamaki. First-order compiler: A deterministic logic program synthesis algorithm. *Journal of Symbolic Computation* 8(6):605–627, 1989.

[75] Iztok Savnik and Peter Flach. Bottom-up induction of functional dependencies from relations. In *Proceedings of the AAAI'93 Workshop on Knowledge Discovery in Databases*.

[76] Ehud Y. Shapiro. *Algorithmic Program Debugging*. Ph.D. Thesis, Yale University, New Haven (CT, USA), 1982. Published under the same title by The MIT Press, 1983.

[77] Douglas R. Smith. The synthesis of LISP programs from examples: A survey. In Alan W. Biermann, Gérard Guiho, and Yves Kodratoff (editors). *Automatic Program Construction Techniques*, pages 307–324. Macmillan, 1984.

[78] Douglas R. Smith. Top-down synthesis of divide-and-conquer algorithms. *Artificial Intelligence* 27(1):43–96, 1985.

[79] Douglas R. Smith. KIDS: A semiautomatic program development system. *IEEE Transactions on Software Engineering* 16(9):1024-1043, September 1990.

[80] Leon S. Sterling and Marc Kirschenbaum. Applying techniques to skeletons. In [50], pages 127–140.

[81] Phillip D. Summers. A methodology for LISP program construction from examples. *Journal of the ACM* 24(1):161–175, January 1977.

[82] William R. Swartout and Robert Balzer. On the inevitable intertwining of specification and implementation. *Communications of the ACM* 25:438–440, 1982.

[83] Yukihide Takayama. Writing programs as QJ proof and compiling into Prolog programs. In *Proceedings of SLP'87*, pages 278–287.

[84] Nancy L. Tinkham. *Induction of Schemata for Program Synthesis*. Ph.D. Thesis, Duke University, Durham (NC, USA), 1990.

[85] Axel van Lamsweerde. Learning machine learning. In A. Thayse (editor), *From Natural Language Processing to Logic for Expert Systems*, pages 263–356. John Wiley, 1991.

[86] Geraint Wiggins. Synthesis and transformation of logic programs in the Whelk proof development system. In K. Apt (editor), *Proceedings of the Joint International Conference and Symposium on Logic Programming'92*, pages 351–365. The MIT Press, 1992.

Transforming Specifications of Observable Behaviour into Programs

David Gilbert[1], Christopher Hogger[2], Jiří Zlatuška[3]

[1] City University, Northampton Square, London EC1V 0HB, U.K.
drg@cs.city.ac.uk
[2] Imperial College, 180 Queens Gate, London SW7 2BZ, U.K.
cjh@doc.ic.ac.uk
[3] Masaryk University, Burešova 20, 602 00 Brno, Czech Republic
zlatuska@informatics.muni.cz

A methodology for deriving programs from specifications of observable behaviour is described. The class of processes to which this methodology is applicable includes those whose state changes are fully definable by labelled transition systems, for example communicating processes without internal state changes. A logic program representation of such labelled transition systems is proposed, interpreters based on path searching techniques are defined, and the use of partial evaluation techniques to derive the executable programs is described.

1 Motivations

Our methodology provides a means for deriving executable programs from specifications of the observable behaviour of a restricted class of systems. The systems which are tractable by this methodology are those whose observations are discrete fine-grained steps which progressively construct data objects, expressed as terms in our approach. We give the characterisation of these systems by the use of a language based on labelled transition systems, and identify a class of interpreters derived from rewriting and path searching algorithms on the graphs induced by the labelled transition systems. Furthermore, we are able to derive programs by partially evaluating such interpreters with respect to the rules in the language of the labelled transition systems. We also provide a formalism which permits the transformation of the labelled transition systems into the target programs within the framework of computational logic.

The class of computations in which we are interested contains those whose result is incrementally constructed, whilst at the same time the partial results are being output as *observations* which are accessible to the environment of the computing agent. For some of these computations it is natural also to *specify* such processes in terms of their observable behaviour. We believe this may be closer to the user's understanding of the system to be programmed, when the external behaviour of the system and the sequence in which the result is produced (i.e. a trace history) is an essential part of the activity of the process. For the sake of simplicity, we assume that the observation of the external behaviour is tightly linked to the internal computation steps of such a system, i.e. each step of the computation strictly extends the resulting data structure which it constructs.

Our method regards observations of the progress of a computation as *extrinsic specifications* which can be represented as directed acyclic graphs. Each computation that the system can perform is represented by a path through the graph, which in turn can be described in first order logic. Logic programs can be derived from these first order logic descriptions by standard transformations.

We view the long-term objective of our method to be the construction of reactive concurrent systems. As a first step towards this objective we present a working framework for sequential systems. From a more abstract point of view, we can understand the procedure for constructing programs from graphs representing observable behaviour as a compiler of a graph-based production language. A particular strategy for the generation of computations from such a graph can be linked to the particular strategy for the search of the tree representing the trace history of the execution of the corresponding program. In general, the method as presented in this paper makes no assumptions about the sequential or concurrent behaviour of the programs which have been generated; such behaviour is a result of the execution mechanism for these programs.

2 Summary of the Approach

The outline of our approach can be given as follows: first, give a specification of all the possible sequences of the observable behaviour of the system to be constructed. This is done by determining the elementary transitions between the states of the program which produce the observables, and taking these transitions as a definition of a labelled transition system which can generate all the possible state changes. Since we assume that every change of state of a program is immediately reflected in production of an observable (i.e. an output visible to the external observer), we identify internal states with their associated observations.

Second, take a general *interpreter* of the resulting labelled transition system expressed in a suitable language. This interpreter is a path searching algorithm which explores the graph of transitions generated by the system. The structure of observations, labelled transitions, and the path searching procedure can all be formalised in the language of first-order Horn clause logic. Labels of the initial transformation system are identified with suitable variables, and partial objects built as a result of partial execution of the program. These objects can be identified with terms containing variables in places at which the structure will be later extended during further program execution. Within the language of logic, generation of a particular observable corresponds to substituting the term representing this observable for the free variable at the location where the new observable occurs. Hence the term which is being constructed during a program run corresponds to a tree of a particular history of observable state transitions, represented as a term.

When specifying path-searching procedures working over the state space of a particular labelled transition system represented in logic, *substitution* plays the the rôle of the basic operation performed. Each transition of the initial labelled transition system thus corresponds to an atomic substitution, and sequences of transitions correspond to compositions of atomic substitutions of this kind. Based on this, the path-searching procedure works over compositions of substitutions. Therefore the resulting program defined by such a system amounts to a generator of substitutions. These substitutions in turn correspond to changes of program observables, i.e. to program state transitions. The substitution generator therefore works as an interpreter working over the labelled transition system.

Finally, generate a program which implements the labelled transition system by partially interpreting the path searching algorithm applied to the set of transition rules. We employ a partial evaluator for logic programs for this and hence use logic programming for all three steps.

The scheme outlined above depends on the feasibility of realising each of the steps involved so that the goal of generating the program from the specification can in fact be achieved. In this paper we describe a particular formal framework which permits this goal to be accomplished. We start with a simple definition of a labelled transition system defined as a system for synchronously rewriting several labels during one transition step. The intuitive meaning of this definition is that several processing agents can act synchronously within the computational environment. We then embed these systems into clauses defining transitions of the system. From this point on, all of the construction is performed in a logic programming language, Prolog in our test implementation, starting from data structures, path-searching algorithms and generation of state-change histories i.e. terms generated by subsequent applications of substitutions corresponding to observable changes. The partial interpretation needed is therefore just a general logic programming partial interpreter (Mixtus [21] in our case).

Within each of the steps we discuss the data structures involved and the simplifications which can be employed. Note that because of the meta-programming features of our approach which is based on a path-searching interpreter, we need to specify substitutions as operations at the meta-level, rather than to rely on substitutions performed by the underlying engine of the logic programming language used for implementation. If the method is to be practically usable, the implementation of the manipulation of substitutions has to be substantially simplified in order that the generation of the final programs by partial evaluation terminates. Special discussion is therefore devoted to using the general properties of the substitutions which can possibly occur during the process of path generation, and to designing a modified definition of substitution suitable for this step.

3 Specifying Observable Changes

We use a labelled transition system (LTS) to describe possible changes of observables in the system. Such a system is given by a set of transition rules of the

form
$$(x_1, \ldots, x_n) \mapsto (t_1, \ldots, t_n) \qquad \text{where } n \geq 1$$

Note that we permit more than one label on the left-hand side of the transition, enabling us to describe systems where more than one observables may change concurrently. $x_1, \ldots, x_n$ are the labels, or identifiers representing observables, and $t_1, \ldots, t_n$ are general expressions built over the labels and other atoms. These latter denote the resulting configuration after observable change, and may include the observable identifiers again as a proper subpart of any of them. The expressions on the right-hand side correspond to fragments of the trees (terms) of trace histories associated with observable data generation.

An example of a labelled transition rule which describes the generation of a list is
$$(x) \mapsto (a.x)$$

We may extend this to the description of a system which counts the number of items in a list:
$$(x, y) \mapsto (a.x, succ(y))$$

The informal motivation is to consider the labels as states, and to take each transition rule as a definition of a state change, possibly acting synchronously over several processes (if $n > 1$). The expressions on the right-hand side permit the definition of both the observable output and the resulting change of the state, including termination or splitting into several processes. One can think of the expressions generated by systems of this kind as snapshots of trace histories of processes which are represented by labels. Transitions can be applied to any of those labels in order to expand the structure representing the current partial trace-history. Observables produced by the system correspond to functors (atoms) occurring in the expressions generated by the LTS. (In the logic programming representation, these will be functors of the language.)

The descriptive power of the formalism is most easily understood by considering the class of processes which can be determined by a LTS as a language generated by a grammar derived from it. On an abstract level, any LTS corresponds to a *grammar* whose nonterminal symbols represent the labels of the LTS, and whose terminal symbols correspond to data structures. Thus the nonterminals actually correspond to states of a computation (sequential or parallel) represented by expanding the starting state. Even in the sequential case, the resulting pattern is different from just recursive descent due to the treatment of all the non-terminals produced in an expansion step as a partial process output. This reflects our interest in focussing primarily on generating/specifying *traces* of computations as sequences of process outputs, not just the resulting (data) structure given by the words generated by the grammar.

If the transitions of the LTS transitions only have one label expanded at each step the resulting grammar is at most a context free grammar, with all the inherent limitations of CFGs, which for example cannot represent the concurrent update of more than one label. The treatment of the class of systems which we consider within our framework contains transition rules which can concurrently transform several labels at the same time, permitting us to describe concurrent systems, and leads to a sub-class of context grammars which is strictly larger than CFGs.

4 Target Program Structure

The processes specified by this class of LTS can be represented in various ways, depending on the actual programming paradigm selected. In our approach, representation as a logic program is chosen, because of the declarative nature of this paradigm. This permits us to develop a framework for program synthesis which is independent of the particular implementation of the processes it defines, either sequential or concurrent. When the resulting logic programs are coupled with a corresponding evaluation strategy, this is effectively equivalent to a program in a procedural programming language, yet the particular level of abstraction permits a more succinct representation of the problem.

The observables of a logic program are the logical variables in the initial goal associated with it. Unification is the finest level of granularity which is useful to observe, and thus unification steps are taken to be the atomic events which are observable. Communication in a logic programming system occurs via bindings made to *shared* variables, and our assumption is that an observer can detect the *incremental* bindings made to the variables in the initial goal (i.e. to external variables). The observations made are posets of binding sets; we can represent these posets as directed acyclic graphs, due to the write-once nature of the logic variable. The bottom element of such a set represents the initial unbound state of the observable variables. Each path through the graph from the minimum vertex to a maximum vertex comprises the observations of one computation and the union of the sets associated with all such paths comprises the instantiation set of the observational variable(s).

An example is the instantiation set of the following directed graph for the variable x. Nodes are labelled with the term to which x is bound, and an arc from node A to node B is labelled with the substitution which when applied to the term at A results in the term at B.

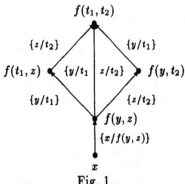

Fig. 1.

$$\{x/f(y,z), x/f(t_1,z), x/f(t_1,t_2)\} \cup \{x/f(y,z), x/f(y,t_2), x/f(t_1,t_2)\}\cup$$

$$\{x/f(y,z), x/f(t_1,t_2)\} = \{x/f(y,z), x/f(t_1,z), x/f(y,t_2), x/f(t_1,t_2)\}$$

Given the choice of logic variables as observables, the actual data items produced by a running program are represented by the functors of the language of terms. Therefore changes of a process state are manifested by assignment of data to the tuple of logic variables representing the process. The particular syntactic structure is just a consequence of our use of logic terms to represent data structures. During the transformation of terms into prefix/postfix notation, the non-variable symbols in them correspond straightforwardly to terminal symbols of the grammar associated with the original LTS. The use of logic variables permits the representation of trace histories as simple variable bindings, with no additional formalism being needed.

An example of an LTS is the pair of transition rules:

$$(x) \mapsto (a.x)$$
$$(x) \mapsto (nil)$$

These rules represent an LTS which describes the behaviour of a process which binds a variable to list and whose trace history is itself a list of the binding states of that variable.

There is a regular grammar corresponding to this system,

$$X \to aX$$
$$X \to \epsilon$$

characterizing the set of process behaviour as the corresponding regular set of sequences of data items a of arbitrary length.

The logic-variable representation uses just one type, X, for the variables of the LTS. The tuples on the right-hand sides of the production rules can be represented by terms $a(x)$, where x is a variable of the type X and a a unary function symbol, and by a nullary functor symbol nil, respectively.

5 Logic Program Representation

When specifying a logic program, we choose to identify the observables by logic variables. As far as these correspond to distinct labels of the LTS, or distinct non-terminals of the grammar, there is a need to ensure that the correspondence of every variable with its associated label is preserved. Without such correspondence substitutions may be applied to incorrect variables. This correspondence can be achieved by partitioning the sets of labels and identifiers using tags. On the level of specification, we choose to work in a *multi-sorted* logic, where types can be used to perform this tagging function. Each variable is therefore associated with a unique fixed *type*. All the usual properties of logic programs are preserved within this, and the only change to the underlying machinery required is that of modifying unification so that it fails whenever an attempt to bind a variable of certain type to a term of *different* type is made.

Thus, for example, in the rule

$$(x, y) \mapsto (a.x, succ(y))$$

we consider that x and y are of different types.

The type scheme resulting from the use of labels of the LTS as types of the system permits simple static type checking, for example that of Gödel [11]. Note that with static type checking it is sufficient to verify type constraints at the level of the source program code, and so at run-time it is possible to use type-less logic programming language, such as Prolog, and hence not to refer to types. In our case this corresponds to the need to ensure proper type constraints when writing the interpreter, as proposed by Hill and Lloyd [10], but the actual programs generated by partial evaluation are ordinary type-less logic programs.

The idea of using typed terms is just a syntactic means for avoiding the use of dynamic predicate-based type checking. The untyped predicate logic is expressible enough to define all that is needed for this, but requires the use of a more complex clause structure for the representation of the transition rules. Specifically we need to introduce predicates for *dynamic* type checking into each of the clauses of the interpreter. The framework of typed terms seems more natural in our context for two reasons. Firstly because of the example of the successful use of types in logic programming which has been set by Gödel, and secondly because the use of types simplifies the representation of transition rules as clauses by effectively moving the type-checking predicates out of such clauses into the code of a general-purpose interpreter.

For the representation of the LTS, the left-hand side of a rule becomes a tuple of logic variables, and the right-hand side is represented by a tuple of terms containing new versions of the variables, all of the variables being typed by the appropriate LTS label types. The version of the above example would be

$$(x, y) \mapsto (a.x', succ(y'))$$

where x and x' are of the same type, and so are y and y'.

In order to implement the above process in Prolog, we choose a representation of variables in which each variable carries its source observable id as a type associated with it. This observable id tag controls the possible variable occurrences which may or may not match with the variables resulting from a labelled transition rule. Obviously, when using a typed logic programming language such as Gödel, the representation could be made simpler.

Observable changes can now be described by the successive instantiation of variables, a characteristic feature being the possibility of binding variable to a term containing yet more variables. Non-linear structures can be generated in such a way, with several new observables being generated as a result, for example the generation of tree structures.

Instantiations of variables are carried out by substitutions, defined as morphisms on terms, fully described by their result on variables. In the case of *finite* substitutions (which only change a finite number of variables), the usual notation

$$\theta = [x_1/t_1, \ldots, x_n/t_n]$$

describes a mapping defined as

$$t\theta = \begin{cases} t_i & \text{if } t = x_i \text{ for } x_i/t_i \in \theta; \\ t & \text{if } t \text{ is a variable, not occurring as } t/u \in \theta \text{ for any } u; \\ f(t_1\theta, \ldots, t_n\theta) & \text{for } t = f(t_1, \ldots, t_n), n \geq 0. \end{cases}$$

Substitutions define state-changing operations on the processes, and the program-generating process developed later in this paper is based on building a meta interpreter which combines substitutions in a suitable way.

The process of observables transformation leads to the composition of substitutions defined as function compositions. On finite substitutions this this gives the following standard definition:

$$\theta\sigma = [x/t\sigma | x/t \in \theta \text{ and } x \neq t\sigma] \cup [y/s \in \sigma \text{ and for every } t, y/t \notin \theta]$$

Note that the second operand of the union allows us to eliminate those changes to variables defined by σ which are ineffective because of a previous elimination of suitable variable occurences by θ. We will employ this fact in the following section to simplify our working definition of composition.

At this point, the LTS can be transformed into substitutions: for each rule

$$(x_1, \ldots, x_n) \mapsto (t_1, \ldots, t_n)$$

generate a set of substitutions of the form

$$[x_1/t'_1, \ldots, x_n/t'_n]$$

with identifiers expressed as logic variables. Moreover, within each pair x_i/t'_i, t'_i is formed from t_i by renaming all variables into fresh ones. As noted above, we assume the existence of typed variables, and hence the framework of a multi-sorted language. When actually implementing this operation in a language lacking strict type discipline (such as Prolog), some extra care must be taken in the actual code to ensure that the types of the variables are preserved when renaming them.

Now the program specification part can be viewed as the set of rules describing the accumulation of substitutions: input substitution is composed with the observable-changing substitution in order that the resulting substitution is a new configuration of the system.

6 Instantiation Steps

Our method describes computations as ones which progressively instantiate variables to terms. We represent terms explicitly by substitution sets, and describe the instantiation of a term t to a more specialised form t' by the relation compose(x, y, z) where

x is the substitution set associated with t

z is the substitution set associated with t'

y is the substitution set whose composition with x results in z.

For example, consider the following set of possible instances of a variable x

$$\{x, f(y, z), f(t_1, z), f(y, t_2), f(t_1, t_2)\}$$

which corresponds to the set of *atomic substitutions* illustrated by Figure 1 above. From this poset we may extract, by closure over the arcs, one of the possible binding histories, e.g.

$$[x/f(y, z)]..[x/f(t_1, z)]..[x/f(t_1, t_2)]$$

which we illustrate in Figure 2 below

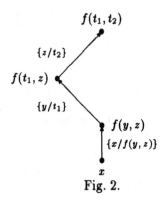

Fig. 2.

We can then associate the following compositions with the binding history:

compose$([x/f(y, z)], [y/t_1], [x/f(t_1, z), y/t_1])$,

compose$([x/f(t_1, z), y/t_1], [z/t_2], [x/f(t_1, t_2), y/t_1, z/t_2])$

We will need to add some restrictions to the standard definition of substitution in order to refine our technique. The first is linked to our basic assumptions about the class of systems we are interested in. The acyclicity of the underlying LTS reflects the intuition that during the process of computation there is always some visible non-empty output after each step. In terms of bindings, this means that there is a *progression* in the binding sequence which prohibits simple renaming occurring as local instantiation steps. This permits the elimination of useless compositions such as compose$([x/y], [y/z], [x/z, y/z])$ and hence prohibits the inclusion of x/y where x and y are variables in the substitution set at the second argument of compose/3

Our top-level definition is

compose(A, B, C) ← progress(A, B, C) , full-compose(A, B, C)

where full-compose/3 is the relation corresponding exactly to the general mathematical definition of substitution composition without any additional restrictions.

Here progress(A, B, C) means that C is not a variant of A, i.e. B must be some non-trivial binding corresponding to an observable state change.

The above-outlined model is too general and contains excessive checks on the substitutions which result from composition. The checks for idempotent substitution pairs and for the elimination of variables from the standard definition of substitution make the partial evaluation of the target program unnecessarily complicated. Pragmatically we found the definition to be too general, and the partial evaluation did not terminate under Mixtus. The picture can be simplified by the consideration of the constraints result from the source LTS structure and the way substitutions are generated from it.

First, variables on the right-hand side of substitution pairs can be renamed, so that $x_i \notin t_i$ for any x_i/t_i. Also, for any $x_i/t_i \in \sigma$ there is no $y_i/u_j \in \sigma$ such that $x_i \in u_j$ for any substitution pair θ, σ, which can possibly be considered a result from composing substitution θ with substitution σ resulting from LTS. This is because variables in u_j have been generated as fresh, not yet occurring elsewhere. As a result, the idempotency check can be omitted from the definition of composition.

Second, in substitutions θ and σ, for any $x_i/t_i \in \theta$ there is no $y_j/u_j \in \sigma$ such that $x_i = y_i$. Again, this is a result of state-changing substitution being generated from LTS rules. Consequently, the elimination check can be omitted as well.

The resulting definition of substitution needed for describing observable changes therefore reads as follows:

$$\theta\sigma = [x_i/t_i\sigma | x_i/t_i \in \theta] \cup \sigma.$$

This allows us to simplify the partial-evaluation phase significantly (see Section 9), and to provide for a manageable program generator out of the source LTS.

7 Constraining Instantiation Steps

The definition of composition as given above is too general for our purposes and does not constrain instantiation to any particular discrete steps. The concept of expanding the underlying LTS-related grammar corresponds to performing computations using the LTS-based specification which leads to the identification of computational steps with the expansion of non-terminals, i.e. variable instantiation. Such an instantiation is limited to terms of the structure which correspond to the left-hand sides of transformation rules, i.e. such an instantiation step typically replaces variables either by constants or by trees of a restricted depth. Hence we are only interested in systems which instantiate terms in such a minimal manner, and thus introduce the notion of a (non-null) atomic substitution set Y whose application to a substitution set X by compose(X,Y,Z) satisfies the following constraints:

- There is at least one non-ground t_i of $v_i/t_i \in X$ which is further instantiated by $v_j/t_j \in Y$ (where t_j is not a variable).
- v_j/t_j is minimal in some sense to the particular application

We can define atomic/1 which is part of the meta-interpreter by:

atomic(Y) ← lts(P ↦ Q), transform(P ↦ Q, Y)

where lts/1 is the object level program and transform/2 performs the transformation operation from the LTS into substitutions referred to in Section 5. For example, we may take for lists the pair of transition rules represented by lts/1:

lts((x) ↦ ($a.x$))
lts((x) ↦ (nil))

which are transformed to [$x/a.x'$] and [x/nil] respectively.

We now can give a definition of a new predicate inst(X,Z) which relates a substitution set X, about a term T, to a substitution set Z, about the immediate successor T' of T, as determined by some applicable substitution set Y:

inst(X,Z) ← atomic(Y), compose(X,Y,Z)

8 Node Traversal of Instantiation Graphs

The inst/2 predicate allows us to describe only one individual step, or edge, in the instantiation graph which represents the graph of transitions, whereas we ultimately intend to describe the graph as a whole. More specifically, having provided a definition or program for 'inst' in respect of some particular application, we want to incorporate it within some encompassing program which *traverses* the DAG determined by 'inst'. The classic path traversal method, for example as shown by Kowalski [13] relates nodes (in this case substitution sets) to their subsequent states. Consider any node N already generated; after one or more atomic steps, various paths will have been developed to some further node N. We can define the following path-finding programs by the transitive closure of inst/2:

Program A1
path(N, F) ← inst(N, F)
path(N, F) ← inst(N, N') , path(N', F)
inst(N, F) ← compose(N, Y, F) , atomic(Y)

Program A2
path(N, F) ← inst(N, F)
path(N, F) ← inst(N', F) , path(N, N')
inst(N, F) ← compose(N, Y, F) , atomic(Y)

Which program is used is determined by the input-output mode with which path/2 is queried; the entire graph can be traversed by inputting N as the bottom node and using path1 to seek all reachable nodes F, or vice-versa.

We can now follow the method of Gilbert and Hogger [8] which derived path exploration programs which computed only the *differences* between nodes. In the context in which compose/3 appears, constrained together with atomic/1, it cannot be used in any way which would not satisfy the following conditions, as defined by Brough and Hogger [4]:

(1) $(\forall N \ \forall Y \ \exists F)(\text{compose}(N,Y,F))$

Some result F be *defined* for any Y applied to any N.

(2) $(\exists I \ \forall Y \ \forall F)(\text{compose}(I,Y,F) \leftrightarrow Y=F)$

compose to have at least one *left-identity* I.

(3) $(\forall N \ \forall Y \ \forall Y' \ \forall F) \ ((\text{compose}(N,Y',F) \leftrightarrow Y'=Y) \leftarrow \text{compose}(N,Y,F))$

$(\forall N \ \forall Y \ \forall F \ \forall F') \ ((\text{compose}(N,Y,F') \leftrightarrow F'=F) \leftarrow \text{compose}(N,Y,F))$

compose to be *functional* in two of its modes.

(4) $(\forall N \ \forall N' \ \forall Y \ \forall Y' \ \forall Y'' \ \forall F) \ ((\text{compose}(N,Y,F) \leftrightarrow \text{compose}(N',Y',F))$
$$\leftarrow (\text{compose}(Y'',Y',Y) \wedge \text{compose}(N,Y'',N')))$$

compose to be *associative*.

This new path exploration relation path/1 is defined by

Program B1

path(Y) ← atomic(Y)
path(Y) ← compose(Y″, Y′, Y) , atomic(Y″) , path(Y′)

Program B2

path(Y) ← atomic(Y)
path(Y) ← compose(Y″, Y′, Y) , atomic(Y′) , path(Y″)

We should note that all the four path searching programs above act as interpreters for programs defined by the lts/1 relation.

9   Program Derivation by Partial Evaluation

Programs A1, A2, B1 and B2 can be used as general templates to describe a class of programs which incrementally instantiate observable variables during the course of their execution; specific instances of programs are determined by the definition of atomic/1, determined by lts/1. We have coded the relations for compose/3, atomic/1 and both path/1 and path/2 in SICStus Prolog in order to explore the possibilities of transforming the generic path programs into specialised forms for given LTS's. A design decision was taken early on to distinguish between meta-level and object-level variables in the Prolog code by using the ground term representation, in order to preserve the semantics of the definitions.

We then use partial evaluation in order to produce a program which incorporates the path searching interpreter and our compose/3 relation with a specific labelled transition system as defined by lts/1. Partial evaluation of logic programs is an optimisation technique which has been described in logic programming terms by Lloyd and Shepherdson [15] as follows: "Given a program P and a goal G, partial evaluation produces a new program P' which is P 'specialised' to the goal G. The intention is that G should have the same (correct and computed) answers w.r.t. P and P', and that G should run more efficiently for P' than for P". Both folding and unfolding are techniques used in partial evaluation:

- *logical folding* is the replacement of a goal that is an instance of the body of a clause by the corresponding instance of the head of the clause;
- *logical unfolding* of the goal X_i in the clause
$H \leftarrow X_1, \ldots, X_{i-1}, X_i, X_{i+1}, \ldots, X_n$
where X_i is defined by $X \leftarrow B_1, \ldots, B_m$ is defined by the following transformation:
$$\{(H \leftarrow X_1, \ldots, X_{i-1}, B_1, \ldots, B_m, X_{i+1}, \ldots, X_n)\theta \mid \mathrm{mgu}(X, X_i) \wedge \theta \neq \text{false}\}$$

Although partial evaluation can be done by hand, we have used Mixtus [21], the excellent partial evaluator for Prolog developed by Dan Sahlin, and have obtained good initial results.

The choice of Prolog permits the simplification of the relationship between the data structures used, the meta-interpreter of the substitution-changing relation, and the resulting synthesized code. This is because the same language is used to represent the observables, the program in the LTS language and the interpreter based on graph searching. The methodology itself is nonetheless applicable to any implementation language, but the amount of code actually needed may be significantly greater.

In the case of an implementation in logic programming, the method naturally does not provide any universal mechanism for synthesizing arbitrary logic programs. By the nature of the initial assumptions chosen, only programs whose state changes result in bindings to variables in an initial query are expressible by this method. On the general level of process description this corresponds to systems whose change of state is always visible to the outside environment, e.g. via observable communication between processes.

10 Comparison with other work

The algebraic structure of atomic formulae, including the lattice properties of the instantiation order relation, were described independently in the early seventies by Reynolds [18] and Plotkin [16]. Both authors were also interested in mechanical theorem proving. Reynolds showed in his paper that the refutation of "transformational systems" (sets of clauses containing only unit clauses and clauses with one positive and one negative literal) was in effect path searching, but that there was no decision procedure for such systems. Plotkin discussed the use of induction to find least generalizations of clauses or literals and showed that there is an algorithm to find the least generalization of any pair of literals or terms. It is interesting to note that in his paper Plotkin considered the possibilities of automated induction; the path searching algorithms described in our work relate to induction, but it is not our goal to try to enhance the specification by generalisation.

Belia and Occhiuto [2] have developed an explicit calculus of substitutions which extends Reynold's gci [18] and combines Robinson's unification [19] and term instances. Their work aims to avoid the drawback of the gap between the theory and current implementations of logic programming languages represented by the use of metalevel structures, like substitutions, and mechanisms, like mgu and instantiation, to deal with the substitution rule. Their calculus of c-expressions permits structures to be dealt with explicitly at the object level which would otherwise typically be hidden at the metalevel. They put the substitution rule as an additional operator, mgi, of the language of terms, and provide c-expressions as programs. We prefer to work initially at the metalevel, and to take a classical approach based on interpretation [13, Chapter 12] and partial evaluation (see for example [15, 3]). C-expressions which use integers as tuple indexes do not exploit the tree structure of terms; Belia and Occhiuto are investigating a different calculus using paths instead of integers, which may be closer to our approach.

The language of Associons [17] was developed by Martin Rem as a program notation without variables; the motivation was to develop a language model which employed more concurrency than traditional languages based on assignment. An associon is a tuple of names defining a relation between entities represented by these names. The state of a computation can be changed by a forward chaining process based on the "closure statement", which creates new associons that represent new relations deduced from the already existing ones. The language of associons is essentially deterministic and is based on sets. In fact the language does have logical variables and also universal quantifiers over closure statements. These statements are effectively normal program clauses (i.e. they can contain negated conditions) [14], and also contain guards. Our programs are expressed as *definite* program clauses, without negated conditions, and in contrast to Rem's language our approach is based on the backward-chaining principle of logic programming and permits the expression of all-solutions non-determinism since our language does not contain guards. A similarity between our approach and that of Rem is that both formalisms permit the construction of programs which are inherently concurrent.

Banâtre and Le Métayer have developed the Gamma language [1] which also permits the construction of programs which are inherently parallel in their operation. A Gamma program is essentially a multiset transformer operating on all the data at once; it 'reacts' on multisets of data by replacing a subset whose elements satisfy a given property by the result of the application of a function on that subset. Gamma is an intermediate language between specifications in first order logic and traditional programming languages; programs in Gamma describe the logic of an algorithm and are transformed into executable programs by expressing lower-level choice such as data representation and execution order. Central to the Gamma language is non-determinism, expressed as the choice between several subsets which are candidates for reaction, and the locality principle, which permits independent and simultaneous reactions on disjoint subsets. The multiset is seen as a representation of the state of a system in Gamma; although our approach is based on sets rather than multisets, there is a similarity in that the set of bindings is regarded as the state of the system. There is no recursive data structure definition in Gamma, and data has to be represented as flat multisets of items; for example trees are represented by nodes and leaves associated with parenthood information. We preserve the tree structure of terms due to our use of term substitution, but do employ types to indicate the position of subterms within a term. In both Gamma and our formalism this means that all components of a data structure are directly accessible, independently of their position in the structure. However, in contrast with Gamma, our technique does not have direct equivalents to the operations of data expansion and data reduction.

The work of Gilbert and Hogger [8, 9] described the instantiation of a given term X by a given substitution Y to give a new term Z by a relation subst(X,Y,Z) where any substitution Y was constructed by applying a tupling <> to some set $\{Y_1, \ldots, Y_n\}$ of simpler substitutions. Y was represented by means of a *substitution-tree*, and a special symbol Δ used exclusively to describe variables. A major drawback of this method is that it was only applicable systems which generated lists, since the non-inclusion of S-trees in the Herbrand universe required the transformation of the subst/3 relation into one which does operate over Herbrand terms, but which is not associative.

Our work is closely related to the area of partial evaluation, in particular as applied to logic programs. Work in this field has been carried out by Lloyd et al. [15, 3] amongst others, who have given a strong theoretical foundation for partial evaluation in logic programming, and Dan Sahlin, who has constructed a robust partial evaluator for Prolog [20, 21]. Partial evaluation for concurrent logic languages has been explored by Fujita et al. [6] for GHC programs, and by Huntbach [12] for Parlog programs.

Furthermore, our method based on a graph traversal template can be regarded as an interpreter for graph-based computations, and in this sense is related to the work by Gallier et al. [7] on graph-based interpreters for general Horn clauses.

11   Conclusions

The work reported in this paper reconstructs the method of Gilbert and Hogger for deriving logic programs from the expected observations of program behaviour. We replace their concept of substitution trees (S-trees) by binding sets, and show that a more general method can be developed based on substitution mappings as the basis for the theory. We have formulated path exploration programs which act as generalised program schemata for certain classes of systems, and have derived specialised instances of these programs by partial evaluation of the schemata and specifications of program behaviour given in terms of labelled transition systems. The partial evaluation stage has been successfully mechanised using Mixtus [21], a partial evaluator for Prolog.

Acknowledgements

This work was partially supported by PECO Fellowship grants CT931675 (David Gilbert) and CT926844 (Jiří Zlatuška) provided by the European Community under the scheme for Scientific and Technical Cooperation with Central and Eastern Europe.

References

1. J-P. Banâtre and D. Le Métayer. Programming by Multiset Transformation. *Communications of the ACM*, 36(1):98–111, 1993.
2. M. Belia and M. E. Occhiuto. C-expressions: a variable–free calculus for equational logic programming. *Theoretical Computer Science*, 107:209–252, 1993.
3. K. Benkerimi and J. W. Lloyd. A partial evaluation procedure for logic programs. In Debray and Hermenegildo [5], pages 343–358.
4. D. R. Brough and C. J. Hogger. Compiling associativity into logic programs. *The Journal of Logic Programming*, 4(4):345–360, December 1987.
5. S. Debray and M. Hermenegildo, editors. *Proceedings of the 1990 North American Conference on Logic Programming*, Austin, 1990. ALP, MIT Press.
6. H. Fujita, A. Okumura, and K. Furukawa. Partial evaluation of GHC programs based on the UR-set with constraints. In R. A. Kowalski and K. A. Bowen, editors, *Proceedings of the Fifth International Conference and Symposium on Logic Programming*, pages 924–941, Seatle, 1988. ALP, IEEE, The MIT Press.
7. J. H. Gallier and S. Raatz. Hornlog: A graph-based interpreter for general Horn clauses. *The Journal of Logic Programming*, 4(2):119–156, June 1987.
8. D. R. Gilbert and C. J. Hogger. Logic for representing and implementing knowledge about system behaviour. In V Mařík, O Štěpánková, and R Trappl, editors, *Proceedings of the International Summer School on Advanced Topics in Artificial Intelligence*, pages 42–49, Prague, Jul 1992. Springer Verlag Lecture Notes in Artificial Intelligence No. 617.
9. D. R. Gilbert and C. J. Hogger. Deriving logic programs from observations. In Jean-Marie Jacquet, editor, *Constructing Logic Programs*. John Wiley, 1993.
10. P. M. Hill and J. W. Lloyd. Analysis of Meta-programs. Technical Report CS-88-08, Department of Computer Science, University of Bristol, Bristol, UK, June 1988.
11. P. M. Hill and J. W. Lloyd. *The Gödel programming language*. MIT Press, 1993.
12. M. Huntbach. Meta-interpreters and partial evaluation in parlog. *Formal Aspects of Computing*, 1(2):193–211, 1989.
13. R. A. Kowalski. *Logic for problem solving*. North Holland, 1979.
14. J. W. Lloyd. *Foundations of Logic Programming*. Spinger-Verlag, Berlin, second edition, 1987.
15. J. W. Lloyd and J. C. Sheperdson. Partial evaluation in logic programming. *The Journal of Logic Programming*, 11(3 & 4):217–242, October/November 1991.
16. G. D. Plotkin. A note on inductive generalization. In B. Meltzer and D. Mitchie, editors, *Machine Intelligence*, pages 153–165, 1970.
17. M. Rem. Associons: A Program Notation with Tuples instead of Variables. *ACM Transactions on Programming Languages and Systems*, 3(3):251–262, Jul 1981.
18. J. C. Reynolds. Transformational systems and the algebraic structure of atomic formulas. In B. Meltzer and D. Mitchie, editors, *Machine Intelligence*, pages 135–151, 1970.
19. J. A. Robinson. A machine-orientated logic based on the resolution principle. *Journal of the ACM*, 12(1):23 – 49, Jan 1965.
20. D. Sahlin. The mixtus approach to automatic partial evaluation of full Prolog. In Debray and Hermenegildo [5], pages 377–398.
21. D. Sahlin. *An Automatic Partial Evaluator for Full Prolog*. PhD thesis, Swedish Institute of Computer Science, Mar 1991.

On Specification Frameworks and Deductive Synthesis of Logic Programs*

Kung-Kiu Lau[1] and Mario Ornaghi[2]

[1] Department of Computer Science, University of Manchester
Oxford Road, Manchester M13 9PL, United Kingdom
kung-kiu@cs.man.ac.uk
[2] Dipartimento di Scienze dell'Informazione
Universita' degli studi di Milano
Via Comelico 39/41, Milano, Italy
ornaghi@imiucca.csi.unimi.it

1 Introduction

Logic program synthesis methods can be roughly divided into those that use a formal specification and a formal synthesis process, and those that use an informal specification and an informal synthesis technique. A taxonomy and a survey of logic program synthesis methods can be found in [5].

In formal methods for logic program synthesis, the issue of what specification language is best has never loomed large, because first-order logic is the obvious candidate and hence the natural choice. This, in our view, has resulted in an over-simplistic view of logic program specifications.

In our own area of *deductive synthesis* of logic programs, for example, a specification is often just an *if-and-only-if* first-order definition, or a set of such definitions. Such a simplistic approach to specification has, in our experience, two major drawbacks:

- It cannot capture the entire body of knowledge needed for and used in the synthesis process.
  For example, our formalisation of deductive synthesis of logic programs in [9] reveals that the synthesis process has to make use of background knowledge not captured by the *if-and-only-if* definition of the specified relation. Such knowledge includes relevant theories, e.g. of data types, with induction schema, etc.
- It does not provide an adequate means to specify properties other than correctness.
  For example, in [10], we show that if we want to synthesise *steadfast* programs, i.e. programs that are modular in the sense that they define program units that remain correct (and hence unchanged) when integrated into larger

* The first author was partially supported by the EC HCM project on Logic Program Synthesis and Transformation, contract no. 93/414. The second author was partially supported by MURST.

programs, then we need to use axiomatisations which allow us to reason about program modularity and composition.

In our work [9, 10], therefore, we have used *specification frameworks* to provide the background for deductive synthesis of logic programs. In this paper, we take a closer look at such frameworks. We shall explain what they are, and how they can be used to specify properties such as correctness and modularity (and hence reusability).

Moreover, we shall show that there is a close two-way relationship between specification frameworks and deductive synthesis. In particular, a deductive synthesis process can provide a useful feedback mechanism which can not only check for desirable properties in the specification framework, but also improve the framework (with regard to such properties) using the result of the synthesis.

In our approach to modularity, we borrow many of the basic ideas developed in the algebraic approach (e.g. [6, 7, 14]). We shall briefly contrast the two approaches in Section 2.3.

2 Specification Frameworks and Specifications

In this section, we give an informal introduction to specification frameworks, and show why and how we distinguish between them and *specifications*. We also briefly explain the background and motivation for our approach.

2.1 Specification Frameworks

We take the view that program synthesis should take place in a general framework within which we can (a) specify, and (b) synthesise programs, for many different computational problems in that framework. The idea is that for (a) a framework should be *sufficiently expressive*, and for (b) it should act as a repository that contains all the knowledge that is needed for reasoning about the problem domain of interest.

The following are two examples of specification frameworks:

Example 1. A well-known example is Peano Arithmetic, which we will denote by $\mathcal{PA}$. Its language contains identity '=', successor 's', sum '+' and product '$\cdot$'. In $\mathcal{PA}$ every natural number n is represented by a *numeral n*.[3] Every computable function $f(x)$ can be expressed by a formula $F(x, z)$, such that for every pair of natural numbers m and n, m $= f(\mathrm{n})$ iff $F(m, n)$, where m and n are the numerals corresponding to m and n respectively, is provable in $\mathcal{PA}$. Thus, in $\mathcal{PA}$ we can *specify* any computable function by a formula; and therefore $\mathcal{PA}$ is *sufficiently expressive*.

Example 2. Another example is the following framework $\mathcal{LIST}$ for lists. We assume that the framework $\mathcal{PA}$ has already been defined. It contains the language

---

[3] Numerals are $0, s(0), s(s(0)), \ldots$

and the axioms of Peano Arithmetic and is enriched by the most useful primitive recursive functions and predicates. We use (i.e. we "import") $\mathcal{PA}$ in the following framework specifying lists, to introduce the *function len(L)* (meaning the length of L) and the *relation elemi(L, i, e)* (meaning the element e occurs at position i in L). Here, the specification of lists leaves undefined the list elements (namely the domain of the sort *Elem*) and the ordering $\leq$ on them, and thus *Elem* and $\leq$ act as *parameters*. We have a many-sorted language with (polymorphic) identity '='. The 'statement' **Import** $\mathcal{PA}$ incorporates the signature and the axioms of $\mathcal{PA}$, while the subsequent 'statements' add new sort, function and relation symbols to the signature, and introduce new axioms.

Framework $\mathcal{LIST}$;
Import $\mathcal{PA}$;

SORTS: *Elem*,*List*;

FUNCTIONS: $nil : \rightarrow List$;
              $| : (Elem, List) \rightarrow List$;
              $len : List \rightarrow Nat$;

RELATIONS: $\leq$ $: (Elem, Elem)$;
              $elemi : (List, Nat, Elem)$

AXIOMS: $\forall a : Elem, B : List. \neg nil = a|B$
             $\forall a_1, a_2 : Elem, B_1, B_2 : List. (a_1|B_1 = a_2|B_2 \rightarrow a_1 = a_2 \wedge B_1 = B_2)$
             $\forall e : Elem, L : List. (elemi(L, 0, e) \leftrightarrow \exists B : List. L = e|B)$
             $\forall e : Elem, L : List, i : Nat.$
             $(elemi(L, s(i), e) \leftrightarrow \exists a : Elem, B : List. (L = a|B \wedge elemi(B, i, e)))$
             $H(nil) \wedge \forall x : Elem, J : List. (H(J) \rightarrow H(x|J)) \rightarrow \forall L : List. H(L)$

End $\mathcal{LIST}$.

The last axiom is the first-order induction schema on lists, where H is any (first-order) formula of the language of lists.

Natural numbers are not needed to specify list. However, the presence of *elemi* yields a more expressive language, which allows us to define in an easy and natural way many different relations on lists, as we will see later.

Thus we define a *specification framework* as a consistent first-order theory with a many-sorted signature, and with identity. We allow full first-order language for the sake of expressiveness. We will denote frameworks by $\mathcal{F}, \mathcal{G}, \ldots$

2.2 Specifications

A framework is thus a general theory, wherein many different computational problems can be specified in a clear, declarative manner. To specify a problem declaratively, we explicitly define the relation to be computed. We also list the goals to be solved by the synthesised program(s).

Definition 1. A *specification* in a framework $\mathcal{F}$ with language $\mathcal{L}_\mathcal{F}$, consists of:

- a *definition axiom* D_r:

$$\forall (r(x) \leftrightarrow R(x))$$

  where r is a new predicate symbol and R is a formula in $\mathcal{L}_{\mathcal{F}}$.
- a set G of *goals* of the form $\exists r(t_0), \exists r(t_1), \ldots$

The link between specifications and programs is stated by the following definition.

Definition 2. Let $\langle D_r, G \rangle$ be a specification within a framework $\mathcal{F}$. A program P is *totally correct* in a model M of $\mathcal{F}$ wrt $\langle D_r, G \rangle$ if, for every goal $\exists r(t_k) \in G$:

- if $\exists r(t_k)$ is true in M, then there is at least a computed answer substitution[4] for $P \cup \{\leftarrow r(t_k)\}$;
- if σ is a computed answer substitution for $P \cup \{\leftarrow r(t_k)\}$, then $r(t_k)\sigma$ is true in M.

We define correctness in a model, since we will consider both closed frameworks, which single out one intended model, and open frameworks, which characterise a class of models.

Example 3. In the framework $\mathcal{LIST}$ of Example 2, we can specify many problems on lists. Firstly, we can specify many new relations on lists, using *definition axioms*. For example, the new relation symbols

- $sub(A, B)$, meaning A is a sub-list of B, i.e. every element of A is an element of B;
- $equiv(A, B)$, meaning A and B are equivalent, i.e. they contain the same elements; and
- $ord(L)$, meaning L is ordered,

can be introduced by the definition axioms D_{sub}, D_{equiv} and D_{ord}:

D_{sub} : $\forall A, B : List. (sub(A, B) \leftrightarrow$
$\qquad \forall e : Elem. (\exists i : Nat.elemi(A, i, e) \rightarrow \exists i : Nat. elemi(B, i, e)))$
D_{equiv} : $\forall A, B : List. (equiv(A, B) \leftrightarrow sub(A, B) \wedge sub(B, A))$
D_{ord} : $\forall L : List. (ord(L) \leftrightarrow$
$\qquad \forall i : Nat, e_1, e_2 : Elem. (elemi(L, i, e_1) \wedge elemi(L, s(i), e_2) \rightarrow e_1 \leq e_2))$

Each relation symbol so defined, together with the corresponding definition axiom, can be added to the framework, thus enriching our language.

We can now specify, for example, the sorting problem by $\langle D_{sort}, G_{sort} \rangle$, with:

$$D_{sort} : \forall L, S : List. (sort(L, S) \leftrightarrow equiv(L, S) \wedge ord(S))$$
$$G_{sort} = \{\exists S : List. sort(l, S) \mid l \text{ is ground}\}$$

The relations *ord* and *sort* depend on the ordering $\leq: List \times List$; hence this specification specifies a class of sorting programs, depending on the interpretations of $\leq: List \times List$.

---

[4] As defined in [11].

2.3 Background and Motivation

Various axiomatisations of data and programs can be found in the literature. For logic programming, Clark and Tärnlund [4] formulated a first-order theory of programs and data. Other examples of first-order theories of general interest to Computer Science are presented in [12].

However, a systematic use of axiomatisations to specify data and programs has been studied mainly in the field of algebraic specifications of abstract data types (ADT's). We refer in particular to the so-called *initial algebra approach*, which has been popularised by many authors (e.g. [8, 6, 7, 14]). In our approach, we incorporate many of their basic ideas (and their terminology): in particular, modularity (and parametric modules, like $\mathcal{LIST}$), and intended-model semantics.

However, there are some major and fundamental differences. In this section, we briefly outline them and thus contrast our approach with that of initial algebra.

We are concerned with program synthesis, and we have a different point of view on what a specification should be, for the purpose of synthesis. We distinguish between specification frameworks and specifications, and we do not impose any restrictions on the form of the axioms. In contrast, the initial algebra approach only allows restricted classes of axioms. Consequently, we cannot exploit the strengths of the initial algebra approach, namely that:

(a) Equational theory admits an initial model [8], unique up to isomorphism, and this result can be easily extended to Horn axioms (as in EQLOG[6]).

(b) There is a correspondence between truth in the initial model and provability for positive formulas. This property provides a kind of goal-solving completeness: a positive existential formula $\exists x A(x)$ (where A contains only $\land, \lor$) is true in the initial model iff there is an instance $A(t)$ provable in the theory.

However, as soon as we introduce negation, for example, we get into the following problems:

(a') The existence of an initial model is lost,[5] and there are no general and effective criteria to isolate those theories which do admit an initial model.

(b') Even if a theory $\mathcal{F}$ has an initial model I, a negated atomic formula may be true in I, but not provable; this destroys the goal-solving completeness property.

Both these points are critical in our approach: (a') because we do not want particular restrictions on the form of the axioms, and (b') because in a definition axiom $\forall x (r(x) \leftrightarrow R(x))$, $R(x)$ is any formula (using negation we may loose goal-solving completeness wrt the goals $\exists r(t)$).

Therefore we need some effective criterion for the existence of the intended model. If we use *isoinitial semantics* instead of initial semantics, we get such a criterion

---

[5] As is well-known in logic programming, the completion of a normal program may be inconsistent, or it may have many incomparable models.

Isoinitial semantics was introduced in [1], as a semantic characterisation of the computable ADT's. In this semantics, the intended model of a theory $\mathcal{F}$ is the *isoinitial model* (when it exists), i.e. a model J of $\mathcal{F}$ such that for every other model M of $\mathcal{F}$, there is a unique *isomorphic embedding* of J in M.

In other words, whereas initial semantics uses homomorphisms, isoinitial semantics uses isomorphic embeddings, which preserve relations and their negations. As a consequence, isoinitial models turn out to be always recursive, whilst this property is not guaranteed, in general, for initial models. An extensive comparison of the two kinds of semantics can be found in [1].

We will make use of the following criterion for the existence of isoinitial models.

Definition 3. We say that a model M of a theory $\mathcal{F}$ is *reachable* if every element of the domain of M is represented by some closed term of the language of $\mathcal{F}$.

We say that $\mathcal{F}$ is *atomically complete* if, for every closed atomic formula A, either $\mathcal{F} \vdash A$, or $\mathcal{F} \vdash \neg A$.

Theorem 4. *If a theory $\mathcal{F}$ has at least one reachable model, then $\mathcal{F}$ admits an isoinitial model if and only if it is atomically complete.*

This theorem gives atomic completeness as the key condition for the exsitence of isoinitial models. In our work, we will show that program synthesis can be used to test for atomic completeness, by the provability of the completion of synthesised programs.

3 Closed and Open Specification Frameworks

We will distinguish between *closed* and *open* frameworks. Informally speaking, an *open* framework leaves open the possibility of different interpretations of the symbols of its language, corresponding to different intended models, while a *closed* framework has a unique *intended model*, determining a unique interpretation.

For example, $\mathcal{PA}$ is a *closed* framework, since the meaning of 's', '$+$', '$\cdot$', '$=$' is completely defined and the intended model is the standard structure of natural numbers. In contrast, $\mathcal{LIST}$ is an *open* framework, since the meaning of *Elem* and $\leq : (Elem, Elem)$ is left open. Indeed, no axiom is given to characterise the domain of the sort *Elem* or the ordering $\leq$ on the elements.

In an open framework, we can distinguish between two kinds of symbols:

- *defining symbols*, namely symbols whose interpretation is left open;
- *defined symbols*, namely symbols whose meaning is defined by the axioms of the framework, possibly in terms of the defining symbols.

In Example 2, *Elem* and $\leq$ are defining symbols, while all the other symbols of the language are defined symbols.

We will write $\mathcal{F} : d_1, \ldots, d_n \Leftarrow p_1, \ldots, p_m$ to indicate that $\mathcal{F}$ is an open framework with *dependency* $d_1, \ldots, d_n \Leftarrow p_1, \ldots, p_m$, i.e. in $\mathcal{F}$, $d_1, \ldots, d_n$ are *defined* symbols and $p_1, \ldots, p_m$ are *defining* symbols. Thus $\mathcal{F} : d_1, \ldots, d_n \Leftarrow$ denotes a closed framework, i.e. all the symbols are defined by the axioms of the framework.

Thus defining symbols act as parameters, and our idea of an open framework is similar to a parametric (algebraic) ADT's. However, unlike the algebraic approach, we use the full first-order language, and therefore we need criteria for discriminating between 'good' and 'bad' frameworks, as we will now discuss.

The two key attributes of a good specification framework are *consistency* and *adequacy*. Consistency has the usual meaning of not containing contradictions, whereas adequacy is related to the dependency of the framework.

Intuitively speaking, a framework is *adequate* wrt a given set of dependencies, if whenever we completely specify the defining symbols, we obtain a complete specification of the defined symbols without adding any new axioms for them. This notion of adequacy wrt a dependency is the first point that we will develop for open frameworks.

3.1 Operations on Frameworks

Since the effort of writing adequate and consistent frameworks, as well as synthesising correct programs in such frameworks, may be considerably high, reusability of both specifications and programs is a very important issue: once we have produced a meaningful, consistent and adequate specification framework and correct and good software for it, we should be able to use them in a wide family of problems. Moreover, we should be able to accumulate and increase our knowledge of the more interesting problem domains, and to enlarge the class of reusable programs, leading to increasingly more complete and useful frameworks. An object-oriented approach, like the one developed in algebraic ADT's, would be very useful in this respect.

So, the second point of our study is the introduction of methods suitable for object-oriented specification and synthesis. We will consider the following operations:

- *Expansion.* Given an adequate and consistent open framework $\mathcal{F} : d_1, \ldots, d_n \Leftarrow p_1, \ldots, p_m$, we can *expand* $\mathcal{F}$, by adding a new (defined) relation symbol r by means of a definition axiom D_r, into a new framework $\mathcal{F} \cup D_r : r, d_1, \ldots, d_n \Leftarrow p_1, \ldots, p_m$.
  Since definition axioms give rise to conservative extensions, the expansion operation preserves consistency, but it may not preserve adequacy.
- *Consolidation.* An adequate open framework $\mathcal{F} : d_1, \ldots, d_n \Leftarrow p_1, \ldots, p_m$ can be *consolidated* in *any* closed framework $\mathcal{G} : p_1, \ldots, p_m, \ldots \Leftarrow$ to obtain the closed framework $\mathcal{F} \cup \mathcal{G} : d_1, \ldots, d_n, p_1, \ldots, p_m, \ldots \Leftarrow$. Such an operation thus enriches $\mathcal{G}$ by the defined symbols of $\mathcal{F}$. We shall call $\mathcal{F} \cup \mathcal{G}$ an *instance* of $\mathcal{F}$ (or, more precisely, an *instantiation* of $\mathcal{F}$ by $\mathcal{G}$).[6]
  This operation preserves adequacy, but it may not preserve consistency.

---

[6] Rather than the more clumsy alternative of 'a consolidation of $\mathcal{F}$ in $\mathcal{G}$'.

Therefore it is necessary to find conditions under which adequacy is preserved by expansions, and consistency by consolidations. For the latter, it is possible to find some sufficient conditions, but we will not discuss them here.

More interestingly, logic program synthesis is strictly related to the above operations. Firstly, it can be used to expand frameworks while preserving adequacy, as we will show in Section 6. Secondly, in an open framework $\mathcal{F}$, it is possible to synthesise *open* programs, which can be *reused* in every instantiation of $\mathcal{F}$, as we will explain in Section 6.3.

4 Closed Frameworks

Definition 5. A *closed* framework $\mathcal{F}$ with language $L_{\mathcal{F}}$ is an axiomatisation (i.e. a set of axioms $\mathcal{F} \subseteq \mathcal{L}_{\mathcal{F}}$) which satisfies the following properties:

- *Reachability.* There is at least one model of $\mathcal{F}$ reachable by a subset of the constant and function symbols of $L_{\mathcal{F}}$, called the *construction symbols*. The ground terms containing only construction symbols will be called *constructions*.
- *Freeness.* $\mathcal{F}$ proves the *freeness axioms* [13] for the construction symbols.
- *Atomic completeness.* $\mathcal{F}$ is atomically complete.

Thus, by Theorem 4, a closed framework is a theory admitting an isoinitial model. The existence of a set of construction symbols satisfying the freeness axioms is not strictly necessary (reachability suffices). It is required here, since we deal with synthesis of logic programs and use Clark's equality theory[7] [3] for construction symbols. Of course, in logic programs only construction symbols can be used for constant and function symbols.

Example 4. As we have already seen, $\mathcal{PA}$ is a closed framework. The construction symbols in $\mathcal{PA}$ are '0' and 's', and the constructions are the numerals $0, s(0), s(s(0)), \ldots$ Note that sum '+' and product '$\cdot$' are not construction symbols, so they need not be considered in freeness axioms. Finally, $\mathcal{PA}$ is atomically complete. Indeed the closed atomic formulas of $\mathcal{PA}$ are of the form $t_1 = t_2$, where t_1, t_2 are closed terms, and, as is well-known, $\mathcal{PA} \vdash t_1 = t_2$ (if t_1 and t_2 denote the same natural number), or $\mathcal{PA} \vdash \neg t_1 = t_2$ (if they denote different numbers).

All the isoinitial models of a framework are isomorphic. For every closed framework $\mathcal{F}$, using the constructor symbols, we can build a particular recursive, isoinitial model I, that we will call the *canonical model*, in the following way:

- For every sort S, the domain of S is the set of constructions of sort S.
- A function symbol f is interpreted in I as the function f_I defined thus: for every tuple a of constructions (of the appropriate sorts), the value of $f_I(a)$ is the construction b such that $\mathcal{F} \vdash f(a) = b$.

---

[7] Containing freeness axioms.

– Every relation symbol r is interpreted in I as the relation r_I such that, for every tuple a of constructions, a belongs to r_I iff $\mathcal{F} \vdash r(a)$. In particular, '=' is the identity of constructions.

Example 5. In the canonical model of $\mathcal{PA}$, the domain (of its unique sort) is the set of numerals $\{0, s(0), s(s(0)), \ldots\}$, '+' is the sum of numerals, e.g. $\mathcal{PA} \vdash s(0) + s(0) = s(s(0))$; '·' is the product of numerals, e.g. $\mathcal{PA} \vdash s(0) \cdot s(0) = s(0)$; and '=' is the identity of numerals ($\mathcal{PA} \vdash s^n(0) = s^m(0)$ iff $n = m$).

Let $\mathcal{F}$ be a closed framework and I its canonical model. By atomic completeness, for every relation symbol r and every tuple a of constructions, $r(a)$ is true in I iff it is true in every model M of $\mathcal{F}$, and $\neg r(a)$ is true in I iff it is true in every model M of $\mathcal{F}$. Therefore the canonical model is representative of any other model wrt the closed atomic and negated atomic formulas. That is, it yields the *intended semantics* of the function and relation symbols axiomatised by the framework.

4.1 Adequate Expansions of Closed Frameworks

As shown by the examples of the previous sections, definition axioms play a central rôle in our approach. In closed frameworks they are adequate when they completely characterise the defined relations, according to the following definition:

Definition 6. Let $\mathcal{F}$ be a closed framework and D_r

$$\forall \underline{x}(r(\underline{x}) \leftrightarrow R(\underline{x}))$$

be a definition axiom. We say that D_r *completely defines* r in $\mathcal{F}$ iff $\mathcal{F} \cup D_r$ satisfies the atomic completeness property.

If D_r completely defines r in $\mathcal{F}$, then $\mathcal{F} \cup D_r$ is in turn a closed framework. It is thus an *adequate expansion* of $\mathcal{F}$ by a new relation r, in the following sense:

– Let I be the canonical model of $\mathcal{F}$, and I_r that of $\mathcal{F} \cup D_r$. Then I_r is an *expansion* of I by r.
– The interpretation of the new symbol r in the expansion I_r is the following: for every tuple t of constructions (of the appropriate sorts), t belongs to r_{I_r} iff $\mathcal{F} \cup D_r \vdash r(t)$.

We recall that in an expansion of a model (see [2]) by new symbols, the interpretation of the old ones remains unchanged. Thus an adequate expansion $\mathcal{F} \cup D_r$ *preserves the semantics* of the symbols already present in $L_{\mathcal{F}}$.

This nice property holds only for completely defined relations, however. If D_r does not completely define a relation, then $\mathcal{F} \cup D_r$ is no longer a closed framework and a canonical model no longer exists. This means that, in general, we have to possibly reconsider the correctness of programs that have already been synthesised. Thus it is important to recognise the axioms D_r which completely define new relations.

5   Open Frameworks

From a general point of view, an open framework is a consistent set $\mathcal{F}$ of first-order axioms, for which we fix a dependency $d_1, \ldots, d_m \Leftarrow p_1, \ldots, p_m$. We require that the dependency satisfies the following properties:

(i) For every *defined* sort symbol S, identity '$=$' on S is a defined symbol. Moreover, there is (in the language of the framework) a non-empty set of constant and function symbols of sort S such that the corresponding freeness axioms are provable in $\mathcal{F}$; these are the *construction symbols* of S.
(ii) For every *defining* sort symbol D, identity '$=$' on D is a defining symbol. Moreover, the language of the framework does not contain constant and function symbols of sort D.

As in the previous section, the closed terms built by construction symbols are called *constructions*. Note that, by (ii), if D is any defining sort symbol, then the set of constructions of sort D is empty.

Example 6. The framework $\mathcal{LIST}$ has the dependency:

$$Nat, List, 0, nil, s, +, \cdot, |, len, elemi \Leftarrow Elem, \leq$$

We have the identities $=: (Nat, Nat)$, $=: (List, List)$, $=: (Elem, Elem)$; where the first two are defined symbols, whilst the third is a defining symbol.

The construction symbols of $List$ are the constant $nil :\to List$ and the function $| : (Elem, List) \to List$, and the only construction of sort $List$ is nil. Note that the only ground term of sort $List$ is nil, since the other ground terms depend on the ground terms of the defining sort $Elem$, and therefore will be known only in particular instances of the framework.

The construction symbols of the sort Nat are '0' and 's' and the constructions of sort Nat are the numerals.

The defining sort $Elem$ has no construction symbols and no constructors.

Let $\mathcal{F} : d_1, \ldots, d_m \Leftarrow p_1, \ldots, p_n$ be an open framework. If $n = 0$ (i.e. no defining symbol is given), then $\mathcal{F}$ is adequate iff it is a closed framework. Otherwise adequacy is defined as follows.

Definition 7. $\mathcal{F} : d_1, \ldots, d_m \Leftarrow p_1, \ldots, p_n$ is *adequate* wrt $d_1, \ldots, d_m \Leftarrow p_1, \ldots, p_n$ iff, for every closed framework $\mathcal{G} : p_1, \ldots, p_n, q_1, \ldots, q_h \Leftarrow$, $\mathcal{F} \cup \mathcal{G}$ is a closed framework such that its canonical model I^* is an expansion of the canonical model I of $\mathcal{G}$ by the new symbols $d_1, \ldots, d_m$.

Let $\mathcal{G} : p_1, \ldots, p_n, q_1, \ldots, q_h \Leftarrow$ be any closed framework that is consistent with $\mathcal{F} : d_1, \ldots, d_m \Leftarrow p_1, \ldots, p_n$; according to Section 3.1, we will say $\mathcal{F} \cup \mathcal{G}$ is an *instance* of $\mathcal{F}$ (or an *instantiation* of $\mathcal{F}$ by $\mathcal{G}$). Adequacy means that every (consistent) instantiation of $\mathcal{F}$ by $\mathcal{G}$ gives rise to an expansion of $\mathcal{G}$ by the symbols defined by $\mathcal{F}$, while preserving the semantics of the old symbols of $\mathcal{G}$.

5.1 Adequate Expansions of Open Frameworks

Using *adequate* open frameworks we have the following advantage: all the work which can be done directly in the open framework is done once and for all, i.e. it can be *reused* in all the instances of the framework. In particular, this holds for *reusable* programs (see Section 6.3) that can be synthesised in such frameworks. This explains why it is interesting to investigate the possibility of performing synthesis in open frameworks. We will show that synthesis of reusable open programs is possible for relations introduced by *adequate definition axioms*, defined as follows.

Definition 8. A definition axiom D_r is *adequate* in an open framework $\mathcal{F}$: $d_1, \ldots, d_m \Leftarrow p_1, \ldots, p_n$ iff, for every instance $\mathcal{F} \cup \mathcal{G}$, D_r completely defines r in $\mathcal{F} \cup \mathcal{G}$.

Axioms adequate in an open framework have the following nice property: they give rise to adequate expansions of the instances of the framework. More precisely, they add to every instance of the framework a new relation r, while preserving the semantics of the old symbols. Thus, if D_r is an adequate definition axiom, we say that $\mathcal{F} \cup D_r$ is an *adequate expansion* of $\mathcal{F}$ by r.

Note that in adequate expansions, the definition r depends only on the interpretation of the defining symbols in the various instances. Therefore the dependency of an expansion $\mathcal{F} \cup D_r$ is : $r, d_1, \ldots, d_m \Leftarrow p_1, \ldots, p_n$.

Finally, it will be convenient to relate adequacy of frameworks and of their expansions (by adequate definition axioms) to restricted classes of instances. Let Γ be a class of closed frameworks which can be used to instantiate an open framework $\mathcal{F}$. To define *adequacy* wrt Γ, it is sufficient to consider in Definition 7 and Definition 8 only the instantiations by closed frameworks $\mathcal{G} \in \Gamma$.

6 Deductive Synthesis and Specification Frameworks

In this section, first we briefly describe deductive synthesis of logic programs in a specification framework. Then we show how synthesis can be used to determine the adequacy of any expansion of a closed or open framework. For an adequate open framework, we also introduce the notion of open, reusable programs.

6.1 Deductive Synthesis of Logic Programs

First, as in [10], for a program P, we define:

- *free*(P) to be the freeness axioms for the constant and function symbols of P.
- $Comp^+(P) = free(P) \cup Ax(P)$, where $Ax(P)$ is the *if*-part of $Comp(P)$, the completion of P [3].

- $Comp^-(P, r)$ is the *only-if*-part of the completed definition of a predicate r in P:

$$\forall (r(x) \rightarrow E_1 \vee \cdots \vee E_k)$$

For convenience, we shall also write $Comp^-(P, r_1, \ldots, r_n)$ for multiple predicate symbols.

Now, let $\mathcal{F}$ be a general (open) framework and consider a specification consisting of a definition axiom D_0

$$\forall (r(x) \leftrightarrow R(x))$$

and a set G of goals.

A *deductive synthesis* process (for this specification) generates a sequence of programs $P_1 \subseteq P_2 \subseteq \cdots \subseteq P_k \subseteq \cdots$, where, for every P_j, there is a corresponding set D_j of definition axioms, introducing the new predicate symbols of P_j .

We assume that the synthesis process ensures that at each step:

(i) $\mathcal{F} \cup D_j \vdash Comp^+(P_j)$;

(ii) P_j terminates , i.e. it either finitely fails or its *SLD*-tree has at least one success node, for every goal in G.

(i) ensures *partial correctness*. However, (i) and (ii) together do not ensure *total correctness*. In Sections 6.2 and 6.3, we will define the notion of total correctness in closed and open frameworks precisely. In the meantime, we state a criterion for halting synthesis with a totally correct program, which we presented in [10].

Definition 9. For a program P, let $r_1, \ldots, r_k$ be the relation symbols occurring in the head of at least one clause of P, and $q_1, \ldots, q_m$ be the ones occurring only in the bodies. Following our terminology for frameworks (without causing confusion), we say that P has the *dependency* $r_1, \ldots, r_k \Leftarrow q_1, \ldots, q_m$, and we write $P : r_1, \ldots, r_k \Leftarrow q_1, \ldots, q_m$. We call $r_1, \ldots, r_k$ the *defined predicates* and $q_1, \ldots, q_m$ the *defining predicates* of P.[8]

In every synthesis step j, we consider the relation symbols introduced by D_j as *defined symbols* of $\mathcal{F} \cup D_j$. Moreover, we require that the dependency of P_j *agrees* with that of $\mathcal{F} \cup D_j$ wrt the defined symbols, namely that the defined predicates of P_j are defined symbols of $\mathcal{F} \cup D_j$. A criterion for halting synthesis is then the following:

A Criterion for Halting Synthesis.
Let $P_j : r_1, \ldots, r_k \Leftarrow q_1, \ldots, q_m$ be the program synthesised in the current step j.

(i) If, for a defined predicate r_h, $\mathcal{F} \cup D_j \vdash Comp^-(P_j, r_h)$, then stop searching for clauses with head r_h.

(ii) If (i) holds for every defined predicate of P_j, and all the defining predicates $q_1, \ldots, q_m$ are defining symbols in $\mathcal{F}$, then halt synthesis.

---

[8] A defined predicate may occur both in the head and (*recursively*) in the body of a clause, whereas a defining one may occur *only* in the body.

As we mentioned in Section 3.1, an adequate (open) framework $\mathcal{F} : d_1, \ldots, d_n \Leftarrow p_1, \ldots, p_m$ completely characterises the defined symbols $d_1, \ldots, d_n$ in terms of the possible interpretations for the defining symbols $p_1, \ldots, p_m$. Correspondingly, a correct open program P should be a program computing some defined relation symbols of the framework, depending on the interpretation of other defining relation symbols. When (i) holds, the completed definition of r_h has been proved in the framework. Roughly speaking, this implies that we have already synthesised all the clauses to compute r_h, depending on the interpretation of the other relation symbols occurring in their bodies.

When (ii) holds, the open program P_j contains *all* the clauses to compute *all* its defined relations, depending *only* on the interpretation of the defining symbols of the framework. This implies *reusability*: for every model M of the framework, it is sufficient to add to P_j the clauses that correctly compute the defining relations in M, to obtain a complete program that is totally correct in M.

On the other hand, if such a P_j can be synthesised, then, as we will see in Section 6.3, the framework is adequate, i.e. synthesis can be used to analyse the adequacy of specification frameworks, and thus we have a useful feedback mechanism between specification and synthesis.

Synthesis in closed frameworks is a limiting case, where we halt synthesis only with programs with dependency $r_1, \ldots, r_k \Leftarrow$, since there are no defining symbols in the framework. We will prove that in this case we obtain (closed) programs which are totally correct (in the usual sense) in the intended model of the framework. Thus in a closed framework, program synthesis can be used both to find correct programs and to study adequate expansions, as we now show more formally.

6.2 Synthesis and Closed Frameworks

Let $\langle D_r, G \rangle$ be a specification in a closed framework $\mathcal{F}$ with canonical model I. We are interested in *adequate specifications*, namely specifications giving rise to adequate expansions. In an adequate specification, D_r completely defines r in $\mathcal{F}$, and $\mathcal{F} \cup D_r$ is a closed framework with a canonical model I_r. Thus the following definition is sensible:

Definition 10. For a given specification $\langle D_r, G \rangle$ and a given framework $\mathcal{F}$, a program P is *totally correct* in the *expansion* $\mathcal{F} \cup D_r$ iff it is totally correct in the canonical model I_r.

For the synthesis of totally correct programs, we can prove the following theorems.

We say that a program P *ground terminates* if, for every relation symbol r of P, and every ground goal $r(t)$, the *SLD*-tree for $P \cup \{\leftarrow r(t)\}$ contains at least one success node or is finitely failed.[9] The following theorem holds (see [10]):

---

[9] We assume a fair computation rule.

Theorem 11. *Let $\langle D_0, G \rangle$ be a specification in a closed framework $\mathcal{F}$, where r_0 is the relation defined by D_0. In a synthesis process, if a ground terminating program $P_j : r_0, \ldots r_k \Leftarrow$ can be synthesised such that $\mathcal{F} \vdash \text{Comp}^-(P, r_0, \ldots r_k)$, then we have:*

(i) $\mathcal{F} \cup D_j$ is an adequate expansion of $\mathcal{F}$ by the symbols introduced in D_j;

(ii) P_j is totally correct in the canonical model of $\mathcal{F} \cup D_j$ wrt the set of ground goals.

This theorem is useful in a synthesis process where we do not know whether the specification $\langle D_0, G \rangle$ we start from is adequate or not. In such a case, we can use synthesis both to obtain a program answering the goals in G and to prove the adequacy of the specification. For every step j, we prove $Comp^+(P_j)$ and the termination of P_j, both for the goals in G and for all the ground goals $r_h(t)$.[10] We halt the synthesis process when we obtain a program $P_j : r_0, \ldots, r_k \Leftarrow$ such that $\mathcal{F} \cup D_j \vdash Comp^-(P_j, r_0, \ldots, r_k)$. If this process halts successfully, then we obtain an adequate expansion of our framework (by the relation symbols defined in D_j) and a totally correct program.

However, it may happen that during this process, we reach the following situation of uncertainty: the current P_j is such that

neither $\mathcal{F} \cup D_j \vdash \neg Comp^-(P_j, r_0, \ldots, r_k)$ nor $\mathcal{F} \cup D_j \vdash Comp^-(P_j, r_0, \ldots, r_k)$.

The following theorem[11] allows us to handle such a situation. In the theorem, 'P_j is totally correct in $\mathcal{F} \cup D_j$ wrt ground goals' means that, for every relation symbol r_i of P_j and every ground atom $r_i(t)$, $r_i(t)$ is true in the canonical model of $\mathcal{F} \cup D_j$ iff the SLD-tree for $P \cup \{\leftarrow r_i(t)\}$ contains at least one success node.

Theorem 12. *Let $P_j : r_0, \ldots, r_k \Leftarrow$ be the current program, and let D_j be definition axioms that completely define $r_0, \ldots, r_k$ in $\mathcal{F}$. If P_j is totally correct in $\mathcal{F} \cup D_j$ wrt ground goals, then $\text{Comp}^-(P_j, r_0, \ldots, r_k)$ is consistent with $\mathcal{F} \cup D_j$, and can be added to it whilst preserving the canonical model.*

According to this theorem, in case we reach the above situation of uncertainty, then we study the definition axioms D_j. If we can prove (in some way) that D_j completely define the corresponding relations, then by Theorem 12 we can add $Comp^-(P_j, r_0, \ldots, r_k)$ as a new axiom, and we obtain both a totally correct program and an adequate expansion. Note that in this case we also have an improvement of the framework, since we add a stronger axiom.

On the other hand, if we discover that some definition axiom in D_j does not completely define the corresponding relation, we have to reconsider both the synthesis process and the framework. Indeed, it may be that in some synthesis step we made a bad 'choice', or that our framework is too weak to define the relations we want to compute.

---

[10] Ground termination is needed to apply Theorem 11.

[11] We omit the proof here for brevity.

6.3 Synthesis and Open Frameworks

To treat the correctness (i.e. the reusability) of an open program in an open framework, it is useful to compare the various Herbrand models of the program with the canonical models of the various instances of the framework. To this end, we introduce the following definitions.

Let M be a model for a language L_M and let C be a set of constant and function symbols.

We say that M is C-generable if it is reachable by C, and the freeness axioms of the function and constant symbols of C are satisfied in M. The closed terms built from C are the *constructions*.

Let M be a C-generable model with language L_M and let r be a relation symbol of L_M. The *positive diagram* of r in M is defined by:

$$diag^+(M, r) = \{r(t) \mid t \text{ are constructions and } M \models r(t)\}$$

For many relations, $diag^+(M, r_1, \ldots, r_k)$ is defined in the obvious way.

Now we can compare the positive diagrams with the Herbrand models of programs, and we introduce and define *steadfastness* (as in [10]) as follows:

Definition 13. Let $\mathcal{M}$ be a class of models, and assume that, for every model $M \in \mathcal{M}$, there is a set C_M of construction symbols of L_M such that M is C_M-generable. We say that

- M is a *steadfast model* of a program $P : r_1, \ldots, r_k \Leftarrow q_1, \ldots, q_m$ iff the constant and function symbols of P belong to C_M, and $diag^+(M, r_1, \ldots, r_k, q_1, \ldots, q_m)$ is the minimum Herbrand model of $Comp^+(P) \cup diag^+(M, q_1, \ldots, q_m)$;
- P is *steadfast* in $\mathcal{M}$ of models iff every $M \in \mathcal{M}$ is a steadfast model of P.

Steadfast programs are *reusable* in the following sense.

Property 14. *Let the program* $P : r_1, \ldots, r_k \Leftarrow q_1, \ldots, q_m$ *be steadfast in a class* $\mathcal{M}$ *of models, and let* M *be any model belonging to* $\mathcal{M}$. *For every program* Q_M *totally correct wrt* $q_1, \ldots, q_m$ *in* M, *if* $P \cup Q_M$ *terminates, then* $P \cup Q_M$ *is totally correct in* M *wrt* $r_1, \ldots, r_k, q_1, \ldots, q_m$.

Thus steadfastness of P works as a kind of 'parametric correctness' in a class of models, where P assumes in each model M the appropriate behaviour that depends on Q_M.

Now we introduce the notion of steadfastness in a class of instances of a framework (as defined in the previous subsection), as follows.

Let $\mathcal{F}$ be an open framework that is adequate wrt a class Γ of closed frameworks. Thus, for every $\mathcal{G} \in \Gamma$, the instance $\mathcal{F} \cup \mathcal{G}$ is a closed framework with a canonical model $I_{\mathcal{F} \cup \mathcal{G}}$. Let $\mathcal{I}_\Gamma$ be the class of canonical models $I_{\mathcal{F} \cup \mathcal{G}}$ such that $\mathcal{G} \in \Gamma$: we say that a program P is steadfast in the class of Γ-instances of $\mathcal{F}$ iff it is steadfast in $\mathcal{I}_\Gamma$.

Synthesis can be used to inform on the adequacy of expansions of open frameworks by new relation symbols, as stated by following theorem.

Theorem 15. *Let $\mathcal{F} : d_1, \ldots, d_n \Leftarrow p_1, \ldots, p_m$ be an open framework, and consider a class Γ of closed frameworks. Let $P_j : r_0, \ldots, r_k \Leftarrow q_1, \ldots, q_h$ be a synthesised program, and assume that all the defining predicates of P_j are defining symbols of $\mathcal{F}$. If*

(i) $\mathcal{F} \cup D_j \vdash Comp^-(P_j, r_1, \ldots, r_k)$; and
(ii) for every $\mathcal{G} \in \Gamma$, there exists at least one program $Q : q_1, \ldots, q_m, \ldots \Leftarrow$, such $\mathcal{G} \vdash Comp^-(Q, q_1, \ldots, q_m)$ and $P_j \cup Q$ terminates at least for the ground goals in $r_1, \ldots, r_k$;

then

(i) P_j is steadfast in the class of Γ-instances of $\mathcal{F}$; and
(ii) the definition axioms D_j are adequate in that class.

The proof is very simple. It follows from the fact that $\mathcal{P} \cup \mathcal{Q} \vdash Comp^-(P_j \cup Q, r_1, \ldots, r_k, q_1, \ldots, q_m)$, and from Theorem 11. More general facts about steadfast programs could be proved (see [10]), but they are not directly applicable to the methodology outlined here.

By Theorem 15, if we halt synthesis according to the criterion given in Section 6.1, then we obtain programs which are steadfast (and hence reusable) in the class of instances where a terminating $P_j \cup Q$, satisfying the hypotheses of the theorem, exists. This class may be a subclass of the class of all the instances. However, the theorem makes sense, since it applies to the class of instantiations by any framework $\mathcal{G}$ that can be obtained by our methodology and that (hence) contains a synthesised program $Q : q_1, \ldots, q_m, \ldots \Leftarrow$, such $\mathcal{G} \vdash Comp^-(Q, q_1, \ldots, q_m)$.

To obtain reusable programs P in this sense, a crucial point is the possibility of stating the termination of $P \cup Q$, for the various possible Q's, once and for all in the open framework. That is, we have to state termination of $P \cup Q$ without knowing Q, but only knowing some possible properties which can be expressed by the axioms of the open framework.

An example of parametric termination is given by the following program $P : p \Leftarrow q$:

$$p(X, 0, 0) \quad \leftarrow$$
$$p(X, s(Y), W) \leftarrow p(X, Y, Z), q(Z, X, W)$$

where the set of goals is $G = \{\exists Z. p(m, n, Z) \mid m, n \text{ are numerals}\}$. To obtain parametric termination, it is sufficient that the programs $Q : q, \ldots \Leftarrow$ used in the instantiations do not contain the predicate p. This requirement is sensible in our general approach, where p is to be computed by the reusable program $P : p \Leftarrow q$, and not by $Q : q, \ldots \Leftarrow$.

Another example is given by parametric sorting programs on lists $P : sort \Leftarrow \leq$. Many other examples can be given, coming from everyday programming practice. We intend to study in future general techniques to state parametric termination in relevant classes.

7 Conclusion

In this paper, we have described specification frameworks and their relationship with deductive synthesis of logic programs. We have shown that such frameworks are necessary to formalise not only the background knowledge needed for synthesis, but also notions related to reusable programs and their synthesis.

Hitherto, research in logic program synthesis has made use of *closed* specification frameworks in which all relation symbols are completely defined. However, such frameworks cannot capture the notion of reusable or steadfast programs. To do so, it is necessary to use *open* frameworks in which the meaning of some relation symbols is open, i.e. it can vary depending on a chosen model of the framework. Open frameworks thus provide a suitable backdrop to the synthesis of steadfast programs, and are therefore vital for an object-oriented approach to specification and synthesis. We have described such an approach, similar to the one developed in algebraic ADT's, for progam synthesis using the full first-order language and isoinitial semantics.

The relationship between synthesis and specification frameworks is very much a two-way affair. As well as reasoning 'forwards' from a specification (in a framework) via synthesis to the specified program, we can use the result of synthesis to inform on and improve the specification framework if necessary, i.e. we can reason 'backwards' from the program via synthesis to the specification framework.

Synthesis in a closed framework can tell us if the initial specification completely defines the specified relation. It can also enhance the framework by adding stronger axioms to it that correspond to the synthesised program.

Similarly, synthesis in an open framework can be used to determine if the initial specification is adequate, i.e. if it defines an adequate expansion of the framework.

In summary, the results presented and discussed in this paper should provide important foundations for our future work on object-oriented deductive synthesis of logic programs.

Acknowledgements

We are very grateful to the referees for their valuable comments and constructive suggestions which have enabled us to vastly improve this paper.

References

1. A. Bertoni, G. Mauri and P. Miglioli. On the power of model theory in specifying abstract data types and in capturing their recursiveness. *Fundamenta Informaticae* **VI**(2):127–170, 1983.
2. C.C. Chang and H.J. Keisler. *Model Theory*. North-Holland, 1973.
3. K.L. Clark. Negation as failure. In H. Gallaire and J. Minker, editors, *Logic and Data Bases*, pages 293-322. Plenum Press, 1978.

4. K.L. Clark and S.-Å. Tärnlund. A first order theory of data and programs. *Proc. IFIP 77*, pages 939–944. North-Holland, 1977.

5. Y. Deville and K.K. Lau. Logic program synthesis. To appear in *J. Logic Programming*, special issue on "Ten Years of Logic Programming", 1994.

6. J.A. Goguen and J. Meseguer. EQLOG: Equality, types,and generic modules for logic programming. In D. DeGroot and G. Lindstrom, editors, *Logic Programming: Functions, Relations, and Equations*, pages 295–363. Prentice-Hall, 1986.

7. J.A. Goguen and J. Meseguer. Unifying functional, object-oriented and relational programming with logical semantics. In B. Shriver and P. Wegner, editors, *Research Directions in Object-Oriented Programming*, pages 417–477. MIT Press, 1987.

8. J.A. Goguen, J.W. Thatcher and E. Wagner. An initial algebra approach to specification, correctness and implementation. In R. Yeh, editor, *Current Trends in Programming Methodology, IV*, pages 80-149. Prentice-Hall, 1978.

9. K.K. Lau and M. Ornaghi. An incompleteness result for deductive synthesis of logic programs. In D.S. Warren, editor, *Proc. 10th Int. Conf. on Logic Programming*, pages 456–477, MIT Press, 1993.

10. K.K. Lau, M. Ornaghi, and S.-Å. Tärnlund. The halting problem for deductive synthesis of logic programs. In P. Van Hentenryck, editor, *Proc. 11th Int. Conf. on Logic Programming*, pages 665–683. MIT Press, 1994.

11. J.W. Lloyd. *Foundations of Logic Programming*. Springer-Verlag, 2nd edition, 1987.

12. Z. Manna and R. Waldinger. *The Deductive Foundations of Computer Programming*. Addison-Wesley, 1993.

13. J.C. Shepherdson. Negation in Logic Programming. in J. Minker, editor, *Foundations of Deductive Databases and Logic Programming*, pages 19-88. Morgan Kaufmann, 1988.

14. M. Wirsing. Algebraic specification. In J. Van Leeuwen, editor, *Handbook of Theoretical Computer Science*, pages 675–788. Elsevier, 1990.

Partial Evaluation of the "Real Thing"

Michael Leuschel

K.U. Leuven, Department of Computer Science
Celestijnenlaan 200A, B-3001 Heverlee, Belgium
e-mail: michael@cs.kuleuven.ac.be

Abstract. In this paper we present a partial evaluation scheme for a "real life" subset of Prolog. This subset contains first-order built-in's, simple side-effects and the operational predicate if-then-else. We outline a denotational semantics for this subset of Prolog and show how partial deduction can be extended to specialise programs of this kind. We point out some of the problems not occurring in partial deduction and show how they can be solved in our setting. Finally we provide some results based on an implementation of the above.

1 Introduction

Partial evaluation has been established as an important research topic, especially in the functional and logic programming communities. The topic has been introduced to logic programming in [10] and has later been called *partial deduction* when applied to pure logic programs. A sound theoretical basis of partial deduction is given in [12].

Although a lot of papers address partial deduction there are few approaches addressing partial evaluation of "real life" programs written for instance in some "real life" subset of Prolog. Even fewer papers discuss the theoretical implications of making the move from partial deduction to partial evaluation. This is what we propose to examine in this paper. First though we have to choose an adequate subset of Prolog. We have chosen a subset encompassing:

1. first-order[1] built-in's, like var/1, nonvar/1 and =../2,
2. simple side-effects, like print/1,
3. the operational if-then-else construct.

Our choice tries to strike a balance between practical usability, complexity of the semantics and the potential for effective (self-applicable) partial evaluation. The most important decision was the inclusion of the if-then-else. As we will see in Sect. 3 the first-order built-in's and the side-effects do not add much complexity to the semantics. In later sections we will also see that the if-then-else lends itself quite nicely to partial evaluation and is much easier to cope with than the "full blown" cut. Using the if-then-else instead of the cut was already advocated in [14] and performed in [21].

---

[1] As opposed to "second order" built-in's which are predicates manipulating clauses and goals, like call/1 or assert/1.

In this paper we first formally define the subset of Prolog in Sect. 2 and give it a semantics in Sect. 3. In the sections 4 through 8 we adapt partial deduction such that it can cope with this subset of Prolog and study some of the added complications. Notably we will show that freeness and sharing information, although completely uninteresting in partial deduction, can be vital to produce efficient specialised programs. In Sect. 9 we extend the partial evaluation technique such that a Knuth-Morris-Pratt like search algorithm can be obtained by specialising a "dumb" search algorithm for a given pattern. We conclude with a discussion of related work and summarise our results.

2 Definition of RLP

The syntax of "Real-life Logic Programming" or simply RLP is based on the syntax of definite logic programs with the following extensions and modifications.

The concept of *term* remains the same. The set of *predicates* P is partitioned into the set of "normal" predicates P_{cl} defined through clauses and the set of built-in predicates P_{bi}. A *normal atom* (respectively a *built-in atom*) is an atom which is constructed using a predicate symbol $\in P_{cl}$ (respectively $\in P_{bi}$).

A *literal* is either an atom or it is an expression of the form (If $\rightarrow$ Then; Else) where If, Then, Else are lists of literals. We will denote by *Goals* the set of all lists of literals. We will represent a list of n literals by $(L_1, \ldots, L_n)$. Sometimes we will also use the notation $\leftarrow L_1, \ldots, L_n$.

A *clause* is an expression of the form $Head \leftarrow Body$ where $Head$ is a normal atom and $Body$ is a list of literals.

We can see from the above that RLP does not incorporate the negation nor the cut, but uses an if-then-else construct instead. This construct will behave just like the Prolog version of the if-then-else which contains a local cut and is usually written as (If -> Then ; Else). Most uses of the cut can be mapped to if-then-else constructs and the if-then-else can also be used to implement the not[2]. The following informal Prolog clauses can be used to define the if-then-else:

```
(If->Then;Else) :- If,!,Then.
(If->Then;Else) :- Else.
```

Thus the behaviour of the if-then-else is as follows:

1. If the test-part succeeds then a local cut is executed and the then-part is entered.
2. If the test-part fails finitely then the else-part is entered.
3. If the test-part "loops" (i.e. fails infinitely) then the whole construct loops.

3 Semantics of RLP

We will now outline a semantics for RLP. This semantics should be preserved by any reasonable partial evaluation procedure for RLP.

---

[2] Both the unsound and the sound version (using a groundness check for soundness).

The inclusion of the if-then-else into RLP has important consequences on the semantic level. As we have already pointed out the if-then-else contains a local cut. It is thus sensitive to the sequence of computed answers of the test-part. An implication being that the computation rule and the search rule have to be fixed in order to give a clear meaning to the if-then-else. From now on we will presuppose the Prolog left-to-right computation rule and the lexical search rule. The two programs hereafter illustrate the above point:

Program P_1	Program P_2
$q(X) \leftarrow (p(X) \rightarrow r(X); fail)$	$q(X) \leftarrow (p(X) \rightarrow r(X); fail)$
$p(a) \leftarrow$	$p(c) \leftarrow$
$p(c) \leftarrow$	$p(a) \leftarrow$
$r(c) \leftarrow$	$r(c) \leftarrow$

Using the Prolog computation and search rules, the query $\leftarrow q(X)$ will fail for program P_1 whereas it will succeed for P_2. All we have done is change the order of the computed answers for the predicate $p/1$. This implies that a partial evaluator which handles the if-then-else has to preserve the sequence of computed answers of all goals prone to be used inside an if-then-else test-part. This for instance is not guaranteed by the partial deduction framework in [12] which only preserves the computed answers but not their sequence.

As noted in Sect. 2 the if-then-else is also sensitive to non-terminating behaviour of the test-part. It is thus vital that our semantics also captures this aspect of RLP programs.

Finally a semantics for RLP has to take into account that the computed answers of built-in's cannot always be specified logically (var/1 for instance) and that built-in's can generate side-effects which have no impact on the computed answers (print/1 for example).

We fulfilled all the above requirements by adapting the denotational semantics of pure Prolog as defined in [2] and [17]. In these papers the semantics sem_P of a pure Prolog program P is a mapping from goals to (possibly infinite) sequences of computed answers. These sequences can be terminated by a least element $\perp$ which captures divergence (non-termination producing no computed answer).

Our semantics is based upon the view that computed answers and side-effects are events which can be observed by a person executing a RLP program. In that sense we will be characterising the observable behaviour of RLP programs.

Definition 1. The *side-effect domain* S_D is a (possibly infinite) set equipped with an equivalence relation $\equiv$. The elements of S_D are called *side-effects*.

In the remaining of this paper we suppose that the side-effect domain S_D is the set of built-in atoms and that the equivalence relation is syntactical identity. For instance the side-effect that "occurs" when $print(a)$ gets executed will be simply represented by the built-in atom $print(a)$. Our approach however makes no assumptions on S_D and it can thus be replaced by a finer structure if required.

Definition 2. An *event* is either a substitution representing a computed answer or a side-effect $\in S_D$. We will denote by *Event* the set of all events.

Definition 3. Two events e_1, e_2 are equivalent with respect to a given goal G, denoted by $e_1 \equiv_G e_2$, if either

1. both e_1 and e_2 are side-effects and $e_1 \equiv e_2$ or
2. both $e_1 = \theta_1$ and $e_2 = \theta_2$ are substitutions and $G\theta_1$ and $G\theta_2$ are variants.[3]

Definition 4. An *event sequence* is an element of one of the following sets

1. *Event**: the set of finite sequences of events
2. *Event** $\times \{\perp\}$: the set of finite sequences of events terminated with $\perp$
3. *Event$^\omega$*: the set of infinite sequences of events

We define the notation *EventSeq* = *Event** $\cup$ *Event** $\times \{\perp\} \cup$ *Event$^\omega$*. Furthermore two event sequences will be equivalent w.r.t. a given goal G ($\equiv_G$) iff they have the same length and the corresponding elements of the sequences are equivalent w.r.t G.

Definition 5. An *ES-semantics sem* for a set of goals $G \subseteq$ *Goals* is a mapping $G \to$ *EventSeq*. If $G =$ *Goals* then *sem* will be called *complete*. An *oracle* is an ES-semantics for $G = \{(A) \mid A$ is a built-in atom $\}$.

The concept of oracle as it is presented here, has nothing to do with the concept as presented in [1] where an oracle is used to abstract away from the sequential depth-first strategy of Prolog. We use an oracle to provide us with the meaning of the built-in's. Starting in the next section we will try to develop a partial evaluation procedure which preserves the semantics of a given RLP program independently of the actual oracle used to model the built-in's.

It is important to note that in Def. 5 the value returned by an oracle depends only on the actual call. In particular it does not depend on the program from which the built-in got called nor from any run-time environment. This makes our approach unsuitable to model built-in's which are not first order (like assert/1 or call/1). But note that unrestricted use of built-in's like assert/1 or retract/1, makes effective partial evaluation almost impossible.[4] Furthermore if the oracle's responses depended on something else it would be impossible to do semantics preserving program transformation without intricate assumptions on how the oracle's responses vary when the program or some run-time environment varies.

Unfortunately due to space restrictions we cannot elaborate on the exact details of our denotational ES-semantics and its fixpoint construction. Let us just state that the equivalence notion $\equiv$ for event sequences induces an equivalence on ES-semantics and we thus obtain an equivalence relation between programs. All further details will be made available in an upcoming technical report.

---

[3] This definition avoids a mistake of [17] which uses equivalence up to a renaming substitution (for which, contrary to what is stated in [17], $\{Y/X_1\}$ and $\{Y/X_2\}$ are not equivalent).

[4] On page 48 of [19] it is stated that "It is a question open to future research whether it is feasible to execute assert/1 and retract/1 by a partial evaluator."

The following table might help in giving the reader an intuition of our ES-semantics. Note that, when abstracting away from the side-effects, all the programs have the same least Herbrand model $M_P = \{q(a)\}$.

RLP Program P	ES-Semantics $sem_P(\leftarrow q(X))$
$q(a) \leftarrow$	$(\{X/a\})$
$q(a) \leftarrow print(a)$	$(print(a), \{X/a\})$
$q(X) \leftarrow q(X)$ $q(a) \leftarrow$	$(\bot)$
$q(a) \leftarrow$ $q(X) \leftarrow q(X)$	$(\{X/a\}, \{X/a\}, \{X/a\}, \ldots)$

4 LDR Trees

In the previous section we have sketched a semantics for RLP. In this section we elaborate on some of the problems when trying to preserve this semantics while performing partial evaluation. It should be clear that "standard" unfolding preserves the ES-semantics of a program as long as it uses the Prolog left-to-right computation rule[5] and it doesn't evaluate any non-logical predicates.

This of course summons the question of what we should do if the left-most literal is non-logical (or if it is logical but further unfolding it would lead us into an infinite loop). The easiest solution would be to just stop unfolding the goal completely. However the loss in specialisation is unacceptable. For instance in the framework of [12] the links (through variable sharing) between the remaining literals of the goal would be lost because the literals will be (even in the best possible case) partially evaluated separately. Worse, all specialisation w.r.t. the remaining literals would be lost if we use a method that constructs just one big SLD tree, as is the case with the partial evaluator developed by the author in [11]. To illustrate the problem let us take a look at the following simple program P_3 and try to specialise it for the query $\leftarrow q(X)$:

Program P_3
$q(f(Y)) \leftarrow print(Y), r(Y), s(Y)$
$r(a) \leftarrow \quad r(b) \leftarrow \quad r(c) \leftarrow$
$s(a) \leftarrow \quad s(b) \leftarrow \quad s(d) \leftarrow$

After the first unfolding step we obtain the goal $\leftarrow print(Y), r(Y), s(Y)$. If we stop unfolding and generate partial evaluations for the uncovered goals $r(Y)$ and $s(Y)$ we get the unmodified and unspecialised program P_3 back.

On the other hand if we do not follow the Prolog left-to-right rule and unfold the goals to the right of $print(Y)$ we obtain the specialised program P_4 which has a different ES-semantics.

---

[5] This is imperative even for purely logical programs. Take for instance the program "$p(X, Y) \leftarrow q(X), q(Y) \quad q(a) \leftarrow \quad q(b) \leftarrow$". If we unfold without following the left-to-right computation rule we change the order of solutions and the ES-semantics. In some cases this might change the (existential) left-termination behaviour.

Program P_4
$q(f(a)) \leftarrow print(a)$
$q(f(b)) \leftarrow print(b)$

For instance we have that $sem_{P_3}(\leftarrow q(f(X))) = (print(X), \{X/a\}, \{X/b\})$ while $sem_{P_4}(\leftarrow q(f(X))) = (print(a), \{X/a\}, print(b), \{X/b\})$. The problem is caused by the backpropagation of substitutions onto the *print* atom.[6]

So we are faced with a dilemma: on the one hand we cannot select the leftmost literal $print(Y)$ but on the other hand we cannot select $r(Y)$ or $s(Y)$ either because the substitutions will backpropagate onto $print(Y)$ changing the ES-semantics. The solution is however quite simple and just requires a minor extension of LD[7] trees: in addition to resolution steps we allow *residualisation* steps in the LD tree. These residualisation steps are not labelled with substitutions but with the selected literal which is thereby removed from the goal and hidden from vicious backpropagations. This extension allows us to follow the Prolog left-to-right computation rule without having to sacrifice neither specialisation nor correctness. This leads to the following definition of LDR trees.

Definition 6. A *complete LDR tree* for a program P is a tree whose nodes consist of goals such that the children for each goal $\leftarrow L_1, \ldots, L_k$ are:

1. either obtained by performing a resolution step with selected literal L_1. In this case the children are ordered according to the lexical ordering of the clauses used in the resolution step and the arcs are labelled with the substitutions of the resolution step.
2. or the goal has just one child which is obtained by performing a residualisation step on selected literal L_1. In this case the arc is labelled with L_1.

For a *(partial) LDR tree* we also allow any node to have no children at all (this corresponds to stopping unfolding). In this case there will be no selected literal.

Figure 1 depicts an LDR tree and shows how the introduction of the residualisation step solves the above dilemma. The generation of residual code from an LDR tree will be addressed in the next section. Note that the selected literal of a residualisation step is not necessarily a built-in. For instance it can be necessary to residualise a purely logical predicate to avoid infinite unfolding (see Sect. 6).

5 Generating Code from LDR trees

It should be quite obvious by looking at Fig. 1 that resultants are no longer sufficient for code generation. In fact we have to avoid the backpropagation of the bindings $\{Y/a\}$ and $\{Y/b\}$ onto the $print(Y)$ atom while ensuring that the

---

[6] The problem of backpropagation seems to have been overlooked in [6], thereby yielding a simple (non-pure) self-applicable partial evaluator whose semantics is however modified by self-application. Also note that under some circumstances backpropagation of single bindings can be allowed, see [19] and [16].

[7] SLD using the Prolog left-to-right computation rule and the lexical search rule.

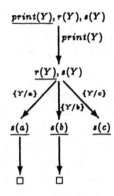

Fig. 1. LDR tree for program P_3

print(Y) atom gets executed only once. The only way to do this in the RLP framework is to generate a new predicate for the goal $\leftarrow r(Y), s(Y)$. The code for this new predicate can then be generated by taking resultants as there are no more residualisation steps in its subtree. Using that code generation strategy we get the following specialised version of P_3:

Program P_8
$q(f(Y)) \leftarrow print(Y), newp(Y)$
$newp(a) \leftarrow$
$newp(b) \leftarrow$

To formalise this process we first need the following notations:

Definition 7. Let τ be a LDR tree. Then $root(\tau)$ denotes the root of the tree. Also $\tau \xrightarrow{\alpha} \tau'$ holds iff τ' is a subtree of τ accessible via a branch whose sequence of labels is α.
We define $children(\tau) = \{(a, \tau') \mid \tau \xrightarrow{(a)} \tau'\}$. We also define $vars(\tau)$ to be the variables occurring in the arcs and leaves (but not the inner nodes) of τ.

In the following definition we present a way to remove a residualisation step from LDR trees. The basic idea is to split an LDR tree at a given residualisation step into two LDR trees which will be linked through the introduction of a new predicate. Figure 2 serves as an illustration of this definition.

Definition 8. Let S be a set of LDR trees. If we can find an LDR tree $\tau \in S$ such that

1. $\tau \xrightarrow{\alpha} \tau' \xrightarrow{R} \tau''$ where R is a residualised literal
2. $root(\tau') = \leftarrow R, G_1, \ldots, G_k$ and $root(\tau'') = \leftarrow G_1, \ldots, G_k$

then $(S \backslash \{\tau\}) \cup \{\tau_1, \tau_2\} \in rsplit(S)$ where

1. τ_1 is obtained from τ by replacing the subtree τ' by a tree consisting of the single node $\leftarrow R, newp(Args)$
2. $children(\tau_2) = \{(\phi, \tau'')\}$ and $root(\tau_2) = newp(Args)$

3. *newp* is a new unused predicate
4. $Args = vars(\leftarrow G_1, \ldots, G_k) \cap vars(\tau'')$.

Note that in a sense we add a clause of the form $newp(Args) \leftarrow G_1, \ldots, G_k$ to the program to be specialised (and use it for folding).

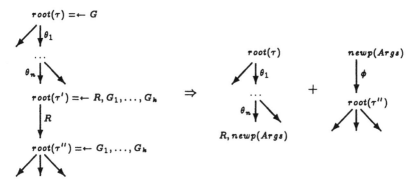

Fig. 2. Illustrating Def. 8 of *rsplit*

Using the above definitions we can now formalise a way to generate code for an LDR tree τ. We will denote by $rsplit^*(\tau)$ the set S_i obtained by the algorithm below. Note that $rsplit^*(\tau)$ is independent of the order in which the residualisation steps are removed.

Algorithm 1: Code Generation
$i = 0, S_0 = \{\tau\}$
while $rsplit(S_i) \neq \phi$ *do*
$\quad S_{i+1} =$ some element $S' \in rsplit(S_i)$
$\quad$ increment i
od
take resultants for the set of partial LD trees S_i

Although the above technique always produces correct code it can generate superfluous predicates: if a newly generated predicate *newp* has less than two defining clauses its introduction was not necessary. We present a post-processing scheme which remedies this (in practice this will be incorporated into *rsplit*).

Definition 9. Let $\theta = \{X_1/T_1 \ldots, X_n/T_n\}$ be a substitution and let *eq* be a binary predicate defined by the following single clause: "$eq(X, X) \leftarrow$".
Then we define $eqcode(\theta) = eq([X_1, \ldots, X_n], [T_1, \ldots, T_n])$. In practice we will write $eqcode(\theta)$ as $X_1 = T_1 \ldots, X_n = T_n$ and $eqcode(\phi)$ as *true*.

Using the above definition we define the post-processing as follows:

1. Transform all clauses "$Head \leftarrow L_1, \ldots, L_{i-1}, newp(Args)$" where *newp* is a new predicate defined by the single clause "$newp(Args\theta) \leftarrow Body$" into the following clause (and then remove *newp* from the program):
   "$Head \leftarrow L_1, \ldots, L_{i-1}, eqcode(\theta), Body$".

2. Transform all clauses "$Head \leftarrow L_1, \ldots, L_{i-1}, newp(Args)$" where $newp$ has no defining clause into "$Head \leftarrow L_1, \ldots, L_{i-1}, fail$".

Note that, by definition of $rsplit$, a newly generated predicate $newp$ can only occur in the last position of a residual clause. Point 2 becomes interesting if we allow "safe" left-propagation of failure (see for instance [16] and [19]).

In [7] it is stated that "only the leftmost choice-point in a goal should be unfolded". Doing otherwise can be harmful for efficiency, for instance it can lead to the situation where an expensive goal has to be re-solved in the specialised program. This is certainly true for partial deduction. But in our case there is no backpropagation of bindings and the introduction of the new predicates not only guarantees that side-effects don't multiply but also that no expensive goal has to be re-solved. So in our framework unfolding (deterministic or not) is never harmful with respect to this problem.

6 Useless Bindings

In this section we will point at a particular problem which occurs when we want to preserve the ES-semantics during partial evaluation. This problem will be linked to "useless bindings" which are informally defined by the following:

A *useless binding* inside a partial SLD tree (respectively LDR tree) is a binding which can be removed without changing the semantics (under consideration) of the resulting residual program.

Let us first show why useless bindings are no problem for partial deduction. When performing partial deduction according to [12] the specialised program is obtained by taking resultants of a partial SLD-tree. The combination of the mgu's, required to reach a leaf, is thus backpropagated on the top-level goal and any useless binding in the combination of mgu's automatically disappears. Suppose for instance that starting from the goal $\leftarrow p(X, Y)$ we reach the leaf $\leftarrow q(Z)$ via the combination of mgu's $\theta = \{X/h(Z), \mathbf{W/h(Z)}, Y/a, \mathbf{V/b}\}$. The resultant will be $p(h(Z), a) \leftarrow q(Z)$ and all the useless bindings in bold have disappeared. In other words backpropagation on the top-level goal ensures that useless bindings have not even an effect on the *syntax* of the residual program.

When we want to preserve the ES-semantics and backpropagation is forbidden these useless bindings can become a big problem. This is even the case for purely logical programs. We will use the following programs to illustrate this.

Program P_6	Program P_7
$p(X, \underline{Z}, a) \leftarrow$	$p(X, \underline{X}, a) \leftarrow$
$p(X, Y, b) \leftarrow p(X, Z, W), q(Z, W)$	$p(X, Y, b) \leftarrow p(X, Z, W), q(Z, W)$
$q(b, b) \leftarrow$	$q(b, b) \leftarrow$
$q(a, a) \leftarrow$	$q(a, a) \leftarrow$

In Fig. 3 we show an LDR tree for the programs P_6 and P_7 in which $p(X, Z, W)$ has been residualised to avoid infinite unfolding. In this LDR tree the problematic

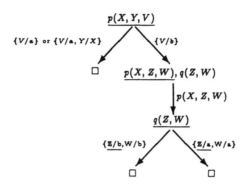

Fig. 3. LDR tree for Programs P_6 and P_7

substitutions $\{Z/a, W/a\}$ and $\{Z/b, W/b\}$ occur. In contrast to partial deduction these substitutions cannot be backpropagated onto the top-level goal.[8] Using the code generation strategy of algorithm 1 in Sect. 5 they will have an effect on the residual code. However for program P_6 the bindings Z/a and Z/b inside these substitutions are useless, i.e. removing them will yield a more efficient residual program with the same ES-semantics. The following table shows the generated residual programs with and without the bindings.

Specialised P_6 with $Z/a, Z/b$	*Specialised P_6 without $Z/a, Z/b$*
$p(X, Z, a) \leftarrow$	$p(X, Z, a) \leftarrow$
$p(X, Y, b) \leftarrow p(X, Z, W), newp(\underline{Z}, W)$	$p(X, Y, b) \leftarrow p(X, Z, W), newp(W)$
$newp(\underline{b}, b) \leftarrow$	$newp(b) \leftarrow$
$newp(\underline{a}, a) \leftarrow$	$newp(a) \leftarrow$

This is just one simple example but our experiments have shown that bad code can be generated when no additional measures are taken. A way to solve this problem is to improve the calculation of the arguments *Args* for the new predicates in Def. 8. For example in Fig. 3 we can detect for P_6 that the variable Z will always be free after a successful call to $p(X, Z, W)$ and that the variable Z cannot possibly share with a variable of the top-most goal $p(X, Y, V)$. Thus Z can be removed from *Args* because no value will ever be transmitted via Z to the subtree for $q(Z, W)$ and the bindings of the subtree which affect Z are of no consequence. By removing Z from *Args* the bindings Z/a and Z/b no longer have an effect on the residual code.[9]

---

[8] Backpropagating $\{Z/a, W/a\}$ and $\{Z/b, W/b\}$ by applying them on the top-level goal and all residualisation steps would generate a program which diverges after the first answer for $p(X, Y, Z)$.

[9] The interested reader might have noticed that for program P_6 in Fig. 3 the entire goal $q(Z, W)$ can be replaced by *true* without changing the semantics. To do this safely however we have to calculate a safe approximation of the possible computed answers for the goal $p(X, Z, W)$. This could be: $\{\{W/b\}, \{W/a\}\}$ allowing us to conclude that one of the branches of $q(Z, W)$ will succeed with the empty computed answer substitution and the other one will fail. Formalising and implementing this technique is still a matter of ongoing research.

Generalising the above we see that we can remove from *Args* all those variables which are guaranteed to be free and guaranteed not to share with any variable of the root of the LDR tree. In order to detect this we have implemented an abstract interpretation scheme which calculates for each node i the following sets of variables:

1. *Free$_i$*, which are variables guaranteed to be free at that node
2. *Share$_i$*, which are potentially sharing with the variables of the root

We can remove a variable V from the set *Args* of Def. 8 if $V \in Free_{root(\tau'')}$ and $V \notin Share_{root(\tau'')}$ where $root(\tau'')$ is the node of the LDR tree following the residualisation step. For example for Fig. 3 we have the following:

Goal of node i	Free$_i$	Share$_i$	Valid for
$\leftarrow p(X, Y, V)$	ϕ	$\{X, Y, V\}$	P_6, P_7
$\leftarrow p(X, Z, W), q(Z, W)$	$\{Z, W\}$	$\{X\}$	P_6, P_7
$\leftarrow q(Z, W)$	$\{Z\}$	ϕ	P_6
$\leftarrow q(Z, W)$	ϕ	$\{Z\}$	P_7

Thus for program P_6 we can remove the variable Z from the arguments *Args* because $Z \in Free_{root(\tau'')}$ and $Z \notin Share_{root(\tau'')}$ while for program P_7 we cannot (where $root(\tau'')$ is the node with goal $\leftarrow q(Z, W)$).

In our implementation we also calculate for each root of an LDR tree (except possibly for the query goal) a set of "unused" argument variables. As we perform renaming and argument filtering (see [8], [3]) these unused variables can be removed from the residual code. More importantly this may diminish the number of variables sharing with the topmost goal and may allow us to generate much better code. A further improvement lies in generating multiple versions of a predicate call according to varying freeness information of the arguments. In our implementation we have accomplished this by integrating the freeness and sharing analysis into the partial evaluation process and using more refined notions of "instance" and "variant". Our system might thus generate two versions for a given goal $\leftarrow p(X)$: one version where X is guaranteed to be free and one where it is not. Under some circumstances this can substantially improve the quality of the generated code (especially when useless if-then-else constructs come into play, see Sect. 8).

We go even further and let the user of our system specify freeness for his top-level goal. In that case the set *Free$_i$* for the top-level goal $\leftarrow p(X, Y, V)$ in the above table would no longer be ϕ but some user specified set. Note that now we are no longer performing partial evaluation in the usual sense. The specialised program will not be correct for all instances of the top-level goal, but only for those instances respecting the freeness information provided by the user.

From this section we can draw the conclusion that, as soon as we want to preserve the sequence of computed answers or the existential termination behaviour of programs (w.r.t. a given static computation rule), partial evaluation should be combined with additional analysis (i.e. abstract interpretation) to give good results.

7 Unfolding the If-then-else

Until now we have not touched the specialisation of the if-then-else predicate. Using the method presented so far we could perform a residualisation step on a selected if-then-else construct. However this technique performs no specialisation whatsoever inside the if-then-else. To remedy this problem we adapt residualisation such that, in addition to removing a selected if-then-else from a goal, it specialises the construct. The technique is based on partially evaluating the test-part and then based on this generating specialised then- and else-parts.

Suppose that the construct (If $\rightarrow$ Then; Else) is selected by a residualisation step. First we partially evaluate If yielding an LDR tree τ. We now calculate $S = rsplit^*(\tau)$ to remove all residualisation steps. We can now take the tree $\tau' \in S$ which has the goal If as root and obtain its sequence of branches β. The sequence β is useful because it provides us with substitutions which can be applied to the then-part. The process of generating the specialised if-then-else construct using β is detailed in algorithm 2 below and we obtain the label of the residualisation step by calling $code(\beta, \text{Then}, \text{Else})$. We also have to add $S \backslash \{\tau'\}$ to the global set of LDR trees determining our partial evaluation.

Algorithm 2: $code(\beta, \text{Then}, \text{Else})$
if β is empty **then**
Partially evaluate Else yielding LDR tree τ_1
Add τ_1' to the global set of LDR trees where
$root(\tau_1') = newp_1(var(\text{Else}))$ and $children(\tau_1') = \{(\phi, \tau_1)\}$
return $newp_1(var(\text{Else}))$
else
Let $\beta = (b_1, \ldots, b_k)$ and let θ be the composition of substitutions of b_1
Partially evaluate Thenθ yielding LDR tree τ_2
Add τ_2' to the global set of LDR trees where
$root(\tau_2') = newp_2(var(\text{Then}))$ and $children(\tau_2') = \{(\phi, \tau_2)\}$
Let G be the leaf goal of b_1 and let $R = code((b_2, \ldots, b_k), \text{Then}, \text{Else})$
return $((eqcode(\theta), G) \rightarrow newp_2(var(\text{Then})) ; R)$
fi

This technique generates a cascade of if-then-else's and an example is given in the next section. Note that to "unfold" the if-then-else we did not have to resort to other non-pure predicates. So in a sense RLP is closed under unfolding.

This unfolding resembles constructive negation (see e.g. [4]) when the if-then-else is used to represent the $not/1$. The definition based on if-then-else seems however much simpler from a practical viewpoint. Also note that our technique is as rather aggressive with respect to specialisation (it can be made even more aggressive by pushing code to the right of an if-then-else inside its then- and else-branches). For instance it specialises more than [9] but there is a risk of code-explosion. The results are however quite similar to the ones obtained by Mixtus in [18] which first transforms the if-then-else using a special form of cut and disjunctions. A similar approach is used in [16] which transforms the if-then-else using "ancestral cuts" and disjunctions. The problems discussed in the next section should also be relevant to these approaches.

8 Useless If-then-else Constructs

The problem of useless bindings discussed in Sect. 6 is also present for the new predicates $newp_1$ and $newp_2$ introduced in algorithm 2 and can be solved in a similar way. The problem tackled in this section usually occurs as soon as predicates with output arguments are present inside the test-part of an if-then-else. For example the variable Y in the following simple program P_8 is used as an output argument to arc.

Program P_8
$arcs\_leave(X) \leftarrow one\_arc(X, Z)$
$one\_arc(X, Y) \leftarrow (arc(X, Y) \rightarrow true; fail)$
$arc(a, b) \leftarrow \quad arc(b, a) \leftarrow \quad arc(a, c) \leftarrow$

Without freeness information we can only come up with the following specialisation for the goal $\leftarrow arcs\_leave(a)$:

$$arcs\_leave(a) \leftarrow ((Z = b) \rightarrow true; ((Z = c) \rightarrow true; fail))$$

Our experiments showed that such useless if-then-else constructs can become very frequent for programs with output arguments. Again through the use of freeness information, we can remedy this problem. For instance, knowing that Z is guaranteed to be free and and hence that $Z = b$ will always succeed exactly once at run-time, we can come up with the following specialisation for $arcs\_leave(a)$:

$$arcs\_leave(a) \leftarrow$$

Note that it is not sufficient to just detect existential variables in residual clauses because output arguments can still be present in the residual program. Again a freeness and sharing analysis is needed to solve the problem. In our implementation we have used the same analysis which was used to solve the problem of useless bindings in Sect. 6. Using this information we can refine algorithm 2 by improving the calculation of arguments to $newp_1$ and $newp_2$ and by detecting when a test-branch (b_1 in Algorithm 2) is guaranteed to succeed exactly once. When this is the case we do not have to create a residual if-then-else construct (a conjunction suffices) and we can drop the else-part.

9 Results

All the ideas of the previous sections have been incorporated into a running system (based on [11]) whose results were very satisfactory. The removal of useless bindings and useless if-then-else constructs turned out to be vital for a lot of non-toy examples, especially for self-application. The system also performs well on the benchmarks used in [18]. We also performed tests with a regular expression parser taken from [13] yielding similar results, i.e. we obtain a deterministic automaton.

We now take a look at the famous match example. We will see that our treatment of the if-then-else is not yet optimal in the sense that we need one further optimisation to get a Knuth-Morris-Pratt (KMP) like search algorithm out of match.

Program P_9
$match(\text{Pat}, \text{T}) \leftarrow match1(\text{Pat}, \text{T}, \text{Pat}, \text{T})$
$match([], \text{Ts}, \text{P}, \text{T}) \leftarrow$
$match([\text{A}|\text{Ps}], [\text{B}|\text{Ts}], \text{P}, [\text{X}|\text{T}]) \leftarrow$
  $(A = B \rightarrow match1(\text{Ps}, \text{Ts}, \text{P}, [\text{X}|\text{T}]) \ ; \ match1(\text{P}, \text{T}, \text{P}, \text{T}))$

We do not obtain a KMP algorithm using the unfolding method presented so far because we do not use the information that the test has failed in the else-branch. This problem can been solved in several ways. For instance in [5],[20] constraints similar to $A \neq B$ are propagated. The SP system presented in [7] derives a KMP algorithm by left-propagation of bindings and the execution of ground not's. In this case the non-ground negative literals play the role of the constraints and the left-propagation of bindings takes care of testing the constraints. The partial evaluators Mixtus (see [18]) and Paddy (see [15]) are also able to derive KMP algorithms in what seems to be a quite similar way.

The question is how do we achieve KMP without left-propagation of bindings nor the explicit introduction of constraints. The solution is quite simple and consists in checking whether there are any apparent incompatibilities between the else-part and the fact that the test-part has to fail in order to reach the else-part. Formalising this we can transform constructs of the form $(\text{If} \rightarrow \text{Th} \ ; \ (E_1, P, E_2))$ where the success of P implies the success of If and where If is purely logical into the simplified construct $(\text{If} \rightarrow \text{Th} \ ; \ (E_1, fail))$. For instance we can transform $(X = Y \rightarrow p(X); (X = a, Y = a, q(Z)))$ into $(X = Y \rightarrow p(X); fail)$.[10] This simple transformation is sufficient to obtain a KMP-like string matcher out of P_9. Program P_{10} below was obtained by partially evaluating P_9 for the query $\leftarrow match([a, a, b], X)$. The above technique can be quite powerful because it does not require that bindings are actually left-propagated on E_1 and because it can extract information out of non-ground not's (when implemented using the if-then-else). Also freeness and sharing information can again be important to obtain good results (to detect success of If).

Program P_{10}
$match\_1(\text{T}) \leftarrow match1\_2(\text{T})$
$match1\_2([\text{H}|\text{T}]) \leftarrow (H = a \rightarrow match1\_3(\text{T})) \ ; \ match1\_2(\text{T}))$
$match1\_3([\text{H}|\text{T}]) \leftarrow (H = a \rightarrow match1\_4(\text{T})) \ ; \ match1\_2(\text{T}))$
$match1\_4([\text{H}|\text{T}]) \leftarrow (H = b \rightarrow true \ ; \ (H = a \rightarrow match1\_4(\text{T})) \ ; \ match1\_2(\text{T})))$

---

10 Discussion and Conclusion

The aspect of trying to preserve the sequence of computed answers (and side-effects) has been dealt with in some of the approaches based on the unfold/fold transformations of [22]. For instance [19] talks about the dangers of left-propagation of bindings and solves this by transforming the bindings into explicit code based on the =/2 predicate and by introducing disjunctions. But as stated in [7]

---

[10] Which can be further simplified (using another simple optimisation based on the fact that a substitution will succeed at most once) into "$X = Y, p(X)$".

such an approach can have a harmful effect by removing indexing information. The method used in [16] relies on the introduction of new predicates in a similar way to ours.[11] However the problems of useless bindings or useless if-then-else constructs are not dealt with in these papers and the semantics that is used stays rather informal and is not denotational.

To our knowledge the preservation of the sequence of computed answers has not been dealt with in the framework of [12] and the concept of a residualisation step is new. Two other partial evaluators handle a significant subset of Prolog. One is Logimix coming from the functional world, described in [13]. The other one is described in [6] but, as noted earlier, does not always preserve the sequence of answers. The primary concern of these papers was self-application and the problems of useless bindings or useless if-then-else constructs are not dealt with.

In this paper we have introduced the ES-semantics for a non-trivial subset of Prolog. We have shown how, through the use of LDR trees, the ES-semantics can be preserved by partial evaluation. An important conclusion of this paper is that using just partial evaluation based on unfold/fold transformations can yield surprisingly bad results, leaving a lot of useless assignments, arguments or if-then-else constructs in the residual program. We have shown how freeness and sharing information, which are of no interest in partial deduction, can be used to solve this problem. Finally our results confirm that the if-then-else is very well suited for partial evaluation and can be used to obtain a KMP like string matcher by partially evaluating the simple "match" algorithm.

Acknowledgements

The author is supported by GOA "Non-standard applications of abstract interpretation." I would like to thank Danny De Schreye and Bern Martens for proof-reading an earlier version of this paper. I also would like to thank them for the stimulating discussions and for their encouragement. I also thank anonymous referees for their helpful remarks and for pointing out related work.

References

1. R. Barbuti, M. Codish, R. Giacobazzi, and M. Maher. Oracle semantics for Prolog. In H. Kirchner and G. Levi, editors, *Proceedings of the Third International Conference on Algebraic and Logic Programming*, Lecture Notes in Computer Science 632, pages 100–114. Springer Verlag, 1992.

2. M. Baudinet. Proving termination of Prolog programs: A semantic approach. *The Journal of Logic Programming*, 14(1 & 2):1–29, 1992.

3. K. Benkerimi and P. M. Hill. Supporting transformations for the partial evaluation of logic programs. *Journal of Logic and Computation*, 3(5):469–486, October 1993.

4. D. Chan and M. Wallace. A treatment of negation during partial evaluation. In H. Abramson and M. Rogers, editors, *Meta-Programming in Logic Programming, Proceedings of the Meta88 Workshop, June 1988*, pages 299–318. MIT Press, 1989.

---

[11] There is however a small mistake w.r.t. the arguments for new predicates.

5. H. Fujita. An algorithm for partial evaluation with constraints. Technical Memorandum TM-0367, ICOT, 1987.

6. H. Fujita and K. Furukawa. A self-applicable partial evaluator and its use in incremental compilation. *New Generation Computing*, 6(2 & 3):91–118, 1988.

7. J. Gallagher. Tutorial on specialisation of logic programs. In *Proceedings of PEPM'93, the ACM Sigplan Symposium on Partial Evaluation and Semantics-Based Program Manipulation*, pages 88–98. ACM Press, 1993.

8. J. Gallagher and M. Bruynooghe. Some low-level transformations for logic programs. In M. Bruynooghe, editor, *Proceedings of Meta90 Workshop on Meta Programming in Logic*, pages 229–244, Leuven, Belgium, 1990.

9. C. A. Gurr. *A Self-Applicable Partial Evaluator for the Logic Programming Language Gödel*. PhD thesis, Department of Computer Science, University of Bristol, January 1994.

10. J. Komorowksi. *A Specification of an Abstract Prolog Machine and its Application to Partial Evaluation*. PhD thesis, Linköping University, Sweden, 1981. Linköping Studies in Science and Technology Dissertations 69.

11. M. Leuschel. Self-applicable partial evaluation in Prolog. Master's thesis, K.U. Leuven, 1993.

12. J. W. Lloyd and J. C. Shepherdson. Partial evaluation in logic programming. *The Journal of Logic Programming*, 11:217–242, 1991.

13. T. Mogensen and A. Bondorf. Logimix: A self-applicable partial evaluator for Prolog. In K.-K. Lau and T. Clement, editors, Logic Program Synthesis and Transformation. *Proceedings of LOPSTR'92*, pages 214–227. Springer-Verlag, 1992.

14. R. O'Keefe. On the treatment of cuts in Prolog source-level tools. In *Proceedings of the Symposium on Logic Programming*, pages 68–72. IEEE, 1985.

15. S. D. Prestwich. The PADDY partial deduction system. Technical Report ECRC-92-6, ECRC, Munich, Germany, 1992.

16. S. D. Prestwich. An unfold rule for full Prolog. In K.-K. Lau and T. Clement, editors, Logic Program Synthesis and Transformation. *Proceedings of LOPSTR'92*, Workshops in Computing, University of Manchester, 1992. Springer-Verlag.

17. M. Proietti and A. Pettorossi. Semantics preserving transformation rules for Prolog. In *Proceedings of the ACM Symposium on Partial Evaluation and Semantics based Program Manipulation, PEPM'91*, Sigplan Notices, Vol. 26, N. 9, pages 274–284, Yale University, New Haven, U.S.A., 1991.

18. D. Sahlin. *An Automatic Partial Evaluator for Full Prolog*. PhD thesis, Swedish Institute of Computer Science, Mar. 1991.

19. D. Sahlin. Mixtus: An automatic partial evaluator for full Prolog. *New Generation Computing*, 12(1):7–51, 1993.

20. D. A. Smith. Partial evaluation of pattern matching in constraint logic programming languages. In N. D. Jones and P. Hudak, editors, *ACM Symposium on Partial Evaluation and Semantics-Based Program Manipulation*, pages 62–71. ACM Press Sigplan Notices 26(9), 1991.

21. A. Takeuchi and K. Furukawa. Partial evaluation of Prolog programs and its application to meta programming. In H.-J. Kugler, editor, *Information Processing 86*, pages 415–420, 1986.

22. H. Tamaki and T. Sato. Unfold/fold transformations of logic programs. In S.-Å. Tärnlund, editor, *Proceedings of the Second International Conference on Logic Programming*, pages 127–138, Uppsala, Sweden, 1984.

Schema-Based Top-Down Design of Logic Programs Using Abstract Data Types

E. Marakakis and J.P. Gallagher

Department of Computer Science, University of Bristol
Queen's Building, University Walk, Bristol BS8 1TR, U.K.
e-mail: manolis@compsci.bristol.ac.uk, john@compsci.bristol.ac.uk

Abstract. This paper presents a set of schemata that support stepwise top-down design of logic programs using abstract data types (ADTs). There are thus three main components to this approach: top-down design, program schemata and ADTs, all of which are already well established notions. Our aim is to combine these ideas in a framework, amenable to support by design tools and allowing the use of existing logic program transformation techniques to optimise the final programs.

1 Introduction and Motivation

This paper presents a set of schemata that support stepwise top-down design of logic programs using abstract data types (ADTs). There are thus three main components to this approach: top-down design, program schemata and ADTs, all of which are already well-established notions. Our aim is to combine these ideas in a framework, amenable to support by design tools and allowing the use of existing logic program transformation techniques to optimise the final programs. The choice of the set of available schemata and ADTs is very important, and is based on a thorough survey in the literature of the use of these techniques in logic program construction.

The main aims of this paper are

- to justify the approach,
- to show a small set of schemata and ADTs with which we are currently working,
- to discuss future directions of this work.

The main components of our approach are now discussed.

1.1 Top-Down Design

Top-down design is a process of successive refinement of a design. At each stage in the process there is a number of "undefined" parts of the design. A refinement step replaces some undefined part by a more detailed construct, which itself may contain further undefined subparts as well as constructs from some executable target language L. The process is finished when the design is entirely in L.

The main justification of top-down design is that it gives a goal-oriented structure to the design process, and suggests a "compositional" style of program in which each refinement step can be independently justified or proved correct. Apart from this very little is implied by the phrase "top-down design". The actual refinement steps may take many forms.

Our belief is that, if tools to support top-down design are to be practical, refinement steps should correspond to the set of basic constructs in the target language L. That is, each refinement step should introduce one language construct. For example, if L is a Pascal-like language we could insist that each refinement would introduce only one **while, repeat, if-then-else** or statement composition construct. The target language L therefore plays a very important role in top-down design since each design decision has to be expressed by a single construct. The constructs available in L should be sufficiently high-level to be regarded as part of the design rather than implementation.

In our approach each refinement step introduces either one instance of a program schema or an ADT operation. In a loose sense, therefore, the choice of schemata and ADTs defines a "design language" whose basic constructs consist of the set of schemata and the set of ADT operations.

The separation of schemata from ADTs is an important aspect of our approach since each schema is independent of any particular types. Design decisions therefore need not confuse control with data. This is unlike most schemata in the logic programming literature, which are associated with specific types (such as lists and trees).

1.2 Program Design Schemata

A program design schema for a language M, or simply schema for M, is an expression that represents a set of programs in M having a common form. Each such program is called an instance of the schema. For logic programs, schemata are defined in Section 4. A schema can be identified with the set of its instances. Schemata are abstractions of program structure, analogous to abstractions of data by ADTs.

Our problem is to define a relatively small set of schemata, each member of which captures a useful form of program. Schemata should contain enough structure to be useful in the design process, but be independent of specific data types. This choice is quite difficult. At one extreme, a single schema could be used to construct all programs [24]. On the other hand the literature contains a large number of possible schemata (variously called schemata, clichés, skeletons, and other things). Too many schemata would be confusing and choosing among them would inevitably obscure the design process. Our current set, shown in this paper, consists of five schemata. Although we do not regard this as final by any means, it is important to stress that we do not envisage a growing library of schemata.

Another requirement on our choice of schemata is that they should have both a clear declarative and a procedural reading. Instantiations of schemata should be executable, but have a clear logical structure.

1.3 Abstract Data Types (ADTs)

The concept of type is used to classify objects with distinct characteristics. In programming, a type or data type is a class of data objects or values, such as integers, reals, sets, graphs, cartesian products, and so on. A number of operations is associated with each type which are applied to the values of the type. Abstract data types give a representation-independent specification of a data type. That is, the operations of the data type are defined in terms of axioms that any representation of the type should satisfy.

We also need to introduce the notion of a parameterised ADT. This is an ADT whose components may be of some other unspecified type. For instance a sequence may consist of integers, sets, other sequences, and so on. Thus we could define a type $sequence(\tau)$ where τ is a type variable indicating the type of elements of the sequence.

In our design approach, abstraction is applied to both the algorithm and the data in a problem. In this way the design is separated from any particular representation of the problem. The use of data abstractions is common, but the use of algorithm abstractions is rarer. Yet we maintain that the experienced program designer makes much use of algorithm abstractions or forms.

The result of the top-down design is an executable design or prototype. To summarise our approach, we might replace the well-known equation [23]

$$\text{Algorithms} + \text{Data Structures} = \text{Programs}$$

by

$$\text{Schemata} + \text{ADTs} = \text{Executable Designs}$$

As a next step in this approach program transformation techniques can be applied to the executable design. These could have the effect of removing the design layer by intertwining it with implementation, or of introducing other standard optimisations. The intention of this step is to improve the performance of the final program.

2 Related Work

2.1 Program Abstractions

Approaches for constructing logic programs based on some kind of abstraction have been followed by many researchers [2], [5], [6], [9], [12], [19], [22]. Their difference apart from technical details is on how they have conceived and used the concept of abstraction. These approaches are classified and discussed based on the kind of abstraction that they use.

Program Skeletons and Clichés. Skeletons are the primary control flow of programs [12], [19], [21]. They are used as the starting point for program development. Techniques are standard Prolog programming practices, e.g. passing back results from computations. Logic programs are constructed by successively enhancing skeletons by applying techniques to them.

Clichés in [2] are defined as commonly occurring program forms that can be reused. Formally clichés are defined to be second-order sentences. The construction process involves instantiation of a cliché with argument values supplied from the programmer.

Abstraction in these approaches is based on control flow similarities between logic programs. These tend to be constructed from commonly occurring program idioms such as processing all elements of a list, and in fact these skeletons tend to be closely related to certain data structures. Experience seems to be the main requirement needed to select such schemata.

Program Schemata. In [9] and [22] a schema is formally defined to be the most specific generalisation [18] of a class of programs. The program construction process in [22] is seen as searching a well-founded ordering of schemata up to certain depth guided by a set of positive and negative examples. Schema instantiation is not discussed in [9]. Schemata for list processing only are shown. Schema classification is based on syntactic similarities.

This approach depends very much on the set of programs that are used for finding the most specific generalisation. This kind of abstraction may result in program schemata with very little structure in them to support the design process. The schemata that can be constructed by that approach are as many as there are programs, that is, infinite. Only small programs have been constructed by this approach. Verification is not discussed in [9] while in [22] program correctness is proved with respect to a set of positive and negative instances that have to be covered and not covered respectively by the predicate under construction.

Logic Description Schemata. A method for constructing logic programs based on structural induction and the generalization paradigm is proposed in [5], [6]. A logic description (LD) is a formula in typeless first-order logic. Two kinds of generalisations are used for the construction of LDs, structural and computational generalisation. Structural induction is the basis of the construction method. The constructed LD is in turn transformed into a logic program. Logic description skeletons are produced automatically from logic description schemata. Undefined predicates in logic description skeletons are defined either automatically using object knowledge or interactively by the user. It is shown that logic descriptions are correct by construction. A logic description schema, i.e. Divide-and-Conquer, is used in [8] to guide the synthesis of logic programs from specifications which consist of examples and properties.

The structural and computational generalisations in [6] correspond to ADTs and to design schemata respectively, from the point of view that both are abstractions over the data and the procedural aspects of a program but they are different kinds of abstractions. The construction of LD and therefore the whole method is data representation dependent. The Divide-and-Conquer schema in [8] is hardwired in the synthesis mechanism while in our approach the schemata are input to the system. Specifications based on examples have the advantage of simplicity but on the other hand the selection of appropriate examples is not always simple. The synthesis of programs which require multiple applications of the Divide-and-Conquer schema are not discussed in [8]. Construction of small programs is demonstrated in [6] and [8]. The construction of large programs does not seem to be a straightforward task. The use of typeless logic is one reason for their limitation.

2.2 Types

The extension of Prolog programs with mode and type declarations is suggested in [3] in order to enhance their readability and obtain more reliable programs. Types are used to detect errors in cases like mistyping arguments or misplaced arguments. Gödel [11] is a strongly typed logic programming language. Its type system is based on many sorted-logic with parametric polymorphism. A polymorphic type checking system for Prolog is proposed in [17]. The type system in [17] is extended with subtype relations in [7] in order to improve its software engineering properies.

Types in [7], [11] and [17] as in our approach are treated as terms over a special alphabet disjoint from the program's alphabet. Unification and type substitution are over the extended set of type symbols which are treated as ordinary terms. Types in our approach are both inferred during program development and for type checking of calls to ADT predicates. The type systems in [7], [11], [17] perform type checking in constructed programs. They use type declarations to check if the programs are type correct.

2.3 Modes

Modes in [3] are used to validate the flow of data through a clause. The problem of automatically inferring modes for predicates in Prolog programs is studied in [4] for optimization purposes. Modes in [16] are used to optimize Prolog programs. The mode analysis in the literature [4], [16] is mainly used for optimization purposes.

In our approach intended modes are declared for the schemata, the ADT operations and the initial predicate specification. During the design, modes are inferred and can be checked for compatibility with the intended modes.

3 Data Types and Modes

3.1 Data Types

The data types that we consider in this approach are polymorphic types. The type variable τ ranges over types. The alphabet of variables V, function symbols

F and predicate symbols P of the underlying logic language has been extended by the type variables V^τ and the type constructors F^τ such that $V \cap V^\tau = \{\}$ and $F \cap F^\tau = \{\}$. The theoretical basis of the types in this approach is *many sorted logic with parametric polymorphism* [10], [11].

A data type term may consist of *type variables, basic types* and *type constructors*. The syntax of data types in BNF production rules is the following

Type ::= τ |
      B |
      C(Type,..,Type)

τ stands for *type variables*. B stands for the *basic types*, or *constructors of arity 0*. C($Type, .., Type$) stands for constructors of arity ≥ 1. Type terms are defined similarly to first-order terms. That is, *type variables* correspond to *individual variables, basic types* correspond to *constants* and *type constructors* to *functions*. A *variable typing* [10], [11], is defined to be a set of the form $\{X_1/\tau_1, .., X_n/\tau_n\}$ where X_i ($1 \leq i \leq n$) are distinct variables and τ_i are types. Each schema individual variable has been assigned a type variable.

3.2 Abstract Data Types

Unfortunately, programming languages do not have facilities for direct modelling of all real world problems. The problem of expressing in a programming language a situation of the real world can be split in two subproblems:

1. Construct a model of the real world situation.
2. Implement that model using the facilities of a programming language.

The relationships of the real world objects are complex and their modelling requires a powerful language. Mathematics seems to be the most appropriate. Once the modelling has been completed, it can be implemented using the facilities of a programming language. An *abstract data type or data type specification* is a representation-independent specification of data types. Each data type is associated with a set of operations which are defined in terms of axioms that any representation of the type should satisfy. A mathematical model of a real world problem can be constructed by using abstract data types. They provide a valuable way to express real world entities and to organize their relationships.

We have implemented several abstract data types such as *sets, ordered sets, stacks, graphs, cartesian product etc.* The operations have been implemented as predicates.

3.3 Modes

The *mode* of a predicate in a logic program indicates how its arguments will be instantiated when that predicate is called. The mode of arguments of a predicate can be seen in two possible states, that is before the call and after the success of the predicate. The mode of an argument in these two states is called *call-mode* and *success-mode* respectively. Call modes are meaningful in a logic program only when the control strategy has been specified. The program schemata in this approach have been constructed with the left-to-right depth-first computation rule of Prolog in mind. The modes i and d are used. They stand for "input" and "don't know" respectively. i identifies the arguments which are expected to be ground. These terms are characterized as "input". An argument with any instantiation can be assigned mode d. Each schema individual variable has been assigned an "intended" mode of use. Modes in this approach are used to guide the application of schemata to undefined predicates and in the refinement of undefined predicates by ADT operations or equality.

Given a partial design and a moded initial goal, a mode inference procedure computes call and success modes as far as possible. These inferred modes are used to check the modes of calls to ADT predicates for compatibility with the intended modes. Our mode inference procedure is an abstract interpretation similar in many respects to other mode inferrence procedures such that in [4]. It is beyond the scope of this paper to describe it in detail.

4 Program Design Schemata

Algorithm design techniques are important in constructing programs [1], [13]. It has been observed that different problems are solved by using the same algorithm. If we take a collection of solution programs for such problems and factor out the differences relating to the problem representation, we obtain the algorithm common to all of them. We call such a problem-independent algorithm a *program design schema*, or *schema*. Such design schemata contain a fixed set of subparts or subgoals, that have different realisations/instantiations for each problem that they are applied to.

4.1 The Design Schema Language

In this section we consider a small set of program design schemata that we find useful for constructing logic programs. Their classification is based on the design strategy that each schema represents.

The schemata are formally defined as expressions in a meta-language of first-order logic. The meaning of the symbols of the meta-language is the following.

- the capital letters X, Y, Z, W, U, V, T, S are schema individual variables. They range over object-language variables or tuples of such variables.
- Other capitalized words are predicate variables. They range over predicate symbols.
- The meta-language logical symbols $\land$, $\neg$ and $\leftarrow$ have their usual meaning.

For each schema individual variable its type and mode are shown as well. For example, (X^τ, i) stands for the schema individual variable X whose type is τ and mode i. In the following, τ_1, τ_2, τ_3 stand for type variables.

4.2 Five Design Schemata

We consider schemata named as follows, and discuss each in turn.

1. Incremental
2. Subgoal
3. Divide-and-conquer
4. Case
5. Backtracking

The predicate variables are given names that indicate their intended meaning.

Incremental. The *Incremental* schema processes one by one each piece of the input data and composes the results for each one to construct the solution. The schema is as follows.

$$Incr((X^{\tau_1}, \mathbf{i}), (Y^{\tau_2}, \mathbf{d})) \leftarrow$$
$$Terminating((X^{\tau_1}, \mathbf{i})) \wedge$$
$$Initial\_result((X^{\tau_1}, \mathbf{i}), (Y^{\tau_2}, \mathbf{d}))$$

$$Incr((X^{\tau_1}, \mathbf{i}), (Y^{\tau_2}, \mathbf{d})) \leftarrow$$
$$\neg Terminating((X^{\tau_1}, \mathbf{i})) \wedge$$
$$Elem\_by\_elem((X^{\tau_1}, \mathbf{i}), (V^{\tau_3}, \mathbf{d}), (W^{\tau_1}, \mathbf{d})) \wedge$$
$$Incr((W^{\tau_1}, \mathbf{i}), (Z^{\tau_2}, \mathbf{d})) \wedge$$
$$Non\_initial\_result((X^{\tau_1}, \mathbf{i}), (V^{\tau_3}, \mathbf{i}), (Z^{\tau_2}, \mathbf{i}), (Y^{\tau_2}, \mathbf{d}))$$

Subgoal. The *Subgoal* schema reduces the problem into a sequence of two sub-problems. The solutions of the simpler problems are composed into a solution of the original problem. This is, an *AND* reduction. Note that V could be empty as well. The schema is as follows.

$$SubGoal((X^{\tau_1}, \mathbf{i}), (Y^{\tau_2}, \mathbf{d})) \leftarrow$$
$$SubGoal1((X^{\tau_1}, \mathbf{i}), (V^{\tau_3}, \mathbf{d}), (Y^{\tau_2}, \mathbf{d}))$$
$$SubGoal2((X^{\tau_1}, \mathbf{i}), (V^{\tau_3}, \mathbf{i}), (Y^{\tau_2}, \mathbf{d}))$$

Divide-and-conquer. The *Divide-and-conquer* schema decomposes the problem in two subproblems. The solution of the problem is constructed by composing the solutions of the subproblems together. The schema is of the following form.

$$DivConq((X^{\tau_1}, \mathbf{i}), (Y^{\tau_2}, \mathbf{d})) \leftarrow$$
$$Terminating((X^{\tau_1}, \mathbf{i})) \wedge Initial\_result((X^{\tau_1}, \mathbf{i}), (Y^{\tau_2}, \mathbf{d}))$$
$$DivConq((X^{\tau_1}, \mathbf{i}), (Y^{\tau_2}, \mathbf{d})) \leftarrow$$
$$\neg Terminating((X^{\tau_1}, \mathbf{i})) \wedge$$
$$Decomposition((X^{\tau_1}, \mathbf{i}), (W^{\tau_1}, \mathbf{d}), (Z^{\tau_1}, \mathbf{d})) \wedge$$
$$DivConq((W^{\tau_1}, \mathbf{i}), (U^{\tau_2}, \mathbf{d})) \wedge DivConq((Z^{\tau_1}, \mathbf{i}), (V^{\tau_2}, \mathbf{d})) \wedge$$
$$Composition((U^{\tau_2}, \mathbf{i}), (V^{\tau_2}, \mathbf{i}), (Y^{\tau_2}, \mathbf{d}))$$

Case. The *Case* schema reduces the problem in two independent subproblems. Each subproblem corresponds to a different case of the original problem. This is, an *OR* reduction. The schema is as follows.

$$Case((X^{\tau_1}, i), (Y^{\tau_2}, d)) \leftarrow$$
$$Case1((X^{\tau_1}, i), (Y^{\tau_2}, d))$$
$$Case((X^{\tau_1}, i), (Y^{\tau_2}, d)) \leftarrow$$
$$Case2((X^{\tau_1}, i), (Y^{\tau_2}, d))$$

Backtracking. The *Backtracking* schema does an organized exhaustive search in the solution space. Solutions are constructed in a stepwise fashion. Note that the solution returned may be composed from the complete search path. The schema is as follows.

$$Backtr((X^{\tau_1}, i), (Y^{\tau_2}, d)) \leftarrow$$
$$Initial\_search\_node((X^{\tau_1}, i), (W^{stack(\tau_3)}, d), (U^{\tau_2}, d)) \wedge$$
$$Backtr1((X^{\tau_1}, i), (W^{stack(\tau_3)}, i), (U^{\tau_2}, i), (Y^{\tau_2}, d))$$

$$Backtr1((X^{\tau_1}, i), (W^{stack(\tau_3)}, i), (U^{\tau_2}, i), (Y^{\tau_2}, d)) \leftarrow$$
$$Solution\_complete((W^{stack(\tau_3)}, i), (U^{\tau_2}, i)) \wedge$$
$$Solution\_assignment((U^{\tau_2}, i), (Y^{\tau_2}, d)$$
$$Backtr1((X^{\tau_1}, i), (W^{stack(\tau_3)}, i), (U^{\tau_2}, i), (Y^{\tau_2}, d)) \leftarrow$$
$$\neg Solution\_complete((W^{stack(\tau_3)}, i), (U^{\tau_2}, i)) \wedge$$
$$Forward\_move((X^{\tau_1}, i), (W^{stack(\tau_3)}, i), (U^{\tau_2}, i),$$
$$(Z^{stack(\tau_3)}, d), (V^{\tau_2}, d)) \wedge$$
$$Backtr1((X^{\tau_1}, i), (Z^{stack(\tau_3)}, i), (V^{\tau_2}, i), (Y^{\tau_2}, d))$$
$$Backtr1((X^{\tau_1}, i), (W^{stack(\tau_3)}, i), (U^{\tau_2}, i), (Y^{\tau_2}, d)) \leftarrow$$
$$\neg Solution\_complete((W^{stack(\tau_3)}, i), (U^{\tau_2}, i)) \wedge$$
$$\neg Forward\_move((X^{\tau_1}, i), (W^{stack(\tau_3)}, i), (U^{\tau_2}, i),$$
$$(Z^{stack(\tau_3)}, d), (V^{\tau_2}, d)) \wedge$$
$$Backtracking((W^{stack(\tau_3)}, i), (U^{\tau_2}, i),$$
$$(T^{stack(\tau_3)}, d), (S^{\tau_2}, d)) \wedge$$
$$Backtr1((X^{\tau_1}, i), (T^{stack(\tau_3)}, i), (S^{\tau_2}, i), (Y^{\tau_2}, d))$$

5 Schema Instantiation

A *program design schema* is instantiated with respect to an undefined predicate.

Let $p(X_1, .., X_n)$ be an undefined predicate with variable typing

$$\Omega = \{X_1/\tau_1, ..., X_n/\tau_n\}.$$

In the following, *mode(X)* stands for the mode of argument X. Let

$$\overline{I} = (Y_1, .., Y_{n1}) : mode(Y_j) = i, \; Y_j \in \{X_1, .., X_n\}$$

$$\overline{D} = (Z_1, .., Z_{n2}) : mode(Z_j) = d, \ Z_j \in \{X_1, .., X_n\}$$

$\{Y_1, .., Y_{n1}\} \cup \{Z_1, .., Z_{n2}\} = \{X_1, .., X_n\}$. The variables in the tuples $\overline{I}$ and $\overline{D}$ respectively have the same order that they have in $p(X_1, .., X_n)$.

Tupling is an operation that is applied on the undefined predicate p/n as follows:

$$tupling(p(X_1, .., X_n)) = p(\overline{I}, \overline{D})$$

Let $\{C_1, .., C_r\}$ be a variant of schema Σ. Each clause C_i $(1 \leq i \leq r)$ has the form $H_i \leftarrow B_i$, where H_i is an atom and B_i is a conjunction of literals.

A *schema substitution* Ψ_i for clause C_i of Σ with respect to $p(X_1, .., X_n)$ is defined as the composition of the following substitutions.

$$\Psi_i = \Theta_i \circ \Delta_i$$

where

- $\Theta_i = mgu(p(\overline{I}, \overline{D}), H_i)$. That is, Θ_i is a substitution of schema predicate variables and individual variables that occur in H_i.
- Δ_i is a substitution of the schema predicate variables in B_i that do not occur in H_i.

Schema predicate variables are substituted by fresh predicate symbols that have not been used before. The instance I' of Σ which defines $p(\overline{I}, \overline{D})$ is defined to be

$$I' = \{C_1\Psi_1, .., C_r\Psi_r\} = \{C'_1, ..., C'_r\}$$

Let A be an atom and Ω a variable typing of A. $A\Omega$ denotes the application of variable typing Ω on the atom A. Similarly let C be a clause and Ξ a variable typing of C. $C\Xi$ denotes the application of variable typing Ξ on the clause C.

Let Ξ_i be the variable typing of clause C_i in schema Σ. Let

$$\phi_i^\tau = mgu(p(\overline{I}, \overline{D}) \ \Omega, H_i \ \Xi_i)$$

Then the *derived variable typing* for C'_i $(1 \leq i \leq r)$ is defined to be

$$\Pi_i = \{X/\sigma\phi_i^\tau : X/\sigma \in \Omega \cup \Xi_i\}$$

Any variable X in C'_i of I' with variable typing of the form $(\tau_1, .., \tau_k)$ is replaced by the tuple of variables $(X_1, .., X_k)$. The variable typing Π_i of C'_i is extended by such variable typings, i.e. $\{X_1/\tau_1, .., X_k/\tau_k\}$. The new variable typing Π'_i for C'_i consists from the union of Π_i and such variable typings.

detupling(A) where A is an atom is defined to be detupling(A) = $p(X_1, .., X_r)$ where $\{X_1, .., X_r\} = vars(A)$ and $X_1, .., X_r$ have the same order as in A.

A detupling operation on clause C of the form $H \leftarrow L_1, .., L_l$ is defined to be

detupling$(H \leftarrow L_1, .., L_l)$ = detupling(H) $\leftarrow$ detupling(L_1),..,detupling(L_l)
detupling(L_i) = $\neg$ detupling(A_i), $(1 \leq i \leq l)$

where A_i is the atom of L_i.

The instance I of schema Σ that defines the predicate $p(X_1, .., X_n)$ is defined to be

$$I = \{\text{detupling}(C_1'),..,\text{detupling}(C_r')\}$$

Example:

Suppose that the *Incremental* schema Σ is applied to the undefined predicate p(X1, X2, X3). The variable typing of p(X1, X2, X3) is $\Omega = \{X1/\text{set(str)}, X2/\tau, X3/\text{orderedSet(str)}\}$ where τ is a type variable, and the "intended" modes of its arguments are mode(X1) = i, mode(X2) = i, mode(X3) = d.

$$tupling(\text{p(X1, X2, X3)} = \text{p((X1,X2), (X3))} = p(\overline{I},\overline{D})$$

Assuming that a renamed substitution for *Incremental* schema is

$$\{\ C_1, C_2\ \} =$$
$$\{\quad Incr(X^{\tau_1}, Y^{\tau_2})\ \leftarrow$$
$$\qquad\qquad Terminating(X^{\tau_1}) \wedge Initial\_result(X^{\tau_1}, Y^{\tau_2})$$

$$Incr(X^{\tau_1}, Y^{\tau_2}) \leftarrow$$
$$\qquad\qquad \neg Terminating(X^{\tau_1}) \wedge Elem\_by\_elem(X^{\tau_1}, V^{\tau_3}, W^{\tau_1}) \wedge$$
$$\qquad\qquad Incr(W^{\tau_1}, Z^{\tau_2}) \wedge Non\_initial\_result(X^{\tau_1}, V^{\tau_3}, Z^{\tau_2}, Y^{\tau_2})\ \}$$

where, τ_1, τ_2, τ_3 are the types corresponding to each schema individual variable. The variable typing for C_1 is $\Xi_1 = \{X/\tau_1, Y/\tau_2\}$. The variable typing for C_2 is $\Xi_2 = \{X/\tau_1, Y/\tau_2, V/\tau_3, W/\tau_1, Z/\tau_2\}$

The *schema substitution* $\Psi_i = \Theta_i \circ \Delta_i$ $i \leftarrow \{1, 2\}$ with respect to $p(\overline{I},\overline{D})$ is the following

$$\Theta_1 = \Theta_2 = \text{mgu}(p(\overline{I},\overline{D}), Incr(X,Y)) = \{\ Incr/p,\ X/(X1,X2),\ Y/(X3)\ \}$$
$$\Delta_1 = \{\ Terminating/q1,\ Initial\_result/q2\ \}$$
$$\Delta_2 = \{\ Terminating/q1,\ Elem\_by\_elem/q3,\ Non\_initial\_result/q4\ \}$$

$$I' = \{C_1\Theta_1\Delta_1, C_2\Theta_2\Delta_2\} = \{C_1', C_2'\}$$
$$= \{\text{p((X1,X2), (X3))} \leftarrow$$
$$\qquad\qquad \text{q1((X1,X2))} \wedge$$
$$\qquad\qquad \text{q2((X1,X2), (X3))}$$

$$\qquad \text{p((X1,X2), (X3))} \leftarrow$$
$$\qquad\qquad \neg\ \text{q1((X1,X2))} \wedge$$
$$\qquad\qquad \text{q3((X1,X2), V, W)} \wedge$$
$$\qquad\qquad \text{p(W, Z)} \wedge$$
$$\qquad\qquad \text{q4((X1,X2), V, Z, (X3))}\ \}$$

Let Φ_1^τ and Φ_2^τ be two type substitutions such that

$$\Phi_1^\tau = \text{mgu}((p(\overline{I},\overline{D})\ \Omega), (\text{Incr(X,Y)}\ \Xi_1)) =$$
$$\{\ \tau_1/(\text{set(str)},\tau),\ \tau_2/(\text{orderedSet(str)})\ \}$$
$$\Phi_1^\tau = \Phi_2^\tau = \text{mgu}((p(\overline{I},\overline{D})\ \Omega), (\text{Incr(X,Y)}\ \Xi_2)) =$$
$$\{\ \tau_1/(\text{set(str)},\tau),\ \tau_2/(\text{orderedSet(str)})\ \}$$

The variable typings of the clauses $\{C_1\Theta_1\Delta_1, C_2\Theta_2\Delta_2\}$ are the following

$$\Pi_1 = \{\Omega\Phi_1^\tau\} \cup \{\Xi_1\Phi_1^\tau\}$$
$$\Pi_2 = \{\Omega\Phi_2^\tau\} \cup \{\Xi_2\Phi_2^\tau\}$$

$$\Pi_1 = \{\ \text{X1/set(str), X2/}\tau\text{, X3/orderedSet(str),}$$
$$\text{X/(set(str),}\tau\text{), Y/(orderedSet(str))}\}$$
$$\Pi_2 = \{\ \text{X1/set(str), X2/}\tau\text{, X3/orderedSet(str),}$$
$$\text{X/(set(str),}\tau\text{), Y/(orderedSet(str)),}$$
$$\text{V/}\tau_3\text{, W/(set(str),}\tau\text{), Z/(orderedSet(str))}\}$$

$$I\ = \{\ \text{detupling}(C_1'),\ \text{detupling}(C_2')\ \} =$$
$$= \{\text{p(X1, X2, X3)} \leftarrow$$

$$\text{q1(X1, X2)} \wedge$$
$$\text{q2(X1, X2, X3),}$$

$$\text{p(X1, X2, X3)} \leftarrow$$
$$\neg\ \text{q1(X1, X2)} \wedge$$
$$\text{q3(X1, X2, V, W1, W2)} \wedge$$
$$\text{p(W1, W2, Z1)} \wedge$$
$$\text{q4(X1, X2, V, Z1, X3)}\ \}$$

$$\Pi_1' = \Pi_1 \cup \{\}$$
$$\Pi_2' = \Pi_2 \cup \{\text{W1/set(str), W2/}\tau\text{, Z1/orderedSet(str)}\}$$

6 Refinement by ADT or Equality Predicates

The definition of an undefined predicate $p(X_1, ..X_n)$ in terms of an ADT or equality predicate will create a new clause of either form

$$p(X_1, ..X_n) \leftarrow R(Y_1, .., Y_m)\ m \le n$$

$$p(X_1, ..X_n) \leftarrow \neg R(Y_1, .., Y_m)\ m \le n$$

where R is an ADT operation predicate symbol or the equality symbol and $\{Y_1, ..Y_m\} \subseteq \{X_1, ..X_n\}$. The variables in the clause are typed according to the declared types of ADTs.

The call-success modes generated by the mode inference procedure are used to validate the call modes of $R(Y_1, .., Y_m)$. That is, the call mode of $R(Y_1, .., Y_m)$ is subsumed by its intended mode.

7   The Design Process

Initially, the programmer gives a specification of the predicate that he wants
to construct. That specification consists of the *data type* and the *mode* of each
argument. Goals are expected to have ground the arguments with mode i, arguments with mode d can be either ground or nonground or partially ground. The
initial refinement is applied to that predicate. The next refinements are applied
to undefined subgoals of the clauses created. Each refinement step involves either
of the following actions.

- The application of a design schema.
- The definition of the predicate being refined in terms of an ADT or equality
    predicate or its negation.

The construction process is a successive top-down application of refinements
until the complete construction of an executable design of the system. A design
is considered to be complete when all of its predicates are defined.

  The above discussion is illustrated by applying an example with five refinements. That is, three applications of design schemata and two predicate definitions in terms of ADT predicates. The effect of the application of each schema
will be shown, not the instantiation details of the schema.

1. *Predicate specification*: $p((I,i,set(\tau)), (D,d,orderedSet(\tau)))$
2. Refinement 1 applied on p/2: *Incremental schema*.

```
p(I, D) :-
      q1(I),
      q2(I, D).
p(I, D) :-
      not q1(I),
      q3(I, X, RI),
      p(RI, RD),
      q4(I, X, RD, D).
```

3. Refinement 2 applied on q2/2: *Subgoal schema*.

```
q2(I, D) :-
      t1(I, Y, D),
      t2(I, Y, D).
```

4. Refinement 3 applied on q3/3: *Subgoal schema*.

```
q3(I, X, RI) :-
      r1(I, X, RI),
      r2(I, X, RI).
```

5. Refinement 4 applied on r1/3: *Abstract data type operation get_elem/2*.

```
r1(I, X, RI) :- get_elem(I, X).
```

6. Refinement 5 applied on r2/3: *Abstract data type operation delete/3*

```
r2(I, X, RI) :- delete(I, X, RI).
```

7.1 Examples

Insertion sort. insert_sort(S1, S2) is true if S2 is the set S1 with its elements sorted.

- *Predicate specification:* insert_sort((S1,i,set(τ)), (S2,d,orderedSet(τ)))

Refinements.
Schema: Incremental

```
insert_sort(S1, S2) :-
   terminating(S1),
   construct_initial_result(S1, S2).
insert_sort(S1, S2) :-
   \+ terminating(S1),
   get_elem_incr(S1, Elem, RestS1),
   insert_sort(RestS1, RestS2),
   construct_non_initial_result(S1, Elem, RestS2, S2).

terminating(S1) :- empty_set(S1).

construct_initial_result(S1, S2) :- empty_ordSet(S2).
```

Schema: Subgoal

```
get_elem_incr(S1, Elem, RestS1) :-
   get_element(S1, Elem, RestS1),
   delete_element(S1, Elem, RestS1).

construct_non_initial_result(S1, Elem, RestS2, S2) :-
   set_elem_ordSet(Elem, RestS2, S2).

get_element(S1, Elem, RestS1) :- get_elem(S1, Elem).

delete_element(S1, Elem, RestS1) :- delete(Elem, S1, RestS1).
```

Abstract Data Type Operations.
ADT operations for unordered sets

```
- empty_set(S)
- delete(X, S1, S2)
- get_elem(S, Elem)
```

ADT operations for ordered sets

```
- empty_ordSet(S)
- set_elem_ordSet(X, S1, S2)
```

8 Conclusions and Further Work

Five novel features in the development of logic programs are introduced by this appoach. First, the problem domain representation, i.e. the abstract data types, are separated from the structural or control part of the algorithm, i.e. the program schemata. Second, the set of program schemata consists of only five schemata, making them manageable by humans. Third, the application of a program schema to an undefined predicate is directed by modes. Fourth, the constructed program designs are type correct. Fifth, static analysis is performed dynamically during program construction.

So far, we find this small set of schemata surprisingly expressive. Larger examples of constructed programs, e.g. depth-first traversal of a graph, can be found in [15]. Related work in functional programming languages has been followed in [20]. One strategy, i.e. divide and conquer, is used in [20] for automatically constructing programs from formal specifications.

A set of support tools for program development using these ideas is under development. The question of heuristics for selecting the appropriate schema or ADT operation at each refinement step is also a subject that requires study.

The transformation of the executable design into a more efficient program will also be investigated. The fact that executable logic programs are produced means that state of the art transformation methods can be brought to bear. Techniques such as partial evaluation have already been applied to removing the overhead of ADTs [14]. Other deeper transformations include those based on unfold/fold and coroutining between goals.

Our approach produces type-correct designs, and the role of ADTs is important for this. On the question of verification in general, we have not assumed the presence of a full formal specification of the problem (that is, a detailed formalisation of the input/output relation of the program). We have considered only questions of algorithm design and have assumed that some aspects of specification are subsumed in the design process. However, higher level specification languages that can be mapped to the schemata could be investigated.

References

1. A. Aho, J. Hopcroft and J. Ullman, *Data Structures and Algorithms*, Addison-Wesley, 1983.
2. D. Barker-Plummer, *Cliché Programming in Prolog*, Proceedings of the Second Workshop on Meta-programming in Logic, (ed. M. Bruynooghe); April 1990, Belgium, pp.247-256.
3. M. Bruynooghe, *Adding Redundancy to Obtain More Reliable and More Readable Prolog Programs*, Proceedings of the First International Logic Programming Conference, Sept. 1982, Marseille, France, pp.129-133.
4. S. Debray, D. Warren, *Automatic Mode Inference for Prolog Programs*, Proceedings of the Symposium on Logic Programming, Salt Lake City, Utah, Sept. 1986, pp.78-88.

5. Y. Deville, J. Burnay, *Generalization and Program Schemata: A Step Towards Computer-Aided Construction of Logic Programs*, Proc. of the North American Conference on Logic Programming, 1989, (eds. E. Lusk and R. Overbeek); pp.409-425, vol.I,.

6. Y. Deville, *Logic Programming: Systematic Program Development*, Addison-Wesley, 1990.

7. R. Dietrich, F. Hagl, *A Polymorphic Type System with Subtypes for Prolog*, ESOP'88 Proceedings of the 2nd European Symposium on Programming, Nancy, France 1988, Lecture Notes in Comp. Sc. no 300, pp.79-93.

8. Pierre Flener, Yves Deville, *Towards Stepwise, Schema-Guided Synthesis of Logic Programs*, Proceedings of International Workshop on Logic Program Synthesis and Transformation, LOPSTR'91, Manchester, U.K., 1991, pp.46-64.

9. T. Gegg-Harrison, *Basic Prolog Schemata*, Technical Report, CS-1989-20, Dept. of Computer Science, Duke University, Durham, North Carolina, Sept. 1989.

10. P. Hill, and R. Topor, *A Semantics for Typed Logic Programs*, Chapter 1 in *Types in Logic Programming* edited by F. Pfenning, The MIT Press, 1992.

11. P. Hill, J. Lloyd, *The Gödel Programming Language*, The MIT Press, 1994.

12. A. Lakhotia, *A Workbench for Developing Logic Programs by Stepwise Enhancement*, PhD thesis, Case Western Reverse University, Department of Computer Engineering and Science, Aug. 1989.

13. J. Kingston, *Algorithms and Data Structures, Design, Correctness, Analysis*, Addison-Wesley, 1990.

14. J. Komorowski, Towards a Programming Methodology Founded on Partial Deduction, Proc. of the European Conference on Artificial Intelligence, Stockholm, Sweden, August 1990.

15. E. Marakakis, J.P. Gallagher, *Schema-Based Top-Down Design of Logic Programs Using Abstract Data Types*, Technical Report, CSTR-94-02, University of Bristol, Department of Computer Science.

16. C. Mellish, *Some Global Optimizations for a Prolog Compiler*, Journal of Logic Programming, 1, 1985, pp.43-66.

17. A. Mycroft, R. O'Keefe, *A Polymorphic System for Prolog*, Artificial Intelligence, 23, 1984, pp. 295-307.

18. G. Plotkin, *A Note on Inductive Generalization*, Machine Intelligence, 5, 1970, pp.153-163.

19. D. Robertson, *A Simple Prolog Techniques Editor for Novice Users*, Proceedings of the 3rd Annual Conference on Logic Programming, Edinburgh, April 1991, pp.78-85.

20. D. Smith, *Top-Down Synthesis of Divide-and-Conquer Algorithms*, Artificial Intelligence, vol. 27, 1985, pp.43-96.

21. L. Sterling, M. Kirschenbaum, *Applying Techniques to Skeletons*, Chapter 6 in *Constructing Logic Programs* edited by Jean-Marie Jacquet, Wiley, 1993.

22. N. Tinkham, *Induction of Schemata for Program Synthesis*, PhD thesis, Department of Computer Science, Duke University, 1990.

23. N. Wirth, *Algorithms + Data Structures = Programs*, Prentice-Hall, 1976.

24. T. Yokomori, *Logic Program Forms*, New Generation Computing, No.4, 1986, pp.305-319.

Generalizing Extended Execution for Normal Programs

Sophie Renault

INRIA-Rocquencourt
BP 105, 78153 Le Chesnay, France
Sophie.Renault@inria.fr

Abstract

We present a set of inference rules aimed at proving declarative (logical) properties of normal programs. Proofs are goal directed and are performed by means of replacement, simplification and rewriting. This work can be seen as a generalization of *Extended Execution* [16] which is in turn an extension of the prolog interpreter. We show the soundness of our generalization and discuss its completeness. Two extensive examples are given. We conclude on the relevance of our approach within the general framework of verification.

1 Introduction

Checking the correctness of a logic program w.r.t. a specification has been considered from various angles, each of them suggesting its own criteria for an adequate semantics and an appropriate definition of correctness and specification.[1]

In this paper our approach of correctness is to prove that a formula Φ, expected to express a desirable property of a program P, is a logical consequence of its completion [5]. P is a normal program (definite clauses with possibly negative literals in their bodies). Φ is any (closed) first-order formula and may be seen as a partial specification for P. A successful proof is a sequence of goals from Φ to *true*. Our framework is named EE_G (*Extended Execution Generalized*) since it is a generalization of the one proposed by Kanamori et al. in [16, 15] for definite programs, namely *Extended Execution* (EE).

EE is based on an extension of the Prolog interpreter for more general queries named *S-formulas* that mainly consist of implicative formulas of the form $\forall \overline{x}(A \rightarrow \exists \overline{y}B)$, where $\overline{x}$ and $\overline{y}$ denote two sequences of variables and A and B denote conjunctions of atoms whose variables are in $\overline{x}$ and $\overline{x}$ or $\overline{y}$ respectively. EE is complete for the class of S-formulas [15]. Notice that they are closed formulas (no free variable). Therefore EE is aimed at proving and not at computing answers, at the contrary of SLDNF. A derivation step in EE involves unfolding or simplification over literals occurring in the goal.

In EE_G the class of S-formulas is extended to any first-order (closed) formula.[2] The main novelty is the acceptance of queries such as $\exists x \neg p(x)$ for which we need

---

[1] [9] and [7] provide a deep survey and a comparison between several proposals.

[2] In general, a normal clause is not an S-formula.

to refer explicitly to the definitions of the completion and therefore to handle the equality relation. The unfolding is now performed by using the iff definitions of the predicates.

The paper is organized as follows: In section 2 we introduce some notation and recall some basic definitions. In section 3 we informally describe EE_G and give an intuition of the use of its rules by an example. Sections 4, 5 and 6 are devoted to a formal description of each rule. In section 7 we show the soundness of EE_G, while its completeness is studied in section 8. Section 9 provides two extensive examples. Finally, section 10 is devoted to a discussion and a comparison with related works.

2 Preliminaries

For the notations, we choose the following conventions. We will use lower case letters possibly indiced and primed to denote individual objects: x, y for variables, t, u for terms, p, q, r for predicate symbols, a, b, f for function symbols, σ, θ for substitutions. Upper case letters will be preferred for collection of objects: P for programs, G, F, W, D, Q, E, L for first-order formulas. These symbols may also be overlined in order to denote a sequence: $\overline{x}$ denotes a sequence of distinct variables, $\overline{t}$ a sequence of terms (not necessarily distinct), $\overline{t} = \overline{u}$ will stand for $\wedge_{i=1}^n (t_i = u_i)$ when $\overline{t} = (t_1, \ldots, t_n)$ and $\overline{u} = (u_1, \ldots, u_n)$. $\overline{x}.\overline{y}$ will denote the concatenation of $\overline{x}$ and $\overline{y}$. $vars(F)$ will denote the sequence of free variables occurring in a term or a formula F.

A replacement of one (resp. every) occurrence of F_1 by F_2 in G will be denoted by $G[F_1 \leftarrow F_2]$ (resp. $G(F_1 \leftarrow F_2)$). A replacement of every occurrence of a variable x in G by a term t will be denoted by $G(x \leftarrow t)$.

An *atom* has the form $p(\overline{t})$. A *literal* is an atom (*positive literal*) or a negated atom $\neg p(\overline{t})$ (*negative literal*).

A *normal clause* has the form $p(\overline{t}) \leftarrow W$ where W is a conjunction of literals. When W is a conjunction of atoms then the clause is a *definite clause*. A *normal program* (resp. *definite program*) P is a finite set of normal (resp. definite) clauses. From now on the qualifier 'normal' will be omitted, unless some confusion arises.

The *polarity* of a subformula is defined as usual: A subformula F has a *positive polarity* (resp. *negative polarity*) in G (or F is a *positive subformula* (resp. *negative subformula*) of G) when F appears in G within an even number of negation (explicitly, or implicitly in the left-hand-side of an implication).

The *scope* of a quantifier $\forall x$ (resp. $\exists x$) in a formula $\forall x F$ (resp. $\exists x F$) is F.

A *substitution* σ is a function from variables to terms with a finite domain. Its domain will be denoted $dom(\sigma)$. $\sigma(x_i)$ will denote the image of x_i by σ. Let $\{x_1, \ldots, x_n\}$ be the domain of σ, then the *range* of σ is the set of terms $\{\sigma(x_1), \ldots, \sigma(x_n)\}$.

P^* will denote Clark's completion [5] of a program. It consists of the set of the iff definitions of every relation occurring in P and the CET (Clark's Equality

Theory).[3] If $\forall \overline{x} \; (p(\overline{x}) \leftrightarrow def_p(\overline{x}))$ is the iff definition of a predicate p then the *completed definition of atom $p(\overline{u})$ is the formula $def_p(\overline{x})$ in which $\overline{u}$ replaces $\overline{x}$.*

3 General description of EE$_G$

3.1 Normal goals

The formula Φ to be proved is assumed to be initially a *normal goal G*:

Definition 1 normal goal. A normal goal G is a closed formula (no free variable) that can be generated by the following grammar:

$$G ::= Q \mid E$$
$$Q ::= Q \wedge Q \mid \forall \overline{x} \; E \mid L$$
$$E ::= E \vee E \mid \exists \overline{y} \; Q \mid L$$
$$L ::= p(\overline{u}) \mid \neg p(\overline{u}) \mid t = u \mid t \neq u \mid true \mid false$$

where $t \neq u$ stands for $\neg t = u$. Moreover we assume that a variable occurs exactly once in a quantification.

Example 1. $\forall x \; (\exists y_1 \; p(x, y_1) \vee \exists y_2 \; q(x, y_2))$ is a normal goal. It is not the case of the equivalent formula $\forall x \; (\exists y \; p(x, y) \vee \exists y \; q(x, y))$ in which $\exists y$ occurs twice. $\forall x \; (p(x, y) \vee \exists y \; q(x, y))$ is not a normal goal since it is not a closed formula (the first occurrence of y is free).

The following lemmas are immediate from definition 1.

Lemma 2. *Every closed first-order formula is equivalent to a normal goal. This normal goal is not unique.*

Proof: It can be easily verified by firstly converting the formula into an equivalent prenex normal form. The non uniqueness is illustrated by the two logically equivalent normal goals $\forall x (\exists y \; (p(x) \wedge q(y)) \vee r(x))$ and $\forall x \; (p(x) \vee r(x)) \wedge \forall x'(\exists y \; q(y) \vee r(x'))$.

Lemma 3. *A variable in a normal goal occurs only in the scope of its quantification.*

Lemma 4. *F is a negative subformula of a normal goal G iff it is a negated atom $p(\overline{u})$.*

---

[3] See [26] for a deep survey on the CET and the results related to its completeness and decidability.

3.2 Derivation and derivation step

EE_G consists mainly of the *Replacement* and the *Simplification* rules. A successful proof in EE_G is a sequence of derivation steps that derives a normal goal G into *true*. A derivation step replaces a normal goal G_i by a new one G_{i+1}. G_{i+1} is produced from G_i by an application of the Replacement or the Simplification rule and a normalization process. More precisely, an application of one of the two rules yields an intermediary formula which is immediately subject to normalization. G_{i+1} is the normalized form of the intermediary formula.

3.3 An example

The following program $\{r(a).\ r(b) \leftarrow r(b).\ r(b) \leftarrow q(a).\ q(a) \leftarrow q(a).\}$ was first given in [1]. Its completion P^* consists of the two axioms:

$$\forall x\ (r(x) \leftrightarrow x = a \ \lor \ (x = b \land r(b)) \ \lor \ (x = b \land q(a)))$$
$$\forall x\ (q(x) \leftrightarrow x = a \ \land \ q(a))$$

Let G_0 be $\exists x(r(x) \land \neg q(x))$. Then we have $P^* \models G_0$. Actually, it is easy to see that $P^* \models r(a)$ and $P^* \models \neg q(b)$. Then, in any (two-valued) model of P^*, $q(a)$ is always *true* or *false*. If it is *true* so must be $r(b)$ and then $r(b) \land \neg q(b)$ is *true*. If it is false then $r(a) \land \neg q(a)$ is true. Finally, either $(r(a) \land \neg q(a))$ or $(r(b) \land \neg q(b))$ is *true* in any model of P^*.

Nevertheless, there is no successful SLDNF-derivation for the goal $r(x) \land \neg q(x)$.[4] EE_G for its part, provides a proof for G_0.[5] It is summarized in the following proof tree:

$$G_0 : \ \exists x\ (r(x) \land \neg q(x))$$

Replacement of $r(x)$ and normalization

$$\underbrace{\neg q(a) \lor r(b) \lor q(a)}_{G_{10}} \land \underbrace{\neg q(a) \lor r(b) \lor \neg q(b)}_{G_{20}} \land \underbrace{\neg q(a) \lor \neg q(b) \lor q(a)}_{G_{30}} \land \underbrace{\neg q(a) \lor \neg q(b) \lor \neg q(b)}_{G_{40}}$$

The replacement of $r(x)$ by its completed definition yields the formula $\exists x((x = a \lor (x = b \land r(b)) \lor (x = b \land q(a))) \land \neg q(x))$. It is transformed during normalization by distributivity of connectors and elimination of equalities. The conjunction of G_{10}, G_{20}, G_{30} and G_{40} is the result of the normalization. The proof is now divided into four subproofs. G_{10} and G_{30} can be derived into *true* by applying a simplification between $\neg q(a)$ and $q(a)$. G_{20} and G_{40} can be derived into *true* by a replacement of $q(b)$ (its completed definition is equivalent to false).

---

[4] That is why this example is used in [1] to illustrate the incompleteness of SLDNF even in the case of a definite program and a non floundering goal.

[5] The inability of SLDNF to provide a successful derivation for $(r(x) \land \neg q(x))$ comes from its attempt to instantiate x. This problem is avoided by EE_G since it is not aimed at computing answer substitutions to queries (they all are closed formulas).

Remark: Although P is a definite program, this proof cannot be performed by EE. The query $\exists x \, (r(x) \wedge \neg q(x))$ is not an S-formula.

The following three sections aim at a formal description of the rules involved in EE_G.

4 The Replacement rule

Definition 5 the Replacement rule. Let G be a normal goal and $p(\overline{u})$ an atom in G. An application of the Replacement rule to G and $p(\overline{u})$ yields:

$$G[p(\overline{u}) \leftarrow F]$$

where F is the completed definition of $p(\overline{u})$. The variables of F that do not occur in $p(\overline{u})$ must be introduced in G as new fresh variables.

Remark: The replacement of $p(\overline{u})$ can be seen as an unfolding using the iff definition of predicate p. This point of view is adopted by Sato in [24].

Lemma 6 Correctness of the Replacement rule. *If G_2 is a formula produced by an application of the Replacement rule to a normal goal G_1, then $P^* \models G_1 \leftrightarrow G_2$.*

Proof: The proof is obvious since this rule merely replaces an atom by its completed definition which is by construction of P^* an equivalent formula.

5 The Simplification rule

We need first the definition of a *deciding substitution* for a normal goal. Intuitively, this notion is comparable to the $\exists$-introduction rule in natural deduction: if G_{i+1} is the result of applying to G_i a deciding substitution $\sigma = \{x \leftarrow t\}$, then G_i can be restored from G_{i+1} by an $\exists$-introduction of variable x for term t. Some conditions on the variables involved in σ must be stated.

5.1 Deciding substitution

Definition 7 deciding substitution. Let G be a normal goal. A substitution σ is a deciding substitution for G if the following holds:

1. every variable in $dom(\sigma)$ is existentially quantified in G,
2. for every variable x in $dom(\sigma)$ and every variable y in $\sigma(x)$, then y is universally quantified in G and $\exists x$ is in the scope of $\forall y$.

Definition 8 deciding substitution instance. Let G be a normal goal and σ be a deciding substitution for G. Then the instance of G by σ is obtained by replacing every occurrence of x_i in $dom(\sigma)$ by $\sigma(x_i)$ and by removing all the quantifiers $\exists x_i$. It is denoted $G\sigma$.
Note that the empty substitution is a (trivial) deciding substitution.

Example 2. $\sigma = \{y_1 \leftarrow x \; ; \; y_2 \leftarrow f(x,x)\}$ is a deciding substitution for $G = \forall x \, \exists y_1 . y_2 \, (p(x,y_1) \wedge q(x,y_2))$ and $G\sigma$ is $\forall x \, (p(x,x) \wedge q(x,f(x,x)))$.
$\theta = \{x \leftarrow y\}$ is not a deciding substitution for $G' = \exists x \forall y \, p(x,y)$ since y lies in the scope of $\exists x$.

Lemma 9. *If $G\sigma$ is a deciding substitution instance for G then $P^* \models G$ if $P^* \models G\sigma$.*

Proof: It follows from the validity of the $\exists$-introduction rule in natural deduction. Stronger conditions on the variables involved in the substitution are required here in order to preserve, after instantiation, the property of bindings stated by the definition of a normal goal.

Lemma 10. *A deciding substitution instance of a normal goal is a normal goal.*

Proof: $G\sigma$ is G in which some variables have been replaced by terms and some irrelevant quantifiers have been removed. By definition σ ensures that every variable in the range of σ will appear in $G\sigma$ in the scope of its quantification. Notice that in example 2, $G\sigma$ is a normal goal with an empty sequence of existential variables. Such normal goals will be splitted during the normalization process (see the renaming rules in section 6.1).

5.2 The simplification rule

Definition 11 the Simplification rule. Let G be a normal goal. Let $p(\bar{u}_1)$ and $\neg p(\bar{u}_2)$ be respectively a positive and a negative literal[6] in G such that $p(\bar{u}_1)$ and $p(\bar{u}_2)$ are both unifiable to $p(\bar{u})$ by a deciding substitution σ for G. Then an application of the Simplification rule to G, $p(\bar{u}_1)$ and $p(\bar{u}_2)$ with σ yields two new goals G_1 and G_2, where G_1 and G_2 are obtained by replacing in $G\sigma$ every occurrence of $p(\bar{u})$ by *true* and *false* respectively, i.e.:

$$G_1 : (G\sigma)(p(\bar{u}) \leftarrow true)$$
$$G_2 : (G\sigma)(p(\bar{u}) \leftarrow false)$$

Lemma 12. *Let G be a normal goal and L be a literal in G. Then it exists in G a subformula F such that F contains all the occurrences of L in G and such that every variable in L is free in F.*

Proof: Let $x_1, \ldots, x_n$ be the n variables occurring in L. By lemma 3, for every quantifier Q_i, where Q_i stands for $\exists x_i$ or $\forall x_i$ with $i \in [1..n]$, it exists a subformula $Q_i F_i$ such that F_i contains all the occurrences of x_i in G, and consequently, all the occurrences of $p(\bar{u})$. It follows that for all i and j ($i \neq j$), either Q_i occurs in F_j or Q_j occurs in F_i. Hence, it exists a k ($k \in [1..n]$) and a subformula $Q_k F_k$ such that F_k contains all the occurrences of L and no occurrence of a Q_i (for $i \in [1..n]$), i.e. such that every variable in L is free in F_k. Finally F is F_k.

---

[6] We give here a simple formulation of the rule in the case where only two literals are selected.

Lemma 13. *Let $(G\sigma)(p(\overline{u}) \leftarrow true)$ and $(G\sigma)(p(\overline{u}) \leftarrow false)$ be the two formulas produced by an application of the Simplification rule to a normal goal G. Let F be a subformula of $G\sigma$ such that F contains all the occurrences of $p(\overline{u})$ and such that every variable of $p(\overline{u})$ is free in F. Then:*
$$P^* \models G\sigma[F \leftarrow (\neg p(\overline{u}) \vee F)] \text{ if } P^* \models G\sigma[F \leftarrow F(p(\overline{u}) \leftarrow true)], \text{ and}$$
$$P^* \models G\sigma[F \leftarrow (p(\overline{u}) \vee F)] \text{ if } P^* \models G\sigma[F \leftarrow F(p(\overline{u}) \leftarrow false)]$$

Proof: Firstly, the existence of F is ensured by lemma 12. We prove the first part of the lemma. By hypothesis $P^* \models G\sigma[F \leftarrow F(p(\overline{u}) \leftarrow true)]$. F is equivalent to a conjunctive prenex form $\overrightarrow{Q}$ $(\wedge_i D_i)$ where D_i are disjunctions of literals and $\overrightarrow{Q}$ is a vector of quantifiers. Hence $G\sigma[F \leftarrow F(p(\overline{u}) \leftarrow true)]$ is equivalent to

$G\sigma[F \leftarrow \overrightarrow{Q}$ $(\wedge_i D_i)(p(\overline{u}) \leftarrow true)]$, i.e. to $G\sigma[F \leftarrow \overrightarrow{Q}$ $(\wedge_i(D_i$ $(p(\overline{u}) \leftarrow true)))]$

The validity of this last formula implies the validity of $G\sigma[F \leftarrow \overrightarrow{Q}$ $(\neg p(\overline{u}) \vee (\wedge_i$ $(D_i(p(\overline{u}) \leftarrow true))))]$ (because F is a positive subformula of $G\sigma$ by lemma 4) which is equivalent (by de Morgan's laws) to:

$$G\sigma[F \leftarrow \overrightarrow{Q} \quad \wedge_i \quad (\neg p(\overline{u}) \vee D_i(p(\overline{u}) \leftarrow true))] \tag{1}$$

For all i,

1. if $p(\overline{u})$ is a positive literal of D_i then $\neg p(\overline{u}) \vee D_i(p(\overline{u}) \leftarrow true)$ and $\neg p(\overline{u}) \vee D_i$ are both equivalent to *true*.
2. otherwise, if $p(\overline{u})$ is a negated atom in D_i then $\neg p(\overline{u}) \vee D_i(p(\overline{u}) \leftarrow true)$ is also equivalent to $\neg p(\overline{u}) \vee D_i$.
3. otherwise $\neg p(\overline{u}) \vee D_i(p(\overline{u}) \leftarrow true)$ is trivially equivalent to $\neg p(\overline{u}) \vee D_i$.

Since the property holds for all i it follows from (1) the formula $G\sigma[F \leftarrow \overrightarrow{Q} \wedge_i (\neg p(\overline{u}) \vee D_i)]$ which is equivalent by de Morgan's laws to

$$G\sigma[F \leftarrow \overrightarrow{Q} (\neg p(\overline{u}) \vee (\wedge_i D_i))] \tag{2}$$

Finally, because no variable in $p(\overline{u})$ occurs in $\overrightarrow{Q}$, it follows from (2) the formula $G\sigma[F \leftarrow (\neg p(\overline{u}) \vee \overrightarrow{Q} (\wedge_i D_i))]$, i.e. $G\sigma[F \leftarrow (\neg p(\overline{u}) \vee F)]$ Hence, from the hypothesis we have inferred $P^* \models G\sigma[F \leftarrow (\neg p(\overline{u}) \vee F)]$.

Lemma 14 Correctness of the Simplification rule. *If G_1 and G_2 are two formulas produced by an application of the Simplification rule to a normal goal G, then $P^* \models G$ if $P^* \models G_1$ and $P^* \models G_2$.*

Proof: Let G_1 and G_2 be $G\sigma(p(\overline{u}) \leftarrow true)$ and $G\sigma(p(\overline{u}) \leftarrow false)$ respectively. By hypothesis, $P^* \models G_1$ and $P^* \models G_2$. Let F be a subformula in $G\sigma$ defined as in lemma 12. Then by lemma 13, and because G_1 and G_2 are equivalent to $G\sigma[F \leftarrow F(p(\overline{u}) \leftarrow true)]$ and $G\sigma[F \leftarrow F(p(\overline{u}) \leftarrow false)]$ respectively, we get $P^* \models G\sigma[F \leftarrow F \vee \neg p(\overline{u})]$ and $P^* \models G\sigma[F \leftarrow F \vee p(\overline{u})]$. It follows from these two statements that $P^* \models G\sigma[F \leftarrow (F \vee p(\overline{u})) \wedge (F \vee \neg p(\overline{u}))]$. Finally, because $(F \vee p(\overline{u})) \wedge (F \vee \neg p(\overline{u}))$ is equivalent to F, we get $P^* \models G\sigma$. Hence $P^* \models G$ (by lemma 9).

6 The normalization rules

The normalization process involves the *rewrite*, the *equality-decomposition* and the *equality-elimination* rules which apply to the subterms of a formula. Rules are given in their general form, i.e. we assume that F_i may also be empty. An empty conjunction and an empty disjunction are the boolean constants *true* and *false* respectively.

6.1 The rewrite rules

We omit here the rules for boolean simplification, for distributivity of negation over $\wedge$ and $\vee$ and for elimination of irrelevant quantification. The well-known de Morgan's laws are applied in a particular context to ensure the termination of the normalization.

de Morgan's laws

$$F_1 \vee (G_1 \wedge G_2) \vee F_2 \longrightarrow (F_1 \vee G_1 \vee F_2) \wedge (F_1 \vee G_2 \vee F_2)$$
$$\exists \overline{x} \, (F_1 \wedge (G_1 \vee G_2) \wedge F_2) \longrightarrow \exists \overline{x} \, (F_1 \wedge G_1 \wedge F_2) \vee \exists \overline{y} \, (F_1 \wedge G_2 \wedge F_2)$$

where $\overline{y}$ is a proper renaming of the variables $\overline{x}$ in F_1, G_2 and F_2.

renaming

$$\forall \overline{x}(F_1 \wedge F_2) \longrightarrow \forall \overline{x} F_1 \wedge \forall \overline{y} F_2 \qquad \exists \overline{x}(F_1 \vee F_2) \longrightarrow \exists \overline{x} F_1 \vee \exists \overline{y} F_2$$

where $\overline{y}$ is a proper renaming of the non empty $\overline{x}$ in F_2.

($\forall$-widening) and ($\exists$-widening)

if $\overline{y} \cap (vars(F_1) \cup vars(F_2)) = \emptyset$,

$$F_1 \vee \forall \overline{y} \, G \vee F_2 \longrightarrow \forall \overline{y} \, (F_1 \vee G \vee F_2)$$
$$F_1 \wedge \exists \overline{y} \, G \wedge F_2 \longrightarrow \exists \overline{y} \, (F_1 \wedge G \wedge F_2)$$

Lemma 15 Correctness of the rewrite rules. *If G_2 is a formula obtained from G_1 by any sequence of applications of the rewrite rules, then $\models G_1 \leftrightarrow G_2$.*

Proof: The whole rewrite rules preserve, in every model, the equivalence between formulas. They correspond to well-known theorems of predicate logic [27].

6.2 The equality-decomposition rule

It is aimed at transforming an equality among two terms u and v into an equivalent conjunction of the form $\wedge_i(x_i = t_i)$ where t_i is a term and x_i is a variable not occurring in t_i. This can be done by using the *Solved Form Algorithm* given in [19].

An application of the equality-decomposition rule to a formula G consists of:

1. a selection in G of an equality $u = v$ (resp. a disequality $u \neq v$)
2. an application of the Solved form algorithm to $\{u = v\}$
3. a replacement in G of the equality $u = v$ (resp. the disequality $u \neq v$) by:
   - *false* (resp. *true*) when failure has been returned.
   - $\wedge_i(x_i = t_i)$ (resp. $\vee_i(x_i \neq t_i)$) otherwise.

Lemma 16 Correctness of the equality-decomposition rule. *If G_2 is a formula produced by any sequence of applications of the equality-decomposition rule to G_1, then $CET \models G_1 \leftrightarrow G_2$.*

Proof: The correctness (and termination) of the Solved form algorithm is proved in [19] (Theorem 3.1). It follows that an application of the equality-decomposition rule merely replaces in G an equality by an equivalent formula in CET.

6.3 The equality-elimination rules

They are aimed at removing in simple cases an occurrence of the equality symbol.

$$\forall \overline{x} \ (F_1 \vee x' \neq t \vee F_2) \longrightarrow \forall \overline{x} \ ((F_1 \vee F_2)(x' \leftarrow t))$$
$$\exists \overline{x} \ (F_1 \wedge x' = t \wedge F_2) \longrightarrow \exists \overline{x} \ ((F_1 \wedge F_2)(x' \leftarrow t))$$

where x' occurs in $\overline{x}$ and not in t.

In particular, in the case of definite programs and S-formulas, an application of the above rules after a replacement will remove all the equality symbols, such that the resulting formula coincides with the one that would have been produced by the rules of EE (see section 8).

They are also useful since they can lead to further normalization steps or make practical the application of the Simplification rule.

Lemma 17 Correctness of the equality-elimination rule. *If G_2 is a formula produced by any sequence of applications of the equality-elimination rules to G_1, then $CET \models G_1 \leftrightarrow G_2$.*

Proof: See [27].

7 Soundness of EE$_G$

Lemma 18 Conservation of the normal form of goals. *Let F be a formula produced by an application of the Replacement or the Simplification rule to a normal goal G_1. Then an exhaustive application of the rewrite rules, the equality decomposition and elimination rules enables the derivation of F into an equivalent stable formula G_2 to which none of the rules is applicable, and G_2 is a normal goal.*

Corollary 19. *A derivation step preserves the normal form of a goal.*

Proof: We describe the form of the intermediary formula F. Following the cases:

- F is obtained from G_1 by a Simplification with a deciding substitution σ. $G_1\sigma$ is a normal goal by lemma 10. F is a conjunction of F_1 and F_2 respectively produced by replacing in $G_1\sigma$ some atoms by *true* and by *false*. Hence, F_1 and F_2 are also normal goals.

- F is obtained from G_1 by a Replacement. Then F is G_1 in which an atom $p(\overline{u})$ is replaced by a formula of the form $\exists \overline{y}(\overline{t} = \overline{u} \wedge W)$ where $\overline{y} = vars(\overline{t}) \cup vars(W)$ is a sequence of fresh variables and W is a conjunction of literals. Hence F is also a closed formula in which every variable occurs in the scope of its quantification.

  - if $p(\overline{u})$ occurs positively in G_1 in a disjunction, then F is also a normal goal: this case is immediate when considering the grammar of the definition 1. If $\overline{y}$ is empty then the first de Morgan's law (cf. subsection 6.1) is applicable and it yields another normal goal (or a conjunction of two normal goals).

  - if $p(\overline{u})$ occurs positively in G_1 in a conjunction then the $\exists$-*widening* rule (cf. subsection 6.1) turns F into a normal goal.

  - if $p(\overline{u})$ occurs negatively in G_1 in a conjunction then F is turned into a normal goal by distributivity of $\neg$ (the rule for negation are not described in subsection 6.1). If $\overline{y}$ is empty then the second de Morgan's law is also applicable. The two rules can be applied in any order and yield a normal goal.

  - if $p(\overline{u})$ occurs negatively in G_1 in a disjunction, then F is turned into a normal goal G_2 by distributivity of $\neg$ and the $\forall$-*widening* rule if $\overline{y}$ is not empty.

Finally it is easy to verify that the normalization process always terminates (notice that when a conjunction of goals is produced at the higher level, the two goals are considered separately in two subproofs, i.e. the second de Morgan's law is not applicable to this conjunction) and that the whole rewrite rules, the equality-decomposition rule and the equality-elimination rules preserve the form of normal goals and do not violate the property of variables binding. The equivalence between F and G_2 is ensured by lemmas 15, 16 and 17.

Remark: The conservation of the uniqueness of variable bindings is crucial regarding the validity of an EE_G proof. Actually a simplification on $\forall x p(x) \vee \forall x \neg p(x)$ would be possible and would derive *true*. It is not the case for the normal goal $\forall x p(x) \vee \forall y \neg p(y)$.

Theorem 20 Soundness of EE_G. *Let G be a normal goal. If there exists a finite sequence of derivation steps that derives G into true then G is a logical consequence of P^*, i.e. $P^* \models G$.*

Proof: The proof follows from the lemmas 6, 9, 14, 15, 16, 17 and 18, which ensure us that when G_{i+1} is obtained from G_i in one derivation step, i.e. by an application of the Replacement or the Simplification rule and of the normalization process (which involves the rewrite rules, the equality-decomposition rule and the equality-elimination rules) then, if $P^* \models G_{i+1}$ then $P^* \models G_i$.

8 A partial completeness result for EE$_G$

Indeed in our generalization work, we have to be sure that we do not lose any power in comparison with EE. This is the aim of this paragraph. We recall a definition for the class of S-formulas.

Definition 21 (*S-formula*). An S-formula is any closed first-order formula which can be written into the following normal goal:
$\forall \overline{x} \ (F_1 \vee \ldots \vee F_{i-1} \vee \exists \overline{y}_i \ F_i \vee \ldots \vee \exists \overline{y}_n \ F_n)$
where F_j is a conjunction of literals (resp. atoms) when $j \in [1..i-1]$ (resp. $j \in [i..n]$).

Theorem 22. *Let P be a definite program and G be a normal goal in the class of S-formulas. Then if G is a logical consequence of P^*, EE$_G$ provides a successful derivation for G.*

Proof: This result is obtained for the rules of EE and the proof is given in [15]. Hence it amounts for us to verify that every derivation in EE can also be achieved by EE$_G$. We just give here the idea of the proof which actually does not present any difficulty but would require the description of each rule of EE. Let us just describe the case where the *dci* rule (unfolding) is applied. Let $p(\overline{u})$ be the selected atom in a goal $G : \forall \overline{x} \ (F_1 \vee \exists \overline{y} \ (F_2 \wedge p(\overline{u}) \wedge F_3) \vee F_4)$, with p defined in P by the clauses $\{p(\overline{t}_i) \leftarrow W_i\}_i$ and let $\overline{y}_i$ be the variables occurring in t_i or W_i. Let $p(\overline{t}_k) \leftarrow W_k$ be the selected clause for the *dci* and σ be the corresponding deciding substitution. Then G leads to $G' : (G[p(\overline{u}) \leftarrow W_k])\sigma$ i.e. to $\forall \overline{x} \ (F_1 \vee \exists \overline{y}.\overline{y}_k \ (F_2 \wedge W_k \wedge F_3)\sigma \vee F_4)$, i.e. to $\forall \overline{x} \ (F_1 \vee \exists \overline{y}.\overline{y}_k \ (F_2 \wedge W_k \wedge F_3 \wedge \overline{t} = \overline{u}_k) \vee F_4)$. In EE$_G$, a replacement of the same atom $p(\overline{u})$ produces, after normalization, $G'' : \forall \overline{x} \ (F_1 \vee F' \vee \exists \overline{y}.\overline{y}_k \ (F_2 \wedge W_k \wedge F_3 \wedge \overline{t} = \overline{u}_k) \vee F'' \vee F_4)$ for some F' and F'' possibly empty. Finally every successful derivation for G' in EE clearly also holds for G''.

The verification for an application of the rule *nfi* follows the same arguments: this time the two goals G' and G'' are exactly the same.

The verification for an application of the *simplification* rule is immediate since our definition of this rule is the same than the definition given for EE. It amounts to verify that the difference between the two definitions of a deciding substitution is not relevant when the substitution is involved in a simplification, i.e. that every deciding substitution used in a simplification in EE is also a deciding substitution following our definition 7. This holds because a simplification requires the selection of at least one negative literal which by definition of an S-formula cannot contain any occurrence of a variable quantified by $\exists x$. Hence no new variable can be introduced in a goal by a simplification.

Finally we verify that the rules aimed at normalizing the goals in the two frameworks coincide.

Remark: We must point out that the symbol of implication is allowed in EE in the syntax of goals and an advantage is that the formulas are kept in a form which emphasizes an intuition of their meaning. This is not the case at present time in EE_G because at right stage it simplifies the formulation of the rules and their proofs. However it must be considered as a possible improvement of our technique.

9 Examples

9.1 First example

We borrow from [8] the following program, where the relation of list inclusion is defined by means of the negation of non inclusion.

$$includ(l_1, l_2) \leftarrow \neg ninclud(l_1, l_2)$$
$$ninclud(l_1, l_2) \leftarrow elem(e, l_1) \wedge \neg elem(e, l_2)$$

We prove the formula $\forall l_1 l_2 l_3 \ (includ(l_1, l_2) \wedge includ(l_2, l_3) \rightarrow includ(l_1, l_3))$ aimed at expressing the transitivity of the list inclusion. Notice that the proof is performed with no use of a definition for $elem$.

A possible proof is summarized by the tree below where each node is labelled by a goal. When G yields a conjunction of n goals G'_i, then the node for G has n sons, each of them is labelled by some G'_i. Moreover the normalization process is implicitly applied at each derivation step (in the first four derivation steps it produces no effect). When a simplification is applied, σ is the associated deciding substitution.

$$\forall l_1 l_2 l_3 (\neg includ(l_1, l_2) \vee \neg includ(l_2, l_3) \vee includ(l_1, l_3))$$
$$\big|\ replacement$$
$$\forall l_1 l_2 l_3 (ninclud(l_1, l_2) \vee \neg includ(l_2, l_3) \vee includ(l_1, l_3))$$
$$\big|\ replacement$$
$$\forall l_1 l_2 l_3 (ninclud(l_1, l_2) \vee ninclud(l_2, l_3) \vee includ(l_1, l_3))$$
$$\big|\ replacement$$
$$\forall l_1 l_2 l_3 (ninclud(l_1, l_2) \vee ninclud(l_2, l_3) \vee \neg ninclud(l_1, l_3))$$
$$\big|\ replacement$$
$$\forall l_1 l_2 l_3 (\exists e(elem(e, l_1) \wedge \neg elem(e, l_2)) \vee ninclud(l_2, l_3) \vee \neg ninclud(l_1, l_3))$$
$$\big|\ replacement$$
$$\forall l_1 l_2 l_3 e' (\exists e(elem(e, l_1) \wedge \neg elem(e, l_2)) \vee ninclud(l_2, l_3) \vee \neg elem(e', l_1) \vee elem(e', l_3))$$

simplification

$$\forall l_2 l_3 e' (\neg elem(e', l_2) \vee ninclud(l_2, l_3) \vee elem(e', l_3)) \qquad\qquad true \qquad \sigma = \{e \leftarrow e'\}$$
$$\big|\ replacement$$
$$\forall l_2 l_3 e' (\neg elem(e', l_2) \vee \exists e(elem(e, l_2) \wedge \neg elem(e, l_3)) \vee elem(e', l_3))$$

simplification

$$\forall l_3 e' (\neg elem(e', l_3) \vee elem(e', l_3)) \qquad\qquad true \qquad \sigma = \{e \leftarrow e'\}$$

simplification

$$true \qquad\qquad\qquad\qquad true \qquad \sigma = \{\}$$

9.2 Second example

Let P be the following program:

$$int(0) \leftarrow \qquad even(0) \leftarrow \qquad odd(s(x)) \leftarrow even(x)$$
$$int(s(x)) \leftarrow int(x) \qquad even(s(x)) \leftarrow \neg even(x)$$

and G be the formula $\forall x \ (int(x) \rightarrow even(x) \vee odd(x))$ which expresses a partial specification of integers. Then G can be easily derived into $true$ by replacements and a simplification.

Now let us suggest an interesting perspective of complementarity with our technique and the proof method introduced by Deransart and Ferrand in [8] which is aimed at proving properties of atoms in the well-founded model [14]. For instance to prove that every ground atom $int(\overline{u})$ in the denotation verifies the property to be always an even or an odd number, we can prove the *validity* (see [8]) of the (partial) specification $S^{int} = even(x) \vee odd(x)$. Following [8] a specification is valid if assuming that the property holds for the body of a (ground) clause, it holds also for the head. This amounts to prove that the following two statements hold (notice the similarity with computational induction):

$$even(0) \vee odd(0)$$
$$\forall x \ (even(x) \vee odd(x) \rightarrow even(s(x)) \vee odd(s(x)))$$

A possibility to prove that these formulas are valid and therefore to complete the proof of the validity of S^{int} is to use EE_G. EE_G ensures us of the validity of these formulas in every model of the completion and therefore in the well-founded model (which is total since P is locally stratified [14]).

10 Discussion and related works

Our approach of proving properties of a program P is addressed with respect to Clark's completion semantics which is the standard approach for a declarative meaning of negation in logic programming. However it is well-known that the completion is not always consistent and therefore the problem of the completeness of a procedural counterpart for this semantics must be studied for special classes of programs [6].

An alternative is to move into Klenee's three-valued logic [11, 17]. It provides a weaker semantics (every two-valued model is a three-valued one) which ensures the consistency of the (three-valued) completion and which allows to extend the completeness result of SLDNF resolution to a wider class of programs [18]. Many recent works propose solutions to overcome the limitations of SLDNF due to the *floundering* problem [21] which is a major obstacle to its completeness.[7] Most of these works are strongly related to the completion even if the completion is not used explicitly in the proof procedure [4, 2, 25]. In [10] another view based on failed (SLDFA)-trees is preferred. All these techniques are proved sound and complete w.r.t. the three-valued completion.

---

[7] It is not the unique one. As it is shown in section 3.3 the problem with excluded middle remains.

Regarding mechanical verification of programs, the choice of a three-valued semantics, because of its weakness is not in our opinion a relevant approach. For example none of the properties proved in the examples of section 9 are three-valued consequences of the completion. It is also the case for the goal of the example in section 3.3.[8]

The crucial distinction of our framework compared to the above mentioned ones lies in the existence of the Simplification rule (which relates to the excluding middle) and which is very useful in practice. That is why, regarding practical matters in mechanical verification, we believe it is better to deal with a two-valued semantics.

Nevertheless even in a two-valued setting, it is a fact that a semantics based on the completion may be too restrictive since it is a strong requirement to be true in every model of the completion. A possible approach would be to consider some special classes of models of P^* expected to fit better the intended meaning of P.

In [10], Drabent defines SLSFA-resolution, an adaptation of SLDFA for the well-founded semantics [14].[9] He proves the completeness of SLSFA for atomic queries which is naturally extended to normal queries. However the properties considered in section 9 are not normal goals and the techniques of [10] do not apply.

In [20], Lever addresses the problem of proving properties of (locally stratified) logic programs by means of SLS-resolution which is the procedural counterpart of the Perfect Model Semantics (Przymusinski [22]). However his method is limited because of the floundering problem, i.e. SLS-trees under consideration must be unfloundered. As a consequence the property of example 9.1 cannot be handled by his techniques since the construction of an SLS-tree from the denial $\leftarrow includ(l_1, l_2) \wedge includ(l_1, l_2) \wedge \neg includ(l_1, l_3)$ leads to the floundering denial $\leftarrow \neg ninclud(l_1, l_2) \wedge \neg ninclud(l_1, l_2) \wedge \neg includ(l_1, l_3)$. It is still the case when we use an extension of SLS-resolution, as is proposed by Przymusinski in [23], namely *SLSC-resolution* (SLS-resolution with Constructive negation), despite of a refinement in the definition of floundering. For its part, the example of section 3.3 can neither be handled since it is not an implicative property acceptable in his framework. Notice that it is always the case when we use an extension of SLS-resolution, as is proposed by Przymusinski in [23], namely *SLSC-resolution*

---

[8] This can be verified by iterating Fitting's operator Φ_P. Kunen in [18] shows the result that for any sentence it is equivalent to be true in every three-valued model of the completion and to be true in $\Phi_P \uparrow^n$ for a finite n. In example 9.2 the expected property $\forall x \, (int(x) \rightarrow even(x) \vee odd(x))$ is true at stage $\omega + 1$ only (more precisely it is the negation of this formula that turns to false at this stage and the least fixpoint of Φ_P is obtained at stage $\omega + 2$). It is still undefined in $\Phi_P \uparrow^\omega$ and consequently it is not a three-valued consequence of the completion. In the example of section 3.3 the query that is proved is not even in the least fixpoint of Φ_P.

[9] The well-founded semantics is a three-valued semantics but the class of programs possessing a total (two-valued) model properly includes the classes of *stratified* and *locally stratified* programs.

(SLS-resolution with Constructive negation), despite of a refinement in the definition of floundering.

11 Conclusion

We have presented a method aimed at proving logical consequences of the completion of normal programs. This is a significant extension of the framework introduced by Kanamori and Seki [16] for definite programs. It is essentially based on an unfolding rule with the iff definitions of the completion and a simplification rule which operates over opposite literals in a goal.

Through some examples, this resulting enriched framework has been shown to be an interesting contribution in program verification. Its first prototyping is currently under improvement. It must be seen as a tool for the stepwise checking of correctness. More generally, it provides a better understanding of a program and its semantics. It can be naturally applied to the field of program equivalence. Section 9.2 gives some perspective of incorporating induction rules in EE_G. Its application to the field of program synthesis has already been investigated for definite programs in [12, 13, 3].

12 Acknowledgments

I would like to thank Mikhail Boulyonkov and anonymous referees for helpful comments. I am grateful to Pierre Deransart, Gerard Ferrand and Laurent Fribourg for valuable discussions and advices.

References

1. K. Apt. Introduction to Logic Programming. In J. Van Leeuven, editor, *Handbook of Theoretical Computer Science*. North Holland, 1990.
2. A. Bottoni and G.Levi. Computing in the Completion. In *Atti dell Ottavo Convegno sulla Programmazione Logica (Gulp)*, pages 375–389. Mediterranean Press, 1993.
3. A. Bouverot. *Comparaison entre la transformation et l'extraction de programmes logiques*. PhD thesis, Université Paris VII, 1991.
4. P. Bruscoli, F. Levi, G. Levi, and M. C. Meo. Intensional Negation in Constraint Logic Programs. In *Atti dell Ottavo Convegno sulla Programmazione Logica (Gulp)*, pages 359–373. Mediterranean Press, 1993.
5. K. L. Clark. Negation as Failure. In H. Gallaire and J. Minker, editors, *Logic and Databases*, pages 293–322. Plenum Press, New York, 1978.
6. A. Cortesi and G. Filé. Graph Properties for Normal Programs. *Theoretical Computer Science*, 107:277–303, 1993.
7. P. Deransart. Proof Methods of Declarative Properties of Definite Programs. *Theoretical Computer Science*, 118:99–166, 1993.
8. P. Deransart and G. Ferrand. Proof Method of Partial Correctness and Weak Completeness for Normal Logic Programs. *Journal of Logic Programming*, 17:265–278, 1993.

9. Y. Deville. A Correctness Definition for Logic Programming. Technical Report RP 88/8, Namur University, 1988.

10. W. Drabent. What is Failure? An Approach to Constructive Negation. Technical Report LITH-IDA-R-91-23, Linkoping University, 1993. to appear in Acta Informatica.

11. M. Fitting. A Kripke-Kleene Semantics for Logic Programs. *Journal of Logic Programming*, 2:295–312, 1985.

12. L. Fribourg. Extracting Logic Programs from Proofs that Use Extended Prolog Execution and Induction. In *Proceedings of the Seventh Int. Conference on Logic Programming, Jerusalem*, pages 685–699, 1990.

13. L. Fribourg. Generating Simplification Lemmas using Extended Prolog Execution and Proof-Extraction. In *Proceedings of the Int. Logic Programming Symposium, San Diego*, 1991.

14. A. Van Gelder, K. A. Ross, and J. S. Schlipf. The well-founded semantics for general logic programs. *Journal of the ACM*, 38(3):620–650, 1991.

15. T. Kanamori. Soundness and Completeness of Extended Execution for Proving Properties of Prolog Programs. Technical Report TR-175, ICOT, 1986.

16. T. Kanamori and H. Seki. Verification of Logic Programs Using an Extension of Execution. In *Proceedings of the Third International Conference on Logic Programming, London*, pages 475–489, 1986.

17. K. Kunen. Negation in Logic Programming. *Journal of Logic Programming*, 4:289–308, 1987.

18. K. Kunen. Signed Data Dependencies in Logic Programs. *Journal of Logic Programming*, 7:231–245, 1989.

19. J. L. Lassez, M. J. Maher, and K. G. Marriott. Unification Revisited. In J. Minker, editor, *Foundations of Deductive Databases and Logic Programming*, pages 587–625. Morgan Kaufmann, Los Alto, Ca., 1988.

20. J. M. Lever. Proving Program Properties by means of SLS-resolution. In *Proceedings of the Eight Int. Conference on Logic Programming, Paris*, pages 614–628, 1991.

21. J. Lloyd. *Foundations of Logic Programming*. Springer-Verlag, Berlin, 1987.

22. T. C. Przymusinski. Perfect Model Semantics. In R. Kowalski and K. Bowen editors, editors, *Proceedings of the Fifth Int. Logic Programming Symposium*, pages 1081–1096. Association for Logic Programing, 1988.

23. T. C. Przymusinski. On Constructive Negation in Logic Programming. Technical report, University of Texas at El Paso, 1990.

24. T. Sato. Equivalence-preserving first order unfold/fold transformation systems. *Theoretical Computer Science*, 105:57–84, 1992.

25. T. Sato and F. Motoyoshi. A Complete Top-down Interpreter for First Order Programs. In *Proceedings of the Int. Logic Programming Symposium, San Diego*, pages 35–53. MIT Press, 1991.

26. J. C. Shepherdson. Language and Equality Theory in Logic Programming. Technical Report PM-88-08, University of Bristol, 1988.

27. D. van Dalen. *Logic and Structure*. Springer Verlag, Berlin, second edition, 1987.

Partial Deduction of Disjunctive Logic Programs: A Declarative Approach

Chiaki Sakama[1] and Hirohisa Seki[2]

[1] ASTEM Research Institute of Kyoto
17 Chudoji Minami-machi, Shimogyo, Kyoto 600 Japan
sakama@astem.or.jp
[2] Department of Artificial Intelligence and Computer Science
Nagoya Institute of Technology, Gokiso, Showa-ku, Nagoya 466 Japan
seki@ics.nitech.ac.jp

Abstract. This paper presents a partial deduction method for disjunctive logic programs. We first show that standard partial deduction in logic programming is not applicable as it is in the context of disjunctive logic programs. Then we introduce a new partial deduction technique for disjunctive logic programs, and show that it preserves the minimal model semantics of positive disjunctive programs, and the stable model semantics of normal disjunctive programs. Goal-oriented partial deduction is also presented for query optimization.

1 Introduction

Partial deduction or *partial evaluation* is known as one of the optimization techniques in logic programming. Given a logic program, partial deduction derives a more specific program through performing deduction on a part of the program, while preserving the meaning of the original program. Such a specialized program is usually more efficient than the original program when executed.

Partial deduction in logic programming was firstly introduced by Komorowski [Kom81] and has been developed by several researchers from various viewpoints (for an introduction and bibliographies, see [Kom92] and [SZ88], for example). From semantic points of view, Lloyd and Shepherdson [LS91] formalized partial evaluation for normal logic programs and showed its correctness with respect to Clark's program completion semantics. On the other hand, Tamaki and Sato [TS84] showed that partial deduction preserves the least Herbrand model semantics of definite logic programs in the context of unfold/fold transformation. The result is extended by Seki to the perfect model semantics for stratified logic programs [Seki91], and the well-founded semantics for normal logic programs [Seki93].

Recent studies of logic programming extended its framework to include indefinite information in a program. *Disjunctive logic programs* are such extensions of logic programming, which possibly include disjunctive clauses in programs. A disjunctive logic program enables us to reason with indefinite information in a program, and its growing importance in logic programming and artificial intelligence is recognized these days. Disjunctive logic programs increase expressiveness of logic programming on the one hand, but their computation is generally expensive on the other hand. Then optimizations of disjunctive logic programs are important issues for practical usage, however, there have been few studies on the subject.

In this paper, we develop partial deduction techniques for disjunctive logic programs. We first show that standard partial deduction is not useful in the presence of disjunctive information in a program, and introduce new partial deduction for disjunctive logic programs. We prove that the proposed partial deduction method preserves the minimal model semantics of positive disjunctive programs, and the disjunctive stable model semantics of normal disjunctive programs.

The rest of this paper is organized as follows. In Section 2, we introduce notations of disjunctive logic programs. In Section 3, we present new partial deduction for positive disjunctive programs and show its correctness with respect to the minimal model semantics. Section 4 extends the result to normal disjunctive programs containing negation as failure, and shows that proposed partial deduction also works well for the disjunctive stable model semantics. Section 5 discusses some connections between normal partial deduction and the proposed one. In Section 6, partial deduction techniques are applied to goal-oriented partial deduction for query optimization. Section 7 summarizes the paper and addresses future work.

2 Disjunctive Logic Programs

A *normal disjunctive program* is a finite set of clauses of the form:

$$A_1 \vee \ldots \vee A_l \leftarrow B_1 \wedge \ldots \wedge B_m \wedge not\, B_{m+1} \wedge \ldots \wedge not\, B_n \quad (l \geq 0, n \geq m \geq 0) \quad (1)$$

where A_i's and B_j's are atoms and *not* denotes the *negation-as-failure* operator. The left-hand side of the clause (1) is called the *head*, while the right-hand side of the clause is called the *body*. The clause (1) is called *disjunctive* (resp. *normal*) if $l > 1$ (resp. $l = 1$). Else if $l = 0$ and $n \neq 0$, it is called an *integrity constraint*. A normal disjunctive program containing no *not* is called a *positive disjunctive program*, while a program containing no disjunctive clause is called a *normal logic program*. A normal logic program containing no *not* is called a *Horn logic program*, and a Horn logic program without integrity constraints is called a *definite logic program*.

In this paper, when we write $A \vee \Sigma \leftarrow \Gamma$, Σ denotes a disjunction (possibly *false*) in the head, and Γ denotes a conjunction (possibly *true*) in the body.[3]

In logic programming, a program is semantically identified with its *ground program*, which is the set of all ground clauses from the program. Then we consider ground programs throughout this paper. We also assume without loss of generality that a disjunction in the head of a ground clause is already *factored*, that is, each atom in the disjunctive head of a clause is different.

An *interpretation* of a program is a subset of the Herbrand base of the program. For a positive disjunctive program P, an interpretation I is called a *minimal model* of P if there is no smaller interpretation J satisfying the program. A program is *consistent* if it has a minimal model, otherwise a program is *inconsistent*. The *minimal model semantics* [Min82] of a positive disjunctive program P is defined as the set of all minimal models of P (denoted by $\mathcal{MM}_P$).

For a normal disjunctive program P, an interpretation I is called a *stable model* of P if I coincides with a minimal model of the positive disjunctive program P^I obtained from P as follows:

$$P^I = \{A_1 \vee \ldots \vee A_l \leftarrow B_1 \wedge \ldots \wedge B_m \mid \text{ there is a ground clause of the form (1)}$$
$$\text{from } P \text{ such that } \{B_{m+1}, \ldots, B_n\} \cap I = \emptyset \, \}.$$

A normal disjunctive program has no, one, or multiple stable models in general. A program which has no stable model is called *incoherent*.

The *disjunctive stable model semantics* [Prz91] of a normal disjunctive program P is defined as the set of all stable models of P (denoted by $\mathcal{ST}_P$). The disjunctive stable model semantics coincides with Gelfond and Lifschitz's stable model semantics [GL88] in normal logic programs.

3 Partial Deduction of Positive Disjunctive Programs

Partial deduction in logic programming is usually defined as *unfolding* of clauses in a program.[4] For a Horn logic program P, partial deduction is formally presented as follows.

Given a Horn clause C from P:

$$C : H \leftarrow A \wedge \Gamma,$$

---

[3] When we write a clause as $A \vee \Sigma \leftarrow \Gamma$, it does not necessarily mean that A should be the leftmost atom in the head of the clause. That is, any two clauses are identified modulo the permutation of disjuncts/conjuncts in their heads/bodies.

[4] Partial deduction is also called partial evaluation. However, we prefer to use the term partial deduction since partial evaluation often includes non-deductive procedures.

suppose that $C_1, \ldots, C_k$ are all of the clauses in P such that each of which has the atom A in its head:

$$C_i : A \leftarrow \Gamma_i \quad (1 \leq i \leq k).$$

Then *normal partial deduction* of P (with respect to C on A) is defined as the program $\pi^N_{\{C;A\}}(P)$ (called a *residual program*) such that

$$\pi^N_{\{C;A\}}(P) = (P \setminus \{C\}) \cup \{C'_1, \ldots, C'_k\}$$

where each C'_i is defined as

$$C'_i : H \leftarrow \Gamma \wedge \Gamma_i.$$

When we simply say normal partial deduction of P (written $\pi^N(P)$), it means normal partial deduction of P with respect to any clause on any atom.

Example 3.1 Let P be the program:

$$P = \{ a \leftarrow b, \quad b \leftarrow c, \quad b \leftarrow a, \quad c \leftarrow \}.$$

Then normal partial deduction of P with respect to $a \leftarrow b$ on b becomes

$$\pi^N_{\{a \leftarrow b;b\}}(P) = \{ a \leftarrow c, \quad a \leftarrow a, \quad b \leftarrow c, \quad b \leftarrow a, \quad c \leftarrow \}. \quad \Box$$

In the context of unfold/fold transformation of logic programs, Tamaki and Sato [TS84] showed that normal partial deduction preserves the least Herbrand model semantics of definite logic programs.

Lemma 3.1 ([TS84]) Let P be a definite logic program and M_P be its least Herbrand model. Then, for any residual program $\pi^N(P)$ of P, $M_P = M_{\pi^N(P)}$. $\Box$

The result also holds for Horn logic programs, that is, programs containing integrity constraints.

Theorem 3.2 Let P be a Horn logic program and $\pi^N(P)$ be any residual program of P. Then $M_P = M_{\pi^N(P)}$.

Proof: By identifying each integrity constraint $\leftarrow G$ with $false \leftarrow G$, M_P contains $false$ iff $M_{\pi^N(P)}$ contains $false$. In this case, both programs are inconsistent. Then the result follows from Lemma 3.1. $\Box$

Thus, in what follows we do not take special care for the treatment of integrity constraints, that is, they are identified with normal clauses during partial deduction as presented above.

Now we consider partial deduction in disjunctive logic programs. If we consider to extend normal partial deduction to a program possibly containing disjunctive clauses, however, normal partial deduction does not preserve the minimal models of the program in general.

Example 3.2 Let P be the program:

$$P = \{\, a \vee b \leftarrow, \quad a \leftarrow d, \quad c \leftarrow a \,\}$$

where $\mathcal{MM}_P = \{\{a,c\}, \{b\}\}$. On the other hand,

$$\pi^N_{\{c \leftarrow a;a\}}(P) = \{\, a \vee b \leftarrow, \quad a \leftarrow d, \quad c \leftarrow d \,\}$$

where $\mathcal{MM}_{\pi^N_{\{c \leftarrow a;a\}}}(P) = \{\{a\}, \{b\}\}$. □

The problem is that normal partial deduction of logic programs is defined as unfolding between normal clauses. In the above example, however, there is the disjunctive clause $a \vee b \leftarrow$ containing the atom a in its head, so unfolding between $c \leftarrow a$ and $a \vee b \leftarrow$ would be needed.

Then our first task is to extend the normal partial deduction method to the one which supplies unfolding for disjunctive clauses.

Definition 3.1 Let P be a positive disjunctive program and C be a clause in P of the form:

$$C : \Sigma \leftarrow A \wedge \Gamma. \tag{2}$$

Suppose that $C_1, \ldots, C_k$ are all of the clauses in P such that each of which includes the atom A in its head:

$$C_i : A \vee \Sigma_i \leftarrow \Gamma_i \quad (1 \leq i \leq k). \tag{3}$$

Then *disjunctive partial deduction* of P (with respect to C on A) is defined as the program $\pi^D_{\{C;A\}}(P)$ (called a *residual program*) such that

$$\pi^D_{\{C;A\}}(P) = (P \setminus \{C\}) \cup \{C'_1, \ldots, C'_k\}$$

where each C'_i is defined as

$$C'_i : \Sigma \vee \Sigma_i \leftarrow \Gamma \wedge \Gamma_i, \tag{4}$$

in which $\Sigma \vee \Sigma_i$ is factored. □

Disjunctive partial deduction is a natural extension of normal partial deduction. In fact, the clause (4) is a resolvent of the clauses (2) and (3). In Horn logic programs, disjunctive partial deduction coincides with normal partial deduction.

Now we show that disjunctive partial deduction preserves the minimal model semantics of positive disjunctive programs. We first present a preliminary lemma.

Lemma 3.3 Let P be a positive disjunctive program and M be its minimal model. Then an atom A is in M iff there is a clause $C : A \vee \Sigma \leftarrow \Gamma$ in P such that $M \setminus \{A\} \models \Gamma$ and $M \setminus \{A\} \not\models \Sigma$.

Proof: ($\Rightarrow$) Suppose that for some atom A in M, there is no clause C in P such that $M \setminus \{A\} \models \Gamma$ and $M \setminus \{A\} \not\models \Sigma$. Then, for each clause C, $M \setminus \{A\} \not\models \Gamma$ or $M \setminus \{A\} \models \Sigma$, and hence it holds that $M \setminus \{A\} \models \Gamma$ implies $M \setminus \{A\} \models \Sigma$. In this case, since $M \setminus \{A\}$ satisfies each clause C, it becomes a model of P, which contradicts the assumption that M is a minimal model. Hence the result follows.

($\Leftarrow$) Assume that A is not in M. Then $M \setminus \{A\} = M$, and for a clause C in P, $M \models \Gamma$ and $M \not\models \Sigma$ imply $A \in M$, contradiction. $\square$

Theorem 3.4 Let P be a positive disjunctive program and $\pi^D(P)$ be any residual program of P. Then $\mathcal{MM}_P = \mathcal{MM}_{\pi^D(P)}$.

Proof: ($\subseteq$) Let M be a minimal model of P. Since the clause (4) is a resolvent of the clauses (2) and (3) in P, M also satisfies each clause (4) in $\pi^D(P)$. Then M is a model of $\pi^D(P)$. Assume that there is a minimal model N of $\pi^D(P)$ such that $N \subset M$. Since N is not a model of P, N does not satisfy the clause (2). Then $N \models \Gamma$, $N \not\models A$, and $N \not\models \Sigma$. As a minimal model N of $\pi^D(P)$ implies A, it follows from Lemma 3.3 that there is a clause C of the form (3) or (4) in $\pi^D(P)$ such that C contains A in its head. (i) Suppose first that C is of the form (3). Then $N \not\models A$ implies $N \setminus \{A\} \models \Gamma_i$ and $N \setminus \{A\} \not\models \Sigma_i$ (by Lemma 3.3). Here $N \setminus \{A\} \models \Gamma_i$ implies $N \models \Gamma_i$. Besides, the disjunctive head $A \vee \Sigma_i$ is assumed to be already factored, then Σ_i does not include A. Thus $N \setminus \{A\} \not\models \Sigma_i$ also implies $N \not\models \Sigma_i$. In this case, however, N does not satisfy the clause (4). This contradicts the assumption that N is a model of $\pi^D(P)$. (ii) Next suppose that C is of the form (4) such that $\Sigma = A \vee \Sigma'$. Then $N \not\models A$ implies $N \not\models \Sigma$, which contradicts the fact $N \not\models \Sigma$. Hence, M is also a minimal model of $\pi^D(P)$.

($\supseteq$) Let M be a minimal model of $\pi^D(P)$. If M is not a model of P, M does not satisfy the clause (2). In this case, $M \not\models \Sigma$, $M \models A$, and $M \models \Gamma$. Since a minimal model M of $\pi^D(P)$ implies A, it follows from Lemma 3.3 that there is a clause C of the form (3) or (4) in $\pi^D(P)$ such that C contains A in its head. When C is of the form (3), $M \models A$ implies $M \models \Gamma_i$ and $M \not\models \Sigma_i$ (by Lemma 3.3 and the discussion presented above). Thus M does not satisfy the corresponding clause (4), which contradicts the assumption that M is a model of $\pi^D(P)$. Else when C is of the form (4) such that $\Sigma = A \vee \Sigma'$, $M \models A$ implies $M \models \Sigma$, which contradicts the fact $M \not\models \Sigma$. Hence M is a model of P. Next assume that there is a minimal model N of P such that $N \subset M$. By ($\subseteq$), N is also a minimal model of $\pi^D(P)$, but this is impossible since M is a minimal model of $\pi^D(P)$. □

Corollary 3.5 Let P be a positive disjunctive program. Then P is inconsistent iff $\pi^D(P)$ is inconsistent. □

Example 3.3 (cont. from Example 3.2) Given the program P, its disjunctive partial deduction $\pi^D_{\{c \leftarrow a;a\}}(P)$ becomes

$$\pi^D_{\{c \leftarrow a;a\}}(P) = \{\, a \vee b \leftarrow, \quad a \leftarrow d, \quad c \leftarrow d, \quad b \vee c \leftarrow \,\},$$

and $\mathcal{MM}_{\pi^D_{\{c \leftarrow a;a\}}(P)} = \{\{a,c\}, \{b\}\}$, which is exactly the same as $\mathcal{MM}_P$. □

4 Partial Deduction of Normal Disjunctive Programs

In this section, we extend disjunctive partial deduction to normal disjunctive programs.

The definition of disjunctive partial deduction of normal disjunctive programs is the same as Definition 3.1, except that in this case each clause possibly contains negation as failure.

Example 4.1 Let P be the normal disjunctive program:

$$P = \{\, a \vee b \leftarrow not\, c, \quad a \leftarrow d, \quad c \leftarrow a \,\}.$$

Then disjunctive partial deduction of P with respect to $c \leftarrow a$ on a becomes

$$\pi^D_{\{c \leftarrow a;a\}}(P) = \{\, a \vee b \leftarrow not\, c, \quad a \leftarrow d, \quad c \leftarrow d, \quad b \vee c \leftarrow not\, c \,\}. \quad \square$$

As shown in the above example, disjunctive partial deduction is not affected by the presence of negation as failure in a program. Thus we can directly apply previously defined disjunctive partial deduction to normal disjunctive programs and the following result holds.

Theorem 4.1 Let P be a normal disjunctive program. Then $\mathcal{ST}_P = \mathcal{ST}_{\pi^D(P)}$.

Proof: Let M be a stable model of P. Then M is a minimal model of P^M. Since P^M is a positive disjunctive program, by Theorem 3.4, M is also a minimal model of $\pi^D(P^M)$. Now let us consider the clauses:

$$\Sigma \leftarrow A \wedge \Gamma \wedge not\, \Gamma' \quad (*)$$

and

$$A \vee \Sigma_i \leftarrow \Gamma_i \wedge not\ \Gamma'_i \quad (1 \leq i \leq k) \quad (\dagger)$$

in P, where $not\ \Gamma'$ is the conjunction of negation-as-failure formulas in the body.

(i) If $M \not\models \Gamma'$ and $M \not\models \Gamma'_i$ for some i $(1 \leq i \leq k)$, the clauses:

$$\Sigma \leftarrow A \wedge \Gamma \quad (*')$$

and

$$A \vee \Sigma_i \leftarrow \Gamma_i \quad (\dagger')$$

are in P^M. From these clauses, disjunctive partial deduction generates the clauses:

$$\Sigma \vee \Sigma_i \leftarrow \Gamma \wedge \Gamma_i \quad (\ddagger')$$

in $\pi^D(P^M)$. On the other hand, from $(*)$ and $(\dagger)$ in P, there are the clauses:

$$\Sigma \vee \Sigma_i \leftarrow \Gamma \wedge \Gamma_i \wedge not\ \Gamma' \wedge not\ \Gamma'_i \quad (\ddagger)$$

in $\pi^D(P)$, which become $(\ddagger')$ in $\pi^D(P)^M$.

(ii) Else if $M \models \Gamma'$ or $M \models \Gamma'_i$ for any i $(1 \leq i \leq k)$, the clauses $(*)$ or $(\dagger)$ is respectively eliminated in P^M. Then the clauses $(\ddagger')$ are not included in $\pi^D(P^M)$. In this case, each clause $(\ddagger)$ in $\pi^D(P)$ is also eliminated in $\pi^D(P)^M$.

Thus, there is a one-to-one correspondence between the clauses in $\pi^D(P^M)$ and the clauses in $\pi^D(P)^M$, hence $\pi^D(P^M) = \pi^D(P)^M$. Therefore M is also a minimal model of $\pi^D(P)^M$, and a stable model of $\pi^D(P)$.

The converse is also shown in the same manner. $\square$

Corollary 4.2 Let P be a normal disjunctive program. Then P is incoherent iff $\pi^D(P)$ is incoherent. $\square$

The above theorem also implies that in normal logic programs, normal partial deduction preserves Gelfond and Lifschitz's stable model semantics.

Corollary 4.3 Let P be a normal logic program. Then $ST_P = ST_{\pi^N(P)}$. $\square$

The above result is also presented in [Seki90].

5 Connections between Normal and Disjunctive Partial Deduction

In this section, we consider connections between normal and disjunctive partial deduction. We first give a sufficient condition such that normal partial deduction preserves the meaning of disjunctive logic programs.

Theorem 5.1 Let P be a normal disjunctive program and C be a clause of the form $\Sigma \leftarrow A \wedge \Gamma$ from P. If A does not appear in the head of any disjunctive clause in P, then $ST_P = ST_{\pi^N_{\{C;A\}}(P)}$. That is, normal partial deduction of P with respect to C on A preserves the disjunctive stable model semantics.

Proof: In this case, disjunctive partial deduction coincides with normal one, hence the result follows from Theorem 4.1. $\square$

Next we present a method to compute disjunctive partial deduction in terms of normal partial deduction.

Definition 5.1 Let P be a normal disjunctive program. The *nlp-transformation* transforms P into the normal logic program $\eta(P)$ which is obtained from P by replacing each disjunctive clause:

$$C: A_1 \vee \ldots \vee A_l \leftarrow \Gamma \tag{5}$$

with l normal clauses:

$$C_i^- : A_i \leftarrow \Gamma \wedge A_1^- \wedge \ldots \wedge A_{i-1}^- \wedge A_{i+1}^- \wedge \ldots \wedge A_l^- \quad (1 \le i \le l). \tag{6}$$

where each A_j^- is a new atom introduced for each A_j.
In particular, $C = C_i^-$ if $l \le 1$. $\square$

Now we show that disjunctive partial deduction of a normal disjunctive program P with respect to a clause C is obtained through normal partial deduction of $\eta(P)$ with respect to each C_i^-. In the following, the function η^{-1} is the reverse transformation which shifts each atom A_j^- appearing in the body of each clause in a program to the atom A_i in the head of the clause. Also Σ^- means $A_1^- \wedge \ldots \wedge A_l^-$ where $\Sigma = A_1 \vee \ldots \vee A_l$.

Theorem 5.2 Let P be a normal disjunctive program. Then $\pi^D_{\{C;A\}}(P) = \eta^{-1}(\pi^N_{\{C_i^-;A\}}(\eta(P)))$ where $\pi^N_{\{C_i^-;A\}}(\eta(P))$ means normal partial deduction of $\eta(P)$ with respect to each normal clause C_i^- on A.

Proof: Corresponding to the clauses (2) and (3) in P, there are the clauses:

$$A' \leftarrow A \wedge \Gamma \wedge \Sigma'^- \quad (\text{where } \Sigma = \Sigma' \vee A') \quad (*)$$

and

$$A \leftarrow \Gamma_i \wedge \Sigma_i^- \quad (1 \le i \le k) \quad (\dagger)$$

in $\eta(P)$, respectively. Then the clauses:

$$A' \leftarrow \Gamma \wedge \Gamma_i \wedge \Sigma'^- \wedge \Sigma_i^- \quad (\ddagger)$$

are obtained from $(*)$ and $(\dagger)$ by normal partial deduction in $\eta(P)$. In this case, by the reverse transformation η^{-1}, each clause of the form $(\ddagger)$ becomes a disjunctive clause of the form (4). Hence, $\pi^D_{\{C;A\}}(P) = \eta^{-1}(\pi^N_{\{C_i^-;A\}}(\eta(P)))$. $\square$

Example 5.1 Let P be the program:

$$P = \{\, a \vee b \leftarrow, \quad a \leftarrow b, \quad b \leftarrow a \,\}.$$

Then,

$$\pi^D_{\{a \leftarrow b;b\}}(P) = \{\, a \vee b \leftarrow, \quad a \leftarrow, \quad a \leftarrow a, \quad b \leftarrow a \,\}.$$

On the other hand, the nlp-transformation of P becomes

$$\eta(P) = \{\, a \leftarrow b^-, \quad b \leftarrow a^-, \quad a \leftarrow b, \quad b \leftarrow a \,\},$$

and

$$\pi^N_{\{a \leftarrow b;b\}}(\eta(P)) = \{\, a \leftarrow b^-, \quad b \leftarrow a^-, \quad a \leftarrow a^-, \quad a \leftarrow a, \quad b \leftarrow a \,\}.$$

Thus,

$$\eta^{-1}(\pi^N_{\{a \leftarrow b;b\}}(\eta(P))) = \{\, a \vee b \leftarrow, \quad a \leftarrow, \quad a \leftarrow a, \quad b \leftarrow a \,\}.$$

Therefore, $\pi^D_{\{a \leftarrow b;b\}}(P) = \eta^{-1}(\pi^N_{\{a \leftarrow b;b\}}(\eta(P)))$. □

The above theorem presents that disjunctive partial deduction $\pi^D_{\{C;A\}}(P)$ is obtained by the transformation sequence: $P \to \eta(P) \to \pi^N_{\{C^-_i;A\}}(\eta(P)) \to \eta^{-1}(\pi^N_{\{C^-_i;A\}}(\eta(P)))$. That is, together with the nlp-transformation, normal partial deduction can also be used for normal disjunctive programs.

6 Goal-Oriented Partial Deduction

In this section, we present goal-oriented partial deduction in disjunctive logic programs. Goal-oriented partial deduction specializes a program with respect to a given goal, which is useful to optimize programs for query-answering. Lloyd and Shepherdson [LS91] discuss a framework of goal-oriented partial evaluation for normal logic programs and provide conditions to assure its correctness with respect to Clark's completion semantics and SLDNF proof procedures. In our framework, goal-oriented partial deduction is presented as follows.

Let us consider a *query* of the form:

$$Q : Q(\mathbf{x}) \leftarrow B_1 \wedge \ldots \wedge B_m \wedge not\ B_{m+1} \wedge \ldots \wedge not\ B_n \qquad (7)$$

where $Q(\mathbf{x})$ is a new atom not appearing elsewhere in a program and $\mathbf{x}$ represents variables appearing in the body of the clause.

Then, given a normal disjunctive program P, partial deduction of P with respect to Q is defined as $\pi^D_{\{Q;B_i\}}(P_Q)$ where B_i is any atom occurring positively in the body of Q and P_Q is the program $P \cup \{Q\}$. When a query contains variables, we consider partial deduction with respect to its ground instances.

An *answer* to a query is defined as a ground substitution σ for variables in $Q(\mathbf{x})$. When Q contains no variable, σ is the empty substitution.

A query Q is *true* in P under the disjunctive stable model semantics if for every stable model I of P_Q there is an answer σ such that $Q(\mathbf{x})\sigma$ is included in I. Else if for some stable model I of P_Q there is an answer σ such that $Q(\mathbf{x})\sigma$ is included in I, the query is *possibly true*. Otherwise, if there is no such answer, the query is *false*. By Theorem 4.1, the following results hold.

Theorem 6.1 Let P be a normal disjunctive program and Q be a query. Then, under the disjunctive stable model semantics,

(i) Q is true in P iff Q is true in $\pi^D_{\{Q;B_i\}}(P_Q)$.

(ii) Q is possibly true in P iff Q is possibly true in $\pi^D_{\{Q;B_i\}}(P_Q)$.

(iii) Q is false in P iff Q is false in $\pi^D_{\{Q;B_i\}}(P_Q)$. $\square$

Example 6.1 Let P be the program:

$$\{\, p(a) \vee p(b) \leftarrow \,\},$$

in which the query $Q: q(x) \leftarrow p(x)$ is true. Then,

$$\pi^D_{\{Q;p(x)\}}(P_Q) = \{\, q(a)\vee p(b) \leftarrow, \quad q(b)\vee p(a) \leftarrow, \quad q(a)\vee q(b) \leftarrow, \quad p(a)\vee p(b) \leftarrow \,\}$$

and Q is also true in $\pi^D_{\{Q;p(x)\}}(P_Q)$ under the disjunctive stable model semantics.
$\square$

Note that in the above example, we assume that the ground queries $q(a) \leftarrow p(a)$ and $q(b) \leftarrow p(b)$ are unfolded consecutively in the program. That is, $\pi^D_{\{Q;p(x)\}}(P_Q)$ means $\pi^D_{\{Q;p(b)\}}(\pi^D_{\{Q;p(a)\}}(P \cup \{\, q(a) \leftarrow p(a), \quad q(b) \leftarrow p(b) \,\}))$. In this case, the order of unfolding does not affect the result of partial deduction since each partial deduction preserves the stable models of the program P_Q.

7   Summary

This paper presented a method of partial deduction for disjunctive logic programs. We first showed that normal partial deduction is not applicable to disjunctive logic programs in its present form. Then we introduced disjunctive partial deduction for disjunctive logic programs, which is a natural extension of normal partial deduction for normal logic programs. Disjunctive partial deduction was shown to preserve the minimal model semantics of positive disjunctive programs, and the disjunctive stable model semantics of normal disjunctive programs. We also showed a method of translating disjunctive partial deduction into normal partial deduction, and presented an application to goal-oriented partial deduction for query optimization.

The partial deduction technique presented in this paper is also directly applicable to disjunctive logic programs possibly containing classical negation [GL91]. Moreover, since positive disjunctive programs are identified with first-order theories, disjunctive partial deduction has potential application to first-order theorem provers. Recently, Brass and Dix [BD94] independently developed a partial deduction technique for disjunctive logic programs which is equivalent to ours. They discuss several abstract properties of disjunctive logic programs and conclude partial deduction as one of the fundamental properties that logic programming semantics should satisfy.

In this paper, we have mainly concerned with declarative aspects of partial deduction and considered propositional programs as a first step. Then our next step is to apply the partial deduction method to programs containing variables and investigate the procedural aspect of disjunctive partial deduction.

References

[BD94] Brass, S. and Dix, J., A Disjunctive Semantics Based on Unfolding and Bottom-up Evaluation, *Proc. 13th World Computer Congress'94, IFIP, GI-Workshop W2, Disjunctive Logic Programming and Disjunctive Databases*, 1994.

[GL88] Gelfond, M. and Lifschitz, V., The Stable Model Semantics for Logic Programming, *Proc. Joint Int. Conf. and Symp. on Logic Programming*, 1070-1080, 1988.

[GL91] Gelfond, M. and Lifschitz, V., Classical Negation in Logic Programs and Disjunctive Databases, *New Generation Computing* 9, 365-385, 1991.

[Kom81] Komorowski, J., A Specification of an Abstract Prolog Machine and its Application to Partial Evaluation, Technical Report LSST 69, Linköping Univ., 1981.

[Kom92] Komorowski, J., An Introduction to Partial Deduction, *Proc. 3rd Int. Workshop on Meta-programming in Logic*, Lecture Notes in Computer Science 649, Springer-Verlag, 49-69, 1992.

[LS91] Lloyd, J. W. and Shepherdson, J. C., Partial Evaluation in Logic Programming, *J. Logic Programming* 11, 217-242, 1991.

[Min82] Minker, J., On Indefinite Data Bases and the Closed World Assumption, *Proc. 6th Int. Conf. on Automated Deduction*, Lecture Notes in Computer Science 138, Springer-Verlag, 292-308, 1982.

[Prz91] Przymusinski, T. C., Stable Semantics for Disjunctive Programs, *New Generation Computing* 9, 401-424, 1991.

[Seki90] Seki, H., A Comparative Study of the Well-Founded and the Stable Model Semantics: Transformation's Viewpoint, *Proc. Workshop on Logic Programming and Nonmonotonic Logic*, Association for Logic Programming and Mathematics Sciences Institute, Cornell University, 115-123, 1990.

[Seki91] Seki, H., Unfold/Fold Transformation of Stratified Programs, *Theoretical Computer Science* 86, 107-139, 1991.

[Seki93] Seki, H., Unfold/Fold Transformation of General Logic Programs for the Well-Founded Semantics, *J. Logic Programming* 16, 5-23, 1993.

[SZ88] Sestoft, P. and Zamulin, A. V., Annotated Bibliography on Partial Evaluation and Mixed Computation, *New Generation Computing* 6, 309-354, 1988.

[TS84] Tamaki, H. and Sato, T., Unfold/Fold transformation of Logic Programs, *Proc. 2nd Int. Conf. on Logic Programming*, 127-138, 1984.

Avoiding Non-Termination when Learning Logic Programs: A Case Study with FOIL and FOCL

Giovanni Semeraro[1], Floriana Esposito[1] and Donato Malerba[1]
Clifford Brunk[2] and Michael Pazzani[2]

[1] Dipartimento di Informatica,
Universitá degli Studi di Bari
Via Orabona 4, 70126 Bari, ITALY
{semeraro, esposito, malerbad}@vm.csata.it

[2] Department of Information and Computer Science,
University of California, Irvine
Irvine, CA 92717 USA
{brunk, pazzani}@ics.uci.edu

Abstract. Many systems that learn logic programs from examples adopt θ-subsumption as model of generalization and refer to Plotkin's framework in order to define their search space. However, they seldom take into account the fact that the lattice defined by Plotkin is a set of equivalence classes rather than simple clauses. This may lead to non-terminating learning processes, since the search gets stuck within an equivalence class, which contains an infinite number of clauses.

In the paper, we present a task that cannot be solved by two well-known systems that learn logic programs, FOIL and FOCL. The failure is explained on the ground of the previous consideration about the search space. This task can be solved by adopting a weaker, but more mechanizable and manageable, model of generalization, called θ-subsumption under object identity (θ_{OI}-subsumption). Such a solution has been implemented in a new version of FOCL, called FOCL-OI.

1 Introduction

Recently, a new research area, called Inductive Logic Programming (ILP) [21], was born. It lies between logic programming and learning from examples, and aims at inductively learning *logic programs* from examples. ILP is characterized by a formal approach to the problem of *inductive generalization*.

The first formal method developed for inductive generalization is due to Plotkin [25]. Most of the work in the area of ILP refers to Plotkin's framework in order to define the space in which the search for concept descriptions is performed. Unfortunately, only in few cases it is realized that Plotkin's lattice is a set of equivalence classes rather than single clauses.

Helft [12] proposed a logical framework for inductive generalization that aims at reducing the size of the search space by restricting the representation language to function-free (but not constant-free) clausal logic. The two constraints adopted by Helft, namely the *difference links* and the *linkedness* of the Horn clauses, allow his algorithm to infer better generalizations from a practical point of view. Helft's framework, together with Plotkin's, is now commonly recognized as one of the two main frameworks for ILP [1] [14].

More recently, Quinlan [26] has developed a system, called FOIL (First Order Inductive Learner), that learns definite Horn clauses from data expressed as relations. FOIL proved effective and efficient on several learning tasks. Nevertheless, Quinlan himself recognizes that *"it is not difficult to construct tasks on which the current version of FOIL will fail"*. He ascribes these failures to the greedy search performed by FOIL. FOCL (First Order Combined Learner) [24] is an extension of FOIL in several aspects. The main extension allows FOCL to define and exploit background knowledge in the inductive learning process. Both FOIL and FOCL are widely acknowledged as an advance in the area of learning from examples [35] and some authors recently provided FOIL with a theoretical foundation by casting it in Plotkin's framework [1].

In this paper, we point out that both FOIL and FOCL may fall into a non-termination problem when they cope with toy world problems taken from the machine learning literature. The same problems occur on real world tasks [18] [30]. Furthermore, we provide an interpretation of the behaviour of these systems based on a pure analytical approach to the problem of learning logical programs. In Section 2 the model of generalization of θ-subsumption under object identity, called $\theta_{OI}-subsumption$, is formally defined and compared to $\theta-subsumption$. Section 3 presents a brief overview of FOIL and FOCL and describes the representation language used by these two systems. Negative experimental results concerning FOIL and FOCL are shown in Section 4. Moreover, we provide an explanation for these negative results. This explanation straightforwardly suggests a way for overcoming the detected conceptual problem. The general theoretically-founded solution is proposed in Section 5. In the same section, we present FOCL-OI, a new learning system based on FOCL, that replaces the model of generalization based on θ-subsumption with that based on θ_{OI}-subsumption. FOCL-OI has been empirically evaluated on those learning tasks that could not be solved by FOIL and FOCL and proved successful in avoiding the problem of non-terminating learning processes.

2 θ-subsumption and θ_{OI}-subsumption

Subsequently, we refer to [17] for what concerns the basic definitions of *substitution, positive* and *negative literal, clause, definite clause, program clause* and *Horn clause*. In particular, we denote a substitution (or *variable binding*) σ with $\{x_1 \leftarrow t_1, x_2 \leftarrow t_2, \ldots, x_n \leftarrow t_n\}$, where the terms $t_1, t_2, \ldots, t_n$ replace the variables $x_1, x_2, \ldots, x_n$, respectively. Furthermore, $dom(\sigma)$ denotes the set $\{x_1, x_2, \ldots, x_n\}$. Given a first order expression ϕ, $vars(\phi)$ is the set of the variables occurring in ϕ. We will consider a clause as the set of its literals. Henceforth, we denote with $\mathcal{C}$ the set of the clauses of the language and with $\mathcal{H}$ the subset of Horn clauses. The term *logic theory* (or simply *theory*) is used as a synonym of *logic program*, i.e. a set of program clauses.

2.1 θ-subsumption

Definition 1 (θ-subsumption). Let C, D be two Horn clauses. We say that D $\theta-subsumes$ C if and only if (iff) there exists a substitution σ such that $D\sigma \subseteq C$.
Without loss of generality, we always assume that $dom(\sigma) \subseteq vars(D)$.

θ-subsumption is a strictly weaker order relation than implication for first-order logic [12] [22] [25]. Here, we use the same order relation in [25], defined in terms of θ-subsumption, on the set $\mathcal{H}$ of the Horn clauses.

Definition 2 (Generality under θ-subsumption). Let C, D be two Horn clauses. We say that D is *more general than or equal to C* - or that C *is more specific than or equal to D* - and we write $C \leq D$ iff D θ-subsumes C, i.e.

$$C \leq D \text{ iff } \exists \; \sigma \; : \; D\sigma \subseteq C$$

Differently from Plotkin, we write $C \leq D$ to indicate that D is more general than C. When the condition above holds, we say that D is a *generalization* of C and C is a *specialization* of D. We write $C < D$ when $C \leq D$ and not$(D \leq C)$ and we say that D is *more general than C (D is a proper generalization of C)* or C *is more specific than D (C is a proper specialization of D)*. We write $C \sim D$ when $C \leq D$ and $D \leq C$ and we say that C is *equivalent to D*.

θ-subsumption induces a *quasi-ordering* upon the set of Horn clauses, that is, $\leq$ is reflexive and transitive, but not antisymmetric. Here, $\sim$ denotes the equivalence relation induced by $\leq$. Notice that it does not coincide with set equality. Indeed, two equivalent clauses under $\sim$ can be not only alphabetic variants, but even have a different number of literals, as for $\{P(x),\ P(f(y))\}$ and $\{P(f(z))\}$. Thus, as Plotkin [25] pointed out, there is a *reduced member* of any equivalence class under $\sim$ and this member is unique to within an alphabetic variant.

Definition 3 (Reduced clause). A clause C is *reduced* if it is not equivalent to any proper subset of itself. Formally, C is *reduced* iff $D \subseteq C$, $D \sim C$ implies that C = D.

Plotkin [25] reports an algorithm that returns a reduced clause D, given any clause C. The literals in $C \setminus D$ are called *redundant*.

Example 1. Consider the following clause

$$C = \{bicycle(X), \neg wheel(X,Y), \neg wheel(X,Z)\}$$

It is easy to see that C is not reduced. Indeed, the two literals $\neg wheel(X, Y)$, $\neg wheel(X, Z)$ can be unified through the substitution $\sigma = \{Z \leftarrow Y\}$ and the resulting factor [11]

$$C' = C\sigma = \{bicycle(X), \neg wheel(X,Y)\}$$

is logically equivalent to C and represents the reduced member of an equivalence class of Horn clauses.

Henceforth, we will always work on the quotient set $\mathcal{H}/\sim$ and, when convenient, we will denote with the name of a clause the equivalence class it belongs to. Indeed, the definition of $\leq$ straightforwardly extends to equivalence classes under $\leq$ as follows:

Given two clauses C and D, let $[C]_\sim$ and $[D]_\sim$ denote the corresponding equivalence classes under $\sim$. We say that $[C]_\sim \leq [D]_\sim$ iff $C \leq D$.

In [30], it is proved that $(\mathcal{H}/\sim, \leq)$ is a *complete lattice* and contains *ascending* and *descending chains of infinite length.*

An immediate consequence of the existence of infinite both ascending and descending chains is that the search space of any inductive system based on θ-subsumption is inherently infinite, when further language restrictions are not considered. Even more so, both specific-to-general and general-to-specific search strategies may go through an infinite path. This consideration points out the relevance of a proper definition both of the goal of the search - for instance, least general generalization vs. most general generalization - and of the language biases which, together with the adopted model of generalization, may reduce/increase the size of the search space, hopefully without reducing its scope (or at least reducing the scope only on meaningless clauses).

2.2 θ_{OI}-subsumption

We consider two distinct language biases, namely *linkedness* and *object identity*. The former constrains the kind of Horn clauses allowed, the latter limits the kind of substitutions considered. Subsequently, we formally define linkedness [12] and object identity.

Definition 4 (Linkedness). A Horn clause is *linked* if all of its literals are linked. A literal is linked if at least one of its arguments is linked. An argument of a literal is linked if either the literal is the head of the clause or another argument in the same literal is linked.

Linkedness is a restriction on the arguments of a clause that generally prevents a learning system from introducing meaningless literals in the body of the learned clause. Therefore, similar restrictions (sometimes exactly the same restriction) are adopted by many researchers and can be found in [17] with the name of *connected formulas*, in [33] with the name of *association chains*, as well as in [3] [5] [19].

Henceforth, $\mathcal{L}$ will denote the set of the linked Horn clauses. Obviously, $\mathcal{L}/\sim \subset \mathcal{H}/\sim \subset \mathcal{C}/\sim$.

Definition 5 (Object Identity). Variables with different names denote distinct objects (or *units*).

Object identity is a restriction on the kind of substitutions allowed. More precisely, substitutions are necessarily injective functions under such an assumption. This means that only one-to-one variable bindings are allowed, instead of many-to-one variable bindings. Formally, object identity corresponds to an equality theory with an axiom schema of the form $f(t_1, t_2, \ldots, t_n) \neq g(t'_1, t'_2, \ldots, t'_m)$ whenever $f \neq g$, where t_k and t'_k, $k = 1, 2, \ldots$, denote the terms of the language. The same axiom has been used by Clark [4], under the name of *inequality* axiom, and by Reiter [28] under the name of *unique name* axiom. For instance, the clause *arch(X, Y, Z) ← on-top(X, Y), left-of(Y, Z)* is equivalent to *arch(X, Y, Z) ← on-top(X, Y), left-of(Y, Z)*, $[X \neq Y], [X \neq Z], [Y \neq Z]$, when object identity holds. The object identity assumption corresponds to introduce *constraints* in form of *difference links* of the type $[X \neq Y]$ into the Horn clause representation of inductively learned rules. Semantically, object identity assumption corresponds to consider only *discriminant interpretations*. These interpretations have the property that each consistent set of formulas has a discriminant model [2].

Definition 6 (θ_{OI}-subsumption). Let C, D be two Horn clauses. We say that D θ_{OI}-*subsumes* C (D θ-*subsumes* C *under object identity*) iff there exists a substitution σ such that $D\sigma \subseteq C$ and σ is injective.

Definition 7 (Generality under θ_{OI}-subsumption). Let C, D be two Horn clauses. We say that D is *more general than or equal to* C *under object identity* and we write $C \leq_{OI} D$ iff D θ_{OI}-subsumes C, that is,

$$C \leq_{OI} D \text{ iff } \exists \sigma : D\sigma \subseteq C \text{ and } \sigma \text{ is injective}$$

Similarly to Definition 2, we can define the concepts of *(proper) generalization/specialization* and *equivalence* under object identity. The corresponding symbols will be endowed with the subscript OI.

Proposition 8. *Let C, D be two Horn clauses, then $C \leq_{OI} D \Rightarrow C \leq D$, while the opposite is not true in general.*

Therefore, θ_{OI}-subsumption is a model of generalization weaker than θ-subsumption.

Example 2. Consider again the clause of Example 1:

$$C = \{bicycle(X), \neg wheel(X, Y), \neg wheel(X, Z)\}$$

It is easy to see that, under object identity, C cannot be further reduced. Indeed, the two clauses C and C' = $\{bicycle(X), \neg wheel(X, Y)\}$ define two distinct equivalence classes under $\sim_{OI}$.

2.3 The algebraic structure of the search space

An analysis of the algebraic properties of the ordering relation defined by θ-subsumption led Plotkin to state that the partially ordered set $(\mathcal{C}/\sim, \leq)$ forms a lattice:

$$(\mathcal{C}/\sim, \vee_C, \wedge_C)$$

where the lattice operations are defined as follows (C_1 and C_2 are assumed to be variable disjoint) :

$$[C_1]_\sim \vee_C [C_2]_\sim = [lgg(C_1, C_2)]_\sim$$
$$[C_1]_\sim \wedge_C [C_2])_\sim = [C_1 \cup C_2]_\sim$$

By $lgg(C_1, C_2)$, we denote the least general generalization of the clauses C_1 and C_2 under θ-subsumption, that is:

Definition 9 (Least general generalization under θ-subsumption). A *least general generalization (lgg)* under θ-subsumption of two clauses is a generalization which is more specific than or equal to any other such generalization. Formally, given $C_1, C_2 \in \mathcal{C}$, C is a lgg of $\{C_1, C_2\}$ iff:

1. $C_i \leq C, i = 1, 2$
2. $\forall D$ such that $C_i \leq D$, $i = 1, 2 : C \leq D$
   $lgg(C_1, C_2) = \{C | C_i \leq C, i = 1, 2 \text{ and } \forall D \text{ such that } C_i \leq D, i = 1, 2 : C \leq D\}$

Least general generalizations are extremely important in the area of inductive learning, since they are the only generalizations that are guaranteed to be *correct*, whenever correct generalizations exist [13]. Definition 9 can be straightforwardly restricted to the subset $\mathcal{H}$ of Horn clauses. The lgg of any set of clauses is unique under $\sim$, i.e. $| lgg(C_1, C_2) |= 1$, where $| S |$ denotes the cardinality of the set S. This property does not hold when the lgg is computed under θ_{OI}-subsumption. Indeed, under object identity assumption, the set of the Horn clauses is no longer a lattice, but it is simply a *quasi-ordered* set. As a consequence, $(\mathcal{H}/\sim, \leq)$ is a lattice, while $(\mathcal{H}/\sim, \leq_{OI})$ is a partially ordered set. Partially ordered search spaces that are not lattices are *finitary* as to the problem of finding a lgg of two or more clauses, that is, there exists a set of lgg's of any pair of elements in $\mathcal{H}/\sim$ and this set is at most finite (since the only generalization operators are the dropping of a literal and, if needed, the turning of constants into variables), while lattices are unitary by definition (there exists a unique reduced lgg). The loss of the requirement of uniqueness and the possibility to have a set of incomparable lgg's is regarded as a desirable property by some authors [23]. As a consequence, when the generalization model of a learning algorithm is θ_{OI}-subsumption, Definition 9 should be modified as follows:

Definition 10 (Lgg under θ_{OI}-subsumption). A least general generalization under θ_{OI}-subsumption of two clauses is a generalization which is not more general than any other such generalization, that is, it is either more specific than or not comparable to any other such generalization.

Formally, given $C_1, C_2 \in \mathcal{C}$, C is a lgg of $\{C_1, C_2\}$ iff:

1. $C_i \leq_{OI} C$, $i = 1, 2$
2. $\forall D$ such that $C_i \leq_{OI} D$, $i = 1, 2 :\ not(D \leq_{OI} C)$
   $lgg(C_1, C_2) = \{C \mid C_i <_{OI} C,\ i = 1, 2 \text{ and } \forall D \text{ such that } C_i \leq_{OI} D,\ i = 1, 2 : not(D <_{OI} C)\}$

Of course, the same results apply to Horn clause logic, in particular to $(\mathcal{H}/\sim, \leq_{OI})$. The space $(\mathcal{H}/\sim, \leq_{OI})$ has another desirable property for the problem of generalization. In this space, there exist only finite strictly ascending chains, since the set of all generalizations of a clause C corresponds to the power set of C, i.e. the set of all subsets of the literals of C. Thus, each proper generalization of C has a number of literals less than the number of literals of C [32]. In formulae:

Let C be a clause and $GEN_{OI}(C)$ the set of all the generalizations of C under object identity. It holds that:

$$GEN_{OI}(C) = \{D \in \mathcal{H}/\sim \mid D \subseteq C\} = 2^C$$

As a consequence, the cardinality of $lgg_{OI}(C_1, C_2)$ in $(\mathcal{H}/\sim, \leq_{OI})$ is upper bounded by the following condition:

$$\mid lgg_{OI}(C_1, C_2) \mid \ \leq \ \mid GEN_{OI}(C_1) \cap GEN_{OI}(C_2) \mid \ \leq \ 2^n,$$

where $n = min\{\mid C_1 \mid, \mid C_2 \mid\}$ and $\mid A \mid$ denotes the number of literals in the clause A.

On the contrary, when many-to-one variable bindings are allowed, as under θ-subsumption, there exist infinite strictly ascending chains in the lattice of the Horn clauses, as previously stated.

Given two clauses C_1 and C_2, it holds that: $lgg(C_1, C_2) \leq D$, for each D in $lgg_{OI}(C_1, C_2)$. Thus, the lgg's generated by assuming θ_{OI}-subsumption as model of generalization may be overly general w.r.t. the corresponding unique θ-subsumption-based lgg.

Overgeneralization can be dealt with by exploiting the proper refinement operator. A consistency-preserving refinement operator that copes with the problem of finding a *most general specialization (mgs)* in the generalization model defined by θ_{OI}-subsumption can be found in [6] [8].

All the results above extend straightforwardly to the proper subset of the linked Horn clauses.

As shown by Buntine [3], surprisingly generalization and specialization hierarchies possess very different properties. In the following, by *size* of the search space we mean the number of equivalence classes in a specialization/generalization hierarchy. Given a partially ordered set $(S/\sim, \leq)$ and a clause $C \in S/\sim$, by *scope* of the search *in a generalization hierarchy* we mean the set of the generalizations of C, that is:

$$GEN(C) = \{D \in \mathcal{S}/\sim|\, C \leq D\}$$

while by *scope* of the search *in a specialization hierarchy* we mean the set of the specializations of C, that is:

$$SPEC(C) = \{D \in \mathcal{S}/\sim|\, D \leq C\}$$

Linkedness reduces the size of both specialization and generalization hierarchies, as well as the scope of the search. Conversely, object identity increases the size of the hypothesis space, as Example 2 shows, but it reduces the scope of the search.

3   FOIL and FOCL

At the high level, both FOIL and FOCL adopt a *separate-and-conquer* search strategy. The *separate* stage of the algorithm is basically a loop that checks for the completeness of the current logical theory and, if this check fails, begins the search for a new consistent clause, while the *conquer* stage performs a general-to-specific hill-climbing search to construct the body of the new clause. The evaluation function used in the latter search is the information theoretic heuristic called *information gain*. Therefore, the search space for the separate stage is $2^{\mathcal{L}/\sim}$, i.e. the space of logic theories. The conquer stage searches for a consistent clause in the only specialization hierarchy of $(\mathcal{L}/\sim, \leq)$ rooted into the linked Horn clause whose head contains the predicate to be learned and whose body is empty, i.e. $P(X) \leftarrow$, where P is the predicate that denotes the concept to be learned.

In FOIL, each k-ary predicate is associated with a *relation* consisting of the set of k-tuples of constants that satisfy that predicate. Such predicates are called *extensional* or *operational*. The relation associated to the predicate that we want to learn is called *target relation*. Each relation, including the target relation, defines both *positive* and *negative instances* of the predicate. Negative instances of a predicate can be defined either explicitly or according to the *closed-world assumption* (CWA) of Prolog. In this last case, negative instances consist of all constant k-tuples other than those considered positive and they are generated by temporarily making a *domain closure assumption* (DCA) [17]. A peculiarity of FOCL is that it is possible to define and use *intensional* or *non-operational predicates*, besides extensional ones. A predicate is defined intensionally when it appears in the head of an inference rule expressed as a Horn clause.

Given a k-ary predicate P, FOIL and FOCL learn a predicate definition or *rule* for P, that is, a set of definite Horn clauses in which $P(X_1, X_2, \ldots, X_k)$ is the literal in the head. Each clause of this rule should be satisfied only by positive tuples (*consistency*), while all clauses together should cover all positive instances of $P(X_1, X_2, \ldots, X_k)$ (*completeness*). Each literal in the body of a clause takes one of the four forms $X_j = X_k, X_j \neq X_k, Q(V_1, V_2, \ldots, V_t), \neg Q(V_1, V_2, \ldots, V_t)$, where the X_i's are variables already introduced by previous literals (*old* variables) in the body, the V_i's can be both old and *new* variables, and Q is the same predicate associated with a relation. Therefore, FOIL and FOCL learn *function-free* Horn clauses, since terms other than variables cannot occur as arguments of a literal.

Finally, we observe that the hill-climbing search strategy, used by FOIL and FOCL while building a single clause, suffers from the *horizon effect*: A literal that proves to be desirable or even essential from a global perspective (the whole clause) may appear relatively unpromising at a local level (a partially developed clause) and so may be left out. In order to avoid this problem, FOIL introduces *determinate literals* [27], while FOCL exploits some user-defined *relational clichés* [31] that suggest potentially useful combinations of literals to be tested while generating a clause of a predicate definition.

4 Experiments with FOIL and FOCL

This section presents some experimental results obtained by running FOIL and FOCL on a classical task from the machine learning literature. Such results point out that both FOIL and FOCL may give rise to non-terminating learning processes.

The objects in Figure 1a are the training examples supplied to FOIL and FOCL in order to learn the nature of an arch much like the arch which Winston describes [34]. The set $A = \{A_1, A_2\}$ includes examples of arches and the set $N = \{N_1, N_2, N_3\}$ contains examples of objects that are not arches. The predicates used to describe the objects are reported in Figure 1b.

Both FOIL and FOCL generate no clause, that is, they are not able to find a definition for the concept *arch* from the given training set.

The problem with FOIL is that it adopts a *stopping criterion* for the conquer

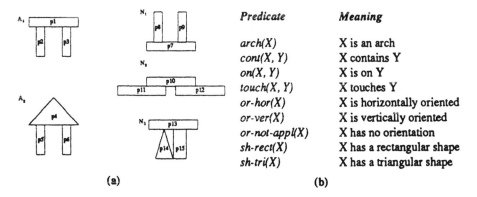

Predicate	Meaning
arch(X)	X is an arch
cont(X, Y)	X contains Y
on(X, Y)	X is on Y
touch(X, Y)	X touches Y
or-hor(X)	X is horizontally oriented
or-ver(X)	X is vertically oriented
or-not-appl(X)	X has no orientation
sh-rect(X)	X has a rectangular shape
sh-tri(X)	X has a triangular shape

(a) (b)

Fig. 1. (a) Training set for the problem *arch*. (b) Description language for the problem *arch*.

stage that is based on Rissanen's Minimum Description Length (MDL) principle [29]. According to this criterion, an inconsistent clause is not extended further if the bits required to encode the new clause, obtained by adding the literal with the maximum information gain, exceed the bits needed to indicate the covered tuples. The elimination of this stopping criterion makes FOIL converge to the following clause:

$$arch(A) : - cont(A,B), or\text{-}ver(B), on(C,B), sh\text{-}rect(B),$$
$$touch(C,D), sh\text{-}rect(D), B <> D$$

which is complete and consistent.

Conversely, FOCL fails because no literal has a positive information gain, and this suggests the use of relational clichés. Indeed, in a second running, we introduced the relational cliché $CONT$, by which FOCL is said to test the pairs of literals $cont(X,Y), Q(V_1, V_2, \ldots, V_t)$, rather than to limit the search by testing separately the literals of the kind $cont(X,Y)$ and $Q(V_1, V_2, \ldots, V_t)$.

Even in this case, FOCL is not able to generate a complete and consistent concept definition and, moreover, the learning process does not converge, that is, FOCL's search does not terminate. An analysis of the ongoing running of FOCL provides a simple explanation of this behaviour. A snapshot of the state of FOCL's search at a time t is the following:

C_1 $arch(?0) : - cont(?0, ?1), or\text{-}not\text{-}appl(?1)$

C_2 $arch(?0) : - cont(?0, ?1), or\text{-}ver(?1), on(?2, ?1), sh\text{-}rect(?1),$
$\qquad\qquad cont(?0, ?3), sh\text{-}rect(?3),$
$\qquad\qquad cont(?0, ?4), sh\text{-}rect(?4), \ldots$

The first clause is consistent and covers only the positive example A_2. The second clause is generated by FOCL in the attempt to achieve completeness. Actually, this clause is a partial one, but it is enough to give us an insight of what is going on: *FOCL's conquer stage searches in the wrong specialization hierarchy. It performs its search in the lattice* $(\mathcal{L}, \leq)$ *of all the linked Horn clauses rather than in the lattice* $(\mathcal{L}/\sim, \leq)$ *of all the equivalence classes* having a linked Horn clause as a reduced member.

Indeed, it is easy to see that the second clause in the rule for *arch* is logically equivalent to the clause:

C_3 $arch(?0) : - cont(?0, ?1), or\text{-}ver(?1), on(?2, ?1), sh\text{-}rect(?1)$

FOCL does not implement an algorithm for testing the equivalence of the newly generated hypothesis with respect to the previous one. As a consequence, the search continues by generating and testing hypotheses that belong to the same equivalence class as the previously generated hypothesis, so remaining inside the same node of the specialization hierarchy and with no means of escape. In other words, the conquer stage in FOCL defines incorrectly its termination condition, therefore there is no guarantee of termination. In fact, the pair of literals *cont(?0, ?n), sh-rect(?n)* turns out to have the maximum information gain at each step of the search for a new literal to add. Moreover, setting one of FOCL's built-in functions, namely the *eql* built-in function that allows the system to test also the literals *eql(X, Y)* and their negations (alphabetic variants of $X = Y$ and $X \neq Y$, respectively), has no effect on the results, since these literals have a low information gain.

It is interesting to observe that Quinlan [26] reports a result obtained by FOIL on the task of learning the concept of an arch. Indeed, FOIL is released with some data files. The file containing the input data for the *arch* problem is *winston.d*. If we compare the experiment performed by Quinlan to the similar experiment presented above, we can easily observe that Quinlan's formulation of the *arch* problem makes the learning process easier. In fact, FOIL learns the predicate *arch(A, B, C)* instead of *arch(A)*. This is apparently a slight difference, but as a matter of fact this representation biases the learning system towards the generation of concept definitions whose clauses are both *domain-restricted* and *range-restricted* [5]. Our formulation of the *arch* problem is more similar to that in [16] and requires that the learning system be able to determine autonomously the number of components (and their characteristics) that are sufficient to discriminate the positive instances of the target concept from the negative ones.

The conclusion that FOCL searches in the lattice $(\mathcal{L}, \leq)$ rather than in $(\mathcal{L}/\sim, \leq)$ applies to FOIL, as well. An empirical proof of this is given in [9]. Furthermore, learning tasks from real-world domains, concerning the field of document understanding [7] [18] [30], showed the same problem of non-termination, as that presented above. This points out that the conceptual deficiency of FOIL and FOCL's conquer stage may become manifested in real world learning tasks, in addition to laboratory-sized ones.

5 FOCL-OI

In this section, we consider the negative results of FOIL and FOCL and propose a theoretical solution that prevents these systems from going through the troubles shown in Section 4. Moreover, a practical solution is proposed and implemented in a new version of FOCL, called FOCL-OI.

The theoretical solution is to change the search space of the conquer stage of FOIL and FOCL from $(\mathcal{L}, \leq)$ to $(\mathcal{L}/\sim, \leq)$. Practically speaking, this means to provide the conquer stage of the high level separate-and-conquer strategy with a procedure that performs a θ-subsumption equivalence test. This procedure should check if the current clause is equivalent (under $\sim$) to the clause obtained by adding the literal (or a combination of literals, if clichés are used) with the highest information gain. In the general case, this is an NP-complete problem, since it involves a test for θ-subsumption, which is NP-complete [10]. In the particular case of FOIL and FOCL, the current clause C is a proper subset of the clause D, obtained by adding the literal with the maximum information gain, since the conquer stage performs a general-to-specific search. Thus, the procedure needs to check only if D θ-subsumes C. In formulae, $C \subset D$ and we want to know if $C \sim D$. But $C \subset D$ implies that $D \leq C$, thus it is enough to check whether $C \leq D$.

However, let us observe that, for the *conquer* stage of FOIL and FOCL, the following proposition holds:

Proposition 11. *Let C and D be two definite Horn clauses:*

$$C \text{ reduced and } C \subset D \text{ and } vars(C) = vars(D) \Rightarrow [C]_\sim \neq [D]_\sim$$

Proof. Let us suppose that C and D are two definite Horn clauses s.t.: C is reduced, $C \subset D$, $vars(C) = vars(D)$ (i.e., D does not introduce new variables) and $[C]_\sim = [D]_\sim$. Let n be the size of C, i.e. $\mid C \mid = n$. By hypothesis, $C \subset D$, thus we can suppose $C = \{L_1, L_2, \ldots, L_n\}$ and $D = \{L_1, L_2, \ldots, L_n, L_{n+1}, \ldots, L_{n+m}\}$, $m \leq 1$.
We prove the proposition by induction on m.
$m = 1$.
Ad absurdum.
$D = C \cup \{L_{n+1}\}$, where L_{n+1} is necessarily a literal in the body of D. Furthermore,

$$vars(C) = vars(D) \Rightarrow vars(\{L_{n+1}\}) \subseteq vars(C) \tag{1}$$

By definition, $[C]_\sim = [D]_\sim$ means $C \leq D$ and $D \leq C$.
$C \leq D \Rightarrow \exists \, \sigma : D\sigma \subseteq C \Rightarrow (C \cup \{L_{n+1}\})\sigma \subseteq C \Rightarrow C\sigma \subseteq C$ and $\{L_{n+1}\}\sigma \subseteq C$.
Let us consider $C\sigma$.
We know that

$$C\sigma \subseteq C \tag{2}$$

$C\sigma \subseteq C \Rightarrow C \leq C\sigma$ and, obviously, it holds also $C\sigma \leq C$ (since $C\sigma = C\sigma \Rightarrow C\sigma \subseteq C\sigma$), thus

$$C\sigma \sim C \tag{3}$$

Now, either $C\sigma \subset C$ or $C\sigma = C$ holds.
If $C\sigma \subset C$ holds, then $C\sigma \neq C$. But (2) and (3) and $C\sigma \neq C$ imply that C is not reduced, which contradicts the first hypothesis.
If $C\sigma = C$ holds, then it suffices to consider $\sigma = \{\}$. But, then $\{L_{n+1}\}\sigma \subseteq C$ is equivalent to say that $\{L_{n+1}\} \subseteq C$ by (1), which contradicts the hypothesis that $C \subset D$.
As a consequence, it holds $[C]_\sim \neq [D]_\sim$
$m > 1$.
Let $D' = C \cup \{L_{n+1}, L_{n+2}, \ldots, L_{n+m-1}\}$, then $D = D' \cup \{L_{n+m}\}$ and $vars(D) = vars(D') \cup vars(\{L_{n+m}\})$. Moreover, by hypothesis, $vars(C) = vars(D)$, then $vars(C) = vars(D') \cup vars(\{L_{n+m}\}) \Rightarrow vars(D') \subseteq vars(C)$.
Thus, $vars(\{L_{n+m}\}) \subseteq vars(C)$.
It follows that:
$vars(C) = vars(D) = vars(D') \cup vars(\{L_{n+m}\}) = vars(D')$.
Now, C is reduced, $C \subseteq D'$, and $vars(C) = vars(D')$. In the trivial case $C = D'$, the proof is reduced to the basic case for $m = 1$ shown above. In the case $C \subset D'$, we can state that $[C]_\sim \neq [D']_\sim$, from the inductive hypothesis. We want to prove

that: $[C]_\sim = [D]_\sim \Rightarrow [C]_\sim = [D']_\sim$, which contradicts the inductive hypothesis. Let us suppose $[C]_\sim = [D]_\sim$, then, by definition, $C \leq D$ and $D \leq C$.

$$C \leq D \Rightarrow \exists \sigma : D\sigma \subseteq C \Rightarrow (D' \cup \{L_{n+m}\})\sigma \subseteq C \Rightarrow D'\sigma \subseteq C \Rightarrow C \leq D' \quad (4)$$

But

$$D = C \cup \{L_{n+1}, L_{n+2}, \ldots, L_{n+m-1}\} \cup \{L_{n+m}\} \quad (5)$$
$$= D' \cup \{L_{n+m}\} \Rightarrow C \subseteq D' \Rightarrow D' \leq C \quad (6)$$

(4) and (5) imply that $[C]_\sim = [D']_\sim$. □

In other words, it is necessary to perform a θ-subsumption equivalence test if and only if the new clause D introduces *new* variables. Therefore, since in FOIL and FOCL we have $D = C \cup \{L_{n+1}\}$, where L_{n+1} is the literal with maximum information gain, and C and D are linked, it is sufficient to search for a unification between L_{n+1} and one of the n literals in C. This search has a linear complexity in the size of the clause. Indeed, it can be proved that there exists no ascending chain of infinite length $\{C_i\}_{i \geq 0}$ such that $C_i \leq C_{i+1}$ and $C_i \subset C_{i+1}$. The trivial case in which $L_{n+1} \in C$ can be easily ruled out by preventing the search from adding a literal that already exists in the body of the currently generated clause. As a consequence, the computational complexity of the procedure that performs a θ-subsumption equivalence test can be reduced to $O(n)$, where n is the size of the current clause C. Indeed, from the literature about unification [16], it is known that the best unification algorithms between terms have a linear complexity in time.

For the sake of completeness, we have to observe that a literal which appears to be redundant in a partially developed clause (*local redundancy*) may no longer be such when further literals are added to the clause. There are two possible solutions to this short-sightedness of the search. A simple one consists in testing pairs (triples, foursomes, ...) of literals, when there is no single literal that helps to jump to a different equivalence class, while the other solution consists in testing pairs of literals $\{L_{n+1}, L_{n+2}\}$ as soon as L_{n+1} turns out to be redundant.

The equivalence test is no longer necessary when we change the underlying model of generalization from θ-subsumption to θ_{OI}-subsumption, which causes a change of FOCL's search space from $(\mathcal{L}, \leq)$ to $(\mathcal{L}, \leq_{OI})$. Indeed, it is easy to see that Proposition 11 holds even under object identity. Moreover, the introduction of a new variable always causes a change of the equivalence class under object identity. Thus, we can conclude that searching into $(\mathcal{L}, \leq_{OI})$ is equivalent to searching into $(\mathcal{L}/\sim, \leq_{OI})$. In order to avoid non-termination problems, we changed the underlying model of generalization of FOCL from θ-subsumption to θ_{OI}-subsumption. This new version of FOCL, called FOCL-OI, implements object identity in a modular way. Indeed, the end-user/teacher can choose the type(s) of variables for which object identity is assumed.

Below, the complete and consistent concept definition produced by FOCL-OI with the same input data - training set and system parameters - as those used in the experiment described in Section 4 is reported.

C_4 arch(?0) : $-$ cont(?0, ?1), or-not-appl(?1)

C_5 arch(?0) : $-$ cont(?0, ?1), or-ver(?1), on(?2, ?1), not(?1 = ?2)
sh-rect(?1), cont(?0, ?3), not(?1 = ?3),
not(?2 = ?3), sh-rect(?3)

In this case, however, the typing information has been properly changed in order to assume object identity for the variables that represent a *part* of an arch. The object identity assumption allows FOCL to prevent the problem of non-termination of the search, when learning a complete and consistent rule for the concept of an *arch*.

6 Conclusions and future work

All learning systems need to clarify the adopted model of generalization and the space in which they perform the search for concept definitions. Formal methods and techniques can be useful to detect potential sources of problems and conceptual shortcomings of the existing empirical learning systems and to suggest the suitable counteractions in order to improve their performance.

This paper constitutes an attempt to apply this analytical approach to two well-known learning systems, FOIL and FOCL. Plotkin's logical framework to inductive generalization [25] proved useful both to point out some lacks that affect the search strategy of these systems and to suggest straightforwardly the adequate correction. This correction has been implemented in a new version of FOCL, called FOCL-OI, and proved effective to overcome non-termination problems. Future work will concern the implementation of a new version of FOCL in which, at first, the θ_{OI}-subsumption generalization model is fully adopted (for any type of variables) and then, the system autonomously learns the *equality literals* in the form $[X_i = X_j]$, if necessary. Theoretically, this new learning strategy should prove more effective and efficient than that in which θ-subsumption and *difference links* $[X_i \neq X_j]$ have to be learned, since θ_{OI}-subsumption is more manageable than θ-subsumption, while learning the equality literals would happen rarely, according to a *when-needed* strategy.

Finally, this analytic approach to concept learning allows us to identify the limits of commonly accepted theories and methodologies for inductive learning and to critically revise and reformulate them on the ground of the cited logical frameworks. For instance, we can now state that, under object identity, adding a literal to a clause is always a specializing rule (unless the added literal already exists in the clause), while, by dropping such assumption, this is no longer true. More precisely, it is true if and only if the new clause is not equivalent to the previous one. In a similar manner, the *dropping condition* generalization rule [20], holds under θ_{OI}-subsumption, but it is no longer a generalization rule under θ-subsumption if the new clause belongs to the same equivalence class as the previous one. The immediate consequence of this observation is that the generalization rules in [20] should be reformulated (indeed, Michalski uses implication as generalization model and the same observation applies also to implication) or, alternatively, it should be stated clearly that they hold under θ_{OI}-subsumption.

References

1. Bell, S., and Weber, S., On the close logical relationship between FOIL and the frameworks of Helft and Plotkin, Proceedings of The Third Int'l Workshop on Inductive Logic Programming ILP'93, Bled, Slovenia, 1-10, 1993.
2. Bossu, G., and Siegel, P., Saturation, Nonmonotonic Reasoning and the Closed-World Assumption, Artificial Intelligence, 25, 13-63, 1985.
3. Buntine, W., Generalized Subsumption and Its Applications to Induction and Redundancy, Artificial Intelligence, 36, 149-176, 1988.
4. Clark, K.L., Negation as failure, in Logic and Databases, H. Gallaire and J. Minker (Eds.), 293-321, Plenum Press, New York, 1978.
5. De Raedt, L., Interactive Theory Revision, Academic Press, San Diego, CA, 1992.
6. Esposito, F., Malerba, D., and Semeraro, G., Specialization in Incremental Learning: The Negation Operator, Proceed. of the AAAI-93 Spring Symp. Series on "Training Issues in Incremental Learning", Stanford, CA, 1993.
7. Esposito, F., Malerba, D., Semeraro, G., and Pazzani, M., A Machine Learning Approach To Document Understanding, Proceed. of the 2nd Int'l Workshop on Multistrategy Learning MSL-93, Harpers Ferry, West Virginia, 276-292, 1993.
8. Esposito, F., Malerba, D., and Semeraro, G., Negation as a Specializing Operator, in Advances in Artificial Intelligence - Proceedings of the Third Congress of the Italian Association for Artificial Intelligence AI*IA 93, Lecture Notes in Artificial Intelligence 728, P. Torasso (Ed.), Springer-Verlag, Turin, Italy, 166-177, 1993.
9. Esposito, F., Malerba, D., Semeraro, G., Brunk, C. and Pazzani, M., Traps and Pitfalls when Learning Logical Definitions, Proceedings of the 8th International Symposium on Methodologies for Intelligent Systems, Charlotte, North Carolina, 1994 (to appear)
10. Garey, M.R., and Johnson, D.S., Computers and Intractability, Freeman, San Francisco, CA, 1979.
11. Genesereth, M.R., and Nilsson, N.J., Logical Foundations of Artificial Intelligence, Morgan Kaufmann, Palo Alto, CA, 1987.
12. Helft, N., Inductive Generalization: A Logical Framework, in Progress in Machine Learning - Proceedings of EWSL 87, I. Bratko & N. Lavrac (Eds.), Sigma Press, Bled, Yugoslavia, 149-157, 1987.
13. Idestam-Almquist, P., Generalization under Implication by Recursive Antiunification, Proceedings of the Tenth International Conference on Machine Learning, Amherst, MA, 151-158, 1993.
14. Kietz, J.U., A Comparative Study Of Structural Most Specific Generalizations Used In Machine Learning, in Logical Approaches to Machine Learning - Workshop Notes: 10th ECAI, Vienna, Austria, 1992.
15. Knight, K., Unification: A Multidisciplinary Survey, ACM Computing Surveys, Vol.21, No.1, 1989.
16. Larson, J.B., Inductive Inference in the Variable Valued Predicate Logic System VL21: Methodology and Computer Implementation, Ph.D. dissertation, Dept. of Computer Science, University of Illinois, Urbana, Illinois, May 1977.

17. Lloyd, J.W., Foundations of Logic Programming, Second Edition, Springer-Verlag, New York, 1987.
18. Malerba, D., Document Understanding: A Machine Learning Approach, Technical Report, Esprit Project 5203 INTREPID, March 1993.
19. Manago, M., and Kodratoff, Y., Model-Driven Learning of Disjunctive Concepts, in Progress in Machine Learning - Proceedings of EWSL 87, I. Bratko & N. Lavrac (Eds.), Sigma Press, Bled, Yugoslavia, 183-198, 1987.
20. Michalski, R.S., A Theory and Methodology of Inductive Learning, Artificial Intelligence, 20, 111-161, 1983.
21. Muggleton, S., Inductive Logic Programming, New Generation Computing, 8(4), 295-318, 1991.
22. Niblett, T., A study of generalization in Logic Programs, Proceedings of the Third European Working Session on Learning, Pitman, London, 131-138, 1988.
23. Nienhuys-Cheng, S.H., van der Laag, P.R.J., van der Torre, L.W.N., Constructing refinement operators by decomposing logical implication, in Advances in Artificial Intelligence - Proceedings of the Third Congress of the Italian Association for Artificial Intelligence AI*IA 93, Lecture Notes in Artificial Intelligence 728, P. Torasso (Ed.), Springer-Verlag, Turin, Italy, 178-189, 1993.
24. Pazzani, M., and Kibler, D., The utility of knowledge in inductive learning, Machine Learning 9, 1, 57-94, 1992.
25. Plotkin, G.D., A Note on Inductive Generalization, in Machine Intelligence 5, B. Meltzer and D. Michie (Eds.), 153-163, Edinburgh University Press, 1970.
26. Quinlan, J. R., Learning Logical Definitions from Relations, Machine Learning 5, 3, 239-266, 1990.
27. Quinlan, J. R., Determinate Literals in Inductive Logic Programming, Proceedings of the 11th International Joint Conference on Artificial Intelligence, Sydney, Australia, 746-750, 1991.
28. Reiter, R., Equality and domain closure in first order databases, Journal of ACM, 27, 235-249, 1980.
29. Rissanen, J., A universal prior for integers and estimation by minimum description length, Annals of Statistics, 11, 1, 416-431, 1983.
30. Semeraro, G., Brunk, C.A., and Pazzani M.J., Traps and Pitfalls when Learning Logical Theories: A Case Study with FOIL and FOCL, Technical Report 93-33, Department of Information and Computer Science, University of California, Irvine, California, July 26, 1993.
31. Silverstein, G., and Pazzani, M., Relational clichés: constraining constructive induction during relational learning, Proceedings of the Eighth International Workshop on Machine Learning, Evanston, Illinois, 203-207, 1991.
32. VanLehn, K., Efficient Specialization of Relational Concepts, Machine Learning 4, 1, 99-106, 1989.
33. Vere, S.A., Multilevel Counterfactuals for Generalizations of Relational Concepts and Productions, Artificial Intelligence, 14, 139-164, 1980.
34. Winston, P.H., Learning Structural Descriptions from Examples, Ph.D. dissertation, Department of Electrical Engineering, Massachussetts Institute of Technology, Cambridge, MA, January 1970.
35. Wirth, R., and O'Rorke, P., Constraints on Predicate Invention, Proceedings of the Eighth International Workshop on Machine Learning, Evanston, Illinois, 457-461, 1991.

Propagation of Inter-argument Dependencies in "Tuple-distributive" Type Inference Systems

Christine SOLNON, Michel RUEHER

I3S, UNSA / CNRS, Route des Colles, B.P. 145,
06903 Sophia Antipolis Cedex, FRANCE
e-mail: {solnon,mr}@mimosa.unice.fr

Abstract. Many type inference systems for Prolog programs are based on the tuple-distributive closure abstraction which ignores inter-argument dependencies. Thus, dependencies specified by head-only shared variables cannot be handled, and the inferred types are often very inaccurate. In this paper, we define an unfolding process which propagates such inter-argument dependencies: each call to a predicate that contains head-only shared variables is replaced by its definition. Hence, dependencies are actually propagated and the accuracy of the inferred types is improved. This unfolding process is repeated until a fix-point is reached in the computation of the type system. Termination is ensured by an abstraction function which limits the depth of recursive structures.

1 Motivations

Prolog is an untyped language. This provides the programmer with flexibility for rapid prototyping, but does not facilitate the debugging, optimization and reuse of programs. Hence, different type inference systems have been proposed for statically approximating the denotation of Prolog programs. Many of these type inference systems are based on the tuple-distributive closure abstraction which ignores inter-argument dependencies. For example, [YS87] proposed to define types with respect to a fix-point of the abstract immediate consequence operator T_P^α. This operator just ignores inter-argument dependencies by computing the distributive closure of tuples (i.e., terms and atoms). [Mis84] proposed to infer set constraints from a Prolog program in order to approximate its denotation, and [HJ92b] showed that a model of these set constraints corresponds to a fix-point of the abstract operator T_P^α. [Hei92b] also proposed a type inference system based on set constraints like the one of [Mis84]. This type system only ignores inter-variable dependencies rather than all inter-argument dependencies and the computed types are a bit more accurate. Tuple-distributive types can also be computed by using regular unary logic (RUL) programs [FSVY91, GdW94].

However, in all these works inter-variable dependencies are ignored. As a consequence, types inferred from programs containing head-only shared variables are often very inaccurate. A *head-only shared variable* is a variable which occurs more than once in the head of a clause but does not occur in its body. Such variables can be seen as type parameters that are universally quantified on the Herbrand universe. Hence, we shall say that a predicate is *parametric* if it contains head-only shared variables in its definition. Parametric predicates express dependencies (i.e., unification constraints between different occurrences of head-only shared variables) that are ignored by tuple-distributive type systems. Consider for example the following program P1:

    eq(X,X).
    p(Y) :- eq(Y,1).

The predicate eq is parametric and the variable X expresses a dependency relation between the two arguments of eq/2. The denotation[1] of P1 is

$$lfp(T_{P1}) = \{ \text{ eq(t,t) } / \text{ t} \in U_{P1} \} \cup \{ \text{ p(1) } \}$$

However, tuple-distributive type inference systems ignore this dependency and the computed approximation for $P1$ is

$$\{ \text{ eq(s,t) } / \text{ t} \in U_{P1} \text{ and } \text{s} \in U_{P1} \} \cup \{ \text{ p(t) } / \text{ t} \in U_{P1} \}$$

Hence, the type associated with the argument of the predicate p is the Herbrand universe and does not bring any significant information[2].

In order to propagate inter-argument dependencies, we propose to *unfold the program to be typed with respect to its parametric predicates*, i.e., each call to a parametric predicate q is replaced by the definition of q. Let us consider again the previous program P1: the predicate p depends on eq which is parametric. Thus, we unfold p with respect to eq, and we obtain the new program $P1'$:

    eq(X,X).
    p(1).

These two programs have a same denotational semantics, but $P1'$ does not contain any call to parametric predicates, and the dependency constraint specified by eq between Y and 1 has been propagated.

The major problem of this approach remains in the termination of the unfolding process: in the case of recursive parametric predicates, each unfolding step may introduce some new clauses that have to be unfolded again. Hence, we propose to perform the unfolding process until the associated type system becomes stable, and we introduce an abstraction function to ensure the stabilization of the type system.

The rest of the paper is organized as follows. In section 2, we define the tuple distributive closure operator α and we present the type inference system which

---

[1] We note U_{P1} the Herbrand universe of P1. On this very short program, U_{P1} only contains the value 1, but if $P1$ is part of a program with functional terms, then the Herbrand universe is infinite.

[2] In a more general way, a shared variable in a clause expresses a dependency relation that is ignored by the abstraction and that implies a loss of accuracy. For example, if the predicate integer defines a set Int of integers, then the type of the predicate "eq_int(X,X) :- integer(X)." is { eq_int(s,t) / s∈Int and t∈Int }, and the type associated with the predicate "p_int(X) :- eq_int(X,1)." is the set Int.

is used to illustrate our approach. In section 3, we define the unfolding process that propagates inter-argument dependencies specified by parametric predicates. In section 4, we show that the new inferred types actually approximate the denotation of the program. In section 5, we define an abstraction function which ensures the termination of the unfolding process. In section 6, we illustrate our approach on two examples. Section 7 discusses related works.

2 Definitions

We shall use the following notations, possibly subscripted:

- f and p respectively denote functional and predicate symbols.
- t denotes a term. A term is either a constant c, a variable X or a functional term $f(t_1, \ldots, t_n)$.
- A denotes a logic atom and D a set[3] of atoms. θA denotes the instance of A obtained by the substitution θ. $(A \leftarrow D)$ denotes a Horn clause.
- P denotes a set of clauses, i.e., a Prolog program. U_P and B_P respectively denote the Herbrand universe and base of P. T_P is the usual immediate consequence operator. $lfp(T_P)$ is the least fix-point of T_P and corresponds to the denotation of P.

2.1 Definition of Positions

Positions are used to refer to sets of terms within sets of atoms, and correspond to types. A position, noted π, is either the empty position $<>$ or a sequence like

$$p(i) \cdot f_1(i_1) \cdot f_2(i_2) \cdot \ \ldots \ \cdot f_n(i_n) \quad (n \geq 0)$$

where $1 \leq i \leq arity(p)$ and $\forall j \in 1..n$, $1 \leq i_j \leq arity(f_j)$. Intuitively, a position π allows one to refer to a sub-term t in an atom A: in the tree that represents A, π denotes the path between the root of A and the root of t. We note $pos(A)$ the set of all the paths between the root of A and all the nodes and leaves of A. If π is a position of A, we note $A_{|\pi}$ the sub-term of A at position π.

Let us consider, for instance, the atom $A = p(c_1, f(c_2))$. This atom defines the positions $pos(A) = \{p(1), p(2), p(2).f(1)\}$. The sub-term of A at position $p(1)$, noted $A_{|p(1)}$ is c_1, the sub-term of A at position $p(2)$, noted $A_{|p(2)}$, is $f(c_2)$ and the sub-term of A at position $p(2).f(1)$, noted $A_{|p(2).f(1)}$, is c_2.

More formally, the set of positions of an atom or a term is:

$pos(p(t_1, \ldots, t_n)) = \bigcup_{i=1}^{n} \{p(i).\pi \ / \ \pi \in pos(t_i)\}$
$pos(f(t_1, \ldots, t_n)) = \{<>\} \cup \bigcup_{i=1}^{n} \{f(i).\pi \ / \ \pi \in pos(t_i)\}$
$pos(t) = \{<>\}$ if t is a variable X or an atomic constant c

We define the sub-terms of a term or an atom as follows:

---

[3] As we are only concerned with fix-point semantics, we assume without loss of generality that the body of a clause is a set (rather than a multi-set) of atoms.

$$p(t_1, ..., t_n)_{|p(i).\pi} = t_{i|\pi} \text{ if } i \in 1..n$$
$$p(t_1, ..., t_n)_{|p'(i).\pi} = \bot \text{ if } p \neq p' \text{ or } i \notin 1..n$$
$$f(t_1, ..., t_n)_{|f(i).\pi} = t_{i|\pi} \text{ if } i \in 1..n$$
$$f(t_1, ..., t_n)_{|f'(i).\pi} = \bot \text{ if } f \neq f' \text{ or } i \notin 1..n$$
$$t_{|<>} = t$$

By extension, the set of the sub-terms at position π in a set D of atoms is

$$D_{|\pi} = \{A_{|\pi} \ / \ A \in D \text{ and } \pi \in pos(A)\}$$

2.2 Definition of the tuple-distributive closure operator

The tuple-distributive closure operator, noted α, approximates a set S of tuples by ignoring dependencies between arguments of tuples as follows:

$$\alpha(S) = \{c \ / \ c \in S\} \cup \{f(t_1, ..., t_n) \ / \ \forall i \in 1..n, \ t_i \in \alpha(S_{|f(i)})\}$$

For example, $\alpha(\{f(1, 2), f(3, 4)\}) = \{f(1, 2), f(1, 4), f(3, 2), f(3, 4)\}$.

The abstract operator T_P^α is defined by the composition of α with the immediate consequence operator T_P, i.e., $T_P^\alpha = \alpha \circ T_P$. The least fix-point of T_P^α, noted $lfp(T_P^\alpha)$, approximates the denotation of the program by ignoring inter-argument dependencies. This set is an interesting basis for defining types [YS87].

2.3 The type inference system

We have defined in [SR94] a type inference system that is more particularly suited to analyse uncomplete Prolog programs. This type inference system ignores inter-argument dependencies and we shall use it to illustrate our approach. This system is based on the inference of a collection of equations that describes set relationships that hold between terms of the program. A set equation looks like $< \pi = E >$ where the position π denotes a type, and E is a set expression:

$$E ::= c \mid \top \mid \bot \mid \pi \mid f(E_1 \times ... \times E_r) \mid E_1 \sqcup E_2 \mid E_1 \sqcap E_2 \mid E_{|\pi}$$

Inference of set equations. Let be $P = \bigcup_{i=1}^n \{A_i \leftarrow D_i\}$ a Prolog program, and $pos(P) = \bigcup_{i=1}^n pos(\{A_i\} \cup D_i)$ the set of positions defined by P. The collection of set equations inferred from P is $type(P)$:

$$type(P) = \{< \pi = \bigsqcup_{i=1}^n exp(\pi, i) > \ / \ \pi \in pos(P) \}$$

where $exp(\pi, i)$ is the set expression associated with the position π in the ith clause $(A_i \leftarrow D_i)$ of P. It is computed by using the following rules:

(1) - $exp(\pi, i) = c$ if $A_{i|\pi} = c$

(2) - $exp(\pi, i) = f(\pi.f(1) \times ... \times \pi.f(r))$ if $A_{i|\pi} = f(t_1, ..., t_r)$

(3) - $exp(\pi, i) = \top$ if $A_{i|\pi} = X$ and $\forall \pi' \in pos(D_i)$, $X \notin D_{i|\pi'}$

(4) - $exp(\pi, i) = \pi_1 \sqcap ... \sqcap \pi_r$ if $A_{i|\pi} = X$ and $\{\pi_1, ..., \pi_r\} = \{\pi' \ / \ X \in D_{i|\pi'}\} \neq \emptyset$

(5) - $exp(\pi, i) = exp(\pi_1, i)_{|\pi_2}$ if $A_{i|\pi} = \bot$ and $\exists \pi_1 \ / \ \pi = \pi_1.\pi_2$ and $A_{i|\pi_1} = X$

(6) - $exp(\pi, i) = \bot$ if $A_{i|\pi} = \bot$ and $\forall \pi_1 \ / \ \pi = \pi_1.\pi_2$, $A_{i|\pi_1} \neq X$

Rules (1) to (4) respectively define the set expression associated with an atomic constant, a functional term, a head-only variable and a head and body variable. Rule (5) states that if there exists a prefix π_1 of π such that the subterm at position π_1 in A_i is a variable, then the set expression associated with π is a projection onto the set expression associated with π_1. For example, the set expression associated with p(1).f(1) in the clause p(X) :- q(X) is $q(1)_{|f(1)}$. Finally rule (6) associates $\perp$ with a position which is not defined in the head atom of the clause.

Solution of a system of equations. An interpretation $\mathcal{I}$ of a system of equations $type(P)$ is a mapping from the set of positions $pos(P)$ to the set of subsets of the Herbrand universe U_P. $\mathcal{I}$ is extended to set expressions as follows:

$$\mathcal{I}(\top) = U_P \qquad \mathcal{I}(f(E_1 \times \ldots \times E_r)) = \{f(t_1,\ldots,t_r) \ / \ t_i \in \mathcal{I}(E_i)\}$$
$$\mathcal{I}(\perp) = \emptyset \qquad \mathcal{I}(E_1 \sqcup E_2) = \alpha(\mathcal{I}(E_1) \cup \mathcal{I}(E_2))$$
$$\mathcal{I}(c) = \{c\} \qquad \mathcal{I}(E_1 \sqcap E_2) = \mathcal{I}(E_1) \cap \mathcal{I}(E_2)$$
$$\mathcal{I}(E_{|\pi}) = \mathcal{I}(E)_{|\pi}$$

A solution of a type system $type(P)$ is an interpretation $\mathcal{I}$ such that $\mathcal{I}(\pi) = \mathcal{I}(E)$ for all equation $< \pi = E >$ of $type(P)$. The least solution $\mathcal{I}$ of a type system $type(P)$ corresponds[4] to the least fix-point of T_P^α, i.e., for each position π of $pos(P)$, $\mathcal{I}(\pi) = lfp(T_P^\alpha)_{|\pi}$.

One can remark that $\sqcup$ is distributive with respect to $\times$ (i.e., $\mathcal{I}(E_1 \sqcup E_2) = \alpha(\mathcal{I}(E_1) \cup \mathcal{I}(E_2))$). Thus, the least solution of a type system exactly corresponds to $lfp(T_P^\alpha)$. This property facilitates some proofs, in particular for characterizing the meaning of an unfolded program. However, one could define $\sqcup$ as a non-distributive operator with respect to $\times$ (i.e., $\mathcal{I}(E_1 \sqcup E_2) = \mathcal{I}(E_1) \cup \mathcal{I}(E_2)$). In this case, the least solution of a type system is a subset of $lfp(T_P^\alpha)$, and the unfolding process can be applied in a same way.

The least solution of a type system can be "computed", i.e., set constraints can be transformed in an explicit form which is a regular tree description of the least model [HJ90, Hei92a]

Example. Let us consider the program P described in Fig. 1.

```
s(a,b).              s*(X,X).
s(b,c).              s*(X,Z) :- s(X,Y), s*(Y,Z).
s(a,d).
                     t(X) :- s*(b,X).
```

Fig. 1. Program P

---

[4] This property only holds for "non-redundant" programs [HJ92a]: a program P is non-redundant if for each clause $(A \leftarrow D)$, there exists a substitution θ such that $\theta D \subseteq lfp(T_P^\alpha)$.

The set of equations inferred from this program is:

$type(P) = \{$ s(1) = a $\sqcup$ b,   s*(1) = $\top$ $\sqcup$ s(1),

           s(2) = b $\sqcup$ c $\sqcup$ d,   s*(2) = $\top$ $\sqcup$ s*(2),

                               t(1) = s*(2) $\}$

The least model of these equations is $\mathcal{I}$:

$\mathcal{I}($s(1)$) = \{$ a, b $\}$ $\mathcal{I}($s*(1)$) = U_P$

$\mathcal{I}($s(2)$) = \{$ b, c, d $\}$ $\mathcal{I}($s*(2)$) = U_P$

                                      $\mathcal{I}($t(1)$) = U_P$

Comparison of set expressions. This type inference system has been designed in order to *analyse uncomplete Prolog programs*. The goal is to extract informations from the definition of some predicates and independently from the definition of some other predicates that are (not yet) defined or that are not relevant. Hence, we have defined a system that deduces subtyping relationships from set equations without resolving them. These relationships do not depend on the interpretation of the undefined predicates [SR94]. For example, from the predicate "p(X) :- q(X), r(X).", and independently from the definition of the predicates q and r, we infer that p(1) is a subtype of both q(1) and r(1).

3 Definition of the unfolding process

The type inference system described in the previous section ignores all inter-argument dependencies. In particular, dependencies expressed by head-only shared variables are ignored. As a consequence, predicates defined with respect to parametric predicates are not accurately approximated at all by this type inference system. For example, on the program of Fig. 1 the dependency between s*(1) and s*(2) is lost and the type associated with t(1) is the Herbrand universe whereas the denotation of t is {t(b), t(c)}. Conversely, one should remark that the types associated with parametric predicates are often very relevant. For instance, in the program of Fig. 1, any term of the Herbrand universe can be instance of the first and the second argument of the predicate s*. Thus the computed types s*(1) and s*(2) exactly correspond to the denotation of s*. Actually, the problem is not to approximate parametric predicates but to approximate predicates defined with respect to parametric predicates.

Thus, we propose to propagate inter-argument dependencies due to head-only shared variables before inferring types. We note P the Prolog program to be typed, and we consider the partition $P = Q \cup R$ so that Q (resp. R) is the set of clauses of P that define parametric (resp. non parametric) predicates. We define $unfold(R)$, the unfolding of R with respect to Q, as follows

$$unfold(R) = \{(A \leftarrow D) \in R \ / \text{ all atoms of } D \text{ correspond to predicates of } R\}$$
$$\cup \ \{(\theta A \leftarrow \theta(D_R \cup D'_1 \cup \ldots \cup D'_n)) \ /$$
$$(A \leftarrow D) \in R \text{ and } D = D_R \cup \{A_1, \ldots, A_n\}$$
$$\text{and } \forall i \in 1..n, \exists (A'_i \leftarrow D'_i) \in Q \text{ and}$$
$$\theta = mgu(A_1, A'_1) \circ \ldots \circ mgu(A_n, A'_n)\}$$

where $D_R \cup \{A_1, \ldots, A_n\}$ is the partition of D such that each atom of D_R is instance of a predicate of R and each A_i is instance of a predicate of Q.
We note R_n the program resulting of n unfolding steps of R with respect to Q:

$$R_0 = R \text{ and } R_n = unfold(R_{n-1}) \ (n > 0)$$

We note R'_n the set of clauses of R_n that do not depend on Q:

$$R'_n = \{(A \leftarrow D) \in R_n \ / \text{ all atoms of } D \text{ correspond to predicates of } R\}$$

Definition. To propagate inter-argument dependencies specified by parametric predicates, the idea is to unfold R with respect to Q until the type system associated with R'_n becomes stable. Hence, *the type system associated with R is* $type(R'_n)$ *if for all positive k, $type(R'_n) = type(R'_{n+k})$.*
   The equality between two systems of equations $type(R'_n)$ and $type(R'_{n+k})$ is a syntactical one. It is established with respect to the idempotence, associativity and commutativity of $\sqcup$ and $\sqcap$.

Example. Let us consider again the program P of Fig. 1. The predicate s* is parametric. We note Q the set of clauses defining s* and R the set of clauses defining s and t. The successive unfolding steps of R with respect to Q are:

```
R'₁ = { s(a,b)., s(b,c)., s(a,d)., t(b). }
R₁ = R'₁ ∪ { t(Z) :- s(b,Y), s*(Y,Z). }

R'₂ = R'₁ ∪ { t(Y) :- s(b,Y). }
R₂ = R'₂ ∪ { t(Z) :- s(b,Y), s(Y,Y1), s*(Y1,Z).}

R'₃ = R'₂ ∪ { t(Y) :- s(b,Y1), s(Y1,Y). }
R₃ = R'₃ ∪ { t(Z) :- s(b,Y), s(Y,Y1), s(Y1,Y2), s*(Y2,Z).}
. . .
```

and the successive inferred set equations for t(1) are

```
type(R'₁) = { t(1) = b } ∪ ...
type(R'₂) = { t(1) = b ⊔ s(2) } ∪ ...
type(R'₃) = { t(1) = b ⊔ s(2) ⊔ s(2) } ∪ ...
         = type(R'₂) = type(R'ₙ) (n>2)
```

Thus, the final type system associated with R is $type(R'_2)$, and the least model $\mathcal{I}$ associated with t(1) is $\mathcal{I}(\text{t(1)}) = \{\text{b, c, d}\}$. This should be compared with the type system $type(P)$ which associates the Herbrand universe to t(1).

4 Semantics of $type(R'_n)$

The termination of the unfolding process is studied in section 5. In this section, we suppose that there exists an n such that for all positive k, $type(R'_n) = type(R'_{n+k})$. The least solution of $type(R'_n)$ corresponds to the set $lfp(T^\alpha_{R'_n})$. In this section, we define this set by means of the initial program P. In particular, we show that $lfp(T^\alpha_{R'_n})$ actually approximates the denotation of P. As the unfolding process does not modify Q, we only focus on the semantics of the types of the predicates of R. We shall note B_R the subset of the Herbrand base B_P corresponding to the predicates of R, and we shall note $lfp(T_P) \cap B_R$ the subset of the denotation of P corresponding to the predicates of R.

Theorem 1. The inferred types approximate the denotation of P, i.e.,

$$\forall k > 0, lfp(T^\alpha_{R'_n}) = lfp(T^\alpha_{R'_{n+k}}) \;\Rightarrow\; lfp(T^\alpha_{R'_n}) \supseteq lfp(T_P) \cap B_R$$

Proof. see appendix A.

Definition of S_P and S_{P_i}. To define the semantics of the inferred set $lfp(T^\alpha_{R'_n})$ with respect to the initial program P, we introduce a new immediate consequence operator S_P which only computes the tuple distributive closure for the non parametric predicates of P (i.e., on R):

$$S_P(I) = T_Q(I) \cup \alpha(T_R(I))$$

In a same way, we define the immediate consequence operator S_{P_i} associated with the ith unfolding step of R with respect to Q:

$$S_{P_i}(I) = T_Q(I) \cup \alpha(T_{R_i}(I))$$

We note $lfp(S_P)$ and $lfp(S_{P_i})$ the least fix-points associated with S_P and S_{P_i}.

Theorem 2. $lfp(S_P)$ better approximates the denotation of P than $lfp(T^\alpha_P)$, i.e.,

$$lfp(T_P) \subseteq lfp(S_P) \subseteq lfp(T^\alpha_P)$$

proof. One can easily check that $\forall I \subseteq B_P$, $T_P(I) \subseteq S_P(I) \subseteq T^\alpha_P(I)$.

Theorem 3. The unfolding process preserves the least fix-point of S_P, i.e.,

$$lfp(S_P) = lfp(S_{P_i}) \;\; \forall i > 0$$

Proof. see appendix B.

Theorem 4. The inferred set $lfp(T^\alpha_{R'_n})$ is a subset of the least fix-point of S_P:

$$\forall k > 0, lfp(T^\alpha_{R'_n}) = lfp(T^\alpha_{R'_{n+k}}) \;\Rightarrow\; lfp(T^\alpha_{R'_n}) \subseteq lfp(S_P)$$

Proof. By definition, $S_P(I) = S_{P_n}(I) = T_Q(I) \cup \alpha(T_{R_n}(I))$ and $R'_n \subseteq R_n$. Thus, $\alpha(T_{R'_n}(I)) \subseteq \alpha(T_{R_n}(I)) \subseteq S_P(I)$ and therefore, $lfp(T^\alpha_{R'_n}) \subseteq lfp(S_P)$.

Example: On the program P of Fig. 1, we compute the following sets:

$lfp(T_P)$ = { s(a,b), s(a,d), s(b,c),
　　　　　　 t(b), t(c),
　　　　　　 s*(a,b), s*(b,c), s*(a,d), s*(a,c) } ∪
　　　　 { s*(X,X) / X∈U_P }

$lfp(S_P)$ = { s(a,b), s(a,c), s(a,d), s(b,b), s(b,c), s(b,d),
　　　　　　 t(b), t(c), t(d),
　　　　　　 s*(a,b), s*(a,c), s*(a,d), s*(b,b), s*(b,c), s*(b,d) } ∪
　　　　 { s*(X,X) / X∈U_P }

$lfp(T_P^\alpha)$ = { s(a,b), s(a,c), s(a,d), s(b,b), s(b,c), s(b,d) } ∪
　　　　 { t(X) / X∈U_P } ∪
　　　　 { s*(X,Y) / X∈U_P, Y∈U_P }

The set $lfp(T_P^\alpha)$ corresponds to the least solution of $type(P)$. Considering this set, the type of t(1) is the whole Herbrand universe. However, after performing the unfolding process, the least solution of $type(R_n')$ is bounded by $lfp(S_P)$. In this new set, the type of t(1) is {b,c,d}, which much better approximates the denotation of t than does the corresponding type in $lfp(T_P^\alpha)$.

5 Stabilization of the unfolding process

The unfolding process defined in section 3 terminates if there exists a step n so that $type(R_n')$ is equal to $type(R_{n+k}')$ for all positive k. In this section, we study the conditions of such a stabilization. In a first time, we show that the type system cannot become stable if the unfolding process introduces terms of increasing depth. Then we define an abstraction function $\mathcal{A}$ which limits the depth of types and ensures a stabilization.

5.1 Definition of recursive positions

The unfolding process is increasing, i.e., for all positive i, $R_i' \subseteq R_{i+1}'$. Hence, one will finitely reach a fix-point if the number of different type systems that can be inferred from the successive unfolded programs $R_1', R_2', \ldots, R_i'$ is finite, i.e., if the number of different constants, functors and positions generated by the unfolding process is finite. However, an infinite number of different positions can be generated by unfolding. Let us consider for example the following program $P = Q \cup R$:

```
Q: mb(X,cons(X,L)).                    R: p(Z) :- mb(a,Z).
   mb(X,cons(Y,L)) :- mb(X,L).
```

The predicate mb defines a recursive structure (i.e., cons/2), and the depth of the argument of the predicate p increases at each unfolding step:

$R'_1 = \{ \text{ p(cons(a,L)). } \}$
$R'_2 = R'_1 \cup \{ \text{ p(cons(X1,cons(a,L))). } \}$
$R'_3 = R'_2 \cup \{ \text{ p(cons(X2, cons(X1,cons(a,L)))). } \}$
. . .

Hence, each unfolding step introduces two new positions[5]:

$pos(R'_1) = \{ \text{ p(1), p(1).cons(1), p(1).cons(2) } \},$
$pos(R'_i) = pos(R'_{i-1}) \cup \{ \text{p(1).\{cons(2)\}}^{i-1}\text{.cons(1), p(1).\{cons(2)\}}^i\}$

and the type system never becomes stable, i.e., $type(R'_i) \neq type(R'_{i+1})$ for all i. In this example, the recursive call occurs on the second argument of the functional term cons. Hence, we shall say that cons(2) is a recursive position.

In a general way, the number of positions can become infinite by unfolding if Q contains a recursive predicate (or mutually recursive predicates) which infinitely introduces terms of increasing depth. The sub-position corresponding to growing parts of terms is called a recursive position.

Static identification of recursive positions. Some recursive positions can be identified by comparing positions of variables that both occur in the head and in the recursive call of a clause. For example, let us consider the predicate rev:

```
rev(nil,L,L).
rev(f(X,f(Y,L)),Res,Acc) :- rev(L,Res,g(X,Acc)).
```

This predicate uses an accumulator g to reverse the elements at even positions in a list f (e.g., rev(f(4,f(3,f(2,f(1,nil)))),nil,g(2,g(4,nil))) is a logical consequence of this program). The variable L occurs in the head atom at position rev(1).f(2).f(2) and in the body at position rev(2). By comparing these two positions, we identify the recursive position f(2).f(2). In a same way, the variable Acc occurs at position rev(3) in the head atom and at position rev(3).g(2) in the body and g(2) is a recursive position. In a more general way, we can identify the recursive positions defined by a recursive parametric predicate p if p does not depend on some other recursive parametric predicates. Otherwise, recursive positions have to be dynamically identified.

Dynamic identification of recursive positions. Recursive positions can be identified during the unfolding process by comparing new positions introduced by unfolding with existing ones. The growing part in a new position may be a recursive position. Hence, a superset of recursive positions can be identified at each unfolding step n as follows:

$$rec - pos = \bigcup_{i=1}^{n} \{\pi \ / \ \pi_1.\pi \in pos(R'_i) \text{ and } \pi_1.\pi \notin pos(R'_{i-1}) \text{ and } \pi_1 \in pos(R'_{i-1})\}$$

rec-pos is a superset of the recursive positions as a position may grow at one step and then becomes stable. To improve the accuracy, this set can be defined by considering sub-positions that have been repeated k times rather than once.

---

[5] Notation: $\{\pi\}^n = \pi.\{\pi\}^{n-1}$ if $n > 1$, $\{\pi\}^1 = \pi$.

5.2 Definition of the $\mathcal{A}$ abstraction

If there is no recursive position, the number of positions is finite by unfolding, and therefore the unfolding process terminates. Otherwise, the unfolding process introduces an infinite number of different positions and no fix-point will be reached. We now define an abstraction function $\mathcal{A}$ which ensures the termination in this case. For example, let us consider again the program $P = Q \cup R$:

```
Q: mb(X,cons(X,L)).                    R: p(Z) :- mb(a,Z).
   mb(X,cons(Y,L)) :- mb(X,L).
```

The unfolding process introduces a recursive position `cons(2)` and therefore an infinite number of different positions

$$p(1).cons(1), \; p(1).cons(2).cons(1), \; \ldots, \; p(1).\{cons(2)\}^i.cons(1), \; \ldots$$

These different positions all correspond to the single notion of "member of the list Z". Hence, to limit the number of positions generated by the unfolding process, the idea is to approximate all these different positions by the single abstract position `p(1).cons(1)`. This abstract position is defined by the union of all set expressions associated with all concrete positions $p(1)\{.cons(2)\}^i.cons(1)$.

Abstract positions. Each concrete position π is approximated by the abstract position $\mathcal{A}(\pi)$ which is obtained by "removing" from π the recursive positions:

$$\mathcal{A}(<>) = <>,$$
$$\mathcal{A}(\pi_1.\pi_2) = \mathcal{A}(\pi_2) \text{ if } \pi_1 \text{ is a recursive position}$$
$$\mathcal{A}(f(i).\pi) = f(i).\mathcal{A}(\pi) \text{ if no recursive position } \pi_1 \text{ is prefix of } f(i).\pi$$

With each abstract position π, we associate the set $\mathcal{C}(\pi)$ of concrete positions:

$$\mathcal{C}(\pi) = \{\pi' \in pos(P) \; / \; \mathcal{A}(\pi') = \pi\}$$

Abstract set expressions. $\mathcal{A}$ is extended to set expressions as follows:

$$\mathcal{A}(\top) = \top \qquad \mathcal{A}(f(E_1 \times \ldots \times E_n)) = f(\mathcal{A}(E_1) \times \ldots \times \mathcal{A}(E_n))$$
$$\mathcal{A}(\bot) = \bot \qquad \mathcal{A}(E_1 \sqcup E_2) = \mathcal{A}(E_1) \sqcup \mathcal{A}(E_2)$$
$$\mathcal{A}(c) = c \qquad \mathcal{A}(E_1 \sqcap E_2) = \mathcal{A}(E_1) \sqcap \mathcal{A}(E_2)$$
$$\mathcal{A}(E_{|\pi}) = \mathcal{A}(E)_{|\mathcal{A}(\pi)}$$

Abstract type systems. The abstract type system $type_{\mathcal{A}}(P)$ that describes all abstract positions associated with a program $P = \cup_{i=1}^n (A_i \leftarrow D_i)$ is:

$$type_{\mathcal{A}}(P) = \{< \pi = \bigsqcup_{\pi' \in \mathcal{C}(\pi)} \sqcup_{i=1}^n \mathcal{A}(exp(\pi', i)) > \; / \; \pi \in \mathcal{A}(pos(P))\}$$

Definition. The type system associated with R is $type_{\mathcal{A}}(R'_n))$ such that for all positive k, $type_{\mathcal{A}}(R'_n) = type_{\mathcal{A}}(R'_{n+k})$.

Theorem 5. The abstract type system is a correct approximation of the concrete one, i.e., if $\mathcal{I}$ and $\mathcal{I}_A$ respectively are the least solutions of $type(P)$ and $type_A(P)$, then for each concrete position π of $pos(P)$, $\mathcal{I}(\pi) \subseteq \mathcal{I}_A(\mathcal{A}(\pi))$.

Proof. In $type_A(P)$, $\mathcal{A}(\pi)$ is defined by the union of all abstract expression $\mathcal{A}(E)$ such that E is the set expression associated with a concrete position $\pi' \in \mathcal{C}(\mathcal{A}(\pi)) \supseteq \{\pi\}$. Then, one can easily check by induction that for any set expression E, we have $\mathcal{I}(E) \subseteq \mathcal{I}(\mathcal{A}(E))$.

Theorem 6. The unfolding process terminates, i.e., there exists n such that for all positive k, $type_A(R'_n) = type_A(R'_{n+k})$.

Proof. The set of different positions that can be generated by the unfolding process is limited by the $\mathcal{A}$ abstraction. Therefore, the number of different abstract type systems associated with the successive unfolding steps is finite.

6 Examples

We illustrate in this section the relevance of the inferred types on two rather small but representative programs.

Example 1: let be the program

```
R: p(X) :- q(L1), s(L2), append(L1,L2,L3), member(X,L3).

    q(cons(a,cons(b,nil))).
    s(cons(c,cons(d,cons(e,nil)))).

Q: append(nil,L,L).
    append(cons(X,L1),L2,cons(X,L3)) :- append(L1,L2,L3).

    member(X,cons(X,L)).
    member(X,cons(Y,L)) :- member(X,L).
```

Q defines the recursive position cons(2). The abstract type system becomes stable at the second unfolding step and is:

```
p(1) = s(1).cons(1) ⊔ q(1).cons(1)
q(1) = cons(q(1).cons(1)×q(1)) ⊔ nil        q(1).cons(1) = a ⊔ b
s(1) = cons(s(1).cons(1)×s(1)) ⊔ nil        s(1).cons(1) = c ⊔ d ⊔ e
```

After resolution of the equations, the type of the argument of p is $\mathcal{I}(p(1)) = \{a, b, c, d, e\}$. However, before unfolding, the type associated with p(1) was the herbrand universe (as it was defined with respect to member(1)).

Example 2: Let be the program:

```
R: p(L1,L2) :- rev(L1,nil,L2).

Q: rev(nil,L,L).
   rev(f(X,f(Y,L)),Acc,Res) :- rev(L,g(X,Acc),Res).
```

Q defines the recursive positions $f(2).f(2)$ and $g(2)$ (see section 5.1). The type system becomes stable at the third unfolding step and is:

$$p(1) = nil \sqcup f(\ p(1).f(1) \times p(1).f(2)\ ) \qquad\qquad p(1).f(1) = \top$$
$$p(1).f(2) = f(\ p(1).f(2).f(1) \times p(1)\ ) \qquad\qquad p(1).f(2).f(1) = \top$$
$$p(2) = nil \sqcup g(\ p(2).g(1) \times p(2)\ ) \qquad\qquad p(2).g(1) = \top$$

The least model of this type system is $\mathcal{I}$:

$$\mathcal{I}(p(1)) = \{\ nil,\ f(U_P,f(U_P,t))\ /\ t \in \mathcal{I}(p(1))\ \}$$
$$\mathcal{I}(p(2)) = \{\ nil,\ g(U_P,t)\ /\ t \in \mathcal{I}(p(2))\ \}$$

Before unfolding, the type associated with $p(2)$ was the herbrand universe (as it was defined with respect to the type of the third argument of **rev**).

7 Related works

Different solutions have been proposed to extend tuple-distributive type inference systems so that they can take into account parametric polymorphism. For example, [Zob87] proposed to parameterize the types by "type variables", and [PR89] proposed to use "conditional types" to express and propagate dependency constraints. However, the resulting types do not allow an intuitive understanding of the underlying structure and they are not easy to handle. Moreover, the comparison of parametric types is still an open problem [DZ92]. Our approach is rather opposite to this one: instead of modifying the type system so that it can handle inter-argument dependencies, we propose to propagate dependencies before inferring types.

A comparable approach to ours has been proposed in [HJ92a]. The idea is to unfold abstract semantic equations inferred from Prolog programs until a fix-point is reached. To obtain a terminating algorithm, this process is curtailed by introducing an abstraction. This algorithm is generic as it can be used by an inference system based on any abstract interpretation (top-down or bottom-up) using set constraints. However, the abstraction function which ensures the termination has to be defined by the programmer. This abstraction function is quite easy to define if one wants to deduce "simple" informations like modes. However, if one wants to infer precise types, the definition of this abstraction requires to know rather precisely what kind of structures are manipulated by the program, and this is just the goal of type inference.

Gallagher and de Wall [GdW94] defined a bottom up algorithm to compute a regular approximation of a logic program. The idea is to combine a tuple distributive closure abstraction (which ignores inter-argument dependencies) with a shortening operator which ensures the termination and the efficiency of the computation. In order to improve the precision of the approximation, top-down computation is simulated through query-answer transformations. This kind of "magic set" style transformations improves the accuracy of the approximation by restricting the calls that can occur with respect to a particular goal and it takes into account some inter-argument dependencies.

Some other type inference systems are based on abstract interpretation of top-down operational semantics of Prolog (e.g., [Bru91, KK93, HCC93]). The abstraction is introduced both to ensure the termination and to improve the efficiency of the computation, and it highly depends on the kind of information to be inferred. In these approaches, argument dependencies can easily be captured by the operational resolution, eventhough they are often ignored for efficiency reasons. The inferred informations are not formally characterized and depend on the choices made in the abstract domain. These approaches usually aim at extracting run time informations (e.g., modes, variable sharing) for a given set of goals in order to design an effective compiler.

8 Discussion

We have defined an unfolding process which propagates inter-argument dependencies specified by shared head-only variables, and thus improves the accuracy of a type inference system based on the α abstraction. In order to ensure the termination of this process, we have introduced another abstraction function $\mathcal{A}$ which approximates different concrete types by a single abstract one. These different concrete types correspond to successive arguments of recursive predicates and they are semantically very closed (e.g., the different elements of a binary list). Hence, the abstract type system is a relevant approximation of the concrete one.

Actually, the two abstractions α and $\mathcal{A}$ are rather complementary in their effects: the α abstraction is used to limit the breadth of types (e.g., $\alpha(\{$ f(a,b), f(a,d), f(c,d) $\})$ is $\{$ f({a,c},{b,d}) $\})$, while the $\mathcal{A}$ abstraction is introduced to limit the depth of types (e.g., $\mathcal{A}(\{$ f(a,f(b,f(c,nil))) $\})$ is X = $\{$ nil, f({a,b,c},X) $\})$. We argue that *these two abstraction functions are both necessary for an accurate and efficient approximation*.

The $\mathcal{A}$ abstraction can be compared with the widening operator used in some type inference systems based on abstract interpretation of top-down execution of logic programs (e.g., [HCC93]) or with the shortening operator used in [GdW94]. Actually, these operators are introduced to ensure the termination of a fix-point computation, and they limit the depth of abstract structures in a similar way than does the $\mathcal{A}$ abstraction.

Acknowledgements: Many thanks to Patrice Boizumault for carefully reading a first version of this paper and to Andreas Podelski for numerous and enriching discussions on this work.

References

[Bru91]   M. Bruynooghe. A practical framework for the abstract interpretation of logic programs. *Journal of Logic Programming*, 10:91–124, 1991.

[DZ92]   P.W. Dart and J. Zobel. A regular type language for logic programs. In Frank Pfenning, editor, *Types in Logic Programming*, pages 157–187, 1992.

[FSVY91] T. Fruhwirth, E. Shapiro, M.Y. Vardi, and E. Yardeni. Logic programs as types for logic programs. In *IEEE-LICS*, pages 300–309, 1991.

[GdW94] J.P. Gallagher and D.A. de Waal. Fast and precise regular approximations for logic programs. In *ICLP*, 1994.

[HCC93] P. Van Hentenrick, A. Cortes, and B. Le Charlier. Type analysis of prolog using type graphs. Technical report, 1993.

[Hei92a] N. Heintze. Practical aspects of set based analysis. In *Joint Int. Conf. & Symp. on Logic Programming*, 1992.

[Hei92b] N. Heintze. Set based program analysis. Phd thesis, School of Computer Science - Carnegie Mellon University, 1992.

[HJ90]   N. Heintze and J. Jaffar. A decision procedure for a class of set constraints. In *IEEE-LICS'90*, pages 42–51, 1990.

[HJ92a] N. Heintze and J. Jaffar. An engine for logic program analysis. In *IEEE-LICS*, 1992.

[HJ92b] N. Heintze and J. Jaffar. Semantic types for logic programs. In Frank Pfenning, editor, *Types in Logic Programming*, pages 141–151, 1992.

[KK93]   T. Kanamori and T. Kawamura. Abstract interpretation based on OLDT resolution. *Journal of Logic Programming*, 15:1–30, 1993.

[Mis84]   P. Mishra. Towards a theory of types in Prolog. In *International Logic Programming Symposium, Atlantic City*, pages 289–298, 1984.

[PR89]   C. Pyo and U.S. Reddy. Inference of polymorphic types for logic programs. In *Logic Programming North American Conference*, pages 1115–1132, 1989.

[SR94]   C. Solnon and M. Rueher. Deduction of inheritance hierarchies from Prolog programs via set constraints. Research report, 1994.

[TS84]   H. Tamaki and T. Sato. Unfold/fold transformation of logic programs. In *2nd ICLP*, pages 127–138, 1984.

[YS87]   E. Yardeni and E. Shapiro. A type system for logic programs. In E. Shapiro, editor, *Concurrent Prolog - Collected papers - Vol. 2*, pages 211–244. MIT Press Series in Logic Programming, 1987.

[Zob87]   J. Zobel. Derivation of polymorphic types for Prolog programs. In *4th ICLP*, pages 817–838, 1987.

A   Proof of theorem 1

Let us show by induction on i that: $\forall i,\ (T_P \uparrow i) \cap B_R \subseteq lfp(T^\alpha_{R'_n})$

• $T_P \uparrow 0 \cap B_R \subseteq lfp(T^\alpha_{R'_n})$ as any clause $(A \leftarrow)$ of R is also a clause of R'_n.

• Let us suppose that $(T_P \uparrow i) \cap B_R \subseteq lfp(T^\alpha_{R'_n})$ and let us show that

$A \in (T_P \uparrow i+1) \cap B_R \Rightarrow A \in lfp(T^\alpha_{R'_n})$:

$A \in (T_P \uparrow i+1) \cap B_R$ implies that there exists a clause $(A_j \leftarrow D_j)$ of R and a substitution θ such that $\theta A_j = A$ and $\theta D_j \subseteq T_P \uparrow i$. Let be $D_j = D_{Q_j} \cup D_{R_j}$ the partition of D_j such that $\theta D_{Q_j} \subseteq B_Q$ and $\theta D_{R_j} \subseteq B_R$.

    – If $D_{Q_j} = \emptyset$ then $(A_j \leftarrow D_j) \in R'_n$. As $\theta D_j \subseteq T_P \uparrow i \cap B_R \subseteq lfp(T^\alpha_{R'_n})$, $A \in lfp(T^\alpha_{R'_n})$.

– Otherwise, $\theta D_{Qj} \subseteq T_P \uparrow i$ implies that there exists an SLD-derivation, from θD_{Qj} to D', that only uses clauses of Q and such that $D' \subseteq (T_P \uparrow i) \cap B_R$.
  As the unfolding process corresponds to a breadth first search, there exists an unfolding step s such that $(\theta A_j \leftarrow D' \cup \theta D_{Rj})$ is an instance of a clause of R'_s.
  · If $s \leq n$ then $R'_s \subseteq R'_n$.
  · Otherwise, as the type system is stable at step n, $lfp(T_{R'_s}^\alpha) = lfp(T_{R'_n}^\alpha)$.
  In both cases, $D' \cup \theta D_{Rj} \subseteq (T_P \uparrow i) \cap B_R \subseteq lfp(T_{R'_n}^\alpha)$, and therefore $A \in lfp(T_{R'_n}^\alpha)$.

B   Proof of theorem 2

• Let us prove by induction on j that: $\forall j \geq 0,\ S_{P_i} \uparrow j \subseteq lfp(S_{P_{i+1}})$

  – $S_{P_i} \uparrow 0 \subseteq lfp(S_{P_{i+1}})$ as any clause $(A \leftarrow)$ of R_i also belongs to $unfold(R_i)$.
  – Let us suppose that $S_{P_i} \uparrow j \subseteq lfp(S_{P_{i+1}})$ and let us show that
    $S_{P_i} \uparrow j+1 \subseteq lfp(S_{P_{i+1}})$, i.e., that $T_Q(S_{P_i} \uparrow j) \cup \alpha(T_{R_i}(S_{P_i} \uparrow j)) \subseteq lfp(S_{P_{i+1}})$.
    $T_Q(S_{P_i} \uparrow j) \subseteq lfp(S_{P_{i+1}})$ as unfolding does not modify clauses of Q.
    Thus, let us first show that $A \in T_{R_i}(S_{P_i} \uparrow j) \Rightarrow A \in lfp(S_{P_{i+1}})$:
    • $A \in T_{R_i}(S_{P_i} \uparrow j)$ implies that there exists a clause $(A' \leftarrow D')$ of R_i and a substitution θ such that $\theta A' = A$ and $\theta D' \subseteq S_{P_i} \uparrow j$. Let be $D' = D'_Q \cup D'_R$ the partition of D' such that $\theta D'_Q \subseteq B_Q$ and $\theta D'_R \subseteq B_R$.
    • If $D'_Q = \emptyset$ then $(A' \leftarrow D') \in unfold(R_i)$, and thus $A \in lfp(S_{P_{i+1}})$
    • Otherwise, let be $D'_Q = \{A'_1, \ldots, A'_n\}$. As $\theta D'_Q \subseteq S_{P_i} \uparrow j$, there exists for each k in 1..n, a clause $(A_k \leftarrow D_k)$ of Q and a substitution θ_k such that $\theta_k A_k = \theta A'_k$ and $\theta_k D_k \subseteq S_{P_i} \uparrow j$. Thus, the clause $(\theta A' \leftarrow \theta D'_R \cup \theta_1 D_1 \cup \ldots \cup \theta_k D_k)$ is an instance of a clause of $unfold(R_i)$ and A belongs to $lfp(S_{P_{i+1}})$.
    Hence, $T_{R_i}(S_{P_i} \uparrow j) \subseteq lfp(S_{P_{i+1}}) \cap B_R$. Then, as $lfp(S_{P_{i+1}}) \cap B_R$ is closed with respect to α and α preserves inclusion relations, $\alpha(T_{R_i}(S_{P_i} \uparrow j)) \subseteq lfp(S_{P_{i+1}}) \cap B_R$.
    Thus, $T_Q(S_{P_i} \uparrow j) \cup \alpha(T_{R_i}(S_{P_i} \uparrow j)) = S_{P_i} \uparrow j+1 \subseteq lfp(S_{P_{i+1}})$

• Let us prove by induction on j that: $\forall j \geq 0,\ S_{P_{i+1}} \uparrow j \subseteq lfp(S_{P_i})$

  – Let us show that $S_{P_{i+1}} \uparrow 0 = T_Q(\emptyset) \cup \alpha(T_{R_{i+1}}(\emptyset)) \subseteq lfp(S_{P_i})$:
    - $T_Q(\emptyset) \subseteq S_{P_i} \uparrow 0 \subseteq lfp(S_{P_i})$
    - $T_{R_{i+1}}(\emptyset) \subseteq lfp(T_P) \cap B_R$ as unfolding preserves the denotation [TS84]. Thus, $\alpha(T_{R_{i+1}}(\emptyset)) \subseteq \alpha(lfp(T_P) \cap B_R) \subseteq lfp(S_{P_i})$.
  – Let us suppose that $S_{P_{i+1}} \uparrow j \subseteq lfp(S_{P_i})$ and let us show that
    $S_{P_{i+1}} \uparrow j+1 \subseteq lfp(S_{P_i})$, i.e., that $T_Q(S_{P_{i+1}} \uparrow j) \cup \alpha(T_{R_{i+1}}(S_{P_{i+1}} \uparrow j)) \subseteq lfp(S_{P_i})$.
    $T_Q(S_{P_{i+1}} \uparrow j) \subseteq lfp(S_{P_i})$ as unfolding does not modify clauses of Q.
    Thus, let us first show that $A \in T_{R_{i+1}}(S_{P_{i+1}} \uparrow j) \Rightarrow A \in lfp(S_{P_i})$:
    • $A \in T_{R_{i+1}}(S_{P_{i+1}} \uparrow j)$ implies that there exists a clause $(A' \leftarrow D')$ of $unfold(R_i)$ and a substitution θ such that $\theta A' = A$ and $\theta D' \subseteq S_{P_{i+1}} \uparrow j$.
    • If $(A' \leftarrow D')$ is also a clause of R_i, then A belongs to $lfp(S_{P_i})$.
    • Otherwise, $(A' \leftarrow D')$ has been generated by unfolding a clause of R_i with respect to Q, i.e., there exists a clause $(A'' \leftarrow D''_R \cup \{A''_1, \ldots, A''_n\})$ of R_i and for each k in 1..n, there exists a clause $(A'_k \leftarrow D'_k)$ of Q and a substitution $\theta' = mgu(A''_1, A'_1) \circ \ldots \circ mgu(A''_n, A'_n)$ such that $\theta' A'' = A'$ and $D' = \theta'(D''_R \cup D'_1 \cup \ldots \cup D'_n)$. As $\theta D' \subseteq S_{P_{i+1}} \uparrow j \subseteq lfp(S_{P_i})$, for each k in 1..n, $\theta \circ \theta' D'_k \subseteq lfp(S_{P_i})$, and therefore, $\theta \circ \theta' A'_k \subseteq lfp(S_{P_i})$. Thus, A belongs to $lfp(S_{P_i})$.
    Hence, $T_{R_{i+1}}(S_{P_{i+1}} \uparrow j) \subseteq lfp(S_{P_i}) \cap B_R$ and $\alpha(T_{R_{i+1}}(S_{P_{i+1}} \uparrow j)) \subseteq lfp(S_{P_i}) \cap B_R$.
    Therefore, $T_Q(S_{P_{i+1}} \uparrow j) \cup \alpha(T_{R_{i+1}}(S_{P_{i+1}} \uparrow j)) = S_{P_{i+1}} \uparrow j+1 \subseteq lfp(S_{P_i})$

Logic Programming and Logic Grammars with First-order Continuations

Paul Tarau[1] and Veronica Dahl[2]

[1] Université de Moncton
Département d'Informatique
Moncton, N.B. Canada E1A 3E9,
tarau@info.umoncton.ca
[2] Logic Programming Group
Department of Computing Sciences
Simon Fraser University
Burnaby, B.C. Canada V5A 1S6
veronica@cs.sfu.ca

Abstract. Continuation passing binarization and specialization of the WAM to binary logic programs have been proven practical implementation techniques in the BinProlog system. In this paper we investigate the additional benefits of having first order continuations at source level. We devise a convenient way to manipulate them by introducing multiple-headed clauses which give direct access to continuations at source-level. We discuss the connections with various logic grammars, give examples of typical problem solving tasks and show how looking at the future of computation can improve expressiveness and describe complex control mechanisms without leaving the framework of binary definite programs. *Keywords:* continuation passing binary logic programs, logic grammars, program transformation based compilation, continuations as first order objects, logic programming with continuations.

1 Introduction

From its very inception, logic programming has cross-fertilized with computational linguistics in very productive ways. Indeed, logic programming itself grew from the automatic deduction needs of a question-answering system in French [7]. Over the years we have seen other interesting instances of this close-relatedness. The idea of continuations, developed in the field of denotational semantics [22] and functional programming [27, 26] has found its way into programming applications, and has in particular been useful recently in logic programming.

In this article we continue this tradition by adapting to logic programming with continuations some of the techniques that were developed in logic grammars for computational linguistic applications. In particular, logic grammars have been augmented by allowing extra symbols and meta-symbols in the left hand sides of rules [6, 17, 9]. Such extensions allow for straightforward expressions of contextual information, in terms of which many interesting linguistic phenomena have been described. We show that such techniques can be efficiently transfered to continuation passing binary programs, that their addition motivates an interesting style of programming for applications other than linguistic ones, and that introducing continuations in logic grammars with multiple left-hand-side symbols motivates in turn novel styles of bottom-up and mixed parsing, while increasing efficiency.

2 Motivation

Several attempts to enhance the expressiveness of logic programming have been made over the years. These efforts range from simply providing handy notational variants, as in DCGs, to implementing sophisticated new frameworks, such as HiLog [5, 29].

In λProlog for instance [14, 15] there is a nice facility, that of scoping of clauses, which is provided by the λProlog incarnation of the intuitionistic rule for implication introduction:

$$\frac{\Sigma; \Gamma, P \vdash C}{\Sigma; P \vdash \Gamma \Rightarrow C} \Rightarrow_R$$

Example 1 *For instance, in the λProlog program* [3]:

```
insert X Xs Ys :-
    paste X => ins Xs Ys.

ins Ys [X|Ys] :- paste X.
ins [Y|Ys] [Y|Zs]:- ins Ys Zs.
```

used to nondeterministically insert an element in a list, the unit clause **paste X** *is available only within the scope of the derivation for* **ins**.

With respect to the corresponding Prolog program we are working with a simpler formulation in which the element to be inserted does not have to percolate as dead weight throughout each step of the computation, only to be used in the very last step. We instead clearly isolate it in a global-value manner, within a unit clause which will only be consulted when needed, and which will disappear afterwards.

Now, let us imagine we are given the ability to write part of a proof state context, i.e., to indicate in a rule's left-hand side not only the predicate which should match a goal atom to be replaced by the rule's body, but also which other goal atom(s) should surround the targeted one in order for the rule to be applicable.

Example 2 *Given this, we could write a program for insert which strikingly resembles the λProlog program given above:*

---

[3] where for instance insert X Xs Ys means insert(X,Xs,Ys) in curried notation

```
insert(X,Xs,Ys):-ins(Xs,Ys),paste(X).
```

```
ins(Ys,[X|Ys]),paste(X).
ins([Y|Ys],[Y|Zs]):-ins(Ys,Zs).
```

Note that the element to be inserted is not passed to the recursive clause of the predicate **ins/2** (which becomes therefore simpler), while the unit clause of the predicate **ins/2** will communicate directly with **insert/3** which will directly 'paste' the appropriate argument in the continuation.

In this formulation, the element to be inserted is first given as right-hand side context of the simpler predicate **ins/2**, and this predicate's first clause consults the context **paste(X)** only when it is time to place it within the output list, i.e. when the fact **ins(Ys,[X|Ys]),paste(X)** is reached.

Thus for this example, we can obtain the expressive power of λProlog without having to resort to an entirely new framework. As we shall see in the next sections, we can simply use any Prolog system combined with a simple transformer to binary programs [24].

3   Multiple Head Clauses in Continuation Passing Binary Programs

We will start by reviewing the program transformation that allows compilation of logic programs towards a simplified WAM specialized for the execution of binary logic programs. We refer the reader to [24] for the original definition of this transformation.

3.1   The binarization transformation

Binary clauses have only one atom in the body (except for some inline 'builtin' operations like arithmetics) and therefore they need no 'return' after a call. A transformation introduced in [24] allows to faithfully represent logic programs with operationally equivalent binary programs.

To keep things simple we will describe our transformations in the case of definite programs. First, we need to modify the well-known description of SLD-resolution (see [11]) to be closer to Prolog's operational semantics. We will follow here the notations of [25].

Let us define the *composition* operator $\oplus$ that combines clauses by unfolding the leftmost body-goal of the first argument.

Definition 1 *Let* $A_0:-A_1,A_2,\ldots,A_n$ *and* $B_0:-B_1,\ldots,B_m$ *be two clauses (suppose* $n > 0, m \geq 0$). *We define*

$$(A_0:-A_1,A_2,\ldots,A_n) \oplus (B_0:-B_1,\ldots,B_m) = (A_0:-B_1,\ldots,B_m,A_2,\ldots,A_n)\theta$$

with $\theta = mgu(A_1,B_0)$. *If the atoms* A_1 *and* B_0 *do not unify, the result of the composition is denoted as* $\perp$. *Furthermore, as usual, we consider* $A_0:-true,A_2,\ldots,A_n$

to be equivalent to $A_0 : -A_2, \ldots, A_n$, *and for any clause* C, $\perp \oplus C = C \oplus \perp = \perp$. *We assume that at least one operand has been renamed to a variant with fresh variables.*

Let us call this Prolog-like inference rule LF-SLD resolution (LF for 'left-first'). Remark that by working on the program P' obtained from P by replacing each clause with the set of clauses obtained by all possible permutations of atoms occurring in the clause's body every SLD-derivation on P can be mapped to an LF-SLD derivation on P'.

Before defining the binarization transformation, we describe two auxiliary transformations.

The first transformation converts facts into rules by giving them the atom **true** as body. E.g., the fact **p** is transformed into the rule **p :- true**.

The second transformation, inspired by [28], eliminates the metavariables by wrapping them in a **call/1** goal. E.g., the rule **and(X,Y):-X, Y** is transformed into **and(X,Y) :- call(X), call(Y)**.

The transformation of [24] (*binarization*) adds continuations as extra arguments of atoms in a way that preserves also first argument indexing.

Definition 2 *Let P be a definite program and Cont a new variable. Let T and $E = p(T_1, \ldots, T_n)$ be two expressions.*[4] *We denote by $\psi(E, T)$ the expression $p(T_1, \ldots, T_n, T)$. Starting with the clause*

(C) $A : -B_1, B_2, \ldots, B_n$.

we construct the clause

(C') $\psi(A, Cont) : -\psi(B_1, \psi(B_2, \ldots, \psi(B_n, Cont)))$.

The set P' of all clauses C' obtained from the clauses of P is called the binarization of P.

Example 3 *The following example shows the result of this transformation on the well-known 'naive reverse' program:*

```
app([],Ys,Ys,Cont):-true(Cont).
app([A|Xs],Ys,[A|Zs],Cont):-app(Xs,Ys,Zs,Cont).

nrev([],[],Cont):-true(Cont).
nrev([X|Xs],Zs,Cont):-nrev(Xs,Ys,app(Ys,[X],Zs,Cont)).
```

These transformations preserve a strong operational equivalence with the original program with respect to the LF-SLD resolution rule which is *reified* in the syntactical structure of the resulting program.

Theorem 1 *([25]) Each resolution step of an LF-SLD derivation on a definite program P can be mapped to an SLD-resolution step of the binarized program P'. Let G be an atomic goal and $G' = \psi(G, true)$. Then, computed answers obtained querying P with G are the same as those obtained by querying P' with G'.*

---

[4] Atom or term.

Notice that the equivalence between the binary version and the original pro- gram can also be explained in terms of fold/unfold transformations as suggested by [18].

Clearly, continuations become explicit in the binary version of the program. We will devise a technique to access and manipulate them in an intuitive way, by modifying BinProlog's binarization preprocessor.

3.2 Modifying the binarization preprocessor for multi-head clauses

The main difficulty comes from the fact that trying to directly access the continu- ation from 'inside the continuation' creates a cyclic term. We have overcome this in the past by copying the continuation in BinProlog's blackboard (a permanent data area) but this operation was very expensive.

The basic idea of the approach in this paper is inspired by 'pushback lists' originally present in [6] and other techniques used in logic grammars to simulate movement of constituents [9]. We will allow a *multiple head notation* as in:

```
a,b,c:-d,e.
```

intended to give, via binarization:

```
a(b(c(A))) :- d(e(A))
```

This suggests the following:

Definition 3 *A multiheaded definite clause is a clause of the form:*
$$A_1, A_2, \ldots, A_m : -B_1, B_2, \ldots, B_n.$$

A multiheaded definite program is a set of multiheaded definite clauses.

Logically speaking, ',',/2 in
$$A_1, A_2, ..., A_m : -B_1, B_2, ..., B_n$$

is interpreted as conjunction on both sides.

The reader familiar with grammar theory will notice (as suggested by [13]) that generalizing from definite to multi-headed definite programs is similar to going from context-free grammars to context-dependent grammars.

The binarization of the head will be extended in a way similar to that of the binarization of the right side of a clause.

Definition 4 *Let P be a multiheaded definite program and Cont a new variable. Starting with the multiheaded clause*
(C) $A_1, A_2..., A_m : -B_1, B_2, ..., B_n.$
we construct the clause
(C') $\psi(A_1, \psi(A_2, ..., \psi(A_m, Cont))) : -\psi(B_1, \psi(B_2, ..., \psi(B_n, Cont))).$

Note that after this transformation the binarized head will be able to match the initial segment of the current implicit goal stack of BinProlog embedded in the continuation, i.e. to look into the immediate future of the computation.

Somewhat more difficult is to implement 'meta-variables' in the left side. In the case of a goal like:

```
a,Next:-write(Next),nl,b(Next).
```

we need a more complex binary form:

```
a(A) :- strip_cont(A,B,C,write(B,nl(b(B,C)))).
```

where `strip_cont(GoalAndCont,Goal,Cont)` is needed to undo the binarization and give the illusion of getting the goal **Next** at source level by unification with a metavariable on the left side of the clause **a/1**.

Notice that the semantics of the continuation manipulation predicate `strip_cont/3` is essentially first order, as we can think of `strip_cont` simply as being defined by a set of clauses like:

```
........
strip_cont(f(X1,...,Xn,C),f(X1,...,Xn),C).
........
```

for every functor **f** occuring in the program.

The full code of the preprocessor that handles multiples-headed clauses is given in **Appendix A**.

4 Programming with Continuations in Multiple-headed clauses

The first question that comes to mind is how well our transformation technique performs compared with handwritten programs.

Example 4 *The following program is a continuation based version of* **nrev/2**.

```
app(Xs,Ys,Zs):-app_args(Xs,Zs),paste(Ys).

app_args([],[X]),paste(X).
app_args([A|Xs],[A|Zs]):-app_args(Xs,Zs).

nrev([],[]).
nrev([X|Xs],R):-
  nrev1(Xs,R),
  paste(X).

nrev1(Xs,R):-
  nrev(Xs,T),
  app_args(T,R).
```

which shows no loss in speed compared to the original program (530 KLIPS with BinProlog 2.20 on a Sparcstation 10-40).

One of Miller's motivating examples for intuitionistic implication in λProlog [5] [14], is the reverse predicate with intuitionistic implication.

Example 5 *By using our multi-headed clauses we can write a such a reverse predicate as follows:*

```
reverse(Xs,Ys):-rev(Xs,[]),result(Ys).

  rev([],Ys), result(Ys).
  rev([X|Xs],Ys):-rev(Xs,[X|Ys]).
```

which gives after binarization:

```
reverse(Xs,Ys,Cont):-rev(Xs,[],result(Ys,Cont)).

  rev([],Ys, result(Ys,Cont)) :- true(Cont).
  rev([X|Xs],Ys,Cont):-rev(Xs,[X|Ys],Cont).
```

with a suitable definition for **true/1** *as:*

```
  true(C):-C.
```

and with a clause like

```
  reverse(Xs,Ys):-reverse(Xs,Ys,true).
```

Notice that such definitions are 'virtual' (i.e. supplied by generic WAM-level operations) in BinProlog, for space and efficiency reasons [23].

By replacing **true(C):-C** an appropriate set of clauses where C=p(X1,...,Xn) for every head of clause **p/n** occurring in the program, we obtain a definite binary program which accurately describes the operational semantics of the original multi-headed program.

Although the existence of such a translation is expected (as binary definite programs are Turing-equivalent) its simple syntactic nature rehabilitates the idea of using a *translation semantics* as an 'internal' means to describe the semantics of multi-headed logic programs, despite the fact that it fixes *an implementation* as the *meaning* of a programming construct.

The availability of such a programming style in the Horn subset of Prolog is also of practical importance as classical Prolog is still about 5-10 times faster than, for instance, the fastest known λProlog implementation [4].

Example 6 *The following program implements a* **map** *predicate with a Hilog-style (see [5]) syntax:*

---

[5] which is, by the way, a motivating example also for Andreoli and Pareschi's Linear Objects, [2]

```
cmap(F),i([]),o([]).
cmap(F),i([X|Xs]),o([Y|Ys]):-G=..[F,X,Y],G,cmap(F),i(Xs),o(Ys).

inc10(X,Y):-Y is X+10.

test:-cmap(inc10),i([1,2,3,4,5,6]),o(Xs),write(Xs),nl.
```

Some interesting optimization opportunities by program transformation can be obtained by using multiple-headed clauses, as can be seen in the following example of elimination of existential variables (this technique has been pioneered by Proietti and Pettorossi, in a fold/unfold based framework, [19]).

Example 7 *Given the program:*

```
r(X,Y):-p(X,N),q(N,Y).

p(a,1).    q(1,10).
p(b,2).    q(2,11).
```

we can rewrite it as:

```
r(X,Y):-p(X),result(Y).

p(a) :-q(1).
p(b) :- q(2).

q(1), result(10).
q(2), result(11).
```

which avoid passing unnecessary information by directly unifying the 2 occurrences of result(_).

Example 8 *Multi-headed clauses can be used to ensure direct communication with the leafs of a tree (represented simply as a Prolog term).*

```
frontier(T,X):-leaf_var(T),result(X).

leaf_var(X), result(X):- var(X).
leaf_var(T) :- compound(T), T=..[_|Xs], member(X,Xs), leaf_var(X).
```

where ?-frontier(X) *returns non-deterministically all the variables in a term seen as a (finite) tree.*

The example also shows that the technique scales easily to arbitrary recursive data-types, and to programs in practice beyond the domain of definite programs.

As it can be seen from the previous examples, the advantage of our technique is that the programmer can follow an intuitive, grammar-like semantics when dealing with continuations and that 'predictability' of the fact that the continuation will be at the right place at the right time is straightforward.

The next section will show some special cases where full access to continuations is more convenient.

5   Multiple headed clauses vs. full-blown continuations

BinProlog 2.20 supports direct manipulation of binary clauses denoted

```
Head ::- Body.
```

They give full power to the knowledgeable programmer on the future of the computation. Note that such a facility is not available in conventional WAM-based systems where continuations are not first-order objects.

Example 9 *We can use them to write programs like:*

```
member_cont(X,Cont)::-
    strip_cont(Cont,X,NewCont,true(NewCont)).
member_cont(X,Cont)::-
    strip_cont(Cont,_,NewCont,member_cont(X,NewCont)).

test(X):-member_cont(X),a,b,c.
```

A query like

```
?-test(X).
```

will return X=a; X=b; X=c; X=*whatever follows from the calling point of* test(X).

```
catch(Goal,Name,Cont)::-
    lval(catch_throw,Name,Cont,call(Goal,Cont)).

throw(Name,_)::-
    lval(catch_throw,Name,Cont,nonvar(Cont,Cont)).
```

where lval(K1,K2,Val) *is a BinProlog primitive which unifies* Val *with a back-trackable global logical variable accessed by hashing on two (constant or variable) keys* K1,K2.

Example 10 *This allows for instance to avoid execution of the infinite* loop *from inside the predicate* b/1.

```
loop:-loop.

c(X):-b(X),loop.

b(hello):-throw(here).
b(bye).

go:-catch(c(X),here),write(X),nl.
```

Notice that due to our translation semantics this program still has a first order reading and that BinProlog's lval/3 is not essential as it can be emulated by an extra argument passed to all predicates.

Although implementation of **catch** and **throw** requires full-blown continuations, we can see that at user level, the multi-headed clause notation is enough.

6 Continuation Grammars

By allowing us to see the right context of a given grammar symbol, continuations can make logic grammars both more efficient and simpler, as well as provide several sophisticated capabilities within a simple framework. We next discuss some of these capabilities.

6.1 Parsing strategy versatility

Changing parsing strategies is usually deemed to require a different parser. We now discuss how we can effect these changes with no more apparatus than continuation grammars.

Take for instance the noun phrase "Peter, Paul and Mary". We can first use a tokenizer which transforms it into the sequence

Example 11 (noun('Peter'), comma, noun('Paul'), and, noun('Mary'))

and then use this sequence to define the rule for list-of-names in the following continuation grammar:

```
names  --> noun('Peter'), comma, noun('Paul'), and, noun('Mary').

noun(X), comma --> [X,','].
noun(X), and, noun(Last) --> [X,'and',Last].
```

This is an interesting way in which to implement bottom-up parsing of recursive structures, since no recursive rules are needed, yet the parser will work on lists of names of any length. It is moreover quite efficient, there being no backtracking. It can also be used in a mixed strategy, for instance we could add the rules:

```
verb_phrase --> verb.
verb --> [sing].
```

and postulate the existence of a verb phrase following the list of nouns, which would then be parsed top-down. The modified rules follow, in which we add a marker to record that we have just parsed a noun phrase, and then we look for a verb phrase that follows it.

```
noun(X), comma --> [X,','].
noun(X), and, noun(Last) --> [X,'and',Last], parsed_noun_phrase.

parsed_noun_phrase --> verb_phrase.
```

DCGs allow further left-hand side symbols provided they are terminal ones (it has been shown that rules with non-leading non-terminal left hand side symbols can be replaced by a set of equivalent, conforming rules [6]). However, because their implementation is based on synthesizing those extra symbols into the strings being manipulated, lookahead is simulated by creating the expectation of finding a given symbol in the future, rather than by actually finding it right away.

With continuations we can implement a more restricted but more efficient version of multiple-left-hand-side symbol-accepting DCGs, which we shall call *continuation grammars*. It is more restricted in the sense that the context that an initial left-hand side symbol is allowed to look at is strictly its right-hand side sisters, rather than any descendants of them, but by using this immediate context right away the amount of backtracking is reduced for many interesting kinds of problems.

Example 12 *For comparison with the length of code, here is a non-continuation based DCG formulation of the "list of names" example which allows for multiple left-hand side symbols* [6]

```
names --> start, start(C), name, end(C), nms, end.

nms --> [].
nms --> start(C), name, end(C), nms.
```

The next two rules are for terminalizing symbols that appear in later rules as non-leading left-hand-side symbols.

```
end --> [end].
start(C) --> [start(C)].

start, [end] --> []
start, [start(C)] --> [].

end(and), end --> [,].
end(comma), start(and) --> [and].
end(comma), start(comma) --> [,].
```

Other possible applications are to procedures to read a word or a sentence, which typically make explicit use of a lookahead character or word, to restrictive relative clauses, whose end could be detected by looking ahead for the comma that ends them, etc.

6.2 Relating of long-distance constituents

One important application in computational linguistics is that of describing movement of constituents. For instance, a common linguistic analysis would state that the object noun phrase in "Jack built the house" moves to the front through relativization, as in "the house that Jack built", and leaves a trace behind in the phrase structure representing the sentence. For describing movement,

---

[6] This is a simplified version of the corresponding code fragment in [8].

we introduce the possibility of writing one meta-variable in the left hand side of
continuation grammar rules. Then our example can be handled through a rule
such as the following:

```
relative_pronoun, X, trace --> [that], X.
```

This rule accounts, for instance, for "the house that jack built", "the house
that the Prime Minister built", "the house that the man with the yellow hat
who adopted curious George built", and so on. A simple grammar containing a
similar rule follows, the reader might try it to derive: "The elephant that Jill
photographed smiles":

Example 13 `sentence --> noun_phrase,verb_phrase.`

```
noun_phrase --> proper_name.
noun_phrase --> det, noun, relative.
noun_phrase --> trace.

verb_phrase --> verb, noun_phrase.
verb_phrase --> verb.

relative --> [].
relative --> relative_marker, sentence.

relative_marker, X, trace --> relative_pronoun, X.

det --> [the].

noun --> [elephant].

relative_pronoun --> [that].

proper_name --> ['Jill'].

verb --> [photographed].
verb --> [smiles].
```

What is noteworthy about this example is that it shows how to use plain Prolog plus binarization to achieve the same expressive power which used to require
more sophisticated grammar formalisms, such as Extraposition or Discontinuous
Grammars [17, 9].

6.3 Computation viewed as parsing

Continuation Grammars are useful not only for linguistic or intrinsically grammatical examples, but can also serve to simplify the description of many problems
which can be formulated in terms of a grammar.

Example 14 *For instance, we can sort a list of numbers by viewing it as an input string to a grammar, which obtains successive intermediate strings in a parallel-like fashion until two successively obtained strings are the same:*

```
sort(Input):- s(V,V,Input,[]), remember(Input).

s(H,[Y,X|Z]) --> [X,Y], {X>Y}, !, s(H,Z).
s(H,[X|Z]) --> [X], s(H,Z).
s(NewString,[]), {remember(OldString)} -->
  {decide(OldString,NewString)}.

decide(Result,Result),output(Result).
decide(OldString,NewString):- sort(NewString).

par_sort(Input,Result):-sort(Input),output(Result).
```

The first clause initializes a difference-list as empty (both the head and tail are represented by a variable V); adds the two extra arguments (input string and output string) that are typically needed by the Prolog translation of the grammar rules; and pastes a reminder of the last string obtained (initially, the input string).

The first grammar rule consumes the next two numbers from the input string if they are unordered, and calls s recursively, keeping track of their ordered version in the difference list represented by the two first arguments of s.

If the next two numbers to be consumed are not ordered, or if there is only one number left, the second grammar rule consumes one number, and keeps track of it in the difference list as it recursively calls s.

The third grammar rule finds no more numbers to consume, so it consults the last string remembered in order to decide whether to print the result (if the new string is no different than the last one) or to start another iteration of the algorithm on the list last obtained.

The last clause returns the answer directly from the continuation. For instance, on the input string L=[8,4,5,3] with the goal par_sort(L,R), we successively obtain: [4,8,3,5], [4,3,8,5] and finally R=[3,4,5,8].

The above example is a variation of the odd-even-transposition parallel algorithm for sorting numbers [1]. A grammatical version of this algorithm has been developed in terms of parametric L-systems [20], an extension of L-systems which operates on parametric words and which can be viewed as a model of parallel computation. It is interesting to note that by using continuation grammars we need no special formalisms to obtain the kind of parallel computation that is provided by more elaborate formalisms such as parametric L-systems.

7 Conclusion

We have proposed the use of continuation-passing binarization both for extending logic programs to allow multiple heads, and for a fresh view of those logic grammars which allow multiple left-hand-side symbols. In both of these areas, the use of continuations has invited interesting new programming (grammar) composition techniques which we have provided examples for.

Other logic programming proposals involve multiple heads, e.g. disjunctive logic programming [12, 21] or contextual logic programming [16, 10]. However, in these approaches the notion of alternative conclusions is paramount, whereas in our approach we instead stress the notion of contiguity that computational linguistics work has inspired. A formal characterization of this stress with respect to logic grammars has been given in [3].

The technique has been included as a standard feature of the BinProlog 2.20 distribution available by ftp from clement.info.umoncton.ca.

Acknowledgment

This research was supported by NSERC Operating grant 31-611024, and by an NSERC Infrastructure and Equipment grant given to the Logic and Functional Programming Laboratory at SFU, in whose facilities this work was developed. We are also grateful to the Centre for Systems Science, LCCR and the Department of Computing Sciences at Simon Fraser University for the use of their facilities. Paul Tarau thanks also for support from the Canadian National Science and Engineering Research Council (Operating grant OGP0107411) and a grant from the FESR of the Université de Moncton.

References

1. S. Akl. *The design and analysis of parallel algorithms.* Prentice Hall, Englewood Cliffs, 1989.
2. J.-M. Andreoli and R. Pareschi. Linear objects: Logical processes with built-in inheritance. In D. Warren and P. Szeredi, editors, *7th Int. Conf. Logic Programming,* Jerusalem, Israel, 1990. MIT Press.
3. J. Andrews, V. Dahl, and F. Popowich. A Relevance Logic Characterization of Static Discontinuity Grammars. Technical report, CSS/LCCR TR 91-12, Simon Fraser University, 1991.
4. P. Brisset. Compilation de λProlog. Thèse, Université de Rennes I, 1992.
5. W. Chen and D. S. Warren. Compilation of predicate abstractions in higher-order logic programming. In J. Maluszyński and M. Wirsing, editors, *Proceedings of the 3rd Int. Symposium on Programming Language Implementation and Logic Programming, PLILP91, Passau, Germany,* number 528 in Lecture Notes in Computer Science, pages 287–298. Springer Verlag, Aug. 1991.
6. A. Colmerauer. *Metamorphosis Grammars,* volume 63, pages 133–189. Springer-Verlag, 1978.
7. A. Colmerauer, H. Kanoui, R. Pasero, and P. Roussel. Un systeme de communication homme-machine en francais. Technical report, Groupe d'Intelligence Artificielle, Universite d'Aix-Marseille II, Marseille, 1973.
8. V. Dahl. Translating spanish into logic through logic. *American Journal of Computational Linguistics,* 13:149–164, 1981.

9. V. Dahl. Discontinuous grammars. *Computational Intelligence*, 5(4):161–179, 1989.

10. J.-M. Jacquet and L. Monteiro. Comparative semantics for a parallel contextual logic programming language. In S. Debray and M. Hermenegildo, editors, *Proceedings of the 1990 North American Conference on Logic Programming*, pages 195–214, Cambridge, Massachusetts London, England, 1990. MIT Press.

11. J. W. Lloyd. *Foundations of Logic Programming*. Springer-Verlag, 1987.

12. J. Lobo, J. Minker, and A. Rajasekar. Extending the semantics of logic programs to disjunctive logic programs. In G. Levi and M. Martelli, editors, *Proceedings of the Sixth International Conference on Logic Programming*, pages 255–267, Cambridge, Massachusetts London, England, 1989. MIT Press.

13. J. Maluszyński. On the relationship between context-dependent grammars and multi-headed clauses., June 1994. Personal Communication.

14. D. Miller. A logic programming language with lambda-abstraction, function variables, and simple unification. *J. Logic and Computation*, 1(4):497–536, 1991.

15. D. A. Miller. Lexical scoping as universal quantification. In G. Levi and M. Martelli, editors, *Proceedings of the Sixth International Conference on Logic Programming*, pages 268–283, Cambridge, Massachusetts London, England, 1989. MIT Press.

16. L. Monteiro and A. Porto. Contextual logic programming. In G. Levi and M. Martelli, editors, *Proceedings of the Sixth International Conference on Logic Programming*, pages 284–299, Cambridge, Massachusetts London, England, 1989. MIT Press.

17. F. Pereira. Extraposition grammars. *American Journal for Computational Linguistics*, 7:243–256, 1981.

18. M. Proietti. On the definition of binarization in terms of fold/unfold., June 1994. Personal Communication.

19. M. Proietti and A. Pettorossi. Unfolding-definition-folding, in this order, for avoiding unnecessary variables in logic programs. In J. Maluszyński and M. Wirsing, editors, *Proceedings of the 3rd Int. Symposium on Programming Language Implementation and Logic Programming, PLILP91, Passau, Germany*, number 528 in Lecture Notes in Computer Science, pages 347–358. Springer Verlag, Aug. 1991.

20. P. Prusinkiewicz and J. Hanan. *L-systems: from formalism to programming languages*. Springer-Verlag, 1992.

21. D. W. Reed, D. W. Loveland, and B. T. Smith. An alternative characterization of disjunctive logic programs. In V. Saraswat and K. Ueda, editors, *Logic Programming Proceedings of the 1991 International Symposium*, pages 54–70, Cambridge, Massachusetts London, England, 1991. MIT Press.

22. J. Stoy. *Denotational Semantics: the Scott-Strachey Approach to Programming Language Theory*. Cambridge, MA. The MIT Press, 1977.

23. P. Tarau. Program Transformations and WAM-support for the Compilation of Definite Metaprograms. In A. Voronkov, editor, *Logic Programming, RCLP Proceedings*, number 592 in Lecture Notes in Artificial Intelligence, pages 462–473, Berlin, Heidelberg, 1992. Springer-Verlag.

24. P. Tarau and M. Boyer. Elementary Logic Programs. In P. Deransart and J. Maluszyński, editors, *Proceedings of Programming Language Implementation and Logic Programming*, number 456 in Lecture Notes in Computer Science, pages 159–173. Springer, Aug. 1990.

25. P. Tarau and K. De Bosschere. Memoing with Abstract Answers and Delphi Lemmas. In Y. Deville, editor, *Logic Program Synthesis and Transformation*, Springer-Verlag, Workshops in Computing, Louvain-la-Neuve, July 1993.

26. P. Wadler. Monads and composable continuations. *Lisp and Symbolic Computation*, pages 1–17, 1993.
27. M. Wand. Continuation-based program transformation strategies. *Journal of the Association for Computing Machinery*, 27(1):164–180, 1980.
28. D. H. D. Warren. Higher-order extensions to Prolog – are they needed? In D. Michie, J. Hayes, and Y. H. Pao, editors, *Machine Intelligence 10*. Ellis Horwood, 1981.
29. D. S. Warren. The XOLDT System. Technical report, SUNY Stony Brook, electronic document: ftp sbcs.sunysb.edu, 1992.

Appendix A

```
% converts a multiple-head clause to its binary equivalent
def_to_mbin((H:-B),M):-!,def_to_mbin0(H,B,M).
def_to_mbin(H,M):-def_to_mbin0(H,true,M).

def_to_mbin0((H,Upper),B,(HC:-BC)) :- nonvar(H),!,
  termcat(H,ContH,HC),
  add_upper_cont(B,Upper,ContH,BC).
def_to_mbin0(H,B,(HC:-BC)) :- !,
  termcat(H,Cont,HC),
  add_cont(B,Cont,BC).

add_upper_cont(B,Upper,ContH,BC):-nonvar(Upper),!,
  add_cont(Upper,ContU,ContH),
  add_cont(B,ContU,BC).
add_upper_cont(B,Upper,ContH,BC):-
  add_cont((strip_cont(ContH,Upper,ContU),B),ContU,BC).

% adds a continuation to a term

add_cont((true,Gs),C,GC):-!,add_cont(Gs,C,GC).
add_cont((fail,_),C,fail(C)):-!.
add_cont((G,Gs1),C,GC):-!,
  add_cont(Gs1,C,Gs2),
  termcat(G,Gs2,GC).
add_cont(G,C,GC):-termcat(G,C,GC).

strip_cont(TC,T,C):-TC=..LC,append(L,[C],LC),!,T=..L.
```

The predicate termcat(Term,Cont,TermAndCont) is a BinProlog builtin which works as if defined by the following clause:

```
termcat(T,C,TC):-T=..LT,append(LT,[C],LTC),!,TC=..LTC.
```

Improving the Whelk System: a type-theoretic reconstruction

Geraint A. Wiggins

Department of Artificial Intelligence
University of Edinburgh
80 South Bridge, Edinburgh EH1 1HN
Scotland
geraint@ai.ed.ac.uk

Abstract. I present a reformulation of the Whelk system [Wiggins 92b], as a higher-order type theory. The theory is based on that of [Martin-Löf 79], adapted to facilitate the extraction of logic programs from proof objects. A notion of normalization is used to ensure that the extracted program is executable by standard logic-programming methods. The extension admits specifications over types and programs, and so allows modularity and the construction of program combinators. In doing so, it demonstrates that logic program synthesis techniques have potential for solving "industrial-strength" problems.

1 Introduction

In this paper, I present a reconstruction of the Whelk proof development and program construction system which was explained in [Wiggins *et al* 91, Wiggins 92a, Wiggins 92b]. The reconstruction uses a higher order logic which admits quantification over types, and, following the style of Martin-Löf Type Theory [Martin-Löf 79], views sentences in the logic as types. Proofs of such sentences are then thought of as members of the types. The addition of the higher order features means we can now synthesise meta-programs and program modules. Making such a gain means that the system is more complicated than before, because type membership is undecidable. However, much of the extra proof obligation entailed may be dealt with automatically, and good heuristics exist for those parts which are not straightforward. The correctness argument for the new system is greatly simplified over the original Whelk, because it can be given largely by reference to Martin-Löf's existing theory.

The paper is structured as follows. In section 2, I remind the reader of the *modus operandi* of the original Whelk system, and illustrate the disadvantages which led to the reconstruction explained here; in section 3 I explain the shift from the first order logic to the type theory and outline the changes necessary. In section 4, I give those inference figures of the reconstructed system necessary for an example. Correctness is shown by reference to Martin-Löf original calculus, but there is not space to cover it in depth here. Section 5 shows the synthesis of the subset/2 predicate for comparison with the original system [Wiggins 92b]. Finally, because of space constraints, I merely sketch the normalization process for generating the runnable programs in section 6.

2 Background

The Whelk system is an attempt to bring ideas used in the domain of type theory and functional program synthesis to bear on the synthesis and/or transformation of logic programs. We wish to give the specification of a program in terms only of the logical relation between its arguments; the construction of a suitable algorithm is then to be left to a process of proof that the specification can be realised. The notion on which this is based is the *proofs-as-programs* paradigm of [Constable 82], in which a constructive proof of an existentially quantified conjecture is encoded as an *extract term*, a function, expressed as a λ-term, embodying an algorithm which implements the input/output relation specified in the conjecture. The proofs-as-programs idea has been adapted [Bundy *et al* 90b] to synthesise logic programs. This is achieved by viewing logic programs as functions on to the Boolean type. The synthesised relations are then Boolean-valued functions (expressed as logic programs, called in the all-ground mode).

In [Wiggins *et al* 91, Wiggins 92b], the original version of the logic system designed for this purpose is explained. While the system does its job effectively, and is in use for teaching and research at various institutions, the background theory is not as tidy as it might be. In particular, the means of specifying what constitutes a *synthesis conjecture* leads us to convoluted correctness proofs, which are less than elegant. Further, the original system is restricted to first order specifications, with no meta-language. Therefore, it is not possible to synthesise program modules, or to specify meta-programs, simply because quantification over mathematical structures other than the natural numbers and parametric lists is not allowed. (In fact, it is not clear that the modularity issue is as important in the context of program synthesis as elsewhere; however, the fact remains that the expressive power of modularity is desirable.)

In Whelk, the fact that we wish to synthesise a program is expressed by means of a *decidability* operator, ∂, which is applied to the specification of the desired program. By proving constructively that a specification is decidable, we synthesise the decision procedure – in the form of a logic program – for that specification. In earlier versions of the system [Wiggins 92b], ∂ was defined by reference to the existence of the extracted program, and not *via* sequent calculus rules, like the other logical connectives. Therefore, it behaved rather like a macro, defined at the meta-level with respect to the sequent calculus. This approach was taken because it allowed certain restrictions to be placed on the form of the programs derived from Whelk proofs, so that no program components would be generated by parts of the proof concerned purely with verification of correctness. In particular, the meta-flavour of the ∂ definition allowed certain restrictions on the applicability of inference rules to be hidden in a way which seems, in retrospect, undesirable. In the reconstruction explained here, these meta-level definitions and restrictions have been abandoned in favour of logical elegance, their respective effects being reproduced by a more conventional sequent calculus definition and application of a simple heuristic during proof construction.

In the following sections of this paper, I will discuss the use of the reconstructed calculus, which I will call WTT, for program manipulation, with reference to the existing example from [Wiggins 92b], showing how the program can be modularised by type. The Whelk conjecture, which clearly has types built in, is as follows:

$$\vdash \forall k : list(\mathbb{N}).\forall l : list(\mathbb{N}).\partial (\forall x : \mathbb{N}.x \in k \rightarrow x \in l)$$

where $\in$ is the usual list membership; it specifies the list inclusion relation. Proof of this synthesis conjecture is fairly straightforward, and will be outlined again in section 5. The extracted program, expressed as a logical equivalence, may be automatically converted the Gödel code, as shown in [Wiggins 92b].

The new higher-order calculus presented here is also designed for describing and manipulating logic programs. However, because of the first order nature of logic programs themselves, it is clear that programs parameterised by module or type will not be expressible as such, in a strictly first-order way. In what follows, therefore, I assume that the parameters of the modules I synthesise will be instantiated *within* WTT, and the resulting Gödel program extracted subsequently, rather than the converse.

3   Reconstruction

There are two main aspects to the type-theoretic reconstruction of Whelk. First, the decidability operator, ∂, on which the construction of logic programs rests, has been integrated into the sequent calculus as an operator in its own right, rather than as a set of operators related to the standard ones, as is the case in Whelk. This results in a considerable reduction in the number of rules (roughly half), which necessarily makes things less complicated.

The second change is the raising of the calculus to higher order. This is carried out by following the rules of Martin-Löf's Type Theory [Martin-Löf 79], and adapting and extending them where necessary. This approach has the advantage that correctness can be argued in terms of the original system. The introduction of higher-order structures raises the possibility that functional terms and variables might appear in the extracted programs, and we will need to be careful that this does not prove to be the case – the programs would not then be logic programs. This point is the main motivation for the building of a new calculus, from the bottom up, as it were, rather than simply implementing the WTT logic on top of an existing type theory. In the latter of these options, it is not clear how (or indeed if) we could guarantee the first order structures we need.

4   Whelk Type Theory

4.1   Introduction

In order to specify our type theory, we must specify what is the syntactic form of each type, and what is the syntactic form of objects of each type. In general, there will be two forms of objects: the *non-canonical* and *canonical* forms – *i.e.* forms which may be further evaluated and forms which may not. The calculus must also supply a means of computation to enable the derivation of canonical forms from non-canonical ones. It will be clear from the expression of the evaluation strategy below, which follows Martin-Löf's description closely, that the theory is extensional – that is to say, equality of functions is expressed in terms of equality of corresponding inputs and outputs.

Like Martin-Löf's type theory, WTT admits four kinds of "judgement" – a judgement being essentially what we can prove is true (*i.e.*, the logical sentence to the right of the sequent symbol). The four kinds of judgement I will use here relate to the formation of *types* and the *objects* which inhabit them. In each case, entities themselves, and equality between entities is covered.

Judgement	Form
A is a type	A type
A and B are equal types	$A = B$
a is an object of type A	$a \in A$
a and b are equal objects of type A	$a = b \in A$

4.2 Preamble

(First order) Whelk uses distinct languages for expression of its specification $(\mathcal{L}_S)$ and of its programs $(\mathcal{L}_P)$. A mapping connects the two, and allows us to say what the synthesised program means in terms of the specification – for more details, see [Wiggins 92b]. In WTT, a similar arrangement pertains: loosely, types correspond with $\mathcal{L}_S$-formulæ and objects with $\mathcal{L}_P$-formulæ, and type membership replaces the interpretation. However, there is a significant difference: in WTT, we have a type of types, known as a *universe*, so it is possible to view a type as an object also. For the purposes of the examples here, however, these higher-order complications are irrelevant. (Incidentally, there is a hierarchy of universes, each containing the one "below" it, which avoids Russell's Paradox, while still giving all the types we need.)

The type language admits the familiar connectives (though for reasons of limited space I have given only those necessary for the example): ∧ (and), ∨ (or), → (implies), ∀ (for all). Contradiction is denoted by {} (the empty type), and negation by implication, so the negation of A is $A \to \{\}$. We also have identity within a type, $\equiv_\tau$, (NB, this is not the same as the object equality judgement) and the decidability operator, ∂. The quantifiers are typed in the usual way, : denoting type membership. The operators of the object language are the constructors of the members of the types. They are best given in terms of the types they construct and destruct. To facilitate comparison, these constructions are given in the style of [Martin-Löf 79].

Type	Canonical Form	Non-canonical Form
$\forall x{:}A.B$	$\sqcap x{:}A.b$	$c(a)$
$A \to B$	$x{:}A \mapsto b$	$c\{a\}$
$A \wedge B$	$a \sqcap b$	$\mathcal{L}_\wedge(a), \mathcal{R}_\wedge(b)$
$A \vee B$	$\mathcal{L}_\sqcup(a), \mathcal{R}_\sqcup(b)$	$\mathcal{D}_\vee(x,y)(c,d,e)$
∂A	$\mathcal{L}_\partial(a), \mathcal{R}_\partial(a)$	$\mathcal{D}_\partial(x,y)(c,d,e)$
$a \equiv_A b$	$c \overset{A}{=} d$	$\mathcal{I}(c \overset{A}{=} d, e)$
$\{\}$		false
A list	$[]_A, [a\,|\,b]_A$	$p(x); p(x{:}A \text{ list}) = \text{List}\mathcal{C}(x, d, x_0, x_1, v, e)$

The non-canonical forms operate as follows:

1. If c has canonical form $⊔x{:}A.b$ and $a \in A$ then $c(a)$ is $b[a/x]$.
2. If c has canonical form $x{:}A \mapsto b$ and $a \in A$ then $c\{a\}$ is b.
3. If c has canonical form $a \sqcap b$ then $\mathcal{L}_\wedge(c)$ is a and $\mathcal{R}_\wedge(c)$ is b.
4. If c has canonical form $\mathcal{L}_\vee(a)$ then $\mathcal{D}_\vee(x,y)(c,d,e)$ is $d[a/x]$; if c has canonical form $\mathcal{R}_\vee(b)$ then $\mathcal{D}_\vee(x,y)(c,d,e)$ is $e[b/y]$.
5. If b has canonical form $\mathcal{L}_\partial(a)$ then $\mathcal{D}_\partial(x,y)(b,c,d)$ is $c[a/x]$; if b has canonical form $\mathcal{R}_\partial(a)$ then $\mathcal{D}_\partial(x,y)(b,c,d)$ is $d[a/y]$.
6. $a \equiv_A b$ expresses the identity relation between a and b, and is syntactically distinct from the equality judgement $a = b \in A$. If c has canonical form $p \overset{A}{=} q$ then $\mathcal{J}(c,d)$ is $d[q/p]$.
7. $\{\}$ is uninhabited; it has no constructors.
8. If x is $[\,]_A$ then $\mathrm{List}\mathcal{C}(x,d,x_0,x_1,v,e)$ is d;
   if x is $[x_0|x_1]$ then $\mathrm{List}\mathcal{C}(x,d,x_0,x_1,v,e)$ is $e[p(x_1)/v]$.

The appearance of the non-canonical forms in synthesised programs corresponds with the application of elimination rules in the proof. Similarly, the application of introduction rules in the proof corresponds with the appearance of object constructors in the program. This will become obvious on inspection of the example inference figures, below.

The sequent symbol is written $\vdash$. Non-empty formulæ are represented by upper case Roman letters (A,B,C,D); variables by lower case (x,y,z); types by lower case Greek letters (τ); and sequences of formulæ by upper case. Hypothesis management is implicit, so the order of the hypotheses is insignificant; unchanging hypotheses are not shown in the figures below. As the calculus is constructive, there is only one formula on the right of $\vdash$, so no management is needed there. Because this is a sequent calculus, the introduction and elimination rules of natural deduction correspond with operations on the right and left of the sequent symbol, respectively.

The construction proving each formula is associated with it by $\in$, which may be thought of as type membership. Thus, a proof/program may be viewed as inhabiting the type which is its specification. Note that program constructions in the inference figures, in particular those associated with hypotheses, may contain uninstantiated (meta-)variables; the construction process may be seen as a process of instantiation. Some cases (e.g. proofs of implications not in the scope of ∂) will lead to finished constructions containing uninstantiated variables. These constructions may be thought of as program transformers: they take an "input", by unification, and return a new structure related to that input in a way described by the conjecture whose proof produced the transformer. These ideas will be explored elsewhere.

In order to verify or synthesise a WTT program, we prove a conjecture of the general form

$$\Lambda \vdash \phi \in \forall \overline{a{:}\tau}.\partial\, S(\overline{a})$$

where Λ is an initial theory providing the definitions and axioms necessary for the proof; S is the specification of the program; $\overline{a{:}\tau}$ is a vector of typed arguments; and ϕ is the (synthesised) program, which includes an initial goal. The difference between verification and synthesis is the instantiation level of the object ϕ – if it is fully ground, we are verifying, if it is a pure variable we are synthesising, and correspondingly for any case in between. This is essentially the same as Whelk. However, given the improved uniformity of the reconstruction, we can now say precisely what an object/program means for any type membership judgement which is part of a proof. If we have a complete proof of the conjecture

$$\Lambda \vdash \phi \in \Phi$$

so that ϕ is ground, then we know that ϕ explicitly encodes a proof of Φ. The computation rules of the calculus allow us to manipulate this proof – instantiating arguments, for example – in a way which models the execution of the program in a higher-order functional programming language. This is relevant to logic programs because, when executed in the all-ground mode, they are equivalent to boolean functions. We can therefore read a subset of the proof objects produced by the system – that subset whose member are members of ∂ types – as logic programs.

4.3   General Rules

Reflexivity

$$\frac{\vdash a \in A}{\vdash a = a \in A} \qquad\qquad \frac{\vdash A \text{ type}}{\vdash A = A}$$

Symmetry

$$\frac{\vdash a = b \in A}{\vdash b = a \in A} \qquad\qquad \frac{\vdash B = A}{\vdash A = B}$$

Transitivity

$$\frac{\vdash a = b \in A \quad \vdash b = c \in A}{\vdash a = c \in A} \qquad\qquad \frac{\vdash A = B \quad \vdash B = C}{\vdash A = C}$$

Equality of Types

$$\frac{\vdash a \in A \quad \vdash A = B}{\vdash a \in B} \qquad\qquad \frac{\vdash a = b \in A \quad \vdash A = B}{\vdash a = b \in B}$$

Substitution

$$\frac{\vdash a \in A \quad x{:}A \vdash B[a/x] \text{ type}}{\vdash B \text{ type}} \qquad\qquad \frac{\vdash a \in A \quad x{:}A \vdash b[a/x] \in B[a/x]}{\vdash b \in B}$$

$$\frac{\vdash a = c \in A \quad x{:}A \vdash B[a/x] = D[c/x]}{\vdash B = D}$$

$$\frac{\vdash a = c \in A \quad x{:}A \vdash b[a/x] = d[c/x] \in B[a/x]}{\vdash b = d \in B}$$

Axiom

$$\frac{}{a{:}A \vdash a \in A}$$

4.4 Dependent Function Type

Formation

$$\frac{\vdash A \text{ type} \quad x{:}A \vdash B \text{ type}}{\vdash \forall x{:}A.B \text{ type}} \qquad \frac{\vdash A = C \quad x{:}A \vdash B = D}{\vdash (\forall x{:}A.B) = (\forall x{:}C.D)}$$

Construction

$$\frac{x{:}A \vdash b \in B}{\vdash (\sqcap x{:}A.b) \in (\forall x{:}A.B)} \qquad \frac{x{:}A \vdash b = c \in B}{\vdash (\sqcap x{:}A.b) = (\sqcap x{:}A.c) \in (\forall x{:}A.B)}$$

Selection

$$\frac{\vdash t \in B \quad b{:}B[t/x] \vdash g \in G}{c{:}(\forall x{:}A.B) \vdash g[c(t)/b] \in G}$$

$$\frac{\vdash c \in (\forall x{:}A.B) \quad \vdash t \in B \quad b{:}B[t/x] \vdash g \in G}{\vdash g[c(t)/b] \in G}$$

Evaluation

$$\frac{\vdash a \in A \quad x{:}A \vdash b \in B}{\vdash ((\sqcap x{:}A.b)(a)) = b[a/x] \in B[a/x]} \qquad \frac{\vdash c \in (\forall x{:}A.B)}{\vdash ((\sqcap x{:}A.c)(x)) = c \in (\forall x{:}A.B)}$$

4.5 Function Type

Formation

$$\frac{\vdash A \text{ type} \quad \vdash B \text{ type}}{\vdash A \to B \text{ type}} \qquad \frac{\vdash A = C \quad \vdash B = D}{\vdash (A \to B) = (C \to D)}$$

Construction

$$\frac{a{:}A \vdash b \in B}{\vdash (x{:}A \mapsto b) \in (A \to B)} \qquad \frac{x{:}A \vdash b = d \in B}{\vdash (x{:}A \mapsto b) = (x{:}A \mapsto d) \in (A \to B)}$$

Selection

$$\frac{\vdash a{:}A \quad b{:}B \vdash g \in G}{c{:}(A \to B) \vdash g[c\{a\}/b] \in G} \qquad \frac{\vdash c \in (A \to B) \quad \vdash a \in A \quad b{:}B \vdash g \in G}{\vdash g[c\{a\}/b] \in G}$$

Evaluation

$$\frac{\vdash a \in A \quad \vdash b \in B}{\vdash ((x{:}A \mapsto b)\{a\}) = b \in B} \qquad \frac{\vdash b \in (x{:}A \mapsto B)}{\vdash ((x{:}A \mapsto b)\{x\}) = b \in (x{:}A \mapsto B)}$$

4.6 Product Type

Formation

$$\frac{\vdash A \text{ type} \quad \vdash B \text{ type}}{\vdash A \wedge B \text{ type}} \qquad \frac{\vdash A = C \quad B = D}{\vdash (A \wedge B) = (C \wedge D)}$$

Construction

$$\frac{\vdash a \in A \quad \vdash b \in B}{\vdash a \sqcap b \in A \wedge B} \qquad \frac{\vdash a = c \in A \quad \vdash b = d \in B}{\vdash (a \sqcap b) = (c \sqcap d) \in (A \wedge B)}$$

Selection

$$\frac{a:A,b:B \vdash g \in G}{c:(A \wedge B) \vdash g[a,b/\mathcal{L}_\wedge(c),\mathcal{R}_\wedge(c)] \in G} \qquad \frac{\vdash c:(A \wedge B) \quad a:A,b:B \vdash g \in G}{\vdash g[a,b/\mathcal{L}_\wedge(c),\mathcal{R}_\wedge(c)] \in G}$$

Evaluation

$$\frac{\vdash a \sqcap b \in A \wedge B}{\vdash \mathcal{L}_\wedge(a \sqcap b) = a \in A} \qquad \frac{\vdash a \sqcap b \in A \wedge B}{\vdash \mathcal{R}_\wedge(a \sqcap b) = b \in B}$$

4.7 Disjoint Union Type

Formation

$$\frac{\vdash A \text{ type} \quad \vdash B \text{ type}}{\vdash A \vee B \text{ type}} \qquad \frac{\vdash A = C \quad \vdash B = D}{\vdash (A \vee B) = (C \vee D)}$$

Construction

$$\frac{\vdash a \in A}{\vdash \mathcal{L}_\vee(a) \in A \vee B} \qquad \frac{\vdash a = c \in A}{\vdash \mathcal{L}_\vee(a) = \mathcal{L}_\vee(c) \in (A \vee B)}$$

$$\frac{\vdash b \in B}{\vdash \mathcal{R}_\vee(b) \in A \vee B} \qquad \frac{\vdash b = d \in B}{\vdash \mathcal{R}_\vee(b) = \mathcal{R}_\vee(d) \in (A \vee B)}$$

Selection

$$\frac{x:A \vdash d:G \quad y:B \vdash e:G}{c:(A \vee B) \vdash \mathcal{D}_\vee(x,y)(c,d,e) \in G} \qquad \frac{\vdash c \in (A \vee B) \quad x:A \vdash d \in G \quad y:B \vdash e \in G}{\vdash \mathcal{D}_\vee(x,y)(c,d,e) \in G}$$

Evaluation

$$\frac{\vdash a \in A \quad x:A \vdash d \in C \quad y:B \vdash e \in C}{\vdash \mathcal{D}_\vee(x,y)(\mathcal{L}_\vee(a),d,e) = d[a/x] \in C}$$

$$\frac{\vdash a \in A \quad x:A \vdash d \in C \quad y:B \vdash e \in C}{\vdash \mathcal{D}_\vee(x,y)(\mathcal{R}_\vee(b),d,e) = e[b/y] \in C}$$

4.8 Decision Type

Formation

$$\frac{\vdash A \text{ type}}{\vdash \partial A \text{ type}} \qquad\qquad \frac{\vdash A = B}{\vdash (\partial A) = (\partial B)}$$

Construction

$$\frac{\vdash a \in A}{\vdash \mathcal{L}_\partial(a) \in \partial A} \qquad\qquad \frac{\vdash a = b \in A}{\vdash \mathcal{L}_\partial(a) = \mathcal{L}_\partial(b) \in \partial A}$$

$$\frac{\vdash a \in (A \to \{\})}{\vdash \mathcal{R}_\partial(a) \in \partial A} \qquad\qquad \frac{\vdash a = b \in (A \to \{\})}{\vdash \mathcal{R}_\partial(a) = \mathcal{R}_\partial(b) \in \partial A}$$

Selection

$$\frac{x{:}A \vdash d \in G \quad y{:}(A \to \{\}) \vdash e \in G}{a{:}\partial A \vdash \mathcal{D}_\partial(x,y)(a,d,e) \in G} \qquad \frac{\vdash a \in \partial A \quad x{:}A \vdash d \in G \quad y{:}(A \to \{\}) \vdash e \in G}{\vdash \mathcal{D}_\partial(x,y)(a,d,e) \in G}$$

Evaluation

$$\frac{\vdash a \in A \quad x{:}A \vdash d \in C \quad y{:}(A \to \{\}) \vdash e \in C}{\vdash (\mathcal{D}_\partial(x,y)(\mathcal{L}_\partial(a),d,e)) = d[a/x] \in C}$$

$$\frac{\vdash a \in A \quad x{:}A \vdash d \in C \quad y{:}(A \to \{\}) \vdash e \in C}{\vdash (\mathcal{D}_\partial(x,y)(\mathcal{R}_\partial(a),d,e)) = e[a/y] \in C}$$

4.9 Identity Relation

Formation

$$\frac{\vdash A \text{ type} \quad \vdash a{:}A \quad \vdash b{:}A}{\vdash a \equiv_A b \text{ type}} \qquad \frac{\vdash A = C \quad \vdash a = c{:}A \quad \vdash b = d{:}A}{\vdash (a \equiv_A b) = (c \equiv_C d)}$$

Construction

$$\frac{\vdash a = b \in A}{\vdash (c \stackrel{A}{=} d) \in (a \equiv_A b)} \qquad \frac{\vdash a = b \in A}{\vdash (c \stackrel{A}{=} d) = (c \stackrel{A}{=} d) \in (a \equiv_A b)}$$

Selection

$$\frac{\vdash e \in G[b/a]}{c{:}(a \equiv_A b) \vdash \mathcal{I}(c,e) \in G} \qquad \frac{\vdash c \in (a \equiv_A b) \quad \vdash e \in G[a/b]}{\vdash \mathcal{I}(c,e) \in G}$$

Evaluation

$$\frac{\vdash a = b \in A \quad \vdash e \in C[a/b]}{\vdash \mathcal{J}(a \stackrel{A}{=} b, e) = e \in C}$$

4.10 Void Type

Formation

$$\vdash \{\} \text{ type}$$

$$\vdash \{\} = \{\}$$

Selection

$$a:\{\} \vdash \text{false} \in G$$

4.11 Parametric Lists

Formation

$$\frac{\vdash A \text{ type}}{\vdash A \text{ list type}}$$

$$\frac{\vdash A \text{ type}}{\vdash (A \text{ list}) = (A \text{ list})}$$

Construction

$$\frac{\vdash A \text{ type}}{\vdash []_A : A \text{ list}}$$

$$\frac{\vdash A \text{ type}}{\vdash []_A = []_A \in A \text{ list}}$$

$$\frac{\vdash a:A \quad \vdash b \in A \text{ list}}{\vdash [a\,|\,b]_A \in A \text{ list}}$$

$$\frac{\vdash a = c \in A \quad \vdash b = d \in A \text{ list}}{\vdash [a\,|\,b]_A = [c\,|\,d]_A \in A \text{ list}}$$

Selection

$$\frac{\vdash d:G[[]_A/x] \quad x_0:A, x_1:A \text{ list}, y:G[x_1/x] \vdash e \in G[[x_0\,|\,x_1])/x]}{x:A \text{ list} \vdash (p(x); p(x:A \text{ list}) = \text{list}\mathcal{C}(x, d, x_0, x_1, v, e)) \in G}$$

$$\frac{\vdash x \in A \text{ list} \quad \vdash d:G[[]_A/x] \quad x_0:A, x_1:A \text{ list}, y:G[x_1/x] \vdash e \in G[[x_0\,|\,x_1])/x]}{\vdash (p(x); p(x:A \text{ list}) = \text{list}\mathcal{C}(x, d, x_0, x_1, v, e)) \in G}$$

Evaluation

$$\frac{\vdash d \in G[[]_A/x] \quad x_0:A, x_1:A \text{ list}, y:G[x_1/x] \vdash e \in G[[x_0\,|\,x_1]/x]}{x:A \text{ list} \vdash (p(x); p(x:A \text{ list}) = \text{list}\mathcal{C}([]_A, d, x_0, x_1, v, e) = d) \in G}$$

$$\frac{\vdash d \in G[[]_A/x] \quad x_0:A, x_1:A \text{ list}, y:G[x_1/x] \vdash e \in G[[x_0\,|\,x_1]/x]}{x:A \text{ list} \vdash (p(x); p(x:A \text{ list}) = \text{list}\mathcal{C}([z_0\,|\,z_1], d, x_0, x_1, v, e)) = e[z_0, z_1, p(x_1)/x_0, x_1, v] \in G}$$

4.12 Universes

Formation

$$\frac{}{\vdash U_n \ type} \qquad\qquad \frac{}{\vdash U_n = U_n}$$

Construction (similarly for all types except U_m)

$$\frac{\vdash A:U_n \quad x:A \vdash B:U_n}{\vdash (\forall x:A.B):U_n} \qquad \frac{\vdash A = U_n \quad x:A \vdash B = D:U_n}{\vdash (\forall x:A.B) = (\forall x:C.D):U_n}$$

Selection

$$\frac{\vdash A \ type}{\vdash A:U_n} \qquad\qquad \frac{\vdash A = B}{\vdash A = B:U_n}$$

$$\frac{\vdash A:U_n}{\vdash A:U_{n+1}} \qquad\qquad \frac{\vdash A = B:U_n}{\vdash A = B:U_{n+1}}$$

5   Example: subset/2

For this example, I use the conjecture which specifies the subset/2 predicate using lists as a representation for sets – that is, the predicate which succeeds when all the members of the list given as its first argument are members of that given as its second. The specification is parameterised by base type (*i.e.*, by the type of the list elements), and, necessarily, by a decision procedure for equality for that type. In the event that such a decision procedure is not supplied, the proof process will not yield a logic program. I assume a background theory defining the $\in_\tau$ (typed member/2), as follows:

$$\exists \tau:U_0. \ \exists d:(\forall a:\tau.\forall b:\tau.\partial (a \equiv_\tau b)).$$
$$\exists x:\tau. \ \exists y:\tau \ list.(m(y);$$
$$m(y:\tau \ list) =$$
$$listC(y,$$
$$\mathcal{R}_\partial(v:\{\} \mapsto false),$$
$$y_0,y_1,v,$$
$$\mathcal{D}_\partial (p_0,p_1)(d(\tau)(x)(y_0), \mathcal{L}_\vee(\mathcal{L}_\partial(p_0)), \mathcal{R}_\vee(v))))$$
$$\in \forall \tau:U_0.\forall d:(\forall a:\tau.\forall b:\tau.\partial (a \equiv_\tau b)).$$
$$\forall x:\tau.\forall y:\tau \ list.\partial (x \in_\tau y) \qquad (1)$$

This definition is the statement that a program implementing member inhabits the appropriate type. It is worth mentioning at this stage that a user of WTT will not normally have to deal with such complex constructions. Instead, we also have two logical equivalences (*i.e.*, type equalities):

$$\forall \tau:U_0.\forall x:\tau.(x \in_\tau []_\tau) = \{\} \qquad (2)$$
$$\forall \tau:U_0.\forall x:\tau.\forall y_0:\tau.\forall y_0:\tau \ list.(x \in_\tau [y_0|y_1]) = (x \equiv_\tau y_0 \lor x \in_\tau y_1) \qquad (3)$$

It is, however, necessary to show that the type defined using $\in_\tau$ is inhabited – so we must construct the object; in fact, this turns out to be done *via* a simple proof.

Returning to the example conjecture, we start with

$$\vdash \forall \tau : \mathsf{U}_0 . \forall d : (\forall a : \tau . \forall b : \tau . \partial\, (a \equiv_\tau b)).$$
$$\forall x : \tau\ \text{list} . \forall y : \tau\ \text{list} . \partial\, (\forall z : \tau . z \in_\tau x \rightarrow z \in_\tau y) \tag{4}$$

The proof proceeds by primitive induction on lists, first on x and then on y. Note that the proof is presented in refinement style, with the rules applied "backwards", and that I have omitted unchanging hypotheses unless they are used in the current proof step. Further, since we begin with a completely uninstantiated proof object, I have omitted it and the $\in$ symbol from the sequents – otherwise the sequents would be completely illegible.

The first move is to introduce the universally quantified type parameters, and the quantifier of x in 4. This yields the conjecture

$$\tau : \mathsf{U}_0, d : (\forall a : \tau . \forall b : \tau . \partial\, (a \equiv_\tau b)),$$
$$x : \tau\ \text{list} \vdash \forall y : \tau\ \text{list} . \partial\, (\forall z : \tau . z \in_\tau x \rightarrow z \in_\tau y) \tag{5}$$

and the program

$$\sqcup \tau : \mathsf{U}_0 . \sqcup d : (\ \sqcup a : \tau . \ \sqcup b : \tau . \partial\, (a \equiv_\tau b)). \sqcup x : \tau\ \text{list} . \phi_5$$

where ϕ_5 is a meta-variable which will be instantiated during the rest of the proof. Note the difference from the original Whelk here: the $\sqcup$s were previously represented by a predicate definition head and arguments. Also, we have two (syntactically detectable) higher-order arguments, whose presence indicates that what we are synthesising here is not a logic program, but a logic-program-valued function - *i.e.*, a module. These, then, are the first two significant differences between this presentation of the example and that of [Wiggins 92b].

Next, we apply list selection to x. This gives us two subconjectures:

$$\vdash \forall y : \tau\ \text{list} . \partial\, (\forall z : \tau . z \in_\tau [\,]_\tau \rightarrow z \in_\tau y) \tag{6}$$

$$x_0 : \tau, x_1 : \tau\ \text{list},$$
$$v : (\forall y : \tau . z \in_\tau x_1 \rightarrow z \in_\tau y) \vdash \forall y : \tau\ \text{list} . \partial\, (\forall z : \tau . z \in_\tau [x_0 | x_1] \rightarrow z \in_\tau y) \tag{7}$$

and the following synthesised program, where ψ_6 and ψ_7 are the program constructions corresponding with (6) and (7), respectively:

$$\sqcup \tau : \mathsf{U}_0 . \sqcup d : (\ \sqcup a : \tau . \ \sqcup b : \tau . \partial\, (a \equiv_\tau b)).$$
$$\sqcup x : \tau\ \text{list} . (p(x); p(x : \tau\ \text{list}) = \ \text{list}\mathcal{C}(x, \psi_6, x_0, x_1, v, \psi_7))$$

Here is another difference between Whelk and WTT: the choice between [] and non-[] lists is built into the language as a type-selector, so no explicit disjunction appears here.

To prove the base case, (6), observe that the expression within the scope of ∂ is true, because the antecedent of the implication is always false. We can now use the new decision type rules to introduce the ∂ and then show that the resulting statment is indeed true. We introduce y and then use the first ∂ construction, to yield the following conjecture:

$$y:\tau \text{ list} \vdash \forall z:\tau.z \in_\tau []_\tau \to z \in_\tau y \tag{8}$$

and the program

$$\boxdot\tau:U_0.\ \boxdot d:(\boxdot a:\tau.\ \boxdot b:\tau.\partial\,(a \equiv_\tau b)).$$
$$\boxdot x:\tau \text{ list}.(p(x); p(x:\tau \text{ list}) = \text{list}\mathcal{C}(x, \mathcal{L}_\partial(\boxdot y:\tau \text{ list}.\psi_8), x_0, x_1, v, \psi_7))$$

where ψ_8 is the program corresponding with (8) as before. It is the $\mathcal{L}_\partial()$ operator, appearing in the $[]$ part of the list$\mathcal{C}$ selector which tells us that, when we reach this point in execution, we have succeeded; the argument of list$\mathcal{C}$ is just verification proof (maybe with some embedded identity relations, which can be extracted easily).

Proof of the remaining conjecture, (8), is trivial (from the definitions of $\in_\tau$ in 2) and is omitted here for lack of space. When it is finished, we have the following program.

$$\boxdot\tau:U_0.\ \boxdot d:(\boxdot a:\tau.\ \boxdot b:\tau.\partial\,(a \equiv_\tau b)).$$
$$\boxdot x:\tau \text{ list}.(p(x);$$
$$p(x:\tau \text{ list}) =$$
$$\text{list}\mathcal{C}(x,$$
$$\mathcal{L}_\partial(\boxdot y:\tau \text{ list}.\ \boxdot z:\tau \text{ list}.v:(z \in_\tau []\tau \mapsto \text{false})),$$
$$x_0, x_1, v, \psi_7))$$

The step case of the induction on x, (7) is harder. First, we use the definition (3) of $\in_\tau$ rule to unfold the leftmost occurrence in the conclusion. This gives

$$\vdash \forall y:\tau \text{ list}.\partial\,(\forall z:\tau.(z \equiv_\tau x_0 \lor z \in_\tau) \to z \in_\tau y) \tag{9}$$

Note here that I am using the type equality rules to perform and prove correct these rewrites, and not the evaluation rules. As such, the rewrites are effectively under equivalence, and the program does not change.

Now, we introduce the universal quantifier of y and rewrite under logical equivalence, again proving correctness by reference to type equality, to get:

$$y:\tau \text{ list} \vdash \partial\,(\forall z:\tau.(z \equiv_\tau x_0 \to z \in y) \land \forall z:\tau.(z \in x_1 \to z \in y))$$

As with Whelk, the rewriting can be performed automatically, via the *rippling* paradigm of [Bundy *et al* 90a, Bundy *et al* 90c]. Again, this step does not change the structure of the synthesised program.

One more logical rewrite gives us

$$\vdash \partial\,(\forall z:\tau.(z \equiv_\tau x_0 \to z \in y)) \land \partial\,(\forall z:\tau.(z \in x_1 \to z \in y))$$

again, not contributing to the program structure. The next step, however, $\wedge$ construction, does contribute. Its sub-sequents are:

$$\vdash \partial\,(\forall z{:}\tau.z \equiv_\tau x_0 \rightarrow z \in y) \tag{10}$$

$$\vdash \partial\,(\forall z{:}\tau.z \in_\tau x_1 \rightarrow z \in y) \tag{11}$$

and the program is:

$$\sqcup\tau{:}U_0.\,\sqcup d{:}(\,\sqcup a{:}\tau.\,\sqcup b{:}\tau.\partial\,(a \equiv_\tau b)).$$
$$\sqcup x{:}\tau\ \mathrm{list}.(p(x);$$
$$p(x{:}\tau\ \mathrm{list}) =$$
$$\mathrm{list}\mathcal{C}(x,$$
$$\mathcal{L}_\partial(\,\sqcup y{:}\tau\ \mathrm{list}.\,\sqcup z{:}\tau\ \mathrm{list}.v{:}(z \in_\tau [\,]\tau \mapsto \mathrm{false})),$$
$$x_0, x_1, v,$$
$$\psi_{10} \sqcap \psi_{11}))$$

We show (10) by first simplifying its conclusion:

$$\vdash \partial\,(x_0 \in_\tau y)$$

and then applying induction on y. This can now be demonstrated by appeal to the definition of $\in_\tau$, whose inhabitant proof object instantiates the program. The program now looks like this:

$$\sqcup\tau{:}U_0.\,\sqcup d{:}(\,\sqcup a{:}\tau.\,\sqcup b{:}\tau.\partial\,(a \equiv_\tau b)).$$
$$\sqcup x{:}\tau\ \mathrm{list}.(p(x);$$
$$p(x{:}\tau\ \mathrm{list}) =$$
$$\mathrm{list}\mathcal{C}(x,$$
$$\mathcal{L}_\partial(\,\sqcup y{:}\tau\ \mathrm{list}.\,\sqcup z{:}\tau\ \mathrm{list}.v{:}(z \in_\tau [\,]\tau \mapsto \mathrm{false})),$$
$$x_0, x_1, v,$$
$$\mathcal{M}(\tau)(x_0)(z) \sqcap \psi_{11}))$$

where $\mathcal{M}$ is the proof object of definition (1).

Finally, we appeal to the induction hypothesis of x (called v) to fill in the uninstantiated ϕ_{11} and complete the program.

$$\sqcup\tau{:}U_0.\,\sqcup d{:}(\,\sqcup a{:}\tau.\,\sqcup b{:}\tau.\partial\,(a \equiv_\tau b)).$$
$$\sqcup x{:}\tau\ \mathrm{list}.(p(x);$$
$$p(x{:}\tau\ \mathrm{list}) =$$
$$\mathrm{list}\mathcal{C}(x,$$
$$\mathcal{L}_\partial(\,\sqcup y{:}\tau\ \mathrm{list}.\,\sqcup z{:}\tau\ \mathrm{list}.v{:}(z \in_\tau [\,]\tau \mapsto \mathrm{false})),$$
$$x_0, x_1, v,$$
$$(M(x_0))(z) \sqcap v))$$

6 Normalization

We now have our WTT module for parameterised subset. However, (12) is clearly a higher-order structure, since it has two arguments which are types, τ and d. It also has embedded non-canonical forms, such as the application of $\mathcal{M}$ to its arguments. Before we can convert this program into one runnable directly as a "normal" logic program, we must evaluate and instantiate these structures, respectively. Evaluation of M, as far as the instantiation of its variables may be performed immediately, in the obvious way, as licensed by the $\forall$ type evaluation rules. To instantiate τ and d, we need another type. I will use $\mathbb{N}$ here, although I have omitted its proof rules for reasons of restricted space. All we need to know is that $\mathbb{N} \in U_0$, and that the type $\forall a \in \mathbb{N}.\forall b \in \mathbb{N}.\partial\,(a \equiv_\mathbb{N} b)$ is inhabited, thus:

$$\boxminus a \in \mathbb{N}.\,\boxminus b \in \mathbb{N}.e(a);$$
$$e(a:\mathbb{N}) = \mathbb{N}C(a,$$
$$f(b);$$
$$f(b:\mathbb{N}) = \mathbb{N}C(b,$$
$$\mathcal{L}_\partial(0 \stackrel{\mathbb{N}}{=} 0),$$
$$b_0, w,$$
$$\mathcal{R}_\partial(u:(0 \stackrel{\mathbb{N}}{=} s(0) \mapsto \text{false})),$$
$$x_0, v,$$
$$g(b);$$
$$g(b:\mathbb{N}) = \mathbb{N}C(b,$$
$$\mathcal{R}_\partial(u:(s(0) \stackrel{\mathbb{N}}{=} 0 \mapsto \text{false})),$$
$$b_0, w,$$
$$v(b_0))$$
$$\vdash \forall a:\mathbb{N}.\forall b:\mathbb{N}.\partial\,(a \equiv_\mathbb{N} b)$$

Again, it is worth emphasising that the production of such an algorithm is a straightforward, and in this case trivially automated, proof. This version of the equality predicate works by counting down the natural numbers; a better version would use a decision procedure built in as a basic type, which would yield a more efficient program.

Given this definition, the evaluation rules of the calculus may be used to reduce the original modular specification to the following:

$$\forall x:\mathbb{N} \text{ list.}\forall y:\mathbb{N} \text{ list.}\partial\,(\forall z:\mathbb{N}.z \in_\mathbb{N} \to z \in_\mathbb{N} y)$$

Such an evaluation results in parallel evaluation of the proof object, so the proof object we end up with is convertible into the subset relation for lists of naturals as required.

7 Conclusion

In this paper I have discussed how a higher-order extension of the Whelk system can help us synthesise modular programs. The example has shown that the system works for programs which are parameterised both in type and in sub-module. I have sketched the outline of an example proof, and shown that it is essentially the same as that for the same example in the Whelk system, though the modularity of WTT makes things slightly easier, in that it is not necessary to re-synthesised the member predicate – we can simply plug in an existing definition as a module.

The consequences of all this are significant. It has been shown elsewhere that logic program synthesis techniques work well for compiling naïve or non-executable specifications into comparatively efficient programs. The question has always been: "what about scaling up?". The modularity of the WTT system is one answer to this important question.

8 Acknowledgements

This work has been supported by ESPRIT basic research project #6810, "Compulog II". I am very grateful to my colleagues in the DREAM group at Edinburgh, in particular Alan Smaill, and in the Compulog project for their advice and support. I also thank Frank Pfenning, Andrei Voronkov and David Basin for their comments on the original version of Whelk.

References

[Bundy et al 90a] A. Bundy, A. Smaill, and J. Hesketh. Turning eureka steps into calculations in automatic program synthesis. In S.L.H. Clarke, editor, *Proceedings of UK IT 90*, pages 221–6, 1990. Also available from Edinburgh as DAI Research Paper 448.

[Bundy et al 90b] A. Bundy, A. Smaill, and G. A. Wiggins. The synthesis of logic programs from inductive proofs. In J. Lloyd, editor, *Computational Logic*, pages 135–149. Springer-Verlag, 1990. Esprit Basic Research Series. Also available from Edinburgh as DAI Research Paper 501.

[Bundy et al 90c] A. Bundy, F. van Harmelen, A. Smaill, and A. Ireland. Extensions to the rippling-out tactic for guiding inductive proofs. In M.E. Stickel, editor, *10th International Conference on Automated Deduction*, pages 132–146. Springer-Verlag, 1990. Lecture Notes in Artificial Intelligence No. 449. Also available from Edinburgh as DAI Research Paper 459.

[Constable 82] R.L. Constable. Programs as proofs. Technical Report TR 82-532, Dept. of Computer Science, Cornell University, November 1982.

[Martin-Löf 79] Per Martin-Löf. Constructive mathematics and computer programming. In *6th International Congress for Logic, Methodology and Philosophy of Science*, pages 153–175, Hanover, August 1979. Published by North Holland, Amsterdam. 1982.

[Wiggins 92a]   G. A. Wiggins.   Negation and control in automatically generated logic programs. In A. Pettorossi, editor, *Proceedings of META-92.* Springer Verlag, Heidelberg, 1992. LNCS Vol. 649.

[Wiggins 92b]   G. A. Wiggins. Synthesis and transformation of logic programs in the Whelk proof development system. In K. R. Apt, editor, *Proceedings of JICSLP-92*, pages 351–368. M.I.T. Press, Cambridge, MA, 1992.

[Wiggins *et al* 91] G. A. Wiggins, A. Bundy, I. Kraan, and J. Hesketh. Synthesis and transformation of logic programs through constructive, inductive proof. In K-K. Lau and T. Clement, editors, *Proceedings of LoPSTr-91*, pages 27–45. Springer Verlag, 1991. Workshops in Computing Series.

A model of costs and benefits of meta-level computation

Frank van Harmelen

SWI
University of Amsterdam
frankh@swi.psy.uva.nl

Abstract. It is well known that meta-computation can be used to guide other computations (at the object-level), and thereby reduce the costs of these computations. However, the question arises to what extent the cost of meta-computation offsetts the gains made by object-level savings. In this paper we discuss a set of equations that model this trade-off between object-savings and meta-costs. The model shows that there are a number of important limitations on the usefulness of meta-computation, and we investigate the parameters that determine these limitations.

1 Introduction

One of the most often stated aims of meta-programming is *search-control*: a meta-program is used to guide the computation at the object-level. Often, this takes the form of a meta-program choosing among multiple applicable object-level computations. A large body of literature exists on this type of meta-programs, in areas like knowledge-representation, (logic-)programming and theorem proving.

Although many other types of meta-programs are both possible and useful, this paper will only consider meta-programs that are used to guide the computation at the object-level. This type of meta-program gives rise to a trade-off situation, in which costs should be compared with benefits. The benefit of meta-computation is that it leads to a better choice among object-computations, and therefore to savings at the object-level, since useless or expensive object-computations can be avoided (see e.g. [1] for results in the area of theorem proving). On the other hand, meta-computations themselves often have a considerable cost, and this cost might offset any savings that are obtained by that very same computation.

This trade-off (between savings made by meta-level choices and the costs of having to make these choices) has been recognised in the literature: [5], [2] and [4, chapter 7] report on *experimental results* on measuring the size of the meta-level overhead, and the large literature on partial evaluation tries to reduce the size of this overhead.

The goal of this paper is to investigate a *theoretical model* of the costs and benefits of meta-computation. After setting out the formal assumptions that underlie this work (Sect. 2), we present in Sect. 3 a quantitative model developed by [6]. In the context of this model, we postulate some reasonable properties of meta-computations (Sect. 4), and illustrate the model with some examples (Sect. 5). In Sect. 6 we extend and generalise the basic model from Sect. 3.

2 Assumptions

We will assume that there are two independent methods for solving a particular object-level problem, and we will call these methods x and y[1]. We also assume that each of these methods has a certain expected cost, which we will denote by c_x and c_y. Furthermore, we assume that x and y are heuristic methods, i.e. they are not guaranteed to solve the object-problem. Instead, we will assume that each method has a specific chance of solving the given object-problem, which we will write as p_x and p_y.

The goal of the meta-computation is to choose among the two object-methods x and y in order to solve a given problem in such a way that the overall cost of the object-computation is minimised. Because in general x and y are not guaranteed to solve the problem (p_x and p_y might be smaller than 1), the meta-computation must not choose between either x or y, but it must choose the ideal ordering of x and y: first try x, and if it fails, then try y, or vice versa. We will write $c_{x;y}$ to denote the expected cost of first executing x followed by y if x fails, and similarly for $c_{y;x}$.

The meta-computation that determines this choice will again have a certain cost, which we will write as c_m. Again, we will assume that this meta-computation is heuristic, i.e. it will make the correct choice of how to order x and y only with a certain chance, which we will write as p_m.

We assume that without meta-level computation, the system will try to use the two methods in a random order to solve the object-problem. The goal of the model will be to compute the savings (or losses, i.e. negative savings) that are obtained by using meta-computation to choose the ordering of x and y instead of making a random choice[2]. These expected savings will be denoted by s.

All of this notation (plus some additional notation used in later sections) is summarised in Table 1.

3 The Savings Function

In this section, we will derive the expression for the expected savings obtained by making a correct meta-level choice concerning the optimal order in which to execute two object-level methods.

Given the assumptions about x, y, c_x, c_y, p_x and p_y, the expected cost of executing x before y, $c_{x;y}$ is:

$$c_{x;y} = c_x + (1 - p_x)c_y \tag{1}$$

---

[1] In Sect. 6, we will show how the model can be extended to deal with an arbitrary number of methods.

[2] In Sect. 6 we will show how the model can be adjusted to accommodate the more realistic assumption that the system will execute the methods in some fixed order if no meta-computation is done.

Table 1. summary of notation

Notation	meaning
x,y	object-methods
c_x, c_y	expected cost of object-methods
$c_{x;y}$	expected cost of first executing x then y
p_x, p_y	chance that object-method will solve problem
c_m	cost of meta-computation
p_m	chance that meta-computation makes the correct choice
s	expected savings made by meta-computation
Δ	difference in expected costs between the two object scenarios
$\phi(x)$	utility of method x

namely the expected cost of executing x plus the expected cost of executing y, but reduced by the chance that y is not executing because x has succeeded in solving the problem. An analogous expression holds for $c_{y;x}$. Notice that (1) reduces to simply the expected cost of c_x when x is a complete method (i.e when $p_x = 1$).

The decision to try x before y should be made when

$$c_{x;y} < c_{y;x} \tag{2}$$

or equivalently, using (1):

$$p_y/c_y < p_x/c_x. \tag{3}$$

The quantity $\phi_x = p_x/c_x$ can be seen as a measure of the *utility* of method x. The utility of a method x increases with its success rate p_x and decreases with its expected costs c_x. The above inequality (3) says that the method with the highest utility should be tried first.

However, the values for success rates and expected costs of x and y (and therefore the values of $\phi(x)$ and $\phi(y)$) will not in general be available to the system, and will have to be computed at the cost of some meta-level effort, c_m. Once the meta-level has estimated $\phi(x)$ and $\phi(y)$, the optimal ordering of the two methods can be determined. We can now derive the expteced savings s made by executing the methods in this optimal order as follows: we assume that without any meta-level effort, the system chooses a random ordering of x and y. The expected savings are then the cost of executing a randomly chosen ordering minus the cost of executing the methods in the optimal ordering, increased with the cost of finding the optimal ordering:

$$\text{savings} = \text{cost-of-random-ordering} -$$
$$(\text{cost-of-chosen-ordering} + \text{meta-level-cost})$$

and the expected cost of executing the methods in a random order is:

$$\frac{c_{x;y} + c_{y;x}}{2}. \tag{4}$$

If the system spends c_m on meta-level effort and then chooses x before y as the optimal ordering, the expected execution cost would be:

$$c_{x;y} + c_m. \tag{5}$$

The expected savings would then be the difference between these two formulae:

$$\frac{c_{y;x} - c_{x;y}}{2} - c_m. \tag{6}$$

This would be the expected savings if the system preferred x over y on the basis of its estimates of $\phi(x)$ and $\phi(y)$. In general, we cannot expect that the meta-level will always succeed in computing the true values of $\phi(x)$ and $\phi(y)$. We can adjust our model to the assumption that the meta-level prefers x over y (i.e. it claims $\phi(x) > \phi(y)$), but that this decision is only correct with a probability p_m. In this case, the expected execution costs would be

$$p_m c_{x;y} + (1 - p_m) c_{y;x} + c_m. \tag{7}$$

If we subtract this from the costs of executing a random ordering, and simplify the result, we get:

$$\frac{(2p_m - 1)(c_{y;x} - c_{x;y})}{2} - c_m \tag{8}$$

These are the expected savings when the meta-level prefers x over y. An analogous expression holds for the reverse case. We can combine these two expressions to obtain the following expression for the expected savings s made by a meta-computation:

$$s = \frac{(2p_m - 1)\Delta}{2} - c_m = (p_m - \tfrac{1}{2})\Delta - c_m \tag{9}$$

where Δ is notation for $|c_{x;y} - c_{y;x}| = |p_y c_x - p_x c_y|$. The intuitive interpretation of Δ is that it represents the difference in costs between the optimal and the non-optimal object-scenario's (executing first x and then, if necessary, y, or vice versa). This concludes the derivation of the savings function (which closely followed [6]).

An alternative derivation for s is as follows: the expected gains made by making the correct choice between $x; y$ or $y; x$ are $|c_{x;y} - c_{y;x}| = \Delta$. If the chance of a correct meta-decision is only p_m, the maximal expected gains are reduced to $p_m \Delta$. The savings made by a random choice would already have been $\frac{1}{2}\Delta$, reducing the expected savings made through meta-computation to $p_m \Delta - \frac{1}{2}\Delta$. The costs of the meta-computation must also be subtracted from this, yielding a total of:

$$s = p_m \Delta - \tfrac{1}{2}\Delta - c_m = (p_m - \tfrac{1}{2})\Delta - c_m \tag{10}$$

From this value for the savings function s we can already draw some conclusions. The form $s = (p_m - \tfrac{1}{2})\Delta - c_m$ shows that even an ideal meta-level computation (with perfect results, $p_m = 1$ and no costs, $c_m = 0$), can at best only save half the difference in expected object-costs ($s = \tfrac{1}{2}\Delta$). This is already a severe limit on what meta-level computations can gain.

This maximal savings of $\frac{1}{2}\Delta$ also puts an upper limit on the maximum amount of meta-effort that we should invest: because the expected savings can never amount to more than $\frac{1}{2}\Delta$, it will certainly never be useful to make c_m larger than this same amount. Thus, any potential losses by spurious meta-computation could be limited given an estimate of Δ.

Furthermore, we see that s is monotonically increasing with Δ, and this makes sense: the larger the difference in expected costs between the different object-computations, the larger the expected savings to be made by meta-computation. In the boundary case, when $\Delta = 0$ and the ordering of object-computations is irrelevant, the expteced savings will be negative (since in that case, $s = -c_m$), and meta-computations will lead to a loss in overall efficiency.

Finally, we see that when $p_m = \frac{1}{2}$, i.e. when the meta-level decision does not improve over a random choice, meta-computation will again only lead to a loss in overall efficiency ($s = -c_m$), as expected.

4 Properties of Meta-computations

In realistic situations, the value of p_m, the probability of making a correct choice between methods, will be dependent on the amount of meta-level effort spent, i.e. $p_m = f(c_m)$. Placing this in (9) above, we get:

$$s(c_m) = (f(c_m) - \tfrac{1}{2})\Delta - c_m \tag{11}$$

Obviously, we want to maximise s as a function of c_m. Exactly what shape $s(c_m)$ will have will depend on how the accuracy of the meta-computation (i.e. p_m) depends on the meta-level effort spent by the system, i.e. the shape of $f(c_m)$. The following are reasonable assumptions to make about $f(c_m)$:

Monotonicity: We expect of a meta-level that the quality of its results does not decrease with increased effort, making $f(c_m)$ non-decreasing:

$$\frac{df}{dc_m}(c_m) \geq 0. \tag{12}$$

Lowerbound: With no meta-level effort, i.e. $c_m = 0$, we assume $p_m = \frac{1}{2}$ (because of the random ordering of x and y):

$$f(0) = \tfrac{1}{2}. \tag{13}$$

Upperbound: Since $f(c_m)$ is a probability, we certainly expect $0 \leq f(c_m) \leq 1$. Together with (12) and (13) this gives:

$$\tfrac{1}{2} \leq f(c_m) \leq 1. \tag{14}$$

This upperbound on $f(c_m)$ immediately makes clear that any meta-level computation is eventually doomed to lead to losses:

$$\lim_{c_m \to \infty} s(c_m) = -\infty \tag{15}$$

since the positive term in (9) is limited to $\frac{1}{2}\Delta$, the negative term $-c_m$ will eventually dominate.

Diminishing returns: Finally, although this is not strictly necessary, we can expect some effect of diminishing returns, giving a smaller increase in p_m for every further increase in c_m:

$$\frac{d^2 f}{d(c_m)^2}(c_m) \leq 0. \tag{16}$$

5 Examples

For illustrational purposes, it is interesting to look at a number of example functions for $f(c_m)$ which have the above properties, and to see what the actual shape of the savings curve $s(c_m)$ would be for these functions. The first two examples are taken from [6].

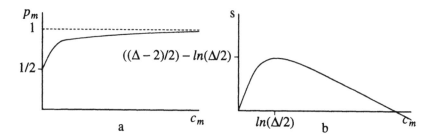

Fig. 1. example of diminishing returns in p_m

Example 1: Figure 1a shows the case for

$$p_m = f(c_m) = 1 - \tfrac{1}{2}e^{-c_m}. \tag{17}$$

The only significance of this function is that it obeys conditions (12)–(16), but otherwise it is arbitrary. It serves to illustrate the case where a relatively small amount of meta-level effort results in substantially improved performance, while there are diminishing returns for subsequent effort. Its convergence to $p_m = 1$ for $c_m \to \infty$ is arbitrary, and does not influence the qualitative shape of the savings curve for $s(c_m)$ shown in Fig. 1b. In this figure, we can see that at a certain point, the expected savings achieved by meta-computation reach a maximum, and any further meta-level effort will only reduce the overall savings. With even more effort spent on meta-computation, the system will eventually behave worse than without any meta-level effort at all.

It is instructive to notice that at the point of maximal expected savings, the value of p_m is $1 - \frac{1}{\Delta}$ (simply substitute $ln(\Delta/2)$ in (17)). This shows that for large Δ, it pays to spend effort trying to get p_m close to 1, whereas for smaller Δ, a small value of p_m is all that we can afford to compute. The costs of computing any better approximation of p_m will then not be paid back by the corresponding savings in object-computations.

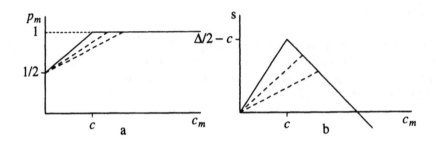

Fig. 2. example of maximised p_m

Example 2: Similarly, Fig. 2a shows an example where increased meta-level effort initially improves the reliability of the choice among object-methods, but after some point ($c_m = c$) the decision is made with maximum reliability (in this case perfect reliability, $p_m = 1$), and any further meta-level effort does not contribute to a more effective control decision. The function used in this figure (shown as a solid line) is:

$$p_m = f(c_m) = \begin{cases} \frac{1}{2}(c_m/c) + \frac{1}{2} & \text{if } c_m \leq c \\ 1 & \text{if } c_m > c \end{cases} \tag{18}$$

Again, the only crucial properties of this example for $f(c_m)$ are conditions (12)–(16),. Other properties, such as the slope of $f(c_m)$ on $[0, c]$, or the fact that $f(c_m) = 1$ on $[c, \infty)$ are irrelevant to the qualitative shape of the saving curve shown as the solid line in Fig. 2b, where again the expected savings reach a maximum at some point, beyond which further meta-level effort will only degrade the performance of the system.

Example 3: Whereas the previous examples showed meta-computations which improved the behaviour of the control decision gradually (in accordance with (16)), a final example shows what happens if meta-level reliability improves suddenly after some initial effort.

Figure 3a shows the case of a meta-level which needs some initial effort to compute a good choice between object-computations, but whose quality does not improve after having made its decision:

$$p_m = f(c_m) = \begin{cases} \frac{1}{2} & \text{if } c_m \leq c \\ 1 & \text{if } c_m > c \end{cases} \tag{19}$$

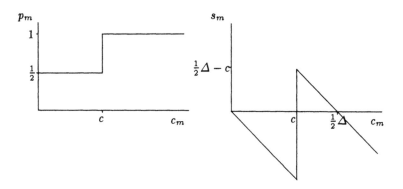

Fig. 3. example of suddenly increasing p_m

The corresponding savings curve is shown in Fig. 3b. The interesting property of this curve is that the expected savings are initially negative ($0 \leq c_m < c$), and only become positive after a required minimum amount of effort. This is an important difference with the previous two examples: there, insufficient meta-effort might lead to non-maximal savings, but would never lead to actual losses, whereas the meta-computation of this third example does indeed yield losses at insufficient meta-effort. This third example shares with the previous two examples the property that too much meta-effort leads to reduced gains, and eventually to losses over no meta-computation at all.

6 Extensions

The simple model above can extended in a number of ways.

6.1 Complete Methods

The first extension is not really an extension, but rather a special case: our current model deals with heuristic methods which would have to be tried in sequence and we therefore have to take into account the expected costs of executing the second method in those cases where the first method failed. We might also consider choosing between complete object-methods, which always succeed. The choice is then not between executing first x and then y or vice versa, but between executing either x or y. The model deals with this situation as a simple special case. We simply take $p_x = p_y = 1$. Then Δ reduces to $|c_x - c_y|$, i.e. the difference between the expected costs of executing *either* x or y as expected .

6.2 Harder Meta-level Problems

We call one meta-level problem *harder* than another if for the same amount of meta-level effort c_m, the system achieves a lower value of p_m (i.e. the choice between the applicable methods is made less reliably).

In the case of the definition for $f(c_m)$ used in example 2, harder meta-level problems are represented by an increasing value of c, and are indicated by the family of dashed lines in Fig. 2a. The corresponding behaviour of the saving function $s(c_m)$ is shown by the dashed lines in Fig. 2b. We see that when the meta-level problem gets harder, the optimum meta-level effort is found for a larger value of c_m, and the corresponding savings are reduced.

This behaviour illustrates a phenomenon often observed in developing meta-level systems, namely that the usefulness of meta-level inference cannot be illustrated adequately on very simple toy problems. For those problems, c will be very small, and any significant amount of meta-level effort is likely to be larger than c, and will thus overshoot the point of maximum expected savings.

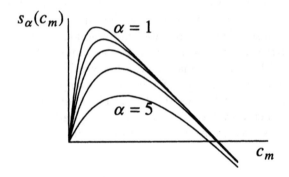

Fig. 4. savings for harder meta-problems

A similar set of curves can be drawn for the example used in Fig. 1. If we take

$$p_m = f(c_m) = 1 - \tfrac{1}{2}e^{-c_m/\alpha} \tag{20}$$

(instead of $p_m = f(c_m) = 1 - \tfrac{1}{2}e^{-c_m}$), then increasing values of $\alpha \geq 1$ will represent harder meta-level problems. Again, the maxima for $s(c_m)$ for different values of α lie at increasing values of c_m, and the corresponding savings are reduced. An example of this behaviour is shown in Fig. 4 which displays different curves $s(c_m)$ for values $\alpha = 1, 1\tfrac{1}{2}, 2, 3, 5$.

The same phenomenon occurs in example 3, where again, increasing values of c represent harder meta-level problems, which again lead to lower maximum expected savings, obtained at higher values of c_m. An interesting additional effect is shown in this example: harder meta-level problems not only lead to reduced maximum expected savings, but the savings function is also positive on a smaller interval, making the balance for meta-computation even more delicate.

6.3 Number of Methods

Rather than modelling just two methods x and y, we can adjust the model to choose between n methods (a suggestion also made by [6]). Expression (1) can be generalised from 2 to n methods, so that the expected cost of executing methods $x_1, \ldots, x_n$ in that order is:

$$c_{x_1; \ldots; x_n} = c_{x_1} + \sum_{i=2}^{n} (\prod_{j=1}^{i-1} (1 - p_{x_j})) c_{x_i} \tag{21}$$

(i.e. the total cost is the sum of the cost of each method multiplied with the chance that all earlier methods failed). The optimal order for executing the methods would of course be some order $x_{i_1}; \ldots; x_{i_n}$ such that for all j, $1 \leq j \leq n - 1 : \phi(x_{i_j}) \geq \phi(x_{i_{j+1}})$

Before we can derive the savings function for this generalised case, we must introduce some notation. The methods $x_1, \ldots, x_n$ can be executed in $n!$ different orders, and we shall denote each such sequence by some σ_j, $1 \leq j \leq n!$. Then c_{σ_j} is the expected cost associated with executing such a sequence (written as $c_{x_k; \ldots; x_m}$ in (21)). Furthermore, let us assume that σ_1 is the optimal order (i.e. it satisfies $\phi(x_k) \geq \phi(x_l)$ whenever x_k occurs earlier in σ_1 than x_l). With this notation, we can derive the savings function analogously to the case $n = 2$ in Sect. 3. The expected cost of executing the methods $x_1, \ldots, x_n$ in a random order (i.e. the generalised version of (4)) is:

$$\frac{\sum_{j=1}^{n!} c_{\sigma_j}}{n!} \tag{22}$$

Equation (7), the expected cost of executing the optimal method (σ_1), chosen by the meta-level with probability p_m, now becomes:

$$p_m c_{\sigma_1} + (1 - p_m) \frac{\sum_{j=2}^{n!} c_{\sigma_j}}{n! - 1} + c_m \tag{23}$$

namely: the cost of σ_1 times the probability that σ_1 was chosen, plus the chance that any of the other methods was chosen times the average cost of one of the other methods, plus the cost of making the meta-level choice.

As before, the expected savings of a meta-computation is the difference between these two formula. Subtracting (23) from (22) gives, after simplification:

$$s = (p_m - \frac{1}{n!})\Delta - c_m \tag{24}$$

where

$$\Delta = \frac{\sum_{j=2}^{n!} c_{\sigma_j}}{n! - 1} - c_{\sigma_1}. \tag{25}$$

Equation (25) shows us that the proper interpretation of Δ is the difference between the cost of the correct choice and the average cost of an incorrect choice.

In the case $n = 2$, this reduces to the difference in expected costs between the two possible sequences, $c_{\sigma_2} - c_{\sigma_1}$, written as $c_{y;x} - c_{y;x}$ in Sect. 3.

From the generalised savings function shown in (24), we see that the maximal expected savings are $(1 - \frac{1}{n!})\Delta$. If we assume that Δ stays more or less constant with increasing n (since there is no reason why the average cost of a set of methods must increase or decrease for larger sets), we can conclude that for increasing n, the maximal expected savings will increase and will rapidly approach Δ (within 1% for $n = 5$). This gives a slightly rosier perspective on the maximal expected savings to be made by meta-computation than originally derived in Sect. 3, where the expected savings were limited to $\frac{1}{2}\Delta$, namely $(1 - \frac{1}{n!})\Delta$ for $n = 2$.

6.4 Initial Ordering of Methods

The model assumed that with no meta-level effort, the system would make an arbitrary choice for the order in which it executed its object-methods x and y. A more realistic assumption would be that the system would apply x and y in some fixed order, say first x and then y, on the basis of some a priori knowledge that the system's designer has about $\phi(x)$ and $\phi(y)$. Suppose that with no meta-effort, the system chooses x before y, and that this choice is indeed the right one with a chance p_0. In other words, the value $f(0)$, the quality of the meta-level decision at no meta-effort, is no longer $\frac{1}{2}$, as specified in (13), but is now p_0. Presumably, $\frac{1}{2} < p_0 \leq 1$, since the fixed ordering programmed into the system will be better than a random ordering. Because (13) has changed to

$$f(0) = p_0 \tag{26}$$

Formula (14) must also change, into

$$p_0 \leq f(c_m) \leq 1. \tag{27}$$

To derive the new value of the savings function, we follow the derivation of (10) above. In this derivation, we subtracted the savings made by a random choice $(\frac{1}{2}\Delta)$. This term must now be replaced by the expected savings made by an informed initial choice, which are $p_0\Delta$. This leads to the following new savings function:

$$s(c_m) = f(c_m)\Delta - p_0\Delta - c_m = (f(c_m) - p_0)\Delta - c_m \tag{28}$$

Notice that this reduces to (9) for the case $p_0 = \frac{1}{2}$ (the random choice case).

Because of the new boundary condition (26), this again implies $s(0) = 0$, as before. Furthermore, if $p_0 = 1$ is (and the initial ordering is already perfect), then (27) implies $f(c_m) = 1$, and the savings function reduces to $s(c_m) = -c_m$, and any meta-computation only leads to losses, as expected.

Finally, the value of (28) is always smaller than the value of (9), because $p_0 > \frac{1}{2}$. For the same reason, the maximum expected savings that can be obtained by a meta-computation are reduced from $\frac{1}{2}\Delta$ to $(1 - p_0)\Delta$. This is just what one would expect, since without meta-effort, the system performs better than it did before, and as a result, the potential savings that can be made by the meta-computation are smaller.

6.5 Cost Dependent on Success or Failure

The assumption above was that a method, say x, had some associated expected cost c_x, which was independent of whether the method succeeded or failed. This assumption can be lifted by introducing two costs, namely c_x^s, the expected cost of x if it succeeds, and c_x^f, the expected cost of x if it fails. The expected cost of executing method x is then:

$$c_x = p_x c_x^s + (1 - p_x)c_x^f \tag{29}$$

namely the cost of x succeeding times the probability that it will succeed plus the cost of x failing, times the probability that it will fail. We can then uniformly substitute this new expression for any c_x in the above, and similarly for y. If we do this in (1), representing the expected cost of executing first method x and then method y, the resulting expression can be rewritten to the following:

$$c_{x;y} = p_x c_x^s + (1 - p_x)(c_x^f + p_y c_y^s + (1 - p_y)c_y^f) \tag{30}$$

which is exactly what we expect, namely the cost of x succeeding times the probability that x succeeds plus the sum of x failing and executing y times the probability that x fails. Notice that this equation reduces to (1) when the distinction between success and failure costs is irrelevant (i.e. when $c_x^s = c_x^f$).

If we substitute (29) into the definition of the utility $\phi(x)$, we obtain:

$$\phi(x) = \frac{p_x}{p_x c_x^s + (1 - p_x)c_x^f} = \frac{1}{c_x^s + \frac{(1-p_x)}{p_x}c_x^f} \tag{31}$$

From this we can see that the utility of x decreases if either failure costs or success costs increase, or when the success chance decreases. This is again as expected, and similar to the old value of $\phi(x)$.

Furthermore, we see that when $p_x > \frac{1}{2}$, then an increase or decrease in c_x^s has a larger effect on $\phi(x)$ than an equal increase or decrease in c_x^f. Thus, for a method which often succeeds, $\phi(x)$ is dominated by the success costs, and similarly the utility for mostly failing methods is dominated by the failure costs.

More surprisingly, (31) also shows that for methods with very low failure costs c_x^f, the utility $\phi(x)$ becomes simply inversely proportional to the success costs, and independent of the success rate. This is surprising because in the case of two methods with zero failure cost (thus: $c_x^f = c_y^f = 0$), but different success rates (say $p_x > p_y$) we would expect x to be preferred over y, but (31) fails to capture this.

A final, and perhaps somewhat counterintuitive implication from (31) would be the following. Suppose that method x fails by running into an infinite loop (e.g. because of a loop in the search space). If we model this by $c_x^f = \infty$, this immediately means $\phi(x) = 0$, implying that x is always to be selected as the last method. The counterintuitive aspect of this is that this conclusion is again independent of the success chance.

7 Discussion

The problem of choosing the correct amount of meta-effort in order to optimise the total savings is very similar to the problems tackled in classical decision theory (e.g. [7]). A number of the effects that we have observed in the models above are also well known in decision theory. In particular, the fact that the maximal savings are limited to $\frac{1}{2}\Delta$ (in the case $n = 2$) is known as the *base-rate effect*: the gains of an informed decision are limited if a random choice already performs well (when the "base-rate" is already high), as in the case of $n = 2$. The reduction of the base-rate and the corresponding increase in maximal savings for increasing n (to $(1 - \frac{1}{n!})\Delta$) is also a known effect.

The *law of diminishing returns* (in our case modelled by assumption (16)) is also familiar from decision theory. Decision theory also gives us an explanation of the counterintuitive results at the end of the previous section for the case of zero or infinite failure costs. In decision theory this is called the *boundary effect*: if one of the dimensions in a multi-dimensional decision problem becomes 0 or ∞, then the model is no longer valid because this single dimension will entirely dominate all the other dimensions. In such cases, we should move to a new model, rather than simply substitute 0 or ∞ in the old model.

An interesting application of our theoretical model can possibly be found in the area of explanation-based learning. A well-known problem in that area is the so-called *utility problem* [3]: learning more control knowledge does not always lead to more efficient problem-solving behaviour, and learning too much may actually adversely affect efficiency. At first sight, it would seem that this effect could be explained by a model similar to ours. Further research is required to shed light on this issue.

Finally, a remark is in order on the usefulness of our model. One might object that the above results are not very useful since in practice it will be very hard, if not impossible, to obtain the actual values for the costs and probabilities involved in a particular system. The main value of our model is therefore a different one: rather than using it to compute numerical values in a concrete system, it teaches us general lessons about the qualitative behaviour of a whole class of meta-systems in general, lessons which are sometimes surprising and perhaps not obvious at first sight.

8 Conclusions

We can summarise the main conclusions from this model as follows:

- The model shows that a *trade-off* does indeed exist between the costs of meta-computation and the savings obtained by it. This results in a situation where savings are maximised at a certain amount of meta-effort. Both more and less meta-effort will lead to non-maximal savings.
- Meta-computation may also lead to a *loss* in efficiency (when compared to doing only object-computation). This will always happen if the meta-effort

becomes greater than a certain maximum (too much meta-effort), but might also happen in certain cases if meta-effort is less than a certain minimum (not enough meta-effort). The loss of overall efficiency at high amounts of meta-effort explains the phenomenon that the usefulness of meta-computation is often hard to illustrate on the basis of small examples.

- An important quantity in the analysis of meta-computation is the difference in the costs of alternative object-computations (Δ in the above). The savings to be made by a meta-computation are proportional to Δ, and the maximum savings (and consequently an upperbound for the amount of meta-effort) are Δ, and even less than that in the case of an informed initial ordering.

The overall conclusion from all of this is that meta-computations are not an unqualified blessing, and that the savings to be made from such computations are often less than expected, and only occur within a narrowly defined interval of meta-effort. Before one decides to use meta-computation as a means to control object-computation, a careful analysis of crucial quantities like Δ, ϕ and $f(c_m)$ is required.

Acknowledgement: I am grateful to Annette ten Teije for her detailed comments and suggestions. She also provided the derivation for the case $n > 2$.

References

1. J. Hesketh A. Bundy, F. van Harmelen and A. Smaill. Experiments with proof plans for induction. *Journal of Automated Reasoning*, 7:303–324, 1991. Earlier version available from Edinburgh as DAI Research Paper No 413.
2. H. Lowe. Empirical evaluation of meta-level interpreters. Master's thesis, Department of Artificial Intelligence, University of Edinburgh, 1988. (M.Sc. Thesis).
3. S. Minton. Quantitative results concerning the utility of explanation-based learning. *Artificial Intelligence*, 42:363–392, 1990.
4. R. A. OKeefe. *The Craft of Prolog.* MIT Press, Massachussetts, 1990.
5. S. Owen. The development of explicit interpreters and transformers to control reasoning about protein topology. Technical Memo HPL-ISC-TM-88-015, Hewlett-Packard Laboratories Bristol Research Centre, Bristol, U.K., 1988.
6. J.S. Rosenschein and V. Singh. The utility of meta-level effort. Report HPP-83-20, Stanford Heuristic Programming Project, March 1983.
7. D. von Winterfeldt and Ward Edwards. *Decision Analysis and Behavioral Research.* Cambridge University Press, 1986.

A Basis for a Multilevel Metalogic Programming Language

Jonas Barklund, Katrin Boberg and Pierangelo Dell'Acqua*

Uppsala University
Computing Science Dept., Box 311, S-751 05 Uppsala, Sweden
E-mail: jonas@csd.uu.se, katrin@csd.uu.se or pier@csd.uu.se

Abstract. We are developing a multilevel metalogic programming language that we call *Alloy*. It is based on first-order predicate calculus extended with metalogical constructs. An Alloy program consists of a collection of theories, all in the same language, and a representation relation over these theories. The whole language is self-representable, including names for expressions with variables. A significant difference, as compared with many previous approaches, is that an arbitrary number of metalevels can be employed and that the object-meta relationship between theories need not be circular.

The language is primarily intended for representation of knowledge and metaknowledge and is currently being used in research on hierarchical representation of legal knowledge. We believe that the language allows sophisticated expression and efficient automatic deduction of interesting sets of beliefs of agents.

This paper aims to give a preliminary and largely informal definition of the core of the language, a simple but incomplete and inefficient proof system for the language and a sketch of an alternative, more efficient, proof system. The latter is intended to be used as a procedural semantics for the language. It is being implemented by an extension of an abstract machine for Prolog.

1 Introduction

We are interested in the representation of complex formalized knowledge, e.g.,

- knowledge containing metaknowledge, e.g., knowledge about how some part of the knowledge should be interpreted or applied;
- knowledge that is too voluminous to organize as a single, one-level collection; or
- knowledge that involves beliefs about other knowledge bases or agents.

Our hypothesis is that such knowledge can favourably be structured in a collection of theories, where some theories contain theorems about some of the other theories. This hypothesis is supported by previous research conducted by, e.g.,

---

* P. Dell'Acqua has been financially supported by both Uppsala Univ. and Univ. degli Studi di Milano.

Bowen [6], Eshghi [19], Aiello, Nardi & Schaerf [1], Costantini & Lanzarone [16], Kim & Kowalski [21], and Brogi & Turini [11].

In order to test our hypothesis we are developing a metalogic programming language, *Alloy*, suited for expressing knowledge of the kind just described. A brief description of the language was recently given by Barklund & Hamfelt [5]. The purpose of this paper is to summarize the language and its properties in further detail.

The language can be characterized as a Horn clause language with resolution [25] that is extended with a representation of the linguistic elements and of its demonstrability. The novel ideas, as we see it, are the following.

- The language has multiple theories; inferences of demonstrability for a certain theory may be made in another theory that represents it.
- The proof systems allows deductions at different metalevels to occur and interact in an intuitively appealing way.
- One proof system allows a correct and efficient interleaving of computations at different metalevels, also in the presence of names of expressions with variables.

The language has been influenced by Prolog [15], Reflective Prolog [18] and unnamed languages proposed by Kowalski [22, 23]. The intended procedural semantics is an extension of that proposed by Barklund, Costantini, Dell'Acqua and Lanzarone [4]. The procedural semantics will be realized through translation to sequences of instructions for an abstract machine, currently under development.

As stated above, we primarily intend this language to be used in applications involving metareasoning and representation of metaknowledge, and secondarily in tools for manipulating programs (such as compilers and program transformers). This intention has influenced the development of the language by making us focus on representation of demonstrability. So far the language has been used for representing legal knowledge [5] and for a solution (not yet published) to the 'Three Wise Men' puzzle, a well-known benchmark problem for metareasoning formalisms [1, 21].

Even though we intend Alloy to become a language with a well-defined concrete syntax we prefer to use an abstract syntax in this report. This is partly because we find it too early to decide on a definitive syntax and partly because the requirements on the concrete syntax of a programming language make it less convenient for a concise presentation of the language and its properties.

Lack of space prevents us from presenting a background of metaprogramming in logic. We therefore trust the reader to have some prior knowledge of metaprogramming and of computational logic.

2 Language Elements

An Alloy program constitutes a definition of a collection of theories, which are here sets of program clauses, closed under inference rules that will be presented in Sect. 5. Such a definition contains theory-tagged program clauses and representation statements, the latter identifying the metatheories for every theory. The language is common for all theories and we describe it in this section.

We let *terms*, *definite program clauses* and *definite goal clauses* be as usual [24] (but in what follows we will simply write "program clause" or "goal clause" for a definite clause).[2] We will refer to expressions composed from atomic sentences (or "atoms") by conjunction and implication operators as *sentences*.

We will employ the standard convention that every free variable in a program clause is implicitly universally quantified. (We make this note here because below there will be nested encodings of tagged program clauses; this rule disambiguates such clauses.)

A *theory term* is simply a term but one that is assumed to denote a theory. We introduce this name only to make the presentation of the syntax and the inference systems easier to follow, because theory terms will formally be treated just as the other terms.

A *tagged atomic sentence* (or "tagged atom") is an expression $\tau \vdash \alpha$, where τ is a theory term and α is an atomic sentence.

A *tagged program clause* is an expression $\tau \vdash \kappa$, where τ is a theory term and κ is a program clause. It states that κ is a theorem of the theory denoted by τ (from here on we will simply write "the theory τ"). Throughout this paper, the symbol '$\vdash$' denotes provability using the inference rules of Sect. 5.

A *tagged goal clause* is like a goal clause except that it contains tagged atoms instead of ordinary atoms. The tagged goal clauses includes the empty tagged goal clause with no atoms and is true in every interpretation.

A *representation statement* is an expression $\tau_M \triangleright \tau_O$. It states that the theory τ_M contains a representation of the theory τ_O. We also say that the theory τ_M is a *metatheory* for τ_O. In order to explain what this means, we first have to explain representation.

For every expression ϕ of the language, there is an encoding of the expression denoted by a ground expression $\ulcorner \phi \urcorner$ that is also in the language. The encodings of terms and clauses are denoted by terms. The encodings of tagged clauses and representation statements are denoted by sentences.[3] We will not say here exactly what the representations are; we can view them here as abstract datatypes. However, we expect that application of substitutions and the unification operation is extended to handle encodings. Moreover, in this presentation we will also assume that the sentences representing tagged clauses and representation statements are atomic, so expressions $\ulcorner \tau_O \vdash \kappa \urcorner$ and $\ulcorner \tau_M \triangleright \tau_O \urcorner$ can appear in any place where an atomic sentence can appear. It might help the reader, when comparing our work with that of others, to assume that $\ulcorner \tau_O \vdash \kappa \urcorner$ is an atomic sentence $Demo(\ulcorner \tau_O \urcorner, \ulcorner \kappa \urcorner)$. We will discuss these issues in more detail in connection with the proof systems.

We can now explain the meaning of a representation statement $\tau_M \triangleright \tau_O$. It states that $\tau_O \vdash \kappa$ is true if and only if $\tau_M \vdash \ulcorner \tau_O \vdash \kappa \urcorner$ is true. Therefore, if we

---

[2] We will use an alphabet where function and predicate symbols begin with uppercase letters while variable symbols begin with lowercase letters.

[3] Employing sentences as encodings is clearly a nonstandard device, but note that they are used only to represent the metalogical part of the language.

have proved $\tau_O \vdash \kappa$ and $\tau_M \rhd \tau_O$ then we may infer $\tau_M \vdash \ulcorner \tau_O \vdash \kappa \urcorner$. Similarly, from $\tau_M \vdash \ulcorner \tau_O \vdash \kappa \urcorner$ and $\tau_M \rhd \tau_O$ we are entitled to deduce $\tau_O \vdash \kappa$.

Representation statements define a representation relation between theories. We may want to restrict this relation in order to provide more efficient proof procedures. Some possible restrictions are:

- Existence of a (irreflexive) partial order on theories (i.e., no cycles).
- A finite number of theories.
- That no theory have more than one metatheory.
- No infinite chains $\cdots \tau_3 \rhd \tau_2 \rhd \tau_1$.

Some of these restrictions will prevent a theory from being its own metatheory. Such a restriction seems to simplify implementation but yields a weaker language without direct self reference. In many situations this is highly desirable, and certain reasonable forms of self-reference can instead be programmed at the metalevel. However, we have not yet investigated the full consequences of the various restrictions.

3 Encodings, Substitutions, Equality and Unification

As stated above, $\ulcorner \phi \urcorner$ is some expression that denotes an encoding of ϕ. We allow the writing of "parameterized" encodings, containing metavariables where there would otherwise have been terms, function symbols or predicate symbols. By "metavariables" we mean variables having a scope which is the program clause in which the current encoding appears as a term; the dot signifies a "hole" in the encoding, where such a variable can appear. The metavariables are thus not themselves part of the encoded expression, but encodings of expressions of the appropriate kind may be substituted for them, yielding proper encodings. (We refrain here from discussing what results from substituting other expressions for the metavariables.)

For example, $\ulcorner F(A, v) \urcorner$ is the encoding of the (nonground) term $F(A, v)$, while $\ulcorner F(A, \dot{z}) \urcorner$, $\ulcorner \dot{x}(A, v) \urcorner$, $\ulcorner \dot{x}(\dot{y}, \dot{z}) \urcorner$ and $\ulcorner \dot{w} \urcorner$ are parameterized encodings. Substituting an encoding of a function symbol for x and encodings of terms for y, z and w yields proper encodings. (The last expression, $\ulcorner \dot{w} \urcorner$, is equivalent to w.)

Furthermore, $\ulcorner T(A) \vdash P(F(v), B) \urcorner$ properly encodes the tagged program clause $T(A) \vdash P(F(v), B)$ while $\ulcorner T(\dot{x}) \vdash \dot{y}(F(v), \dot{z}) \urcorner$ is parameterized with the metavariables x, y and z. Of these, encodings of terms must eventually be substituted for x and z, while an encoding of a predicate symbol must be substituted for y. (In both examples, v is the only variable known to be universally quantified over the program clause. However, in the second example, instantiating z to an encoding of a nonground term might reveal further variables with the scope of the program clause.[4])

---

[4] If this appears to give ambiguous scoping, let us point out that this is what happens when encodings of program clauses are developed through unification. When all "dotted variables" have been instantiated to proper encodings, then the encoding of the program clause is complete and its local variables can be unambiguously identified.

The equality theory of our language is "Herbrand equality", as defined by Clark [14], extended with three axioms for encodings:

$$\ulcorner t_0(t_1,\ldots,t_k)\urcorner = \ulcorner u_0(u_1,\ldots,u_l)\urcorner \rightarrow$$
$$k = l \wedge$$
$$\ulcorner t_0 \urcorner = \ulcorner u_0 \urcorner \wedge \ulcorner t_1 \urcorner = \ulcorner u_1 \urcorner \wedge \cdots \wedge \ulcorner t_k \urcorner = \ulcorner u_l \urcorner$$
$$\ulcorner \tau_1 \vdash \kappa_1 \urcorner = \ulcorner \tau_2 \vdash \kappa_2 \urcorner \rightarrow$$
$$\ulcorner \tau_1 \urcorner = \ulcorner \tau_2 \urcorner \wedge$$
$$\ulcorner \kappa_1 \urcorner = \ulcorner \kappa_2 \urcorner$$
$$\ulcorner \tau_{1,1} \triangleright \tau_{1,2} \urcorner = \ulcorner \tau_{2,1} \triangleright \tau_{2,2} \urcorner \rightarrow$$
$$\ulcorner \tau_{1,1} \urcorner = \ulcorner \tau_{2,1} \urcorner \wedge$$
$$\ulcorner \tau_{1,2} \urcorner = \ulcorner \tau_{2,2} \urcorner$$
$$\ulcorner \dot{t} \urcorner = t.$$

Together these statements ensure compositionality of encodings. For example, we have that $x = \ulcorner t \urcorner \leftrightarrow \ulcorner F(\dot{x}) \urcorner = \ulcorner F(t) \urcorner$.

Moreover, for any pair of distinct symbols σ_1 and σ_2, there is an axiom $\ulcorner \sigma_1 \urcorner \neq \ulcorner \sigma_2 \urcorner$, stating uniqueness of encodings.

The unification algorithm must be extended accordingly. We only give two examples here of what it should produce when given (partially instantiated) encodings.

- Unification of the names $\ulcorner T(\dot{x}) \vdash \dot{y}(F(A),v) \urcorner$ and $\ulcorner T(u) \vdash Q(\dot{z},\dot{w}) \urcorner$ succeeds with the substitution that replaces x with $\ulcorner u \urcorner$, y with $\ulcorner Q \urcorner$, z with $\ulcorner F(A) \urcorner$, and w with $\ulcorner v \urcorner$ as most general unifier.
- The names $\ulcorner F(A,u) \urcorner$ and $\ulcorner F(A,v) \urcorner$ do not unify because that would imply $\ulcorner u \urcorner = \ulcorner v \urcorner$, which is false.

4   Examples

Before explaining the proof system of the language we shall look at two examples that show some of its abilities.

The following program (from a similar Reflective Prolog program [18]) contains three theories denoted by the symbols M, O_1 and O_2, where M is intended to be a metatheory of O_1 and O_2.

$$M \vdash \ulcorner O_2 \vdash \dot{x}(\dot{a},\dot{b}) \urcorner \leftarrow \ulcorner O_1 \vdash \dot{x}(\dot{a},\dot{b}) \urcorner \tag{1}$$
$$M \vdash \ulcorner O_2 \vdash \dot{x}(\dot{b},\dot{a}) \urcorner \leftarrow Symmetric(x) \wedge \ulcorner O_1 \vdash \dot{x}(\dot{a},\dot{b}) \urcorner \tag{2}$$
$$M \vdash Symmetric(\ulcorner As\_good\_as \urcorner) \tag{3}$$
$$O_1 \vdash As\_good\_as(p, Gold) \leftarrow Honest(p) \tag{4}$$
$$O_1 \vdash Honest(Mary) \tag{5}$$
$$M \triangleright O_1 \tag{6}$$
$$M \triangleright O_2 \tag{7}$$

The statement $As\_good\_as(Mary, Gold)$ is a theorem of both O_1 and O_2 but in addition also $As\_good\_as(Gold, Mary)$ is a theorem of O_2. It could be derived as follows. Resolving clause 4 with unit clause 5 yields

$$As\_good\_as(Mary, Gold)$$

in O_1. As M is a metatheory of O_1 (according to 6) we obtain

$$\ulcorner O_1 \vdash As\_good\_as(Mary, Gold)\urcorner \tag{8}$$

as a theorem of M. The sentences $Symmetric(x)$ and $\ulcorner O_1 \vdash \dot{x}(\dot{a}, \dot{b})\urcorner$ of clause 2 can be resolved with unit clauses 3 and 8, respectively, yielding

$$\ulcorner O_2 \vdash As\_good\_as(Gold, Mary)\urcorner$$

as a consequence. Because M is a metatheory also of O_2 (7) we obtain

$$As\_good\_as(Gold, Mary)$$

as a theorem of O_2.

Without clause 6 we would not have had M as a metatheory of O_1. Then there would have been no correspondence between having a theorem ϕ in O_1 and a theorem $\ulcorner O_1 \vdash \phi\urcorner$ in M, and vice versa (similarly for clause 7 and the relationship between M and O_2).

It should be obvious that in the theory O_2 the predicate symbol As_good_as denotes a relation that is the symmetric closure of the relation denoted by As_good_as in the theory O_1. (The same would be true also of any other predicate symbol π for which $Symmetric(\ulcorner\pi\urcorner)$ is a theorem in M.)

Alloy is expressive enough, however, that much more sophisticated relationships between theories can be expressed. For example, the following program makes theory O_2 'inherit' (not necessarily unit) clauses from O_1 and obtains a symmetric relation that depends also on information in theory O_2.

$$M \vdash \ulcorner O_2 \vdash \dot{x}(\dot{a}, \dot{b}) \leftarrow \dot{z}\urcorner \leftarrow \ulcorner O_1 \vdash \dot{x}(\dot{a}, \dot{b}) \leftarrow \dot{z}\urcorner \tag{9}$$

$$M \vdash \ulcorner O_2 \vdash \dot{x}(\dot{b}, \dot{a}) \leftarrow \dot{z}\urcorner \leftarrow Symmetric(x) \wedge \ulcorner O_1 \vdash \dot{x}(\dot{a}, \dot{b}) \leftarrow \dot{z}\urcorner \tag{10}$$

$$M \vdash Symmetric(\ulcorner As\_good\_as\urcorner) \tag{11}$$

$$O_1 \vdash As\_good\_as(p, Gold) \leftarrow Honest(p) \tag{12}$$

$$O_2 \vdash Honest(Mary) \tag{13}$$

$$M \triangleright O_1 \tag{14}$$

$$M \triangleright O_2 \tag{15}$$

The statements $As\_good\_as(Mary, Gold)$ and $As\_good\_as(Gold, Mary)$ are theorems of O_2 also for this program, even though none of them is a theorem of O_1. The second of these could now be derived as follows. The clause

$$As\_good\_as(p, Gold) \leftarrow Honest(p),$$

where p is universally quantified, is apparently a theorem of O_1; M is a metatheory of O_1 (14) so

$$\ulcorner O_1 \vdash As\_good\_as(p, Gold) \leftarrow Honest(p)\urcorner \tag{16}$$

is a theorem of M. The sentences $Symmetric(x)$ and $\ulcorner O_1 \vdash \dot{x}(\dot{a}, \dot{b}) \leftarrow \dot{z} \urcorner$ of clause 10 can be resolved with unit clauses 11 and 16, respectively, yielding

$$\ulcorner O_2 \vdash As\_good\_as(Gold, p) \leftarrow Honest(p) \urcorner$$

as a theorem of M. As M is a metatheory also of O_2 (15) we obtain

$$As\_good\_as(Gold, p) \leftarrow Honest(p) \qquad (17)$$

as a theorem of O_2. Resolving clauses 17 and 13 finally yields the theorem $As\_good\_as(Gold, Mary)$ in O_2.

This expressivity is similar to that obtained by Brogi, Mancarella, Pedreschi and Turini [9] when applying their work on composition operators for theories [8] to modules. In their work, a module can be constructed extensionally or intensionally, corresponding to the subgoals $\ulcorner O_1 \vdash \dot{x}(\dot{a}, \dot{b}) \urcorner$ and $\ulcorner O_1 \vdash \dot{x}(\dot{a}, \dot{b}) \leftarrow \dot{z} \urcorner$ of clauses 2 and 10, respectively, that inspect either the extension or the intension of the theory O_1. In other aspects their work is quite different from ours, e.g., in that they construct an algebra for theories, where a theory, in practice, can be treated much as a powerful kind of data structure. We, on the other hand, write executable *specifications* for theories. The approaches are thus complementary.

As a third example, showing some use of encoding nonground expressions, consider the following program. (The last clause contains the encoding of a nonground program clauses $Irresistible(f) \leftarrow Female(f)$ and $Likes(y, z) \leftarrow Irresistible(z)$.)

$$Beliefs(Mary) \vdash$$
$$Charming(x) \leftarrow \ulcorner Beliefs(\dot{n}) \vdash Likes(\dot{n}, Mary) \urcorner \wedge Name(x, n) \qquad (18)$$
$$Beliefs(Mary) \vdash$$
$$\ulcorner Beliefs(\dot{n}) \vdash Irresistible(f) \leftarrow Female(f) \urcorner$$
$$\leftarrow Male(m) \wedge Name(m, n) \qquad (19)$$
$$Beliefs(Mary) \vdash \ulcorner Beliefs(\dot{x}) \vdash Female(Mary) \urcorner \qquad (20)$$
$$Beliefs(Mary) \vdash Male(John) \qquad (21)$$
$$Beliefs(Mary) \vdash \ulcorner Beliefs(\dot{x}) \vdash Likes(y, z) \leftarrow Irresistible(z) \urcorner \qquad (22)$$

The clauses can be seen as stating that Mary has the following five basic beliefs: anyone who she believes is thinking they like her, she finds charming (clause 18); all men think all women are irresistible (clause 19); everyone think that Mary is a woman (clause 20); John is a man (clause 21); everyone think that if they are irresistible then everyone likes them (clause 22). We have assumed the existence of a predicate $Name$ above, intending that an atomic sentence $Name(t, \ulcorner t \urcorner)$, can be proved for any ground term t, in any theory.

From this program we can prove that Mary finds John charming (actually that she finds anyone charming who she believes is a man); the query can be expressed as a tagged goal clause

$$\leftarrow Beliefs(Mary) \vdash Charming(John). \qquad (23)$$

From clauses 19 and 21 (when solving the *Name* atom) we obtain

$$Beliefs(Mary) \vdash \ulcorner Beliefs(John) \vdash Irresistible(f) \leftarrow Female(f) \urcorner,$$

i.e., that Mary believes that John thinks all women are irresistible. By resolving this clause with clause 20 we obtain

$$Beliefs(Mary) \vdash \ulcorner Beliefs(John) \vdash Irresistible(Mary) \urcorner,$$

i.e., that Mary believes that John thinks she is irresistible. By resolving this clause with clause 22 we obtain

$$Beliefs(Mary) \vdash \ulcorner Beliefs(John) \vdash Likes(y, Mary) \urcorner,$$

i.e., that Mary believes that John thinks that everyone likes her (the variable y in the unit clause $Likes(y, Mary)$ is universally quantified in that unit clause). Note that this resolution step involved substitution of variables at two levels: $\ulcorner John \urcorner$ for x and *Mary* for z. Resolving the clause so obtained with clause 18 and solving the *Name* atom produces a tagged program clause

$$Beliefs(Mary) \vdash Charming(John)$$

that can be resolved with the tagged goal clause 23 yielding an empty tagged goal clause, so the original goal clause is entailed by the program.

5 A Proof System

This proof system is intended to give the reader an understanding of what can be proved from an Alloy program; it is not intended to be efficient. (See the next section for a short description of a more complicated but efficient proof system.)

The following inference rule is a generalized Horn clause resolution rule that is capable of "indirect" reasoning. A theory τ_1 contains a theorem that a theory τ_2 contains a theorem that a theory τ_3 contains a theorem, and so on. Finally, there is a theorem that a theory τ_k contains two procedure clauses $A \leftarrow C$ and $H \leftarrow B$, where H can be unified with a selected atom $\uparrow C$ (we designate the rest of the conjunction by $\downarrow C$). Using the rule we may infer that τ_1 contains a theorem, etc., that τ_k contains a theorem that is the resolvent of $A \leftarrow C$ and $H \leftarrow B$ on $\uparrow C$.

$$\frac{\tau_1 \vdash \ulcorner \tau_2 \vdash \cdots \ulcorner \tau_k \vdash A \leftarrow C \urcorner \cdots \urcorner \quad \tau_1 \vdash \ulcorner \tau_2 \vdash \cdots \ulcorner \tau_k \vdash H \leftarrow B \urcorner \cdots \urcorner}{\tau_1 \vdash \ulcorner \tau_2 \vdash \cdots \ulcorner \tau_k \vdash (A \leftarrow B \wedge \downarrow C)\theta \urcorner \cdots \urcorner} \quad \theta = mgu(H, \uparrow C)$$

As special cases of this rule we obtain a single-level resolution rule

$$\frac{\tau \vdash A \leftarrow C \quad \tau \vdash H \leftarrow B}{\tau \vdash (A \leftarrow B \wedge \downarrow C)\theta} \quad \theta = mgu(H, \uparrow C)$$

and a two-level resolution rule

$$\frac{\tau_1 \vdash \ulcorner \tau_2 \vdash A \leftarrow C \urcorner \quad \tau_1 \vdash \ulcorner \tau_2 \vdash H \leftarrow B \urcorner}{\tau_1 \vdash \ulcorner \tau_2 \vdash (A \leftarrow B \wedge \downarrow C)\theta \urcorner} \quad \theta = mgu(H, \uparrow C)$$

For example, when the theories represent the beliefs of agents, this rule allows an agent to reason that he believes another agent believes in some statement P, because he believes that the second agent believes in the statements Q and $P \leftarrow Q$.

The following rule is a generalized meta-to-object linking rule [7, 28].

$$\frac{\tau_1 \vdash \ulcorner \tau_2 \vdash \cdots \ulcorner \tau_k \vdash \ulcorner \tau_M \triangleright \tau_O \urcorner \urcorner \cdots \urcorner \qquad \tau_1 \vdash \ulcorner \tau_2 \vdash \cdots \ulcorner \tau_k \vdash \ulcorner \tau_M \vdash \ulcorner \tau_O \vdash \kappa \urcorner \urcorner \cdots \urcorner}{\tau_1 \vdash \ulcorner \tau_2 \vdash \cdots \ulcorner \tau_k \vdash \ulcorner \tau_O \vdash \kappa \urcorner \urcorner \cdots}$$

To understand the rule, it may help to first look at some of its special cases. The first is a direct linking rule

$$\frac{\tau_M \triangleright \tau_O \quad \tau_M \vdash \ulcorner \tau_O \vdash \kappa \urcorner}{\tau_O \vdash \kappa}$$

stating: if τ_M is a metatheory for τ_O and contains as a theorem that κ is a theorem of τ_O, then we may infer that κ is indeed a theorem of τ_O.

Another special case is an indirect linking rule

$$\frac{\tau_I \vdash \ulcorner \tau_M \triangleright \tau_O \urcorner \quad \tau_I \vdash \ulcorner \tau_M \vdash \ulcorner \tau_O \vdash \kappa \urcorner \urcorner}{\tau_I \vdash \ulcorner \tau_O \vdash \kappa \urcorner}.$$

If a theory I has a theorem that theory M is a metatheory for theory O and another theorem that M has a theorem that theory O contains κ, then theory I infers that theory O contains κ.

The generalized rule not only allows transfer of information from a metatheory an object theory, but also permits reasoning that is conditional to a hypothesis that such a relationship exists.

There is also a corresponding generalized object-to-meta linking rule

$$\frac{\tau_1 \vdash \ulcorner \tau_2 \vdash \cdots \ulcorner \tau_k \vdash \ulcorner \tau_M \triangleright \tau_O \urcorner \urcorner \cdots \urcorner \qquad \tau_1 \vdash \ulcorner \tau_2 \vdash \cdots \ulcorner \tau_k \vdash \ulcorner \tau_O \vdash \kappa \urcorner \urcorner \cdots \urcorner}{\tau_1 \vdash \ulcorner \tau_2 \vdash \cdots \ulcorner \tau_k \vdash \ulcorner \tau_M \vdash \ulcorner \tau_O \vdash \kappa \urcorner \urcorner \cdots}$$

with corresponding special cases. One might get the impression that the two linking rules together with a single-level resolution rule could replace the generalized resolution rule. For example, to get

$$\tau_1 \vdash \ulcorner \tau_2 \vdash (A \leftarrow B \wedge \downarrow C)\theta \urcorner$$

from

$$\tau_1 \vdash \ulcorner \tau_2 \vdash A \leftarrow C \urcorner \quad \text{and} \quad \tau_1 \vdash \ulcorner \tau_2 \vdash H \leftarrow B \urcorner$$

(where θ is a mgu of H and $\uparrow C$) one would infer $\tau_2 \vdash A \leftarrow C$ and $\tau_2 \vdash H \leftarrow B$ using the meta-to-object linking rule, resolve these clauses (obtaining $\tau_2 \vdash (A \leftarrow B \wedge \downarrow C)\theta$) and finally apply the object-to-meta linking rule. This is correct only if $\tau_1 \triangleright \tau_2$, while the generalized inference rule can always be applied. Expressed in the domain of agents: One may reason about how other agents would reason, without actually asking them to hypothesize the premisses and tell us their conclusions. Expressed from an implementation point of view: We can simulate a computational device instead of actually invoking it.

We should mention that an implementation of this proof system is likely to be incomplete (with respect to a so far unstated, intended semantics) because it may encounter proofs that require applications of the meta-to-object linking rule but where the encoding contains uninstantiated metavariables. The inference rule is then not applicable, unless an arbitrary instantiation is made. For completeness an interpreter would have to try all possible instantiations, a hopeless task. The proof system outlined in the next session is not only more efficient but its linking rules employ constraints on names, rather than mere translation, allowing them to be applied to partially instantiated encodings.

6 Outline of a More Efficient Proof System

As a full description would become too long for this paper, we only outline here the efficient proof system for Alloy on which an implementation is being based. First we should analyse what 'efficient' should mean in this context.

SLD-resolution is goal-directed in the following two ways:

1. Every SLD-resolution step in a successful resolution proof is necessary in the sense that it resolves away some atom of the goal.
2. Unification will immediately stop the exploration of a resolution series, if the selected atom cannot unify with the chosen clause.

Of course, a nondeterministic choice of clauses may lead to unsuccessfully terminated resolution series before a proof is found, or even to nontermination, but that is a separate problem. Moreover:

3. In an SLD-resolution series, the resolvents are always goal clauses. Therefore the set of program clauses (i.e., the left side of clauses sequents) remains constant while a proof is being sought.

This property is exploited in the efficient implementation of Prolog-like languages, as it allows compilation of the program clauses to efficient machine code. Note that the inference rules in the proof system of the previous section created a new program clause. Operationally that would correspond to an addition to the program, preventing or complicating compilation of program clauses. (Even if we were to replace the generalized resolution rule with a generalized SLD-resolution rule, there are no obvious direct replacements of the linking rules.)

Just like RSLD-resolution [18], our proof procedure combines reflection inferences with SLD-resolution steps in order to retain the three properties above. This ensures that if an object level deduction invokes a metalevel deduction, or vice versa, then it is because the result is needed. Moreover, unification (which is extended to a form of constraint satisfaction) will terminate unsuccessful multilevel proofs as soon as possible. Finally, the program remains constant during execution.

Extending unification to constraint satisfaction over the domain of names is an elegant and implementation-wise straightforward way to get correct linking between levels. We employ the scheme that was proposed recently by Barklund, Costantini, Dell'Acqua and Lanzarone for a somewhat more restricted language [3, 4] (see also Christiansen [13]).

First, unification and SLD-resolution is re-expressed in terms of equations instead of substitutions. Each resolution step sets up an equation system that either can be transformed to a solved form, corresponding to a most general unifier, or can be shown to have no solution.

Next, the language of the equations to be solved by unification is extended with the operators 'up' and 'down' (for computing the name of an expression and computing what an expression names, respectively). It is no longer certain that an equation system can be transformed to a solved form or failed, because an 'up' or 'down' operator may be applied to an insufficiently instantiated expression. Therefore, it is necessary to retain some unsolved equations between resolution steps. We can think of these equations as unsolved constraints.

The scheme above only computes names of ground expressions. A further complication when implementing the present language is that the 'up' and 'down' operators must have an additional parameter indicating a scope, so encodings of variable symbols occurring in the encoding of some program clause can be associated consistently with variables. The added complication is the price to be paid for this richer language. We should add that these "scopes" seem to be quite efficiently implementable (cf. the dictionaries of Barklund [2] and the substitutions of Sato [26]).

7 Related Work

In constructing this language we have tried to remain closer to work on reflection and encodings in mathematical logic [27] than most other approaches.

The language was obviously inspired by Reflective Prolog [18], differing from it in mainly three aspects:

- Alloy has names for variables, so nonground expressions can be represented directly. This should make the language suitable for writing program manipulating tools, such as partial evaluators, at least for the language itself. Moreover, knowledge that is naturally represented by quantified formulas can be more directly represented. We have not yet begun to exploit this capability.
- The multiple theories allows a more elaborate structure of metaknowledge to be expressed directly (in Reflective Prolog there is a single amalgamated theory) and it seems possible to trade some expressivity for ease of implementation by preventing theories from being their own metatheories. On the other hand, in principle it should not be too difficult to write in Reflective Prolog an interpreter for a language with this kind of theory structure. However, incorporation of this fundamental construct explicitly in the language seems justified on grounds of efficiency and convenience. (The interesting multiple theory extension proposed for Reflective Prolog [17] was intended

for modelling communicating agents, by describing how theories exchange information, and is therefore quite different from the extension described here.)

- The view of what is actually reflected between metalevels is conceptually quite different in the two languages, although the views coincide when only ground expressions are transferred.

There are also other formalisms where a comparison should be useful:

- Our language is also related to languages proposed by Kowalski [22, 23] in the use of multiple theories. Our language is indeed an attempt to concretize these ideas.
- Brogi, Mancarella, Pedreschi & Turini [8, 9, 10] have investigated a collection of operators for composing theories. The "meta-interpretive" definitions of the operators can be expressed directly in Alloy. Although theories in Alloy are not to be thought of as data structures, the language allows theories to be denoted by arbitrary ground expressions. Alloy could accomodate a programming style where many theories (denoted by compound expressions) are defined, although extensional representations of the theories need not be computed. The operators by Brogi *et al.* could serve as a useful basis for defining more specific operators.
- The encodings in this language are more abstract than the structural names of Barklund [2], Costantini and Lanzarone [18] and of Cervesato & Rossi [12], but less abstract than those of Hill & Lloyd [20]: We do not commit to a particular ground representation but there is a syntax for the representations.
- The truth predicate proposed by Sato [26] is related to Alloy in that it too allows encodings of nonground expressions and reflection. However, there are also important differences, the most important being that the theory structure of Alloy can be used to prevent a theory from introspecting its own theorems. Sato shows how paradoxes can easily be expressed using his definition of the truth predicate; the programs can still be given semantics in three-valued logic. Alloy, as presented here, is a Horn clause language. It might be possible to incorporate some of Sato's ideas into Alloy, in order to extend the language to general clauses.
- Alloy has interesting similarities with provability logics and we are presently investigating what results from this area are applicable, and if this language and its inference systems bring any novel ideas to that area.

Acknowledgments

We are indebted to Stefania Costantini, Gaetano Lanzarone, Andreas Hamfelt and Robert A. Kowalski for valuable discussions. Their ideas have significantly influenced this work. The referees provided many useful comments and suggestions that have improved the paper.

The research reported herein was supported financially by the Swedish National Board for Technical and Industrial Development (NUTEK) under contract No. 92–10452 (ESPRIT BRP 6810: *Computational Logic 2*).

References

1. Aiello, L. C., Nardi, D. and Schaerf, M., Reasoning about Knowledge and Ignorance, in: H. Tanaka and K. Furukawa (eds.), *Proc. Intl. Conf. on Fifth Generation Comp. Sys. 1988*, Ohmsha, Tokyo, 1988.

2. Barklund, J., What is a Meta-Variable in Prolog?, in: H. Abramson and M. H. Rogers (eds.), *Meta-Programming in Logic Programming*, MIT Press, Cambridge, Mass., 1989.

3. Barklund, J., Costantini, S., Dell'Acqua, P. and Lanzarone, G. A., Reflection through Constraint Satisfaction, in: P. Van Hentenryck (ed.), *Logic Programming: Proc. 11th Intl. Conf.*, MIT Press, Cambridge, Mass., 1994.

4. Barklund, J., Costantini, S., Dell'Acqua, P. and Lanzarone, G. A., SLD-Resolution with Reflection, to appear in Proc. ILPS'94, 1994.

5. Barklund, J. and Hamfelt, A., Hierarchical Representation of Legal Knowledge with Metaprogramming in Logic, *J. Logic Programming*, 18:55–80 (1994).

6. Bowen, K. A., Meta-Level Programming and Knowledge Representation, *New Generation Computing*, 3:359–383 (1985).

7. Bowen, K. A. and Kowalski, R. A., Amalgamating Language and Metalanguage in Logic Programming, in: K. L. Clark and S.-Å. Tärnlund (eds.), *Logic Programming*, Academic Press, London, 1982.

8. Brogi, A., Mancarella, P., Pedreschi, D. and Turini, F., Composition Operators for Logic Theories, in: J. W. Lloyd (ed.), *Computational Logic*, Springer-Verlag, Berlin, 1990.

9. Brogi, A., Mancarella, P., Pedreschi, D. and Turini, F., Meta for Modularising Logic Programming, in: A. Pettorossi (ed.), *Meta-Programming in Logic*, LNCS 649, Springer-Verlag, Berlin, 1992.

10. Brogi, A., Program Construction in Computational Logic, Ph.D. Thesis, Dipartimento di Informatica, Università di Pisa, 1993.

11. Brogi, A. and Turini, F., Metalogic for Knowledge Representation, in: J. A. Allen, R. Fikes and E. Sandewall (eds.), *Principles of Knowledge Representation and Reasoning: Proc. 2nd Intl. Conf.*, Morgan Kaufmann, Los Altos, Calif., 1991.

12. Cervesato, I. and Rossi, G. F., Logic Meta-Programming Facilities in 'LOG, in: A. Pettorossi (ed.), *Meta-Programming in Logic*, LNCS 649, Springer-Verlag, Berlin, 1992.

13. Christiansen, H., Efficient and Complete Demo Predicates for Definite Clause Languages, in: P. Van Hentenryck (ed.), *Logic Programming: Proc. 11th Intl. Conf.*, MIT Press, Cambridge, Mass., 1994.

14. Clark, K. L., Negation as Failure, in: H. Gallaire and J. Minker (eds.), *Logic and Data Bases*, Plenum Press, New York, 1978.

15. Colmerauer, A., Kanoui, H., Pasero, R. and Roussel, P., Un Système de Communication Homme-Machine en Français, Technical report, Groupe de Recherche en Intelligence Artificielle, Univ. d'Aix-Marseille, Luminy, 1972.

16. Costantini, S. and Lanzarone, G. A., Analogical Reasoning in Reflective Prolog, in: A. A. Martino (ed.), *Pre-Proc. 3rd Intl. Conf. on Logica Informatica Diritto*, Istituto per la documentazione giuridica, Florence, 1989.

17. Costantini, S., Dell'Acqua, P. and Lanzarone, G. A., Reflective Agents in Metalogic Programming, in: A. Pettorossi (ed.), *Meta-Programming in Logic*, LNCS 649, Springer-Verlag, Berlin, 1992.

18. Costantini, S. and Lanzarone, G. A., A Metalogic Programming Language, in: G. Levi and M. Martelli (eds.), *Proc. 6th Intl. Conf. on Logic Programming*, MIT Press, Cambridge, Mass., 1989.

19. Eshghi, K., Meta-Language in Logic Programming, Ph.D. Thesis, Dept. of Computing, Imperial College, London, 1986.
20. Hill, P. M. and Lloyd, J. W., *The Gödel Programming Language*, MIT Press, Cambridge, Mass., 1994.
21. Kim, J. S. and Kowalski, R. A., An Application of Amalgamated Logic to Multi-Agent Belief, in: M. Bruynooghe (ed.), *Proc. Second Workshop on Meta-Programming in Logic*, Dept. of Comp. Sci., Katholieke Univ. Leuven, 1990.
22. Kowalski, R. A., Meta Matters, invited presentation at Second Workshop on Meta-Programming in Logic, 1990.
23. Kowalski, R. A., Problems and Promises of Computational Logic, in: J. W. Lloyd (ed.), *Computational Logic*, Springer-Verlag, Berlin, 1990.
24. Lloyd, J. W., *Foundations of Logic Programming, Second Edition*, Springer-Verlag, Berlin, 1987.
25. Robinson, J. A., A Machine-oriented Logic Based on the Resolution Principle, *J. ACM*, 12:23–41 (1965).
26. Sato, T., Meta-Programming through a Truth Predicate, in: K. Apt (ed.), *Proc. Joint Intl. Conf. Symp. on Logic Programming 1992*, MIT Press, Cambridge, Mass., 1992.
27. Smorynski, C., The Incompleteness Theorems, in: J. Barwise (ed.), *Handbook of Mathematical Logic*, North-Holland, Amsterdam, 1977.
28. Weyhrauch, R. W., Prolegomena to a Theory of Mechanized Formal Reasoning, *Artificial Intelligence*, 13:133–70 (1980).

Logic Programs with Tests *

Marion Mircheva

Institute of Mathematics & Computer Science
Bulgarian Academy of Sciences
"Acad. G.Bonchev" str. bl. 8, Sofia 1113 , Bulgaria
E-mail: marion@bgearn.bitnet

Abstract. We extend logic programming to deal with logic programs
that include new truth functional connectives called *tests*. Stable Model
Semantics and Three Valued Stable Model Semantics are extended to
give meaning to programs with *tests*.
We consider three possible applications of such programs. It is shown
how to define a particular semantics in terms of another one with the
help of an appropriate transformation of normal programs into programs
with tests. Our approach can be applied for resolving inconsistency in
logic programs with *explicit* negation. Programs with tests can serve as
a promising tool for eliminating or encoding *integrity constraints* within
particular background semantics.

1 Introduction

In this paper we propose a general framework for extending two valued and three
valued stable model semantics to programs with new truth functional connectives
called *tests*. We consider at least three possible applications of logic programs
with such metatools.

For some well known semantics we present a method to calculate a particular
semantics (set of models) of a program in terms of some other semantics of
the program obtained from the initial one after a suitable transformation. The
resulting program however includes a new test connective. For example the stable
models of the program

$$a \leftarrow \sim b$$
$$b \leftarrow \sim a$$

$M_1 = \{a, \sim b\}$ and $M_2 = \{b, \sim a\}$ coincide with the three valued stable
models of the following program

$$a \leftarrow \sim b, \sim t(a)$$
$$b \leftarrow \sim a, \sim t(b)$$

---

* This paper has been developped during my visit at the Institute for Logic, Complexity and Deductionsystems, University of Karlsruhe, Germany. It was supported by
the European Economic Community within the framework of the *Community's Action for Cooperation in Science and Technology with Central and Eastern European
Countries* under the contract number CIPA3510CT924716.

The modal[2] operator $t(a)$ has the meaning:

$$\text{if } a = \mathbf{u} \text{ then } t(a) = \text{true, otherwise } t(a) = \text{false.}$$

This test operator distinguishes undefined value from the Boolean ones. Obviously its meaning can be given by the truth table:

a	$t(a)$	$\sim t(a)$
1	0	1
u	1	0
0	0	1

Along this line we present answer sets [4] via stable models [3], stable models via three valued stable models and three valued stable models for extended programs [11] via three valued stable models for normal programs. In fact the last evidence presents elimination of integrity constraints with the help of a program with a test.

The next application concerns extended logic programs, that are programs with *default* ($\sim$) and *explicit* ($\neg$) negations. We propose a method for resolving inconsistency caused by the interaction between *default* and *explicit* negations within three valued stable semantics. We present an alternative extension of three valued stable models to programs with explicit negation that gives meaning to every logic program. Our models can be calculated within three valued stable semantics over logic programs with a special test connective.

Within our framework, any background semantics could be easily adapted to deal with programs augmented with integrity constraints presented as denials $\leftarrow a_1, \cdots, a_n$. This can be done either by withdrawing the models that brake up the constraints or by revising the models in a satisfactory way. The revision strategy has to be encoded in the meaning of the test connectives that has to be used (witness section 5).

The paper is organized as follows. Section 2 involves the syntax of our framework. In Section3 we formally describe our extension of two valued stable models to programs with *tests* and present a case study example. Section 4 extends Stable3 of Przymusinski to include programs with *tests* and presents two case study examples. Section 5 describes a new extension of three valued stable models to programs with explicit negation. An appropriate transformation into a class of programs with *guard* connective is considered here. Section 6 discusses related papers and gives some conclusive remarks.

2 General Framework: Normal Logic Programs with Test Formulae

In this section we extend the propositional logic programming to deal with programs that include some new truth functional connectives, called *test* operators.

---

[2] We use the term "modal" according to Polish logical tradition: each connective that transforms multi valued domain into Boolean one is called modal.

Some of them have modal flavor as they transform all possible values into Boolean ones. Their meaning is defined with correspondent truth tables.

A normal logic program is a set of formulae of the form

$$c \leftarrow a_1, \cdots, a_m, \sim b_1, \cdots, \sim b_n$$

where $n \geq 0$, $m \geq 0^3$ and all c, a_i and b_j are atoms. The sign $\sim$ stands for default negation. A clause containing variables is treated as standing for the set of all ground instances of that clauses.

We extend ordinary language of logic programs to include a new truth functional connective $t(x_1, \cdots, x_n)$ where $x_1, \cdots, x_n$ are atoms or literals (if the language includes classical negation). The next are examples of test operators:

1. $test(a, b) = true$ if $a = true$ and $b = true$ otherwise $t(a, b) = false$. For a two valued domain this is classical conjunction or in terms of logic programming that is the meaning of an *integrity constraint* of the form $\leftarrow a, b$. So, $\sim test(a, b) = true$ (equivalently $test(a, b) = false$) if and only if $\leftarrow a, b$ is satisfied.
2. $test(a) = true$ if $a = undefined$ otherwise $test(a) = false$. This is a test that will be used in a three valued domain for separating the Boolean models from the others.
3. If $a = true$ then $test(a, b) = b$ otherwise $test(a, b) = false$. This is again a three valued connective and in the sequel we call it *strong guard*. [4] Note that restricted to a two valued domain this operator is equivalent to the test connective from 1.

For the aim of simplicity we restrict our general presentation to binary test connectives.

Definition 2.1. A program with *test* is a set of clauses such that some rules may have default negated test formulae in addition to the other premises. Consequently it might contains rules with test:

$$c \leftarrow c_1, \cdots, c_m, \sim c_{m+1}, \cdots, \sim c_n, \sim test(a, b)$$

---

[3] If $n = m = 0$ we mean the rule $c \leftarrow$ and when $m = 0$ or $n = 0$ we mean $c \leftarrow \sim b_1, \cdots, \sim b_n$ or $c \leftarrow a_1, \cdots, a_m$ correspondingly.

[4] A similar connective has been used by Fitting [2] in connection with logic programming languages based on billatices. Fitting considers the usefulness of *guard* connective "P guards Q" (P:Q) with the meaning:

if $P = true$ then $P : Q = Q$ otherwise $P : Q = undefined$.

Our strong guard is also close to what Fitting refers as *Lisp conjunction* or conjunction of Prolog as actually implemented, with its left-right, top-down evaluation. Obviously, our *strong guard* is also an asymmetric connective. However all Fitting's connectives are regular in Kleene's sense (that is k-monotone), while the *strong guard* is not regular in that sense.

3 Two Valued Stable Model Semantics for Programs with Test

We slightly modify Stable Model Semantics introduced by Gelfond and Lifschitz [3] to deal with programs with a test formulae. In fact we prolong stable semantics to programs with test connective. This is done according to the meaning of a particular test operator.

Definition 3.1 (I_t interpretation). Given a program t-Π with test, let I be an ordinary 2-valued interpretation of t-Π (set of atoms). I_t is obtained from I by adding $test\,(a,b)$ to I for any test formula $test(a,b) \in Lang(t\text{-}\Pi)$ if $test(a, b) = true$ for the current value of a and b in I.

I_t extends an ordinary interpretation I in a natural way to interpretations over the language with test and this extension conforms with the semantics for a test connective.

For the presentation below let us assume t-Π be a logic program with test. The clauses of such programs are allowed to have in addition to ordinary premises only default negated test sentences.

Definition 3.2 (t-Π/I_t program). By t-Π/I_t transformation we mean a program obtained from t-Π by performing the following two operations that concern each default negated atom or *default negated test* in the bodies of the program rules.

1. Remove from t-Π all rules containing $\sim X$, if $X \in I_t$;
2. Remove from all remaining rules their default premises $\sim X$.

The only difference compared to the original GL *modulo transformation* is the new concern with *default negated tests*. The resulting program t-Π/I_t is positive by definition and it has a unique least 2-valued model.

Definition 3.3. An interpretation I is called a Stable model if *least model* $(t$-$\Pi/I_t) = I$. Stable semantics for a program t-Π with test is determined by the set of all stable models of t-Π.

There is no syntactical difference between stable models for programs with tests and ordinary programs. In both cases any stable model is a set of atoms. Moreover t-interpretations play only an auxiliary role during the calculations.

3.1 Case Study Example: Answer Sets via Stable Models

Answer Set Semantics (AS) [4] extends Stable Models [3] to programs that include both explicit ($\neg$) and default ($\sim$) negations.

An extended logic program is a set of formulae of the form

$$l_0 \leftarrow l_1, \cdots, l_m, \sim s_1, \cdots, \sim s_n$$

where $n \geq 0$, $m \geq 0$[5] and all l_i and s_j are literals. The sign $\neg$ stands for a classical (explicit) negation and $\sim$ stands for a default negation. A literal is a formula of the form a or $\neg a$, where a is an atom. By default literal we mean $\sim l$ where l is a literal.

Recall the definition of answer sets, which is in two parts.

Let Π be any set of ground clauses not containing $\sim$, and let Lit be the set of all ground literals in the language of Π. The *answer set* of Π, denoted by $\alpha(\Pi)$, is the smallest subset S of Lit such that

1. for any clause $l_0 \leftarrow l_1, \cdots, l_m \in \Pi$, if $l_1, \cdots, l_m \in S$ then $l_0 \in S$, and
2. if S contains a pair of complementary literals, then $S = Lit$.

Now let Π be any extended logic program which doesn't contain variables. For any subset S of Lit, let Π^S be the set of clauses without $\sim$ obtained

1. by deleting every clause containing a condition $\sim l$, with $l \in S$ and
2. by deleting in the remaining clauses every condition $\sim l$.

S is an answer set of Π if and only if S is the answer set of Π^S, i.e. if $S = \alpha(\Pi^S)$.

According to Answer Set Semantics extended logic programs are divided into three disjoint classes:

1. Π is *as-consistent* if it has a consistent [6] answer set.
2. Π is *as-contradictory* if it has an inconsistent answer set (namely only one answer set Lit).
3. Π is *as-incoherent* if it has no answer set.

It is a well known fact that contradictions (in *as-contradictory* programs) are "stable" in the sense that they do not depend on the contribution of default negation. In *as-contradictory* programs, contradictions can not be removed even if either any clause is added to the program or any clause with $\sim$ is removed from the program.

For the aim of presenting answer sets in terms of stable models we need to separate the class of *as-contradictory programs*.

Definition 3.1.1. (Legal Programs). A program Π is legal if the set $C(\Pi)$ is consistent where $C(\Pi)$ is the smallest set such that:

If $l_0 \leftarrow l_1, \cdots, l_m \in \Pi$ and $l_1, \cdots, l_m \in C(\Pi)$ then $l_0 \in C(\Pi)$.

Intuitively legal programs have consistent *monotonic* part. For instance the program

---

[5] If $n = m = 0$ we mean the rule $l_0 \leftarrow$ and when $m = 0$ or $n = 0$ we mean $l_0 \leftarrow \sim s_1, \cdots, \sim s_n$ or $l_0 \leftarrow l_1, \cdots, l_m$ correspondingly.

[6] A set of literals is consistent if it does not contain both a and $\neg a$ for some atom a.

$$\neg a \leftarrow$$
$$a \leftarrow$$

is not legal and it is not in the scope of our interests.

Statement 3.1.1. A program Π is not legal if and only if Π is as-contradictory. Consequently legal programs are as-incoherent or as-consistent.

Notations:

- For an arbitrary set of clauses R we denote $Head(R) = \{l \mid l \leftarrow \cdots \in R\}$.
- $\Pi^\neg$ denotes the renamed version of Π, that is the next transformation is applied to every literal occurring in Π.

$$l^\neg = \begin{cases} a, & \text{if } l = a \text{ for some atom } a \\ a^\neg, & \text{if } l = \neg a \text{ for some atom } a \end{cases}$$

Extended logic programs could be considered as a particular case of normal logic programs together with a set of integrity constraints of the form $\leftarrow a, a^\neg$ for any atom a. Thus instead of answer sets of a given extended program Π we can calculate the stable models of its renamed version in which the correspondent constraints are properly encoded.

Definition 3.1.2.

- ($\wedge$-test) For any two atoms let $\wedge(a, b)$ be a new formula with the following meaning:
  $\wedge(a, b) = true$ if $a = true$ and $b = true$, otherwise $t(a, b) = false$.
  This is the first test connective from the examples of tests in section 2.
- ($\wedge$-Π) For any extended program Π let $\wedge$-Π be a program with $\wedge$-test obtained from Π by adding a default negated formula $\sim \wedge(a, \neg a)$ to the premises of each rule with a head a and $\neg a$.

Theorem 3.1.1.
For any extended legal program Π

$$\text{Answer Sets}(\Pi) = \text{Answer Sets}(\wedge\text{-}\Pi) = \text{Stable Models}^7 (\wedge\text{-}\Pi)^\neg$$

Proof: Follows from the fact that answer sets of Π and $\wedge$-Π are always consistent.

Example.
$$\Pi_1: \quad a \leftarrow \sim b$$
$$b \leftarrow \sim a$$
$$\neg a \leftarrow$$

The program Π_1 has one answer set $M = \{\neg a, b\}$. Then $M_t = \{\neg a, b\}$, so $\sim \wedge(a, \neg a)$, $\sim \wedge(b, \neg b)$, $\sim a$ are true and $\sim b$ is false in M_t (def 3.1). It is easy to calculate (def 3.2, def 3.1.2.):

---

[7] We mean stable models after reversing the renamed literals.

$$\land\text{-}\Pi_1 : \quad a \leftarrow \sim b, \sim \land(a, \neg a)$$
$$b \leftarrow \sim a, \sim \land(b, \neg b)$$
$$\neg a \leftarrow \sim \land(a, \neg a)$$
$$\land\text{-}\Pi_1/M_t : \quad b \leftarrow$$
$$\neg a \leftarrow$$

Then M is also a stable model of $\land\text{-}\Pi_1$ because the least model$(\land\text{-}\Pi_1/M_t) = M$ (def.3.3).

4   Three Valued Stable Model Semantics for Programs with Test

In this section we extend Three-valued Stable Semantics (Stable3) in style of Przymusinski [12,15] to programs with test formulae. For a normal logic program the F-least[8] stable3 model determines the so called Well Founded Semantics, originally introduced by Van Gelder at al. [15].

We use the standard notion of the three-valued (partial) interpretation $I = T \cup \sim F$ where T is the set of true atoms and F is the set of false atoms. The rest are undefined.

Definition 4.1 (I_t interpretation). Given a program $t\text{-}\Pi$ with test, let I be an ordinary 3-valued interpretation of $t\text{-}\Pi$ (a set of true and false atoms). I_t is obtained from I in the following way:

- add $test(a, b)$ to I for any test formula $test(a, b) \in Lang(t\text{-}\Pi)$ if $test(a, b) = true$ for the current value of a and b in I;
- evaluate $test(a, b)$ as undefined if $test(a, b) = undefined$ for the current value of a and b in I;
- add $\sim test(a, b)$ otherwise.

I_t extends an ordinary interpretation I in a natural way to interpretations over the language with a test and this extension conforms with the semantics for the test connective.

For the presentation below let us assume $t\text{-}\Pi$ be a logic program with test. Remind that the clauses of such programs might have in addition to ordinary premises only default negated test sentences.

Definition 4.2 ($t\text{-}\Pi/I_t$ program). By $t\text{-}\Pi/I_t$ transformation we mean a program obtained from $t\text{-}\Pi$ by performing the following three operations that concern each default negated atom or *default negated test* in the bodies of the program rules.

1. Remove from $t\text{-}\Pi$ all rules containing $\sim X$, if $X \in I_t$;
2. Remove from all remaining rules their default premises $\sim X$, if $\sim X \in I_t$;
3. Replace all the remaining default premises by proposition **u**.

---

[8] F stands for Fitting ordering, though Fitting call it k(nowledge)-ordering.

The only difference compared to the original *modulo transformation* [def. 6.23,14] concerns the inclusion of *default negated test* formulae. The resulting program Pi/I_t is by definition non-negative and it has a unique least 3-valued model.

Definition 4.3. An interpretation I is called a Stable3 model if *least model* $(t\text{-}\Pi/I_t) = I$. Stable3 semantics for a program $t\text{-}\Pi$ with test is determined by the set of all Stable3 models of $t\text{-}\Pi$.

Similarly to the case with two valued stable models, there is no syntactical difference between Stable3 models for programs with tests and ordinary programs. In both cases any Stable3 model is a set of atoms and default atoms. However not every program with test possesses Stable3 models (witness $t\text{-}\Pi_3$ bellow).

4.1 Case Study Example: Stable Models via Stable3 Models

We aim to define two valued stable models of a normal program Π in terms of 3-valued stable models of a correspondent program with test $t\text{-}\Pi$. The test operator must distinguish exact values from undefinedness.

Definition 4.1.1.

– (test for **u**). For any atom a let $t(a)$ be the following three valued connective:

$$\text{if } a = \mathbf{u} \text{ then } t(a) = \text{true, otherwise } t(a) = \text{false.}$$

This test operator distinguishes undefined values from the Boolean ones. Obviously its meaning can be given by the truth table:

a	$t(a)$	$\sim t(a)$
1	0	1
u	1	0
0	0	1

– For any normal program Π, $t\text{-}\Pi$ denotes the program with test for **u** obtained from Π by adding a default negated formula $\sim t(a)$ to the premises of each rule in Π with a head a (for every head a).

Theorem 4.1.1.

For any normal program Π let $t\text{-}\Pi$ be a program with test for **u** obtained from Π by adding a default negated formula $\sim t(a)$ to the premises of each rule in Π with a head a. Then

$$\text{Stable Models}(\Pi) = \text{Stable3 Models}(t\text{-}\Pi)$$

Proof: Follows from the observation that in every Stable3 model of $t\text{-}\Pi$ the set of undefined atoms is empty.

Examples.

$\Pi_2 : a \leftarrow \sim b$
$\qquad b \leftarrow \sim a$

The program Π_2 has two stable models, $M_1 = \{a, \sim b\}$ and $M_2 = \{b, \sim a\}$. It is easy to check that M_1 and M_2 are the only three valued stable models of $t\text{-}\Pi_2$.

$t\text{-}\Pi_2 : a \leftarrow \sim b, \sim t(a)$
$\qquad b \leftarrow \sim a, \sim t(b)$

Consider the program Π_3:

$\Pi_3 : a \leftarrow \sim a$

Then the correspondent program (def.4.1.1) $t\text{-}\Pi_3$ is as follows:

$t\text{-}\Pi_3 : a \leftarrow \sim a, \sim t(a)$

Neither Π_3 possesses any stable models nor $t\text{-}\Pi_3$ admits Stable3 models.

4.2 Case Study Example: Stable3 Models for Extended Programs via Stable3 Models for Normal Programs

As we have already mentioned Przymusinski extended Stable3 semantics to programs with explicit negation [11]. This was done in a very simple way. Przymusinski takes all Stable3 models of the renamed program $\Pi^\neg$ and throws out all "inconsistent" models, i.e. those which contain both l and $l^\neg$ for some atom l. Thus the program

$\Pi_4 : \neg a \leftarrow \sim b$
$\qquad a \leftarrow$

hasn't any Stable3 models.

In the sequel we use Stable3 to denote three-valued stable models for normal and extended programs, as well as for programs with test (section 3) when the context makes the background language clear.

We can easily notice that Stable3 semantics for an extended logic program is in fact Stable3 semantics for the normal program $\Pi^\neg$ with $\wedge$-test (see def.3.1.2. for $\wedge$-test).

a	b	$\wedge(a,b)$	$\sim\wedge(a,b)$
1	1	1	0
1	u	0	1
⋮	⋮	⋮	⋮
0	0	0	1

Statement 4.2.1. Let Π be an extended program and let $\Pi^\neg$ be its renamed version. Let $\wedge\text{-}\Pi^\neg$ be a program with $\wedge$-test obtained from $\Pi^\neg$ by adding a default negated formula $\sim\wedge(a, a^\neg)$ to the premises of each rule with head a and $a^\neg$. Then

$$\text{Stable3}(\Pi) = \text{Stable3}(\wedge\text{-}\Pi^\neg)$$

Example. For the program Π_4 above, the correspondent program $\wedge\text{-}\Pi_4^\neg$ is as follows:

$\wedge\text{-}\Pi_4^\neg : a^\neg \leftarrow \sim b, \sim\wedge(a, a^\neg)$
$\qquad a \leftarrow \sim\wedge(a, a^\neg)$

The program $\wedge\text{-}\Pi_4^\neg$ doesn't have any Stable3 models.

5 Resolving Inconsistency in Extended Logic Programs

As we already pointed out Stable3 semantics for an extended logic program is in fact Stable3 semantics for the normal program $\Pi^{\neg}$ with $\wedge$-test.

For the aim of a better representation of incomplete knowledge, hypothetical reasoning, abduction and other forms of non-monotonic reasoning, classical (explicit) negation ($\neg$) has been proposed to be used in logic programs along with negation as failure ($\sim$) [4,5,6,8]. Though default negation has a clear procedural meaning its application is limited in the sense that $\sim a$ does not refer to the presence of knowledge asserting the falsehood of the atom a but only to the lack of evidence about its truth. In this section we propose an alternative extension of Stable3 semantics to deal with explicit negation. We define models for every legal (see def.3.1.1.) extended program. If an extended program Π possesses Stable3 models of Przymusinski, then our models form a superset of Stable3. In the sequel we show that our semantics for extended logic programs is in fact three valued stable semantics over normal logic programs with a new connective called *strong guard*. Thus in this section we aim to illustrate another application of test connectives for resolving inconsistency within three valued stable model semantics.

There are at least two awkward questions related to the inconsistency of extended logic programs that Stable3 Semantics is faced with:

- First, Stable3 semantics does not distinguish between inconsistency arising as a result of a mutual relationship between Closed World Assumption and explicit negation (witness Π_4) and inconsistency created by the monotonic part of the program.
- There are two different notions of falsehood in extended programs and no satisfactory semantics for explicit negation besides the constraint that a and $\neg a$ create an unwilling situation.

We aim to resolve partly both questions. We propose an alternative extension of Stable3 (for normal programs) to programs with explicit negation, evaluating the explicit negation as Kleene's three valued negation. For normal programs our semantics coincide with Stable3. For a rather general case (Statement 5.1.2) our semantics coincide with Przymusinski's semantics.

The appearance of the two kinds of negations make the use of the notion *false* ambiguous. As usual we refer to *not a* as default negation, that is a is false-by-default in Π if *not a* is provable from Π. We define the meaning of explicit negation as Kleene's three valued negation i.e. there is a *logical falsehood* beside false-by-default. So $\neg a$ (a) is *logically false* (simply false) if and only if a ($\neg a$) is true and $\neg a$ (a) is unknown whenever neither a nor $\neg a$ is true (provable).

We give a meaning to each legal program by resolving inconsistency caused by the interaction between default and explicit negations. For instance in the program Π_4, the rule $\neg a \leftarrow \sim b$ is considered less trustful then the nonnegative rule $a \leftarrow$. Each rule that depends on default negation is considered as context dependent. Though $\neg a$ has an explanation $\sim b$ in Π_4 the meaning of $\neg a$ depends on the value of a in the current context. Since a has more trustful explanation then $\neg a$, then a has to be accepted as *true* and $\neg a$ has to be revised to *undefined*. The intuitive meaning of $\neg a \leftarrow \sim b$ turns out to be: $\neg a$ is true if b is false and a is not true, but if a turns out to be true along with $\sim b$ then a must be revised to undefined. This holds for all rules that depend on default assumptions. Thus we propose $\{a, \sim b\}$ to become a model for Π_1. All of these lead to the next definition.

Definition 5.1 (Split Program). Let Π be a logic program. Π could be uniquely split into two disjoint parts, $\Pi = T \cup H$ where T is the minimal set inductively defined as follows:

- if $l \leftarrow \in \Pi$, then $l \leftarrow \in T$;
- if $l \leftarrow l_1, \cdots, l_m \in \Pi$ and $l_i \in Head(T)$ for each i, then $l \leftarrow l_1, \cdots, l_m \in T$.

The couple (T, H) is called split program.

Definition 5.1 holds for extended logic programs as well as for normal programs. To every program Π we associate the corresponding splitting (T, H). The set T forms the monotonic part of the program and the set H stands for the defeasible part of the program. H also could be treated as a set of hypotheses in the context of Π. In this setting hypotheses are not predominant as it is usually considered in hypothetical reasoning frameworks [10] and sometimes in logic programming frameworks [6]. In our case each program determines by itself the set of defeasible clauses.

Statement 5.1. A program $\Pi = T \cup H$ is legal if T is consistent (T is considered as a set of inference rules).

5.1 Three Valued Stable Semantics for Split Programs: s-Stable3

In this section we formally define our intended models for extended logic programs. We supply with definite meaning every legal program. For normal programs our semantics coincides with Stable3 models.

We resolve the inconsistency caused by the interaction between default and explicit negation. Consider a split program $\Pi = T \cup H$. If a contradiction arises between rules with heads l and $\neg l$, the rules from T override the rules from H. If both clauses have equal priority, that is l and $\neg l$ come from H, then multiple models are adopted – one in which l is true and $\neg l$ is undefined and the other in which $\neg l$ is true and l is undefined. For example the program

$\Pi_5 : \neg a \leftarrow \sim b$

$\qquad a \leftarrow \sim c$

has two s-Stable3 models: $\{a, \sim b, \sim c\}$ and $\{\neg a, \sim b, \sim c\}$.

Definition 5.1.1. Let $I \subseteq Lit(\Pi)$ be an interpretation and $\Pi = T \cup H$ be a split (legal extended) program. By $\Pi/^s I$ we mean the program obtained from Π after performing the following five operations that concern each default literal in the bodies of the program rules.

1. Remove from Π all rules containing condition $\sim l$, if $l \in I$;
2. Remove from Π all rules with a head l if l and $\neg l$ belong to I;
3. Remove from all rules in Π their default premises $\sim l$, if $\sim l \in I$;
4. Add the proposition **u** to the premises of all clauses in H with a head l, if l does not belong to $T \cup F$ and $\neg l \in T$;
5. Replace all the remaining default premises by proposition **u**.

We changed the original *modulo transformation* Π/I definition by adding extra conditions (2) and (4). Item (2) rejects all inconsistent models, item (4) ensures the revision to logical falsehood for those literals l which counterpart $\neg l$ is true.

The resulting program $\Pi/{}^s I$ is by definition non-negative and it has a unique least three-valued model.

Definition 5.1.2. A 3-valued interpretation I of the language of Π is called an s-Stable3 model if $I = least\ model\ (\Pi/{}^s I)$. Three-valued s-Stable semantics is determined by the set of all s-Stable3 models of Π.

Remark. In fact the points 1,2,3,5 from def.5.1.1 together with def.5.1.2 are equivalent to Stable3 semantics of Przymusinski. Provision 4 supplies with models those programs that do not have Przymusinski's semantics. For instance the program $\Pi_4 = T \cup H$

$H : \neg a \leftarrow \sim b$
$T : \quad a \leftarrow$

has an s-Stable3 model $M = \{a, \sim b\}$ because M is indeed the least model of $\Pi_4/{}^s M$.

$\Pi_4/{}^s M : \neg a \leftarrow \sim b,\ \mathbf{u}$
$\qquad\qquad a \leftarrow$

Statement 5.1.1. S-Stable3 models are always consistent (it directly follows from the provision 2, def.5.1.1).

Statement 5.1.2. Let Π be an extended legal program. If all Stable3 models of its renamed version $\Pi^\neg$ are "consistent" then Stable3 models of Π coincide with s-Stable3 models of Π.

Statement 5.1.3. Every extended legal program has at least one s-Stable3 model. (Follows from the observation that the number of s-Stable3 models of Π is no less that the number of the Stable3 models of $\Pi^\neg$.)

Theorem 5.1.1.

Let $\Pi = T \cup H$ be a legal extended logic program. If M is a Stable3 model of Π then M is also an s-Stable3 model of Π.

Proof. Assume M is a Stable3 model of Π, that is $M = least\ model(\Pi/M)$. Then the only difference between Π/M and $\Pi/{}^s M$ concerns some rules $l \leftarrow \cdots$ from Π/M for which neither l nor $\sim l$ belong to M and $\neg l$ is true in M. These rules are replaced with $l \leftarrow \cdots, \mathbf{u}$ in $\Pi/{}^s M$. However these rules don't contribute to the model M as l is also undefined in M. Therefore $M = least\ model(\Pi/{}^s M)$.
#

5.2 Stable3 Semantics for Programs with Strong Guard Coincides with s-Stable3

We need to extend our syntax to include a new truth functional connective - *strong guard* - in order to express our intended meaning of logic programs in terms of Stable3 semantics. Given a program Π we find a correspondent program Π^g such that

$$\text{s-Stable3}(\Pi) = \text{Stable3}(\Pi^g).$$

Π^g is a program with *strong guard*.

Definition 5.2.1 (Strong Guard). We extend the propositional language with a new binary truth functional connective $(a; b)$ (a and bl are literals) and read it "a guards b". It has the following meaning:

if $a = true$ then $(a; b) = b$ otherwise $\sim (a; b) = true$.

This definition shows when a guard formula is true (provable) with respect to some semantics and when it is false-by-default. Now the rule for default negation over literals is again Closed World Assumption, but a guard formula $(a; b)$ is false-by-default if a is not true. In that sense the guard connective performs a metacalculation.

We adopt the notation $(a; b)$ to distinguish from Fitting's *guard* $(:)$ [2].

Now definition 2.1 adapted to a strong guard connective looks like: A program with *strong guard* is an extended program in which some rules may have default guard sentences in addition to the other premises. Thus it might contain rules with guard:

$$l \leftarrow l_1, \cdots, l_m, \sim l_{m+1}, \cdots, \sim l_n, \sim (a; b)$$

Following the content of section 3, Stable3 semantics can easily be prolong to programs with strong guards. This is done in according to the meaning of the test operator — in this case a strong guard connective.

We've already given some hints about how to transform a program Π to a program Π^g with strong guard such that

$$\text{s-Stable3}(\Pi) = \text{Stable3}(\Pi^g).$$

Definition 5.2.2 (The Transformation). Let $\Pi = T \cup H$ be a legal extended logic program. Then $\Pi^g = T \cup H^g$ where

$H^g = \{l \leftarrow \cdots, \sim (\neg l; l) | l \leftarrow \cdots \in H \text{ and } \neg l \in Lit(\Pi)\} \bigcup$
$\{l \leftarrow \cdots | l \leftarrow \cdots \in H \text{ and } \neg l \notin Lit(\Pi)\}.$

In fact this transformation changes only the set H in Π, as we add auxiliary premises to the clauses in H. That is why the transformation preserves the legal property, i.e. if Π is a legal program then Π^g is also a legal program and $T \cup H^g$ is indeed its splitting.

Recall the program $\Pi_5 = T \cup H$, that has two s-Stable3 models: $\{a, \sim b, \sim c\}$ and $\{\neg a, \sim b, \sim c\}$.

$\Pi_5 : T = \emptyset$

$\quad H : \quad a \;\leftarrow \sim b$

$\qquad\qquad \neg a \leftarrow \sim c$

Its transformed version Π_5^g

$\Pi_5^g : T = \emptyset$

$\quad H : \quad a \;\;\leftarrow \sim b, \sim(\neg a; a)$

$\qquad\qquad \neg a \leftarrow \sim c, \sim(a; \neg a)$

has two Stable3 models (for a program with guard):

$\{a, \sim b, \sim c\}$ and $\{\neg a, \sim b, \sim c\}$, that are identical to s-Stable3 models of Π_5.

The following theorem shows that the transformation preserves the meaning of the original program.

Theorem 5.2.1.

Let Π be a legal program and Π^g be the program obtained after the transformation. Then

1) s-Stable3$(\Pi) = $ Stable3(Π^g)

2) Stable3$((\Pi^g)^-) = $ Stable3(Π^g)

Proof: 1) We show more generally that for any 3-valued consistent interpretation I of Π

$$\Pi/^s I = \Pi^g / I_g$$

The only difference between $\Pi/^s I$ and Π^g/I_g concerns those rules from H with heads l for which the correspondent rules from H^g contain a default guard $not\,(\neg l; l)$, i.e. $l \leftarrow \cdots \in H$ and $l \leftarrow \cdots, not\,(\neg l; l) \in H^g$. The nontrivial subcases concern the provision when $\neg l$ is true in I and l is undefined in I, otherwise $\Pi/^s I = \Pi^g/I_g$ directly follows from def.5.1.1. and def.4.2. For such l and $\neg l$ by provision 4, def.5.1.1. we get $l \leftarrow \cdots, \mathbf{u} \in \Pi/^s I$. As $\neg l$ is true in I_g we get $not\,(\neg l; l) = not\, l = \mathbf{u}$. Therefore $l \leftarrow \cdots, \mathbf{u} \in \Pi^g/I_g$.

2) follows from the fact that Stable3 models of $(\Pi^g)^-$ are always consistent (after reversing the renaming).#

5.3 Elimination of Integrity Constraints

Extended logic programs could be considered as a particular case of normal logic programs together with a set of integrity constraints of the form $\leftarrow a, b$ where b is the renamed version of $\neg a$. In this setting the framework proposed in this section can serve as a method for elimination of integrity constraints. For example let us consider a theory that contains the program Π and the constraints $\leftarrow a, b$:

$\Pi : a \leftarrow \sim c$

$\quad\;\; b \leftarrow \sim d$

This theory has two expected models $M_1 = \{a, \sim c, \sim d\}$ and $M_2 = \{b, \sim c, \sim d\}$. These models are exactly the Stable3 models of a program Π^g with strong guard:

$\Pi^g : a \leftarrow \sim c, \sim(b; a)$

$\qquad b \leftarrow \sim d, \sim(a; b)$

Indeed we can easily check that

least model$(\Pi^g/M_1) = M_1$
least model$(\Pi^g/M_2) = M_2$

because the correspondent *modulo* programs are as follows:

$\Pi^g/M_1 : a \leftarrow$
$\qquad b \leftarrow \mathbf{u}$
$\Pi^g/M_2 : b \leftarrow$
$\qquad a \leftarrow \mathbf{u}$

Consider the more general case of integrity constraints like $\leftarrow a_1, \cdots, a_n$. It is read as "$a_1, \cdots, a_n$ are not allowed to be true together". To eliminate these constraints within Stable3 semantics we need n guard connectives: $(a_1, \cdots, a_{i-1}, a_{i+1}, \cdots, a_n; a_i)$, $i = 1, \cdots, n$ with the following meaning:

if $a_1, \cdots, a_{i-1}, a_{i+1}, \cdots, a_n$ are true, then $(a_1, \cdots, a_{i-1}, a_{i+1}, \cdots, a_n; a_i) =$
a_i otherwise $\sim (a_1, \cdots, a_{i-1}, a_{i+1}, \cdots, a_n; a_i) = true$

Then the meaning of a program $\Pi = T \cup H$ with a constraint $\leftarrow a_1, \cdots, a_n$ is equivalent to Stable3 semantics of the program Π^g obtained from Π by adding the guard formula $\sim (a_1, \cdots, a_{i-1}, a_{i+1}, \cdots, a_n; a_i)$ to the premises of each rule from H with a head a_i.

6 Discussion

Our framework was originally motivated by the goal of providing an extension of 3-valued stable semantics that makes the use of explicit negation harmless (section 5). Later we found out that programs with some other metatools like test operators could help for a better understanding the mutual relationships between different semantics as well as to increase the expressive power of logic programs.

The idea to use a transformation between programs (over one and the same language) was widely explored in different contexts. Besides the other advantages this idea is useful as the transformation usually facilitates implementation. Thus for s-Stable3 models for programs with guard logic programming methods for three valued stable models [13,16] can be applied over the transformed programs. A transformation that can eliminate certain restricted forms of integrity constraints from definite clause databases is proposed in the work of Asirelli et al [1]. In [7] a particular case of elimination of constraints in some restricted program databases has been independently considered.

Kowalski and Sadri [6] proposed a technique for a contradiction removal within the context of default reasoning based on answer set semantics. They allow an explicit representation of exceptions, that are rules with heads of the form $\neg x$, in addition to the original program. They restrict the syntax of the original program, so that explicit negation does not appear in the heads of ordinary clauses. They give a semantics for rules and exceptions in a way that exceptions have higher priority then rules. We also use a priority assignment of some rules over the others but in a completely different setting – split programs. To encode the priority relation and to resolve inconsistency we propose a transformation that helps to calculate the intended meaning within Stable3 semantics. Kowalski and Sardi use a transformation that calculates their intended meaning in terms of answer sets. For that aim they propose a *cancellation technique* that is different from ours. We need a new connective in the language as we deal with a three valued domain. Our models are not deductively closed in a sense that it is possible to have $a \leftarrow \sim b$ in a program, b false-by-default and a not true. This holds also for the models of *Logic Programs with Exceptions* of Kowalski and Sardi and for some semantics for inconsistent databases [8].

A recent work by Pereira, at al. [9] independently concerns how to present knowledge for default reasoning by using extended logic programs. They propose to resolve inconsistency within three valued semantics by retracting some negation as failure formulae from the rules of a program. That way they restrict the *Closed World Assumption* assigning value *undefined* instead of *false* to some of the default negated premises. This is done by a rather complicated procedure based on tracing all possible sets of admissible hypotheses (default literals). We on the contrary don't restrict CWA but propose a transformation that helps calculate our models within programming methods for 3-valued stable semantics.

References

1. Asirelli, P., De Santis, M. and M.Martelli. Integrity constraints in logic databases. *J. Logic Programming*, vol 2, number 3, 221-233, 1985.
2. Fitting, M. Kleene's three valued logics and their children. August, 1993. Manuscript.
3. Gelfond, M and V.Lifschitz. The stable model semantics for logic programming. *In R.A.Kowalski and K.A.Bowen, editors, 5th ICLP*, 1070- 1080. MIT Press, 1988.
4. Gelfond, M and V.Lifschitz. Logic programs with classical negation. *In Warren and Szeredi, editors, 7th ICLP*, 579-597, MIT Press, 1990.
5. Kowalski, R. A. and F.Sadri. Logic Programs with exceptions. *In Warren and Szeredi, editors, 7th ICLP*, 588-613. MIT Press, 1990.
6. Inoue, K. Extended Logic programs with default assumptions. *Proceeding of 8th ICLP*, 491-504. MIT Press, 1991.
7. Mircheva, M. Declarative semantics for inconsistent database programs. *In D. Pearce and H.Wansing, editors, Proc. of JELIA '92, LNCS 663*, 252- 263, Springer-Verlag, 1992.
8. Pearce P. and G. Wagner. Logic Programming with strong negation. *In P. Schroeder-Heister, editors, Proceedings of the International Workshop on Extension of Logic Programming*, pages 311-326, Tubingen, Dec. 1989. Lecture Notes in Artificial Intelligence, Springer-Verlag.
9. Pereira, L. M., Aparicio, J.N. and J.J.Alferes. Contradiction removal within well founded semantics. *In A. Nerode, W. Marek and V.S.Subrahmanian, editors, Proceedings of the First International Workshop on Logic Programming and Nonmonotonic Reasoning*, pages 105-119, MIT Press, Cambridge, MA, 1991.

10. Poole, D. A logical framework for default reasoning. *Artificial Intelligence 36*, 27-47, 1988.

11. Przymusinski, T. Extended stable semantics for normal and disjunctive programs. *In Waren and Szeredi, editors, 7th ICLP*, pages 459-477. MIT Press, 1990.

12. Przymusinski, T. Well founded semantics coincides with three valued stable models. *Fundamenta Informaticae 13*, pages 445-463, 1990.

13. Przymusinski, T. and David S. Warren. Well founded semantics: theory and implementation *Technical report,* Dep. of Computer Science, University of California at Riverside and SUNY at Stony Brook, March,1992.

14. Przymusinska, H. and T. Przymusinski. Semantic issues in deductive databases and logic programs. *In Formal Techniqies in Artificial Intelligence. A Sourcebook. R.B.Banerji (editor),* Elsevier Science Publishers B.V. (North-Holland),1990.

15. Van Gelder, A., Ross, K.A. and J.S.Schlipf. The well founded semantics for general logic programs. *Journal of ACM*, pages 221-230, 1990.

16. Warren, D. The XWAM: A machine that integrate prolog and inductive database query evaluation. *Technical report #25, SUNY at Stony Brook*, 1989.

An Architecture with Multiple Meta-Levels for the Development of Correct Programs [1]

Barbara Dunin-Kęplicz

Institute of Informatics
University of Warsaw
02-097 Warszawa, Banacha 2
Poland
e-mail: keplicz@mimuw.edu.pl

Abstract

In this paper we design a multi-meta-level compositional architecture for correct programs development. In this architecture an *object level*, describing an application domain, together with a *meta-level*, representing the semantics of a programming language, and a *meta-meta-level*, reflecting the adopted methodology, provide a specification of a generic system supporting the user in the process of correct programs construction. The ideas reported in this paper are illustrated in a prototype version of the system, designed for Dijkstra's guarded command programming language.

Keywords: multi-level compositional architecture, formal program development, specification and verification, programming in logic.

1   Introduction

Most software systems designed to help the user in a program derivation are so large that they need the support of a rigorous methodology. A precise specification of methodological assumptions enables one to describe a correct program development process in formal terms. Our goal in this paper is to design an architecture which gives rise to a system which is parametric w.r.t. methodology, programming language and application domain. Distinguishing and separate treatment of these three aspects of program construction is the key point of the reusability of the system.

---

[1] This work has been partially supported by grants: KBN 2 1199 9101 (Polish Research Commitee), CRIT-1 (CEC).

The expressive power needed to cover various methodological approaches seems to be hard to realize without research in this field; different methodological frameworks need different tools to formalize them. However it is generally accepted that the most conceptually natural programming style is top-down program development. For this reason our general framework will be characterized through the two main principles, namely:

- the *step-wise refinement* technique.

- *verifying* the correctness of the program *during* its *development*.

The above methodological perspective can be expressed via an adequate parametrization of various aspects of the program construction.

The first, very intuitive, principle is the realization of *an analytic design scheme* characterized in [9] (p.139):

> The analytic design consists of a systematic application of the following reasoning:
>
> 1. Given a problem P, is it possible to express its solution in a reasonably concise fashion using primitive notions of the linguistic level at which we want to program? If yes – write the program, if not – invent notions $P1,...,Pn$, such that
>
>   (a) each of the notions $P1,...,Pn$ is well specified,
>   (b) using the notions according to their specification it is possible to write a satisfactory program for problem P.
>
> 2. Consider each of the notions $P1,...,Pn$ in turn as a new problem and repeat the reasoning.
>
> This process continues until all invented, intermediary notions are implemented in terms of primitive notions.

Applying this scheme we can show that a program may be constructed by creating a sequence of refinements. Each "notion" may be treated as a well-specified module: let the *symbolic module* Π_i correspond to the notion of Pi (for $i = 1,...,n$). In subsequent steps, solving the problem amounts to consecutive development of particular symbolic modules. During the refinement of the program, on any level its correctness may be formally verified. In other words, the notion of symbolic module enables one the on-line verification of the program correctness.

Definition
Let Π be a partially developed program with the symbolic modules $\Pi_1,...,\Pi_n$. We say that Π is *conditionally correct* iff the correct development of the modules $\Pi_1,...,\Pi_n$ guarantees the correctness of the completely developed program. $\square$

Our approach realizes the following mathematical view of programming ([3], p.5):

> "The development of a proof of correctness and the construction of a correct program that satisfies its specification can proceed hand in hand. A program can literally be derived from its specification."

While constructing programs, the user's activity can be viewed as a systematic development of symbolic modules. Our goal is to help the user in those phases of program derivation which can be realized automatically. On the one hand, the system is planned to support, but not to substitute, the user in the program construction. On the other hand, the system can be treated as a tool enforcing a good, systematic programming style, which is essential especially for an inexperienced programmer.

Another point we want to implement is the capability to use a possibly broad class of programming languages. For this purpose we plan to embed the semantics of a programming language in a formal theory. This high-level construction is applicable to imperative programming languages with *axiomatically defined semantics*.

The last premise underlying our system concerns the application domain. The domain knowledge embodies the conceptualization of a domain for a particular application in a form of a *domain theory*. This knowledge is considered as system neutral, i.e. represented in a form that is independent of its use.

The paper is organized as follows. Section 2 characterises a structure of the presented multi-meta-level compositional architecture. Section 3 is an overview of the generic system: a description of three reasoning levels, a communication between them and an inference cycle in the system are discussed. Section 4 illustrates the presented ideas in action; it contains a prototypical version of the system designed for Dijkstra's guarded command programming language. Finally, conclusions and further perspectives are discussed in section 5.

2   Why a Multi-Level Architecture?

2.1   Motivation

Distinguishing three reasoning levels: a theory describing an application domain, a theory representing the semantics of a programming language and a theory reflecting the adopted methodology leads naturally to designing a system in terms of a *multi-level compositional architecture* ([8]). Each of the reasoning levels, namely:

(i) object — application domain;

(ii) meta — semantics of a programming language;

(iii) meta-meta — adopted methodology

is substantially different with respect to:

- goals;
- knowledge;
- reasoning capabilities.

One of the goals of this paper is to argue that designing two meta-levels in this architecture is really well motivated in the field of program construction.

The step-wise program refinement requires a certain mechanism of a *task decomposition*; to be more specific, a precise goal definition on particular reasoning levels. In effect, the essential property of our system is that each reasoning level establishes a goal to be achieved at the lower level. In other words the behaviour of the system can be specified in *a goal directed* manner.

2.2   The User's Role

A translation from specification to implementation cannot be realized automatically but what is possible is the automatic support in some phases of program derivation. In practice this amounts to the formal verification of particular steps of the program construction. To achieve this goal the user interacts with meta-meta-level realizing methodology. At this place he takes strategic decisions both about the shape of the program and, in the case of failure of its construction, about where to start its reconstruction. We will discuss these questions in more detail in subsection 3.3.

3 Overview of the Generic System

3.1 Domain: Object Level

Before starting the program development, the user has to precisely specify what problem is to be solved and to identify any constraints on the input data. This starting specification will then guide the process of program development.

The specification of the program Π has to be defined in terms of an *application domain* of the program. In order to carry out the necessary manipulations of specifications, domain knowledge can be viewed as a declarative theory of the application domain, formalized as the *object theory* T_O. Theorems to be proved in the theory T_O, let say $\gamma_1, \ldots, \gamma_k$, are established on meta-level descri- bing semantics. A communication between object and meta-level is assured by reflection rules defined in subsection 3.5.

The way of describing theory T_O is irrelevant from our point of view — it may be any adequate formal theory with a derivability relation. The only important point is the ability of drawing conclusions based on this theory.

3.2 Semantics: Meta-Level

The assumption about axiomatic semantics allows us to use Hoare's notation. Since we are interested in total correctenss of the program, the triple $\{Q\}I\{R\}$ indicates that the execution of a statement I, which begins in any state satisfy- ing the predicate Q (called *precondition*), is guaranteed to terminate in a state satisfying the predicate R (called *postcondition*).

In our framework the statement I may be :

- At the beginning of program construction — the symbolic module denoting the whole program to be developed. In this case predicates Q and R are the precondition and postcondition for the entire program.

- During the development process — a sequence of programming language statements and/or symbolic modules.

- When the development is completed — the sequence of programming lan- guage statements.

The axiomatic semantics of the programming language can be naturally formu- lated as a *formal meta-theory*: proof rules describing the meaning of particular programming language commands are represented as meta rules in the theory T_{M1}.

The language of T_{M1} contains two predicate symbols THR and SP.

- $THR(Q)$ denotes that Q is a theorem of the object theory T_O.

- $SP(Q, I, R)$ denotes that $\{Q\}I\{R\}$ holds; here Q and R are logical formulas referred to as *precondition* and *postcondition*, respectively and I is a command of the programming language or the name of symbolic module.

We adopt the convention that formulas of the object theory T_O and commands of the programming language will be treated as terms of the theory T_{M1}. To this end we assume a naming function; for convenience, in our presentation we will use the symbols themselves as their names.

The system starts the program construction from the symbolic module corresponding to the program Π in order to develop systematically all symbolic modules occurring in it. Let $\Pi_1, \ldots, \Pi_n$ are symbolic modules in the partially developed program Π. Then Π is *conditionally correct* and the description of the current state of the process of program construction is reflected in the list: $[SP(\alpha_1, \Pi_1, \beta_1), \ldots, SP(\alpha_n, \Pi_n, \beta_n)]$. The verification of the correctness of the development of the symbolic module Π_i amounts to the proof of the theorem $[SP(\alpha_i, \Pi_i, \beta_i)]$.

On the basis of the theory T_{M1} and control information transferred from the meta-meta-level (control-choice(Q,I,R)), pointing out a selected symbolic module (Q), an instruction meant to implement it (I) and the current list of theorems (R), the meta-level is able to generate new list of theorems to be proved: $[THR(\gamma_1), \ldots, THR(\gamma_k), SP(\alpha'_1, \Pi'_1, \beta'_1), \ldots, SP(\alpha'_m, \Pi'_m, \beta'_m)]$ $(k > 0, m \geq 0)$.

Theorems $\gamma_1, \ldots, \gamma_k$, pointed out in the part $THR(\gamma_1), \ldots, THR(\gamma_k)$, are transferred to the object theory T_O — as goals — in order to be proved there.

Depending on the adopted method of formalizing the meta-theory T_{M1} (and of course inference procedure, e.g. programming in logic and resolution, production rules and chaining, etc.) the meta-level reasoning may be realized in various ways.

A communication between meta-meta and meta level is assured by reflection rules defined in subsection 3.4.

3.3 Methodology: Meta-Meta-Level

Formalizing a possibly general methodology of correct program construction is an important research question. Still a lot needs to be done to support the user in this process. While designing the system we are abstracting from a general solution of this problem. The underlying strategy behind our approach is to limit attention to the necessary design decisions. Assuming a stepwise development of programs, the crucial modelling decisions concern:

- Introducing a concept of symbolic module in order to implement a stepwise refinement technique.

- Assuming a certain strategy for symbolic module choice at any moment of the program development.

- Assuming a certain strategy for selecting a program construction to implement a given symbolic module.

All these modelling assumptions will be then translated into control primitives.

Although the formulation of the knowledge sufficient to take all necessary decisions is a promising line of research, we recall that the final choices are always made by the user. Assuming a very strict interaction with the programmer, we will be satisfied with a precise specification of input and output of the meta-meta-level, treating the methodology as a kind of a *black box*. In other words, the adopted methodology amounts to a *highest-level goal selection*. The user, as the one responsible for taking decisions, is obliged to point out:

- the symbolic module at hand.

- the program construction implementing the selected symbolic module.

- the new highest level goal statement.

Thus, the language of the meta-meta-level contains three predicate symbols *selected-goal*, *selected-instruction*, *new-goal-statement*.

- *selected-goal(Q)* denotes that Q is the symbolic module under consideration.

- *selected-instruction(I)* denotes that I is the program construction implementing the selected symbolic module.

- *new-goal-statement(R)* denotes that R is the new goal statement (to be considered at the meta-level).

The problem of establishing a new-goal-statement is more complicated. If the program derivation goes on without any disturbance, the resulting *current-goal-statement* may become, if the user did not change his/her mind in the meantime, the next new-goal-statement in the following reasoning cycle.

Otherwise, when the reasoning cycle fails, i.e. some facts underlying the program costruction turn out not to be true — certain theorems in the object theory T_O remain unproved, the further program derivation should be revised. In order to establish another next-goal-statement on the meta-meta-level, classical backtracking techniques (under the user's control) may be applied. In general, the question where to start the program reconstruction may be answered in many ways, reflecting not only user's taste but also various methodological approaches. So, in future meta-meta-level may be viewed as a sophisticated reasoning level using specialized knowledge, strategies, heuristics, etc.

3.4 Generic Communication Between Meta- and Meta-Meta-Level

The meta-meta-level (M2) communicates with meta-level (M1) via the *downward reflection 2* rule.

$$\textbf{(DR2)} \quad \frac{\begin{array}{c} \textit{selected-goal(Q)} \\ \textit{selected-instruction(I)} \\ \textit{new-goal-statement(R)} \end{array}}{\textit{control-choice(Q,I,R)}}$$

(DR2) transfers control information for meta reasoning.

The next rule assures communication in the opposite direction: between meta-level (M1) and meta-meta-level (M2). The *upward reflection 2* rule is defined as follows:

$$\textbf{(UR2)} \quad \frac{\textit{meta-goal-statement(M)}}{\textit{current-goal-statement(M)}}$$

(UR2) transfers current-goal statement for meta-meta-reasoning.

3.5 Generic Communication Between Meta- and Object Level

The meta-level (M1) communicates with object level (O) via the *downward reflection 1* rule:

$$\text{(DR1)} \quad \frac{meta\text{-}goal\text{-}statement}{object\text{-}goal\text{-}statement([\gamma_1, \ldots, \gamma_k])}$$
$$([THR(\gamma_1), \ldots, THR(\gamma_k), SP(\alpha_1, \Pi_1, \beta_1), \ldots, SP(\alpha_m, \Pi_m, \beta_m)])$$

(DR1) transfers *object-goal-statement*$([\gamma_1, \ldots, \gamma_k])$ for object reasoning.

The *upward reflection 1* rule assures communication in the opposite direction: between object level (O) and meta-level (M1).

$$\text{(UR1)} \quad \frac{T_O \vdash Q}{T_{M1} \vdash THR(Q)}$$

(UR1) transfers information about realized object-goal statements for meta reasoning.

3.6 Global Description of the Reasoning Pattern of the System

During an inference cycle the system tries to develop one symbolic module. This process is realized in eight separate phases, namely:

(A) Reasoning on meta-meta-level: GOAL SELECTION.

(B) Downward reflection from meta-meta-level into meta-level: (DR2).

(C) Reasoning on meta-level: ESTABLISHING a *meta-goal-statement*.

(D) Downward reflection from meta-level to object level: (DR1).

(E) Reasoning on object level: VERIFYING an *object-goal-statement*.

(F) Upward reflection from object level to meta-level: (UR1).

(G) Reasoning on meta-level: ESTABLISHING a new *meta-goal-statement*.

(H) Upward reflection from meta-level to meta-meta-level: (UR2).

The inference cycle in our architecture is constructed in such a way that the reasoning on the object level is realized as soon as possible. Another acceptable strategy is to postpone proving theorems in the object theory, treating it as the last phase of the program development: the constructed program is then considered as *totally correct* under the condition that all the object level theorems generated during the program derivation are provable. In this context our design decision is well motivated. Proving theorems in the object theory T_O implies verification of facts underlying the program construction. If these facts turn out not to be true, program derivation based on them should be blocked as soon as possible.

The choice of the strategy of program construction may be viewed in future as another parameter of the system. Obviously, different strategies imply different imference cycles in this kind of architecture.

4   Prototype System for Dijkstra's Guarded Command Programming Language

We illustrated our multi-meta-level framework by designing the prototype system for Dijkstra's *guarded command programming language* ([2]), providing a very simply and concise notation for expressing algorithms. Since there is only room for very brief presentation of the language, for details please consult [6], [4].

4.1   Dijkstra's Programming Language

4.1.1   The Notion of the Weakest Precondition

To characterize the semantics of his language, Dijkstra introduced the predicate $wp(S, R)$ (a *weakest precondition*) for representation of the weakest precondition of a command S with respect to a postcondition R. The $wp(S, R)$ characterizes the set of all initial states such that if S starts in any one of them it will terminate with R true. We can write $Q \Rightarrow wp(S, R)$ to express that Q implies the weakest precondition for S establishing R. A program S is said *totally correct* with respect to the precondition Q and postcondition R if the relation $Q \Rightarrow wp(S, R)$ holds.

For the purpose of readability, we briefly introduced Dijkstra's programming language.

4.1.2   List of commands

1. The *skip* command. Execution of this command does not change state of any variable.

    $wp(skip, A) = A.$

2. The *abort* command. This command aborts the program execution:

$$wp(abort, A) = False.$$

3. The *multiple assignment* command.

$$wp(x_1, \ldots, x_n := e_1, \ldots, e_n , A) = A[x_1, \ldots, x_n \leftarrow e_1, \ldots, e_n].$$

The variables $x_1, \ldots, x_n$ should be all distinct.

4. The *sequential composition*. This command is equivalent to the sequential execution two commands, in a given order:

$$wp(I_1; I_2, A) = wp(I_1, wp(I_2, A)).$$

5. The *selection* command. This command consists of a set of *guarded commands* and it is executed as follows. A guard which is true is selected, then the command bound to it is executed. If no guard is true, the program execution is aborted.

> **if**
> $$\begin{array}{ll} B_1 \rightarrow I_1 & [] \\ B_2 \rightarrow I_2 & [] \\ & \ldots \\ B_n \rightarrow I_n & \end{array}$$
> **fi**

When more than one guard is true, then the selection of a guarded command is nondeterministic.

6. The *iteration* command. This command also consists of guarded commands: as long as there are guards which are true, one of them is selected and the command bound to it is executed. When no guard is true the iteration terminates.

> **do**
> $$\begin{array}{ll} B_1 \rightarrow I_1 & [] \\ B_2 \rightarrow I_2 & [] \\ & \ldots \\ B_n \rightarrow I_n & \end{array}$$
> **do**

The evaluation of the predicate wp for selection and iteration from a definition may be sometimes very difficult or even impossible, because of the inconstructive character of this definition (for the iteration). To overcome this problem, Dijkstra formulated two theorems for iteration and selection.

Theorem 1 Suppose a predicate A satisfies

(1) $A \Rightarrow (B_1 \lor \ldots \lor B_n)$.
(2) $(A \land B_i) \Rightarrow wp(I_i, B)$, for $1 \leq i \leq n$.

Then (and only then)

$A \Rightarrow wp(IF, B)$. $\square$

Theorem 2 Suppose a predicate A satisfies

(1) $(A \land B_i) \Rightarrow wp(I_i, A)$, for $1 \leq i \leq n$.

Suppose further that an integer function t satisfies the following, where t1 is a fresh identifier

(2) $(A \land (B_1 \lor \ldots \lor B_n)) \Rightarrow (t > 0)$.
(3) $(A \land B_i) \Rightarrow wp((t1 := t; I_i), t < t1)$, for $1 \leq i \leq n$.

Then

$A \Rightarrow wp(DO, A \land \neg(B_1 \lor \ldots \lor B_n))$. $\square$

Because the notion of the weakest precondition for other commands can be immediately transformed into Hoare's notation, the characterization of the semantics of Dijkstra's language may be formulated in terms of preconditions and postconditions. This notation will be applied in formalizing the meta-theory T_{WP}.

4.2 The Object Theory T_Π

Domain knowledge is formalized by a declarative theory of the application domain: the object theory T_Π with derivability relation.

4.3 The Meta-Theory T_{WP}

Proof rules describing the meaning of particular Dijkstra's commands will be now represented as meta rules of the theory T_{WP} (theory of wp). The theory T_{WP} will be formalized as a *Horn clause program* ([1]).

In this theory (meta-rules for) compound statements, e.g. the *sequential composition*, the *selection command* and the *iteration command* are logically related by meta-rules to the specification of simpler constructions and to theorems in the object theory T_Π.

Three simple commands: the *multiple assignment*, *abort* and *skip* are related to the construction of theorems in the object theory T_Π.

Meta-theory T_{WP}

(1) $SP(Q, x_1, \ldots, x_n := e_1, \ldots, e_n, R) \leftarrow THR(Q \Rightarrow R^{x_1, \ldots, x_n}_{e_1, \ldots, e_n})$.

(2) $SP(Q, \mathbf{skip}, R) \leftarrow THR(Q \Rightarrow R)$.

(3) $SP(Q, \mathbf{abort}, R) \leftarrow THR(False)$.

(4) $SP(Q, I_1; I_2, R) \leftarrow SP(Q, I_1, S), SP(S, I_2, R)$.

(5) $SP(Q, \mathbf{if}\ B_1 \rightarrow I_1 [\!] \cdots [\!] B_n \rightarrow I_n\ \mathbf{fi}\ , R) \leftarrow$
$\qquad\qquad THR(B_1 \vee \cdots \vee B_n),$
$\qquad\qquad SP(Q \wedge B_1, I_1, R),$
$\qquad\qquad \ldots$
$\qquad\qquad SP(Q \wedge B_n, I_n, R).$

(6) $SP(Q, I_{INIT}; \mathbf{do}\ B_1 \rightarrow I_1 [\!] \cdots [\!] B_n \rightarrow I_n\ \mathbf{od}, R) \leftarrow$
$\qquad\qquad THR(P \wedge \neg(B_1 \vee \cdots \vee B_n) \Rightarrow R),$
$\qquad\qquad THR(P \wedge (B_1 \vee \cdots \vee B_n) \Rightarrow t > 0),$
$\qquad\qquad SP(Q, I_{INIT}, P),$
$\qquad\qquad SP(P \wedge B_1, I_1, P),$
$\qquad\qquad \ldots$
$\qquad\qquad SP(P \wedge B_n, I_n, P),$
$\qquad\qquad SP(P \wedge B_1 \wedge t1 = t, I_1, t < t1),$
$\qquad\qquad \ldots$
$\qquad\qquad SP(P \wedge B_n \wedge t1 = t, I_n, t < t1).$

4.4 Communication Between Meta-Theory and Object Theory

The meta-theory T_{WP} and the object theory T_Π constitute the two basic levels of reasoning in the system. Because the meta-theory is formalized in a logic programming language, we formulate the *downward reflection 1* rule (between meta and object level) in terms of *goal statements*.

(DR1)
$$\frac{\leftarrow THR(\gamma_1),\ldots,THR(\gamma_k),SP(\alpha_1,\Pi_1,\beta_1),\ldots,SP(\alpha_m,\Pi_m,\beta_m)}{\leftarrow \gamma_1,\ldots,\gamma_k}$$

The meta-goal-statement
$$\leftarrow THR(\gamma_1),\ldots,THR(\gamma_k),SP(\alpha_1,\Pi_1,\beta_1),\ldots,SP(\alpha_m,\Pi_m,\beta_m)$$
is transformed into object-goal-statement
$$\leftarrow \gamma_1,\ldots,\gamma_k.$$

Let us also recall the generic *upward reflection 1* rule.

(UR1)
$$\frac{T_\Pi \vdash Q}{T_{WP} \vdash THR(Q)}$$

$THR(Q)$ is a theorem of the meta-theory T_{WP}, provided that Q is a theorem of the object theory T_Π.

4.5 Methodology

The methodology adopted in the prototype is based on a program development method designed in [2], [7] and [3]. Our framework can be characterized through the following principles that underlie the process of program development:

- Using the concept of symbolic module.

- Assuming that the user is responsible for the selection of the symbolic module to be developed at any moment of the program construction.

- Assuming that the user points out a programming construct implementing the selected symbolic module (and delivers all the necessary information).

4.6 Goal Flow in the System

Goal: To derive program Π from the specification $\{\alpha\}\Pi\{\beta\}$.

The system starts from the symbolic module corresponding to the program Π in order to develop systematically all symbolic modules occurring in the program. (*Step-wise refinement*).

The initial goal-statement is of the form:
new-goal-statement($[SP(\alpha, \Pi, \beta]$).

(A) GOAL SELECTION

Let us assume that the new-goal-statement is of the form
new-goal-statement($[SP(\alpha_1, \Pi_1, \beta_1), \ldots, SP(\alpha_n, \Pi_n, \beta_n)]$) $(n \geq 0)$.

If n=0 the system finishes its work: the developed program is *totally correct*.

Otherwise, the user chooses any module Π_i from $\{\Pi_1, \ldots, \Pi_n\}$ and gives
its partial development, i.e. points out the command (with all necessary
information, e.g. invariant, bound function and a set of guarded commands
in the case of an iterative command). In effect, the meta-meta-level, reflecting the adopted methodology, delivers the following information:
selected-instruction(I)
selected-goal(Π_i)
new-goal-statement($[SP(\alpha_1, \Pi_1, \beta_1), \ldots, SP(\alpha_i, \Pi_i, \beta_i), \ldots,$
$\qquad\qquad\qquad SP(\alpha_n, \Pi_n, \beta_n)]$) $(0 < i \leq n)$.

(B) (DR2)

The system applies the *downward reflection 2* rule in order to transfer the
following control information for the meta-level reasoning:
control-choice($\Pi_i, [SP(\alpha_1, \Pi_1, \beta_1), \ldots, SP(\alpha_n, \Pi_n, \beta_n)], I$) $(0 < i \leq n)$.

(C) RESOLUTION in the META-THEORY T_{WP}

Based on the information obtained from the meta-meta-level, the resolu-
tion rule is applied to the selected literal of the form $SP(\alpha_i, \Pi_i, \beta_i)$ and a
clause of the theory T_{WP} corresponding to the selected instruction I. The
new meta-goal statement is constructed:
$\leftarrow THR(\gamma_1), \ldots, THR(\gamma_k), SP(\alpha'_1, \Pi'_1, \beta'_1), \ldots, SP(\alpha'_m, \Pi'_m, \beta'_m)$
$\qquad\qquad\qquad\qquad\qquad\qquad (k > 0, m \geq 0)$.

(D) (DR1)

The system applies the *downward reflection 1* rule. The new object-goal
statement is of the form:
$\leftarrow \gamma_1, \ldots, \gamma_k$ $(k > 0)$.

(E) THEOREM PROVING in the OBJECT THEORY T_{Π}

The system transfers control to the object theory T_{Π} in order to attempt to
prove the theorems $\gamma_1, \ldots, \gamma_k$. Let us assume that the following theorems
are proved:
$\gamma_1, \ldots, \gamma_l$ $(0 \leq l \leq k)$.

(F) (UR1)

The system applies the *upward reflection 1* rule to transfer information about proved theorems into the meta-level:

$$THR(\gamma_1), \ldots, THR(\gamma_l) \qquad\qquad (0 \leq l \leq k).$$

(G) RESOLUTION in the META-THEORY T_{WP}

The resolution rule is applied to the current meta-goal-statement and to all the literals of the form $THR(\gamma_i)$

The new meta-goal statement is constructed:

$$\leftarrow THR(\gamma_1'), \ldots, THR(\gamma_j'), SP(\alpha_1', \Pi_1', \beta_1'), \ldots, SP(\alpha_m', \Pi_m', \beta_m')$$
$$(0 \leq j \leq k, m \geq 0).$$

(H) (UR2)

The system applies the *upward reflection 2* rule. The new meta-goal-statement is transferred into the meta-meta-level as the:

$$current\text{-}goal\text{-}statement([THR(\gamma_1'), \ldots, THR(\gamma_j'),$$
$$SP(\alpha_1', \Pi_1', \beta_1'), \ldots, SP(\alpha_m', \Pi_m', \beta_m')]) \quad (0 \leq j \leq k, m \geq 0).$$

(A1) GOAL SELECTION

If j=0 (all the object theorems are proved) the current-goal-statement may become the next new-goal-statement. The resulting program is *conditionally correct*.

Otherwise, the program construction was based on a false assumption, so further derivation should be revised. This requires establishing a next new-goal-statement on the meta-meta-level. To realize this point classical backtracking techniques may be applied.

When the new-goal-statement is established, the system transfers control to point (A).

Presented inference cycle for the prototype system gives reasons for a formulation of the following theorem.

Theorem 3 Let Π be a program in Dijkstra's quarded command programming language completely developed from the specification $\{\alpha\}\Pi\{\beta\}$, during some inference cycles of the prototype system. Then the program Π is *totally correct*.
□

5 Conclusions

The idea of supporting the programer in those phases of program derivation which can be realized automatically, deserve further study. In this context, the investigation in sound development methodologies is the most urgent research subject. Possible approaches to program development should be charactedized in terms of the necessary *design decisions.* In order to take adequate choices at any moment of the program construction a highly specialized knowledge (together with reasoning capabilities) has to be precisely formulated.

Treating methodology as a reasoning level responsible for the highest-level goal selection, a classical problem of choosing the next goal statement, especially in the case of failure, arises. From technical viewpoint, the well known techniques of backtracking can be applied here. But let us stress the need of taking decisions at this point as well.

After posing all these methodological questions, answering them in a possible flexible way and planning an adequate interaction with the programmer, methodology can be treated as a full rights reasoning level. Distinguishing methodology as the meta-meta-level in multi-level compositional architecture for correct program derivation suggests a transparent description of this, rather complex, process. The approach reported in this paper illustrates the need of having two meta-levels in order to resolve more complicated problems.

Another line of research is increasing of the expressive power of the system. The methodology may be extended by adopting different program construction techniques, not only restricted to stepwise-refinement. A very natural extention seems to be a combination of top-down program development (using the concept of symbolic module) and/or composing program from already existing components (modules).

The ideas underlying the general-purpose system for program derivation are based on our experiences with CAProDel — a system supporting the user in creating programs in Dijkstra's quarded command programming language. This system was implemented at Institute of Informatics of Warsaw University (for details please consult [5]). A multi meta-levels generic system meant for carrying out experiments in program development is currently under implementation. In this case the methodology for *complex reasoning task* and the modelling language for *compositional architecture* — DESIRE — is applied.

6   Acknowledgements

I would like to thank Witold Lukaszewicz for the fruitful discussion motivating this work.

References

[1] K. Apt. Logic programming. In J. van Leeuven, editor, *Handbook of Theoretical Computer Science*. Elsevier Science Publishers, 1990.

[2] E. W. Dijkstra. *A Discipline of Programming*. Prentice Hall, Englewood Cliffs, 1976.

[3] G. Dromey. *Program Derivation. The Development of Program from Specifications*. Addison Wesley, Reading, Mass., 1989.

[4] B. Dunin-Kęplicz. Formal reconstruction of correct programs development process. Technical report, Institute of Informatics, Warsaw University, 1994.

[5] B. Dunin-Kęplicz, J. Jabłonowski, W. Łukaszewicz, and E. Madalińska-Bugaj. CAProDel: A system for computer aided program development. To appear in *Proceeding of the Sixth International Conference on Software Engineering and Knowledge Engineering*, SEKE'94, Jurmala, Latvia, 1994.

[6] B. Dunin-Kęplicz, J. Jabłonowski, W. Łukaszewicz, and E. Madalińska-Bugaj. Developing programs from specifications: Design of a system. In *Proceedings of the Third International Conference on Information Systems Developers Workbench*, pages 145–168, Gdańsk, 1992.

[7] D. Gries. *The Science of Programming*. Springer, Berlin, 1981.

[8] A. Langevelde, A. Philipsen, and J. Treur. Formal specification of compositional architecture. In *Proccedings of ECAI 92*, pages 272–276, Vienna, 1992.

[9] W. M. Turski. *Computer Programming Methodology*. Heyden, London, 1978.

More on Unfold/Fold Transformations of Normal Programs: Preservation of Fitting's Semantics. *

Annalisa Bossi[1] and Sandro Etalle[1,2]

[1] Dipartimento di Matematica Pura ed Applicata
Università di Padova
Via Belzoni 7, 35131 Padova, Italy

[2] CWI
P.O. Box 4079, 1009 AB Amsterdam, The Netherlands

email: bossi@zenone.math.unipd.it, etalle@cwi.nl

Abstract. The unfold/fold transformation system defined by Tamaki and Sato was meant for definite programs. It transforms a program into an equivalent one in the sense of both the least Herbrand model semantics and the Computed Answer Substitution semantics. Seki extended the method to normal programs and specialized it in order to preserve also the finite failure set. The resulting system is correct wrt nearly all the declarative semantics for normal programs. An exception is Fitting's model semantics. In this paper we consider a slight variation of Seki's method and we study its correctness wrt Fitting's semantics. We define an applicability condition for the fold operation and we show that it ensures the preservation of the considered semantics through the transformation.

1 Introduction

The unfold/fold transformation rules were introduced by Burstall and Darlington [8] for transforming clear, simple functional programs into equivalent, more efficient ones. The rules were early adapted to the field of logic programs both for program synthesis [10, 13] and for program specialization and optimization [2, 16]. Soon later, Tamaki and Sato [26] proposed an elegant framework for the transformation of logic programs based on unfold/fold rules.

The major requirement of a transformation system is its correctness: it should transform a program into an equivalent one. Tamaki and Sato's system was originally designed for definite programs and in this context a natural equivalence on programs is the one induced by the least Herbrand model semantics. In [26] it was shown that the system preserves such a semantics. Afterward, the system was proven to be correct wrt many other semantics: the computed answer substitution semantics [14], the Perfect model semantics [24], the Well-Founded semantics [25] and the Stable model semantics [23, 4].

In [24], Seki modified the method by restricting its applicability conditions. The system so defined enjoys all the semantic properties of Tamaki-Sato's, moreover,

---

* This work has been partially supported by "Progetto Finalizzato Sistemi Informatici e Calcolo Parallelo" of CNR under grant n. 89.00026.69

it preserves the finite failure set of the original program [22] and it is correct wrt Kunen's semantics [21].

However, neither Tamaki-Sato's, nor Seki's system preserve the Fitting model semantics.

In this paper we consider a transformation schema which is similar yet slightly more restrictive to the one introduced by Seki [24] for normal programs. We study the effect of the transformation on the Fitting's semantics [11] and we individuate a sufficient condition for its preservation.

The difference between the method we propose and the one of Seki consists in the fact that here the operations have to be performed in a precise order. We believe that this order corresponds to the "natural" order in which the operations are usually carried out within a transformation sequence, and therefore that the restriction we impose is actually rather mild.

The structure of the paper is the following. In Section 2 we recall the definition of Fitting's operator. In Section 3 the transformation schema is defined and exemplified, and the applicability conditions for the fold operation are presented and discussed. Finally, in Section 4, we prove the correctness of the unfold/fold transformation wrt Fitting's semantics.

2 Preliminaries

We assume that the reader is familiar with the basic concepts of logic programming; throughout the paper we use the standard terminology of [17] and [1]. We consider *normal programs*, that is finite collections of *normal rules*, $A \leftarrow L_1, \ldots, L_m$. where A is an atom and $L_1, \ldots, L_m$ are literals. B_P denotes the Herbrand base and *Ground(P)* the set of ground instances of clauses of a program P. We say that a clause is *definite* if the body contains only positive literals (atoms); a definite program is then a program consisting only of definite clauses. Symbols with a $\sim$ on top denote tuples of objects, for instance $\tilde{x}$ denotes a tuple of variables $x_1, \ldots, x_n$, and $\tilde{x} = \tilde{y}$ stands for $x_1 = y_1 \wedge \ldots \wedge x_n = y_n$. We also adopt the usual logic programming notation that uses "," instead of $\wedge$, hence a conjunction of literals $L_1 \wedge \ldots \wedge L_n$ will be denoted by $L_1, \ldots, L_n$ or by $\tilde{L}$.

Three valued semantics for normal programs. In this paper we refer to the usual Clark's completion definition, $Comp(P)$, [9] which consists of the completed definition of each predicate together with CET, Clark's Equality Theory, which is needed in order to interpret "=" correctly. It is well-known that, when considering normal programs, the two valued completion $Comp(P)$ of a program P might be inconsistent an consequently have no model; moreover, when $Comp(P)$ is consistent, it usually has more models, none of which can be considered the *least* (hence the preferred) one. Following [11], we avoid this problem by switching to a three-valued logic, where the truth tables of the connective are the ones given by Kleene [15]. When working with 3-valued logic, the same definition of completion applies, with the only difference that the connective $\leftrightarrow$ is replaced with $\Leftrightarrow$, Lucasiewicz's operator of "having the same truth value". In this context, we have that a *three valued* *(or partial) interpretation,* is a mapping from the ground atoms of the language $\mathcal{L}$ into the set $\{true, false, undefined\}$.

Definition 1. Let $\mathcal{L}$ be a language. *A three valued (or partial) $\mathcal{L}$-interpretation, I,* is a mapping from the ground atoms of $\mathcal{L}$ into the set $\{true, false, undefined\}$. □

A partial interpretation I is represented by an ordered couple, (T, F), of disjoint sets of ground atoms. The atoms in T (resp. F) are considered to be *true* (resp. *false*) in I. T is the positive part of I and is denoted by I^+; equivalently F is denoted by I^-. Atoms which do not appear in either set are considered to be *undefined*.

If I and J are two partial $\mathcal{L}$-interpretations, then $I \cap J$ is the three valued $\mathcal{L}$-interpretation given by $(I^+ \cap J^+, I^- \cap J^-)$, $I \cup J$ is the three valued $\mathcal{L}$-interpretation given by $(I^+ \cup J^+, I^- \cup J^-)$ and we say that $I \subseteq J$ iff $I = I \cap J$, that is, iff $I^+ \subseteq J^+$ and $I^- \subseteq J^-$. The set of all $\mathcal{L}$-interpretations is then a complete lattice. In the sequel we refer to a fixed but unspecified language $\mathcal{L}$ that we assume contains all the functions symbols and the predicate symbols of the programs that we consider, consequently we will omit the $\mathcal{L}$ prefix and speak of "interpretations" rather than of "$\mathcal{L}$-interpretations".

We now give a definition of Fitting's operator [11]. We denote by $Var(E)$ the set of all the variables in an expression E and we write $\exists y\, B\theta$ as a shorthand for $(\exists y\, B)\theta$, that is, unless explicitly stated, the quantification always applies before the substitution.

Definition 2. Let P be a normal program, and I a three valued interpretation. $\Phi_P(I)$ is the three valued interpretation defined as follows:

- A ground atom A is *true* in $\Phi_P(I)$
  iff there exists a clause $c :\ B \leftarrow \tilde{L}.$ in P whose head unifies with A, $\theta = mgu(A, B)$, and $\exists \tilde{w}\, \tilde\theta$ is *true* in I, where $\tilde{w}$ is the set of local variables of c, $\tilde{w} = Var(\tilde{L})\backslash Var(B)$.

- A ground atom A is *false* in $\Phi_P(I)$
  iff for all clauses $c :\ B \leftarrow \tilde{L}$ in P for which there exists $\theta = mgu(A, B)$ we have that $\exists \tilde{w}\, \tilde{L}\theta$ is *false* in I, where $\tilde{w}$ is the set of local variables of c, $\tilde{w} = Var(\tilde{L})\backslash Var(B)$. □

Recall that a *Herbrand* model is a model whose universe is given by the set of $\mathcal{L}$-terms.

Φ_P is a monotonic operator, that is $I \subseteq J$ implies $\Phi_P(I) \subseteq \Phi_P(J)$, and characterizes the three valued semantics of $Comp(P)$, in fact Fitting, in [11] shows that the three-valued Herbrand models of $Comp(P)$ are exactly the fixpoints of Φ_P; it follows that any program has a *least* (wrt. $\subseteq$) three-valued Herbrand model, which coincides with the least fixpoint of Φ_P. This model is usually referred to as Fitting's model.

Definition 3. Let P be a program, *Fitting's model* of P, $Fit(P)$, is the least three valued Herbrand model of $Comp(P)$. □

We adopt the standard notation: $\Phi_P^{\uparrow 0}$ is the interpretation that maps every ground atom into the value *undefined*, $\Phi_P^{\uparrow \alpha+1} = \Phi_P(\Phi_P^{\uparrow \alpha})$, $\Phi_P^{\uparrow \alpha} = \cup_{\delta < \alpha} \Phi_P^{\uparrow \delta}$, when α is a limit ordinal. From the monotonicity of Φ_P follows that its Kleene's sequence is monotonically increasing and it converges to its least fixpoint. Hence there always exists an ordinal α such that $lfp(\Phi_P) = \Phi_P^{\uparrow \alpha}$. Since Φ_P is monotone but not continuous, α could be greater than ω.

Theorem 4 [11]. *Let P be a program, then, for some ordinal α,*

$$- Fit(P) = \Phi_P^{\uparrow \alpha} \qquad \qquad \Box$$

3 Unfold/fold transformations

3.1 Introduction

Unfold and fold are basic transformation rules but their definition may differ depending on the considered semantics.

Unfolding is the fundamental operation for partial evaluation [18] and consists in applying a resolution step to the considered atom in all possible ways. Usually, it is applied only to positive literals (an exception is [3]).

Folding is the inverse of unfolding when one single unfolding is possible. Syntactically, it consists in substituting a literal L for an equivalent conjunction of literals $\tilde{K}$ in the body of a clause c. This operation is used to simplify unfolded clauses and to detect implicit recursive definitions. In order to preserve the declarative semantics of logic programs, its application must be restricted by some, semantic dependent, conditions. Therefore, the various proposals mainly differ in the choice of such conditions. They can be either a constraint on how to sequentialize the operations while transforming the program [26, 24], or they can be expressed only in terms of (semantic) properties of the program, independently from its transformation history [5, 19]. For normal programs different definitions for folding in a particular transformation sequence are given in [24, 23, 12].

3.2 A four step transformation schema

In this section we introduce the unfold/fold transformation *schema*. All definitions are given modulo reordering of the bodies of the clauses and standardization apart is always assumed.

First we define the unfolding operation, which is basic to all the transformation systems.

Definition 5 Unfold. Let $cl : A \leftarrow \tilde{L}, H.$ be a clause of a normal program P, where H is an atom. Let $\{H_1 \leftarrow \tilde{B}_1, \ldots, H_n \leftarrow \tilde{B}_n\}$ be the set of clauses of P whose heads unify with H, by mgu's $\{\theta_1, \ldots, \theta_n\}$.

- *unfolding an atom H in cl* consists of substituting cl with $\{cl_1', \ldots, cl_n'\}$, where, for each i, $cl_i' = (A \leftarrow \tilde{L}, \tilde{B}_i)\theta_i$.

$unfold\ (P, cl, H) \stackrel{\text{def}}{=} P \backslash \{cl\} \cup \{cl_1', \ldots, cl_n'\}.$ $\qquad \qquad \Box$

Let P be a normal program. A *four step transformation schema* starting in the program P consists of the following steps:

Step 1. Introduction of new definitions. We add to the program P the set of clauses $D_{def} = \{c_i : H_i \leftarrow \tilde{B}_i\}$, where the predicate symbol of each H_i is *new*, that is, it does not occur in P. On the other hand, we require that the predicate symbols found in each $\tilde{B}_i$ are defined in P, and therefore are not *new*. The result of this operation is then

$- P_1 = P \cup D_{def}$ □

Example 1 (min-max, part 1). Let P be the following program

$$P = \{ \quad min([X], X).$$
$$min([X|Xs], Y) \leftarrow min(Xs, Z), inf(X, Z, Y).$$
$$max([X], X).$$
$$max([X|Xs], Y) \leftarrow max(Xs, Z), sup(X, Z, Y).$$
$$inf(X, Y, X) \quad \leftarrow X \leq Y.$$
$$inf(X, Y, Y) \quad \leftarrow \neg(X \leq Y).$$
$$sup(X, Y, Y) \quad \leftarrow X \leq Y.$$
$$sup(X, Y, X) \quad \leftarrow \neg(X \leq Y).$$
$$c_1 : med(Xs, Med) \quad \leftarrow min(Xs, Min),$$
$$max(Xs, Max),$$
$$Med \text{ is } (Min + Max)/2. \quad \}$$

here $med(Xs, Med)$ reports in Med the average between the minimum and the maximum of the values in the list Xs.

We may notice that the definition of $med(Xs, Med)$ traverses the list Xs twice. This is obviously a source of inefficiency. In order to fix this problem via an unfold/fold transformation, we first have to introduce a new predicate *minmax*. Let us then add to program P the following new definition:

$D_{def} = \{c_2 : minmax(Xs, Min, Max) \leftarrow min(Xs, Min), max(Xs, Max). \}$ □

Step 2. Unfolding in D_{def}. We transform D_{def} into D_{unf} by unfolding some of its clauses. The clauses of P are therefore used as unfolding clauses. This process can be iterated several times and usually ends when all the clauses that we want to fold have been obtained; the result of this operation is

$- P_2 = P \cup D_{unf}$ □

Example 1 (min-max, part 2). We can now unfold the atom $min(Xs, Min)$ in the body of c_2, the result is

$$c_3 : minmax([X], X, Max) \quad \leftarrow max([X], Max).$$
$$c_4 : minmax([X|Xs], Min, Max) \leftarrow min(Xs, Y),$$
$$inf(X, Y, Min),$$
$$max([X|Xs], Max).$$

In the bodies of both clauses we can then unfold predicate *max*. Each clause generates two clauses.

$c_5 : minmax([X], X, X).$
$c_6 : minmax([X], X, Max) \qquad \leftarrow max([\,], Z), sup(Z, X, Max).$

$c_7 : minmax([X], Min, X) \qquad \leftarrow min([\,], Y), inf(X, Y, Min).$
$c_8 : minmax([X|Xs], Min, Max) \leftarrow min(Xs, Y),$
$$inf(X, Y, Min),$$
$$max(Xs, Z),$$
$$sup(X, Z, Max).$$

Clauses c_6 and c_7 can then be eliminated by unfolding respectively the atoms $max([\,], Z)$ and $min([\,], Y)$. D_{unf} consists then of the following clauses.

$c_5 : minmax([X], X, X).$
$c_8 : minmax([X|Xs], Min, Max) \leftarrow min(Xs, Y),$
$$inf(X, Y, Min),$$
$$max(Xs, Z),$$
$$sup(X, Z, Max).$$

Still, *minmax* traverses the list Xs twice; but now we can apply a *recursive folding* operation. □

Step 3. Recursive folding. Let $c_i : H_i \leftarrow \tilde{B}_i$ be one of the clauses of D_{def}, which was introduced in *Step 1*, and $cl : A \leftarrow \tilde{B}', \tilde{S}.$ be (a renaming of) a clause in D_{unf}. If there exists a substitution θ, $Dom(\theta) = Var(c_i)$ such that

(a) $\tilde{B}' = \tilde{B}_i\theta;$
(b) θ does not bind the local variables of c_i, that is for any $x, y \in Var(\tilde{B}_i)\backslash Var(\tilde{H}_i)$ the following three conditions hold
  − $x\theta$ is a variable;
  − $x\theta$ does not appear in $A, \tilde{S}, H_i\theta;$
  − if $x \neq y$ then $x\theta \neq y\theta;$
(c) c_i is the only clause of D_{def} whose head unifies with $H_i\theta;$
(d) all the literals of $\tilde{B}'$ are the result of a previous unfolding.

then we can fold $H_i\theta$ in cl, obtaining $cl' : A \leftarrow H_i\theta, \tilde{S}$. This operation can be performed on several conjunctions simultaneously, even on the same clause. The result is that D_{unf} is transformed into D_{fold} and hence

− $P_3 = P \cup D_{fold}$ □

Example 1 (min-max, part 3). We can now fold $min(Xs, Y), max(Xs, Z)$ in the body of c_8. The resulting program D_{fold} consists of the following clauses

$c_5 : minmax([X], X, X).$
$c_9 : minmax([X|Xs], Min, Max) \leftarrow minmax(Xs, Y, Z),$
$$inf(X, Y, Min),$$
$$sup(X, Z, Max).$$

$minmax(Xs, Min, Max)$ has now a recursive definition and needs to traverse the list Xs only once. In order to let the definition of *med* enjoy of this improvement, we need to *propagate* predicate *minmax* inside its body. □

Step 4. Propagation folding. Technically, the difference between this step and the previous one is that now the folded clause comes form the original program P. This allows us to drop condition (d) of the folding operation.

Let c_i : $H_i \leftarrow \tilde{B}_i$ be one of the clauses of D_{def}, which was introduced in *Step 1*, and cl : $A \leftarrow \tilde{B}', \tilde{S}$. be (a renaming of) a clause in the original program P. If there exists a substitution θ, $Dom(\theta) = Var(c_i)$ such that the conditions (a), (b) and (c) defined above are satisfied, then we can fold $H_i\theta$ in cl, obtaining cl' : $A \leftarrow H_i\theta, \tilde{S}$. Also this operation can be performed on several conjunctions simultaneously, even on the same clause. The result is that P is transformed into P_{fold} and therefore

$$- \ P_4 = P_{\mathrm{fold}} \cup D_{\mathrm{fold}} \qquad\qquad\qquad \Box$$

Example 1 (min-max, part 4). We can now fold $min(Xs, Y), max(Xs, Z)$ in the body of c_1, in the original program P. The resulting program is

$$P_{\mathrm{fold}} = P\backslash\{c_1\} \cup \{c_{10} : med(Xs) \ \leftarrow \ minmax(Xs, Min, Max),$$
$$Med \ is \ (Min + Max)/2. \quad \}$$

And then the final program is $P_4 = P_{\mathrm{fold}} \cup D_{\mathrm{fold}} =$

$= \{$ c_5 : $minmax([X], X, X)$.
   c_9 : $minmax([X|Xs], Min, Max) \leftarrow minmax(Xs, Y, Z),$
   $\qquad\qquad\qquad\qquad\qquad\qquad inf(X, Y, Min),$
   $\qquad\qquad\qquad\qquad\qquad\qquad sup(X, Z, Max)$.

   c_{10} : $med(Xs)$ $\qquad\qquad \leftarrow minmax(Xs, Min, Max),$
   $\qquad\qquad\qquad\qquad\qquad\qquad Med \ is \ (Min + Max)/2$.

   $+$ definitions for predicates min, max, inf and sup.$\}$

Notice also that predicates min and max are no longer used by the program. $\Box$

3.3 Semantic considerations

The *schema* (that is, the method we propose) is similar but more restrictive than the *transformation sequence* with *modified* folding[3] proposed by Seki [24]. The (only) limitation consists in the fact that the schema requires the operations to be performed in fixed order: for instance it does not allow a *propagation folding* to take place before a *recursive folding*. We believe that in practice this is not a bothering restriction, as it corresponds to the "natural" procedure that is followed in the process of transforming a program. In fact, in all the papers we cite, all the examples that can be reduced to a transformation sequence as in [24], can also be reduced to the given transformation schema.

Since the *schema* can be seen as a particular case of the transformation *sequence*, it enjoys all its properties, among them, it preserves the following semantics of the initial program: the success set [26], the computed answer substitution set [14], the

---

[3] here we are adopting Seki's notation, and we call *modified* folding the one presented in [22, 24], which preserves the finite failure set, as opposed to the one introduced by Tamaki and Sato in [26], which does not.

finite failure set [24], the Perfect model semantics for stratified programs [24], the Well-Founded semantics [25], the Stable model semantics [23, 4].

However, as it is, the schema suffers of the same problems of the sequence, i.e., Fitting's Models is not preserved. This is shown by the following example.

Example 2. Let $P_1 = P \cup D_{\text{def}}$, where P and D_{def} are the following programs

$$D_{\text{def}} = \{\, p \qquad \leftarrow q(X). \qquad \}$$
$$P \quad = \{\, q(s(X)) \leftarrow q(X), t(0).$$
$$\qquad t(0). \qquad \qquad \}$$

As we fix a language $\mathcal{L}$ that contains the constant 0 and the function $s/1$, we have that $\exists X\, q(X)$ is *false* in $Fit(P_1)$, consequently, p is also *false* in $Fit(P_1)$. Now let us unfold $q(X)$ in the body of the clause in D_{def}; the resulting program is the following. $P_2 = P \cup D_{\text{unf}}$, where

$$D_{\text{unf}} = \{\, p \qquad \leftarrow q(Y), t(0). \,\}$$
$$P \quad = \{\, q(s(X)) \leftarrow q(X), t(0).$$
$$\qquad t(0). \qquad \qquad \}$$

We can now fold $q(Y)$ in the body of the clause of D_{unf}, the resulting program is $P_3 = P \cup D_{\text{fold}}$, where

$$D_{\text{fold}} = \{\, p \qquad \leftarrow p, t(0). \qquad \}$$
$$P \quad = \{\, q(s(X)) \leftarrow q(X), t(0).$$
$$\qquad t(0). \qquad \qquad \}$$

Now we have that p is *undefined* in the Fitting model of P_3. □

So, in order for the transformation to preserve Fitting's model of the original program, we need some further applicability conditions. Therefore the following.

Theorem 6 Correctness. *Let* $P_1, \ldots, P_4$ *be a sequence of programs obtained applying the transformation schema to program* P. *Let also* $D_{\text{def}} = \{H_i \leftarrow \tilde{B}_i\}$ *be the set of clauses introduced in* Step 1, *and, for each* i, $\tilde{w}_i$ *be the set of local variables of* c_i: $\tilde{w}_i = Var(\tilde{B}_i) \backslash Var(H_i)$. *If each* c_i *in* D_{def} *satisfies the following condition:*

A *each time that* $\exists \tilde{w}_i\, \tilde{B}_i\theta$ *is false in some* $\Phi_{P_1}^{\uparrow\beta}$, *then there exists a non-limit ordinal* $\alpha \leq \beta$ *such that* $\exists \tilde{w}_i\, \tilde{B}_i\theta$ *is false in* $\Phi_{P_1}^{\uparrow\alpha}$

Then $Fit(P_1) = Fit(P_2) = Fit(P_3) = Fit(P_4)$.

Proof. The proof is given in the subsequent Section 4. □

On condition A. Condition A is in general undecidable, it is therefore important to provide some other decidable sufficient conditions. For this, in the rest of this Section, we adopt the following notation:

- $D_{\text{def}} = \{c_i : H_i \leftarrow \tilde{B}_i\}$ is the set of clauses introduced in *Step 1*, and, for each i,

- $\tilde{w}_i = Var(\tilde{B}_i)\backslash Var(H_i)$ is the set of local variables of c_i.

First, it is easy to check that if c_i has no local variables, then it satisfies **A**.

Proposition 7. *If $\tilde{w}_i = \emptyset$ then c_i satisfies **A**.*

Proof. It follows at once from the definition of Fitting's operator. □

This condition, though simple, is met by most of the examples found in the literature; if we are allowed an informal "statistics", of all the papers cited in our bibliography, seven contain practical examples in clausal form which can be assimilated to our method ([6, 14, 20, 22, 24, 25, 26]), and of them, only two contain examples where the "introduced" clause contains local variables ([14, 20]). Our Example 1 satisfies the condition as well.

Nevertheless Proposition 7 can easily be improved. First let us consider the following Example[4].

Example 3. Let $P_1 = P \cup D_{def}$, where P and D_{def} are the following programs

$$D_{def} = \{\ c_0 : br(X, Y) \quad \leftarrow\ reach(X, Z), reach(Y, Z).\ \}$$
$$P \quad = \{\qquad reach(X, Y) \leftarrow arc(X, Y).$$
$$reach(X, Y) \leftarrow arc(X, Z), reach(Z, Y).\qquad \} \cup DB$$

Where DB is any set of ground unit clauses defining predicate arc. $reach(X, Y)$ holds iff there exists a path starting from node X and ending in node Y, while $br(X, Y)$ holds iff there exists a node Z which is reachable both from node X and node Y. □

In this Example the definition of predicate br can be specialized and made recursive via an unfold/fold transformation. Despite the fact that clause c_0 contains the local variable Z, it is easy to see that **A** is satisfied. This is due to the fact that P is actually a DATALOG (function-free) program.

We now show that if (a part of) the original program P is function-free (or recursion-free) then **A** is always satisfied.

Let us first introduce the following notation. Let p, q be predicates, we say that p *refers to* q in program P if there is a clause of P with p in its head and q in its body. The *depends on* relation is the reflexive and transitive closure of *refers to*. Let $\tilde{L}$ be a conjunction of literals, by $P|_{\tilde{L}}$ we denote the set of clauses of P that define the predicates which the predicates in $\tilde{L}$ depend on. We say that a program is *recursion-free* if there is no chain $p_1, \ldots, p_k$ of predicate symbols such that p_i refers to p_{i+1} and $p_k = p_1$. With an abuse of notation, we also call a program *function-free* if the only terms occurring in it are either ground or variables.

We can now state the following.

Proposition 8. *For each index i, and each $w \in \tilde{w}_i$, let us denote by $\tilde{L}_w$ the subset of $\tilde{B}_i$ formed by those literals where w occurs. If for every $\tilde{L}_w$, one of the following two conditions holds:*

---

[4] The example is actually a modification of Example 2.1.1 in [22]

(a) $P_1|_{\tilde{L}_w}$ *is recursion-free, or*
(b) $P_1|_{\tilde{L}_w}$ *is function-free;*

then each c_i satisfies **A**.

Proof. First we need the following Observation.

Observation 9 Let Q be a function-free or a recursion-free program, then for some integer k, $Fit(Q) = \Phi_Q^{\uparrow k}$

Proof. Straightforward □

Now fix an index i, and let $\tilde{w}_i = w_1, \ldots, w_m$, and let $\tilde{M}$ be the subset of $\tilde{B}_i$ consisting of those literals that do not contain any of the variables in $\tilde{w}_i$. It is immediate that, for any ordinal α, and for any substitution θ

$$\Phi_{P_1}^{\uparrow \alpha} \models \exists \tilde{w}_i \, \tilde{B}_i \theta \quad \text{iff} \quad \Phi_{P_1}^{\uparrow \alpha} \models \exists w_1 \, \tilde{L}_{w_1} \theta \wedge \ldots \wedge \exists w_m \, \tilde{L}_{w_m} \theta \wedge \tilde{M} \theta \quad (1)$$

Now suppose that, for some ordinal α, and substitution θ, $\exists \tilde{w}_i \, \tilde{B}_i \theta$ is *false* in $\Phi_{P_1}^{\uparrow \alpha}$. By (1), either *(i)* $\tilde{M}\theta$ is *false* in $\Phi_{P_1}^{\uparrow \alpha}$, or *(ii)* there exists an i such that $\exists w_i \, \tilde{L}_{w_i}\theta$ is *false* in $\Phi_{P_1}^{\uparrow \alpha}$; we treat the two cases separately.

(i), $\tilde{M}\theta$ is *false* in $\Phi_{P_1}^{\uparrow \alpha}$, then, by the definition of Φ_{P_1}, there exists a non-limit ordinal $\beta \leq \alpha$ such that $\tilde{M}\theta$ is *false* in $\Phi_{P_1}^{\uparrow \beta}$, and, by (1), $\exists \tilde{w}_i \, \tilde{B}_i \theta$ is *false* in $\Phi_{P_1}^{\uparrow \beta}$.

(ii), $\exists w_i \, \tilde{L}_{w_i}\theta$ is *false* in $\Phi_{P_1}^{\uparrow \alpha}$, since $P_1|_{\tilde{L}_{w_i}}$ is function or recursion-free, by Observation 9 there exists an integer k such that $\exists w_i \, \tilde{L}_{w_i}\theta$ is *false* in $\Phi_{P_1}^{\uparrow k}$; again, by (1), $\exists \tilde{w}_i \, \tilde{B}_i \theta$ is *false* in $\Phi_{P_1}^{\uparrow k}$.

So, in any case, there exists a non-limit ordinal $\beta \leq \alpha$ such that $\exists \tilde{w}_i \, \tilde{B}_i \theta$ is *false* in $\Phi_{P_1}^{\uparrow \beta}$. Since this holds for any index i, the thesis follows. □

Checking A "a posteriori". We now show that condition **A** holds in P_0 iff it holds in any program of the unfold part of the transformation sequence. This gives us the opportunity of providing further sufficient conditions.

First let us restate **A** as follows:

A': For each substitution θ and non-limit ordinal β, if $H_i\theta$ is *false* in $\Phi_{P_1}^{\uparrow \beta+1}$, then $H_i\theta$ is *false* in $\Phi_{P_1}^{\uparrow \beta}$ as well.

Now, let P_1' be a program which is obtained from P_1 by applying some unfolding transformation. It is easy to see[5] that H_i satisfies **A'** in P_1 iff H_i satisfies **A'** in P_1'. So the advantage of **A'** over **A** is that it can be checked a *posteriori* at any time during the unfolding part of the transformation. So Proposition 8 can be restated as follows.

Proposition 10. *Let P_1' be a program obtained from P_1 by (repeatedly) applying the unfolding operation. Let D_{def}' be the subset of P' corresponding to D_{def} in P. If for each clause c of D_{def}', and for every variable y, local to the body of c*

---

[5] This is a direct consequence of Lemma 11, which is given in the next Section

- $P_1'|_{\tilde{L}_y}$ is recursion-free or function-free,

  where $\tilde{L}_y$ denotes the subset of the body of c consisting of those literals where y occurs;

then each c_i satisfies **A** in P_1.

Proof. It is a straightforward generalization of the proof of Proposition 8. $\qquad\square$

4 Correctness of the transformation

The aim of this section is to prove the correctness of the transformation schema wrt Fitting's semantics, Theorem 6.

4.1 Correctness of the unfold operation

First we consider the unfold operation. To prove its correctness we need the following technical Lemma.

Lemma 11. *Let P' be the program obtained by unfolding an atom in a clause of program P. Then for each integer i and limit ordinal β,*

- $\Phi_P^{\uparrow i} \subseteq \Phi_{P'}^{\uparrow i}$ and $\Phi_{P'}^{\uparrow i} \subseteq \Phi_P^{\uparrow 2i}$;
- $\Phi_P^{\uparrow i}(\Phi_P^{\uparrow \beta}) \subseteq \Phi_{P'}^{\uparrow i}(\Phi_{P'}^{\uparrow \beta})$ and $\Phi_{P'}^{\uparrow i}(\Phi_{P'}^{\uparrow \beta}) \subseteq \Phi_P^{\uparrow 2i}(\Phi_P^{\uparrow \beta})$.

Proof. The proof is given in [7]. $\qquad\square$

This brings us to a preliminary conclusion.

Corollary 12 Correctness of the unfold operation. *Let P' be the result of unfolding an atom of a clause in P. Then*

- $Fit(P) = Fit(P')$ $\qquad\square$

It should be mentioned that, because of the particular structure of the transformation sequence, here we never use self-unfoldings (that is, unfoldings in which the same clause is both the unfolded clause and one of the unfolding ones). Consequently the correctness of *Step 2* follows also from a result of Gardner and Shepherdson [12, Theorem 4.1] which states that if the program P' is obtained from P by unfolding (but not self-unfolding), then $Comp(P)$ and $Comp(P')$ are logically equivalent theories[6].

The following is a second, technical result on the consequences of an unfolding operation which will be needed in the sequel.

Lemma 13. *Let P be a normal program, $cl : A \leftarrow \tilde{K}$. be a definite, clause of P. Suppose also that cl is the only clause of P whose head unifies with $A\theta$. If P' is the program obtained by unfolding at least once all the atoms in $\tilde{K}$, then, for each non-limit ordinal α*

---

[6] In [12] this result is stated for the usual two-valued program's completion. By looking at the proof it is straightforward to check that it holds also for the three-valued case

– if $A\theta$ is true (resp. false) in $\Phi_P^{\uparrow\alpha+1}$ then $A\theta$ is true (resp. false) in $\Phi_{P'}^{\uparrow\alpha}$

Proof. Let us first give a simplified proof by considering the case when $\tilde{K}$ consists of two atoms H, J and we perform a single unfolding on them; we will later consider the general case.

Let $\{H_1 \leftarrow \tilde{B}_1., \ldots, H_n \leftarrow \tilde{B}_n.\}$ be the set of clauses of P whose head unify with H via mgu's $\phi_1, \ldots, \phi_n$, and let $\{J_1 \leftarrow \tilde{C}_1., \ldots, J_m \leftarrow \tilde{C}_m\}$ be the set of clauses of P whose head unify with J. Unfolding H in cl and then J in the resulting clauses, will lead to the following program:

$P' = P\backslash\{cl\} \cup \{d_{i,j} : (A \leftarrow \tilde{B}_i, \tilde{C}_j.)\theta_{i,j})\}$

Where $\theta_{i,j} = mgu(J\phi_i, J_j)$. Here some of the clauses $d_{i,j}$ may be missing due to the fact that $J\phi_i$ and J_j may not unify, but this is of no relevance in the proof.

Note that the clauses $d_{i,j}$ are the only clauses of P' whose head could possibly unify with A.

Let $\tilde{y} = Var(H, J)\backslash Var(A)$ be the set of variables local to the body. We have to consider two cases.

a) $A\theta$ is *true* in $\Phi_P^{\uparrow\alpha+1}$. By the definition of Φ_P, $(\exists\tilde{y}\, H, J)\theta$ is *true* in $\Phi_P^{\uparrow\alpha}$. There has to be an extension σ of θ, $Dom(\sigma) = Dom(\theta) \cup \tilde{y} = Var(A, H, J)$ such that $(H, J)\sigma$ is *true* in $\Phi_P^{\uparrow\alpha}$. Let $H_i \leftarrow \tilde{B}_i$ and $J_j \leftarrow \tilde{C}_j$ be the clauses used to prove, respectively, $H\sigma$ and $J\sigma$. Hence there exists a τ such that $\theta_{i,j}\tau|_{Dom(\sigma)} = \sigma$, $H\sigma = H_i\theta_{i,j}\tau$, $J\sigma = J_j\theta_{i,j}\tau$, and $(\tilde{B}_i, \tilde{C}_j)\theta_{i,j}\tau$ is *true* in $\Phi_P^{\uparrow\alpha-1}$. By Lemma 11, $\Phi_P^{\uparrow\alpha-1} \subseteq \Phi_{P'}^{\uparrow\alpha-1}$, hence $(\tilde{B}_i, \tilde{C}_j)\theta_{i,j}\tau$ is *true* in $\Phi_{P'}^{\uparrow\alpha-1}$. It follows that $A\theta_{i,j}\tau = A\sigma = A\theta$ is *true* in $\Phi_{P'}^{\uparrow\alpha}$.

b) $A\theta$ is *false* in $\Phi_P^{\uparrow\alpha+1}$. By the definition of Φ_P, $(\exists\tilde{y}\, H, J)\theta$ is *false* in $\Phi_P^{\uparrow\alpha}$. Hence for all extensions σ of θ, such that $Dom(\sigma) = Dom(\theta) \cup \tilde{y} = Var(A, H, J)$, we have that $(H, J)\sigma$ is *false* in $\Phi_P^{\uparrow\alpha}$.

Hence for all such σ's, and for all i, j and τ such that $\theta_{i,j}\tau|_{Dom(\sigma)} = \sigma$, $H\sigma = H_i\theta_{i,j}\tau$, $J\sigma = J_j\theta_{i,j}\tau$, we have that $(\tilde{B}_i, \tilde{C}_j)\theta_{i,j}\tau$ is *false* in $\Phi_P^{\uparrow\alpha-1}$. By Lemma 11, $\Phi_P^{\uparrow\alpha-1} \subseteq \Phi_{P'}^{\uparrow\alpha-1}$, hence $(\tilde{B}_i, \tilde{C}_j)\theta_{i,j}\tau$ is *false* in $\Phi_{P'}^{\uparrow\alpha-1}$. Since the clauses $d_{i,j}$ are the only ones that define A in P', we have that $A\theta_{i,j}\tau = A\sigma = A\theta$ is *false* in $\Phi_{P'}^{\uparrow\alpha}$.

Now to complete the proof, we have to observe two facts:

- First, that if we perform some further unfoldings on the resulting clauses, then we can only "speed up" the process of finding the truth value of A. In fact, by the same kind of reasoning used above, if $A\theta$ is *true* in $\Phi_{P'}^{\uparrow\alpha}$, and P'' is obtained from P' by unfolding some atoms in the bodies of the clauses $d_{i,j}$, then, for some $\beta \le \alpha$, $A\theta$ is *true* in $\Phi_{P''}^{\uparrow\beta}$.

- Second, that if cl contains just one atom, or more than two atoms, then the exact same reasoning applies. $\square$

4.2 The replacement operation

In order to prove the correctness of the unfold/fold transformation schema we will use (a simplified version of) the results in [6, 7] on the simultaneous replacement operation.

The replacement operation has been introduced by Tamaki and Sato in [26] for definite programs. Syntactically it consists in substituting a conjunction, $\tilde{C}$, of literals with another one, $\tilde{D}$, in the body of a clause.

Similarly, *simultaneous* replacement consists in substituting a set of conjunctions of literals $\{\tilde{C}_1, \ldots, \tilde{C}_n\}$, with another corresponding set of conjunctions $\{\tilde{D}_1, \ldots, \tilde{D}_n\}$ in the bodies of the clauses of program P; here each $\tilde{C}_i$ represents a subset of the body of a clause of P and we assume that if $i \neq j$ then $\tilde{C}_i$ and $\tilde{C}_j$ do not overlap, that is, they are either found in different clauses or they represent disjoint subsets of the same clause.

Note that the fact that each $\tilde{C}_i$ may occur in the body of only one clause of P is not restrictive, as even if $i \neq j$, $\tilde{C}_i$ and $\tilde{C}_j$ may actually represent identical literals.

We now give a simplified version of the *applicability conditions* introduced in [6, 7] in order to ensure the preservation of the semantics through the transformation. Such conditions depend on the semantics we associate to the program. Our first requirement is the semantic equivalence of the replacing and the replaced conjunctions of literals.

Definition 14 Equivalence of formulas. Let E, F be first order formulas and P be a normal program.

- *F is equivalent to E wrt $Fit(P)$, $F \sim_P E$, if for each ground substitution θ $E\theta$ is true (resp. false) in $Fit(P)$ iff $F\theta$ is.* □

Note that $F \sim_P E$ iff $Fit(P) \models \forall(F \Leftrightarrow E)$.

Example 4. Let P be the program in Example 1. We have that

$$med(Xs, Med) \sim_P \exists X, Y \; min(Xs, X) \wedge max(Xs, Y) \wedge Med = (X + Y)/2 \quad \square$$

With many respects, and with some caution, two equivalent (conjunctions of) literals can be used interchangeably; for example, if q is a new predicate we want to give a definition to, and we know that $A \sim_P B$ then defining q by introducing the new clause $q \leftarrow A$ is, from Fitting's semantics viewpoint, equivalent to doing it by introducing $q \leftarrow B$.

Notice that the formula in Example 4 we had to specify X and Y as existentially quantified variables. When we want to replace the conjunction $\tilde{C}$ with $\tilde{D}$, in the clause cl the first requirement of those applicability conditions is the equivalence of $\exists \tilde{x} \; \tilde{C}$ and $\exists \tilde{x} \; \tilde{D}$, where $\tilde{x}$ is a set of "local variables", that is, variables which appear in $\tilde{C}$ and/or $\tilde{D}$, but which do not occur anywhere else in the clause that we are transforming. The equivalence is required as it would make no sense to replace $\tilde{C}$ with something which has a different meaning. Unfortunately this is not enough, in fact we need the equivalence to hold also after the transformation. The equivalence can be destroyed when $\tilde{D}$ depends on cl, in which case the operation may introduce an infinite loop.

In order to prove that no fatal loops are introduced, we make use of a further concept. Here we say that the (closed) formula G is *defined* in the interpretation I, if the truth value of G in I is not *undefined*.

Definition 15 not-slower. Let P be a normal program, E and F be first order formulas. Suppose that $F \sim_P E$. We say that

- F is not-slower that E if for each ordinal α and each ground substitution θ: if $E\theta$ is defined in $\Phi_P^{\uparrow\alpha}$, then $F\theta$ is defined in $\Phi_P^{\uparrow\alpha}$ as well. □

So F is not-slower that E if, for each θ, computing the truth value of $F\theta$ never requires more iterations that computing the one of $E\theta$. In a way we could then say that the definition of F is at least as efficient as the one of E.

The following Theorem shows that if the replacing conjunctions are *equivalent to* and *not-slower than* the replaced ones, then the replacement operation is correct.

Theorem 16. *Let P' be a program obtained by simultaneously replacing the conjunctions $\{\tilde{C}_1, \ldots, \tilde{C}_n\}$ with $\{\tilde{D}_1, \ldots, \tilde{D}_n\}$ in the bodies of the clauses of P. If for each $\tilde{C}_i$, there exists a (possibly empty) set of variables $\tilde{x}_i$ such that the following three conditions hold:*

(a) [locality of the variables in $\tilde{x}_i$]. $\tilde{x}_i$ is a subset of the variables local to $\tilde{C}_i$ and $\tilde{D}_i$, that is, $\tilde{x}_i \subseteq Var(\tilde{C}_i) \cup Var(\tilde{D}_i)$ and the variables in $\tilde{x}_i$ don't occur in $\{\tilde{D}_1, \ldots, \tilde{D}_{i-1}, \tilde{D}_{i+1}, \ldots, \tilde{D}_n\}$ nor anywhere else in the clause where $\tilde{C}_i$ is found.
(b) [equivalence of the replacing and replaced parts]. $\exists \tilde{x}_i\, \tilde{D}_i \sim_P \exists \tilde{x}_i\, \tilde{C}_i$
(c) [the D_i's are not-slower than the C_i's]. $\exists \tilde{x}_i\, \tilde{D}_i$ is not-slower than $\exists \tilde{x}_i\, \tilde{C}_i$.

then $Fit(P) = Fit(P')$.

Proof. This is a particular case of Corollary 3.16 in [7]. □

A property we will need in the sequel is the following.

Proposition 17. *Suppose that $A \leftarrow \tilde{C}, \tilde{E}$ is a clause of P and that P' is obtained from P by replacing $\tilde{C}$ with $\tilde{D}$ in such a way that the conditions of Theorem 16 are satisfied (so that $Fit(P) = Fit(P')$). Then*

- *Each time that $A\theta$ is true (resp. false) in $\Phi_P^{\uparrow\alpha}$ then $A\theta$ is true (resp. false) in $\Phi_{P'}^{\uparrow\alpha}$*

Proof. This is a consequence of the fact that the replacing conjunction is not-slower than the replaced one. The formal proof is omitted here, it can be inferred by analyzing the proof of Theorem 3.15 in [7] □

Before we provide the proof of the correctness of the four step schema, we need to establish some further preliminary results. The first one states that the converse of **A** holds in any case.

Proposition 18. *Each time that $\exists \tilde{w}\, \tilde{B}\theta$ is true in some $\Phi_{P_1}^{\uparrow\beta}$, then there exists a non-limit ordinal $\alpha \le \beta$ such that $\exists \tilde{w}\, \tilde{B}\theta$ is true in $\Phi_{P_1}^{\uparrow\alpha}$.*

Proof. It follows at once from the definition of Fitting's operator. □

The following important transitive property holds:

Proposition 19. *Let P and P' be normal programs, E and F be first order formulas;*

- *If $E \sim_P F$ and $Fit(P) = Fit(P')$, then $E \sim_{P'} F$.* □

Now we can provide the details of the proof.

4.3 Correctness of the four step schema

We now prove the correctness of the four step schema. For the sake of simplicity we restrict ourselves to the case in which *Step 1* introduces only one clause. The extension to the general case is straightforward.

Let $P_1, \ldots P_4$ be the sequence of programs obtained via the four step schema: P_1 is the initial program, i.e. the one that contains D_{def}. P_2, P_3 and P_4, are the programs obtained by applying steps *Step 2* through *Step 4*. In order to show that the Fitting's models of programs $P_1, \ldots P_4$ coincide, we proceed as follows:

By the correctness of the unfolding operation, Corollary 12 we have that $Fit(P_1) = Fit(P_2)$.

We perform some further unfolding on some atoms of P_2, obtaining a new program that we will call P_{2u}, again by Corollary 12 we have that $Fit(P_2) = Fit(P_{2u})$; then we produce a "parallel sequence" of programs P_{3u}, P_{4u} by applying the simultaneous replacement operation, miming, to some extent, the original transformation. By applying Theorem 16 we will show that $Fit(P_{2u}) = Fit(P_{3u}) = Fit(P_{4u})$.

Finally we show that programs P_{3u} and P_{4u} are obtainable respectively from P_3 and P_4 by appropriately applying the unfold operation, and hence, by Corollary 12, that $Fit(P_3) = Fit(P_{3u})$ and that $Fit(P_4) = Fit(P_{4u})$. This will end the proof. Fig.1 illustrates both the original transformation and its parallel sequence.

Initial program. Let us establish some notation: $P_1 \ldots P_4$ are the programs obtained by applying the four step schema to program P, and $c_0 : H \leftarrow \tilde{B}$. is the (only) clause added to program P in *Step 1*. We also denote by $\tilde{w}$ the set of the local variables of c_i, $\tilde{w} = Var(\tilde{B}) \backslash Var(H)$. For the moment, let us make the following restriction:

- till the end of 4.3, we assume that $\tilde{B}$ doesn't contain negative literals.

Later, in subsection 4.4, we will prove the general case.

A simple consequence of the fact that c_0 is the only clause defining the predicate symbol of H is the following.

Observation 20

- $H \sim_{P_1} \exists \tilde{w} \, \tilde{B};$ □

P_2 and P_{2u}. P_2 is obtained by unfolding some of the atoms in $\tilde{B}$, so $P_2 = P \cup \{A_i \leftarrow \tilde{U}_i, \tilde{N}_i\}$, where the atoms in $\tilde{N}_i$ are those that have not been unfolded during *Step 1* (N stands for Not unfolded, while U for Unfolded), so $\tilde{N}_i$ is equal to a subset of an instance of $\tilde{B}$ and each A_i is an instance of H. We obtain P_{2u} from P_2 by further unfolding all the atoms in each $\tilde{N}_i$. We denote by $\{c_{i,j} : (A_i \leftarrow \tilde{U}_i)\gamma_{i,j}, \tilde{D}_{i,j}\}$ the set of clauses of P_{2u} obtained from clause c_i by unfolding the atoms in $\tilde{N}_i$. By the correctness of the unfolding operation, Corollary 12, we have that

$$Fit(P_1) = Fit(P_2) = Fit(P_{2u}) \qquad (2)$$

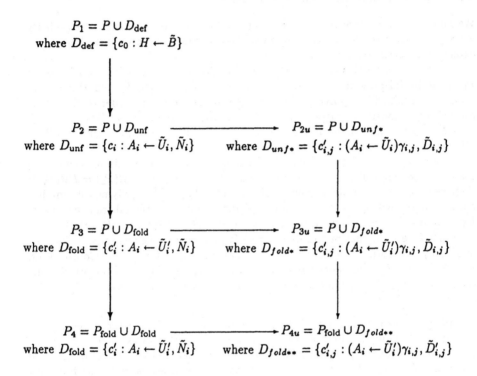

Fig. 1. Diagram of the transformation (left) together with the "parallel sequence" (right).

Moreover, the following properties hold:

Observation 21

- $H \sim_{P_{2u}} \exists\tilde{w}\, \tilde{B}$;
- H *is not-slower than* $\exists\tilde{w}\, \tilde{B}$ *in* P_{2u}.

Proof. From Observation 20 we have that $H \sim_{P_1} \exists\tilde{w}\, \tilde{B}$. The first statement follows then from (2) and Proposition 19. For the second, fix θ and let β be the least ordinal such that $\exists\tilde{w}\, \tilde{B}\theta$ is *true* (or *false*) in $\Phi_{P_{2u}}^{\uparrow\beta}$. The clauses defining the atoms in $\tilde{B}$ are the same in P_1, P_2 and P_{2u}, so $\exists\tilde{w}\, \tilde{B}$ is *true* (resp. *false*) in $\Phi_{P_1}^{\uparrow\beta}$ as well. From condition **A** and Proposition 18 we have that β is a non-limit ordinal. Hence, by the definition of Φ, $H\theta$ is *true* (resp. *false*) in $\Phi_{P_1}^{\uparrow\beta+1}$, and, by Lemma 13 $H\theta$ is *true* (resp. *false*) in $\Phi_{P_{2u}}^{\uparrow\beta}$. $\square$

P_3 and P_{3u}. P_{3u} is obtained from P_{2u} as follows.

Suppose that in *Step 2* we performed a recursive folding on the clause c_i : $A_i \leftarrow \tilde{B}\theta, \tilde{R}_i, \tilde{N}_i$ of P_2, obtaining c_i' : $A_i \leftarrow H\theta, \tilde{R}_i, \tilde{N}_i$ in P_3. In the diagram we denote by $\tilde{U}_i'$ the conjunction of literals resulting from the application of the recursive folding on the conjunction $\tilde{U}_i$ (so $\tilde{U}_i = \tilde{B}\theta, \tilde{R}_i$ and $\tilde{U}_i' = H\theta, \tilde{R}_i$).

On P_{2u} we then perform the following. In each of the clauses $c_{i,j}$ we transform $\tilde{U}_i\gamma_{i,j}$ into $\tilde{U}_i'\gamma_{i,j}$ by replacing conjunctions of literals of the form $\tilde{B}\theta\gamma_{i,j}$ with $H\theta\gamma_{i,j}$ wherever needed; we call the resulting clauses $c_{i,j}'$. It is easy to see that if we unfold all the atoms in $\tilde{N}_i$ in the body of clause c_i' in P_3, then the resulting clauses are exactly the $c_{i,j}'$ in P_{3u}; this is best shown by the diagram. Hence P_{3u} is obtainable from P_3 by appropriately applying the unfolding operation. From Corollary 12 it follows that

$$Fit(P_3) = Fit(P_{3u}) \qquad (3)$$

Now we show that $Fit(P_{2u}) = Fit(P_{3u})$. First we need the following.

Proposition 22. *Let Q be a program, A, B be atoms and $\tilde{y}$ be a set of variables, such that $A \sim_Q \exists\tilde{y}\, B$. Suppose also that η is a renaming over $\tilde{y}$ and that for each variable z that occurs in A or B, but not in $\tilde{y}$, $Var(z\eta) \cap Var(\tilde{y}\eta) = \emptyset$. Then*

$$- A\eta \sim_Q \exists(\tilde{y}\eta)\, B\eta$$

Proof. Straightforward. $\square$

Since $\gamma_{i,j}$ results from unfolding the atoms in $\tilde{N}_i$, we have that $Dom(\gamma_{i,j}) \cap Var(c_i) \subseteq Var(\tilde{N}_i)$. Hence, by the conditions on θ in *Step 2*, $Dom(\gamma_{i,j}) \cap \tilde{w}\theta = \emptyset$ and $\tilde{w}\theta\gamma_{i,j} = \tilde{w}\theta$; so $\theta\gamma_{i,j}$ is a renaming over $\tilde{w}$, and the variables in $\tilde{w}\theta\gamma_{i,j}$ do not occur anywhere else in $c_{i,j}$. From Observation 21 and Proposition 22 we have that

- $H\theta\gamma_{i,j} \sim_{P_{2u}} \exists(\tilde{w}\theta\gamma_{i,j})\, \tilde{B}\theta\gamma_{i,j}$;
- $H\theta\gamma_{i,j}$ is not-slower than $\exists(\tilde{w}\theta\gamma_{i,j})\, \tilde{B}\theta\gamma_{i,j}$ in P_{2u}.

Since we obtained P_{3u} from P_{2u} by simultaneously replacing conjunctions (of the form) $\tilde{B}\theta\gamma_{i,j}$ with $H\theta\gamma_{i,j}$, by Theorem 16

$$Fit(P_{2u}) = Fit(P_{3u}). \qquad (4)$$

Moreover, the following properties hold:

Observation 23

- $H \sim_{P_{3u}} \exists\tilde{w}\, \tilde{B}$;
- H is not-slower than $\exists\tilde{w}\, \tilde{B}$ in P_{3u}.

Proof. The first statement follows from Observation 21, (4) and Proposition 19. For the second first note that going from P_{2u} to P_{3u} we have affected only clauses that define the predicate *new*, moreover no other predicates definition depends on these clauses, in particular the atoms in $\tilde{B}$ are independent from them, hence, since H is not-slower than $\exists\tilde{w}\, \tilde{B}$ in P_{2u}, the statement follows from Proposition 17. $\square$

P_4 and P_{4u}. P_4 is obtained from P_3 by transforming some of the clauses of P of the form $A \leftarrow \tilde{B}\theta, \tilde{E}$ into $A \leftarrow H\theta, \tilde{E}$.

Now we want to obtain P_{4u} from P_{3u} in such a way that P_{4u} is obtainable also from P_4 by unfolding the atoms in the conjunctions $\tilde{N}_i$.

Let $d : A \leftarrow \tilde{B}\theta, \tilde{E}$ be one of the clauses of P_3 that are transformed in *Step 4*. First note that d belongs both to P_3 and P_{3u}, in fact d was already present it the original program P, and never modified. We can then apply the same operations to the clauses of P_{3u}. Observe that for the conditions on θ given in *Step 4*, and by Observation 23 we have that

Observation 24

- $H\theta \sim_{P_{3u}} \exists(\tilde{w}\theta) \, \tilde{B}\theta$
- $H\theta$ *is not-slower than* $\exists(\tilde{w}\theta) \, \tilde{B}\theta$ *in* P_{3u} $\qquad\qquad\qquad\qquad\qquad\qquad$ □

Second, notice that in case that d was used as unfolding clause for going from P_2 to P_{2u}, then some instances of $\tilde{B}\theta$ were propagated into P_{3u}. Using the notation of the diagram, this is the case when some $\tilde{N}_i$ (in P_2) is of the form $A', \tilde{F}_i$ where A and A' are unifiable atoms, then one of the $\tilde{D}_{i,j}$ (in P_{2u}) is of the form $\tilde{D}_{i,j} = (\tilde{B}, \tilde{F}_i)\theta'$. However, if we unfold N_i in P_4, what we get is $\tilde{D}'_{i,j} = H\theta', \tilde{F}_i$, that has $H\theta'$ instead of $\tilde{B}\theta'$. By the same argument used for $\theta\gamma_{i,j}$ in 4.3, we have that

Observation 25

- $H\theta' \sim_{P_{3u}} \exists(\tilde{w}\theta') \, \tilde{B}\theta'$
- $H\theta'$ *is not-slower than* $\exists(\tilde{w}\theta') \, \tilde{B}\theta'$ *in* P_{3u} $\qquad\qquad\qquad\qquad\qquad$ □

So in order to obtain P_{4u} from P_{3u} we have then to do two things: First, replace $\tilde{B}\theta$, with the corresponding $H\theta$ in all the clauses d that are transformed in *Step 4*. Second, replace $\tilde{B}\theta'$ with $H\theta'$ in the $\tilde{D}_{i,j}$ so that P_{4u} contains $\tilde{D}'_{i,j}$ instead of $\tilde{D}_{i,j}$. This tantamounts to the application of a simultaneous replacement.

From Observations 24 and 25, and Theorem 16 we have that

$$Fit(P_{3u}) = Fit(P_{4u}) \qquad\qquad (5)$$

Moreover P_{4u} is obtainable from P_4 by unfolding all the atoms in the conjunctions $\tilde{N}_i$ in the clauses where they occur. Hence

$$Fit(P_4) = Fit(P_{4u}). \qquad\qquad (6)$$

So far, because of (1), (2), (3), (4) and (5), we have the following

Proposition 26. *If condition* **A** *holds and* $\tilde{B}$ *does not contain negative literals, then*

- $Fit(P_1) = Fit(P_2) = Fit(P_3) = Fit(P_4)$ $\qquad\qquad\qquad\qquad\qquad\qquad$ □

4.4 The general case

We can finally prove Theorem 6. Let us state it again.

Theorem 6. *Let $P_1, \ldots, P_4$ be a sequence of programs obtained applying the transformation schema to program P, Let also $D_{\text{def}} = \{H_i \leftarrow \tilde{B}_i\}$ be the set of clauses introduced in Step 1, and, for each i, $\tilde{w}_i$ be the set of local variables of c_i: $\tilde{w}_i = Var(\tilde{B}_i)\backslash Var(H_i)$. If each c_i in D_{def} satisfies the following condition:*

A *each time that $\exists \tilde{w}_i \; \tilde{B}_i \theta$ is false in some $\Phi_{P_1}^{\uparrow \beta}$, then there exists a non-limit ordinal $\alpha \leq \beta$ such that $\exists \tilde{w}_i \; \tilde{B}_i \theta$ is false in $\Phi_{P_1}^{\uparrow \alpha}$*

Then $Fit(P_1) = Fit(P_2) = Fit(P_3) = Fit(P_4)$.

Proof. We consider here the simplified case in which *Step 1* introduces only one clause which in turn contains only one negative literal in the body, i.e. $D_{\text{def}} = \{c_0 : H \leftarrow \neg l(\tilde{y}), \tilde{B}'\}$. The generalization to the case of multiple clauses and multiple negative literals is straightforward and omitted here. Notice that if c_0 contained no negative literals, then the result would following directly from Proposition 26.

We now perform a double transformation on P_1: first, we enlarge it with the following new definition: $d : notl(\tilde{y}) \leftarrow \neg l(\tilde{y})$; then, we replace each instance $\neg l(\tilde{t})$ of $l(\tilde{y})$ that occurs in the body of a clause with the corresponding instance $notl(\tilde{t})$ of $notl(\tilde{y})$. This replacement operation clearly preserves Fitting's model of the programs, in fact it can be undone by unfolding. Let us call P_1' the program so obtained. We have that

$$Fit(P_1) = Fit(P_1')|_{B_{P_1}} \tag{7}$$

Where $Fit(P_1')|_{B_{P_1}}$ denotes the restriction of $Fit(P_1')$ to the atoms in the Herbrand base of P_1.

Now P_1' contains, instead of clause c_0, the following: $c_0' = H \leftarrow notl(\tilde{y}), \tilde{B}'$. which is a *definite* clause.

Now notice that, since the unfold operation is defined only for positive literals, then $\neg l(\tilde{y})$ is never unfolded in the transformation $P_1 \ldots P_4$. It follows that, by performing the same operations used for going from P_1 to P_4, we can obtain another "parallel sequence" $P_1' \ldots P_4'$ that starts with program P_1'. By the same arguments used to prove (7), we have that, for $i \in [1 \ldots 4]$,

$$Fit(P_i) = Fit(P_i')|_{B_{P_1}} \tag{8}$$

Moreover, by Proposition 26,

$$Fit(P_1') = Fit(P_2') = Fit(P_3') = Fit(P_4') \tag{9}$$

From (8) and (9) the thesis follows. $\qquad\square$

References

1. K. R. Apt. Introduction to Logic Programming. In J. van Leeuwen, editor, *Handbook of Theoretical Computer Science*, volume B: Formal Models and Semantics. Elsevier, Amsterdam and The MIT Press, Cambridge, 1990.
2. L. Aiello, G. Attardi, and G. Prini. Towards a more declarative programming style. In E. J. Neuhold, editor, *Proceedings of the IFIP Conference on Formal Description of Programming Concepts*, pages 121–137. North-Holland, 1978.
3. C. Aravidan and P. M. Dung. Partial deduction of logic programs w.r.t. well-founded semantics. In H. Kirchner G. Levi, editor, *Proceedings of the Third International Conference on Algebraic and Logic Programming*, pages 384–402. Springer-Verlag, 1992.
4. C. Aravidan and P. M. Dung. On the correctness of Unfold/Fold transformation of normal and extended logic programs. Technical report, Division of Computer Science, Asian Institute of Technology, Bangkok, Thailand, April 1993.
5. A. Bossi and N. Cocco. Basic Transformation Operations which preserve Computed Answer Substitutions of Logic Programs. *Journal of Logic Programming*, 16(1/2):47–87, 1993.
6. A. Bossi, N. Cocco, and S. Etalle. Transforming Normal Programs by Replacement. In A. Pettorossi, editor, *Meta Programming in Logic - Proceedings META'92*, volume 649 of *Lecture Notes in Computer Science*, pages 265–279. Springer-Verlag, Berlin, 1992.
7. A. Bossi, N. Cocco, and S. Etalle. Simultaneous replacement in normal programs. Technical Report CS-R9357, CWI, Centre for Mathematics and Computer Science, Amsterdam, The Netherlands, August 1993. Available via anonymous ftp at ftp.cwi.nl, or via xmosaic at http://www.cwi.nl/cwi/publications/index.html.
8. R.M. Burstall and J. Darlington. A transformation system for developing recursive programs. *Journal of the ACM*, 24(1):44–67, January 1977.
9. K. L. Clark. Negation as failure rule. In H. Gallaire and G. Minker, editors, *Logic and Data Bases*, pages 293–322. Plenum Press, 1978.
10. K.L. Clark and S. Sickel. Predicate logic: a calculus for deriving programs. In *Proceedings of IJCAI'77*, pages 419–120, 1977.
11. M. Fitting. A Kripke-Kleene semantics for Logic Programs. *Journal of Logic Programming*, 2(4):295–312, 1985.
12. P.A. Gardner and J.C. Shepherdson. Unfold/fold transformations of logic programs. In J-L Lassez and editor G. Plotkin, editors, *Computational Logic: Essays in Honor of Alan Robinson*. 1991.
13. C.J. Hogger. Derivation of logic programs. *Journal of the ACM*, 28(2):372–392, April 1981.
14. T. Kawamura and T. Kanamori. Preservation of Stronger Equivalence in Unfold/Fold Logic Programming Transformation. *Theoretical Computer Science*, 75(1/2):139–156, 1990.
15. S.C. Kleene. *Introduction to Metamathematics*. D. van Nostrand, Princeton, New Jersey, 1952.
16. H.J. Komorowski. Partial evaluation as a means for inferencing data structures in an applicative language: A theory and implementation in the case of Prolog. In *Ninth ACM Symposium on Principles of Programming Languages, Albuquerque, New Mexico*, pages 255–267, 1982.
17. J. W. Lloyd. *Foundations of Logic Programming*. Springer-Verlag, Berlin, 1987. Second edition.
18. J. W. Lloyd and J. C. Shepherdson. Partial Evaluation in Logic Programming. *Journal of Logic Programming*, 11:217–242, 1991.

19. M.J. Maher. Correctness of a logic program transformation system. IBM Research Report RC13496, T.J. Watson Research Center, 1987.

20. M. Proietti and A. Pettorossi. Unfolding, definition, folding, in this order for avoiding unnecessary variables in logic programs. In Maluszynski and M. Wirsing, editors, *PLILP 91, Passau, Germany (Lecture Notes in Computer Science, Vol.528)*, pages 347–358. Springer-Verlag, 1991.

21. T. Sato. Equivalence-preserving first-order unfold/fold transformation system. *Theoretical Computer Science*, 105(1):57–84, 1992.

22. H. Seki. Unfold/fold transformation of stratified programs. In G. Levi and M. Martelli, editors, *6th International Conference on Logic Programming*, pages 554–568. The MIT Press, 1989.

23. H. Seki. A comparative study of the Well-Founded and Stable model semantics: Transformation's viewpoint. In D. Pedreschi W. Marek, A. Nerode and V.S. Subrahmanian, editors, *Workshop on Logic Programming and Non-Monotonic Logic, Austin, Texas, October 1990*, pages 115–123, 1990.

24. H. Seki. Unfold/fold transformation of stratified programs. *Theoretical Computer Science*, 86(1):107–139, 1991.

25. H. Seki. Unfold/fold transformation of general logic programs for the Well-Founded semantics. *Journal of Logic Programming*, 16(1/2):5–23, 1993.

26. H. Tamaki and T. Sato. Unfold/Fold Transformations of Logic Programs. In Sten-Åke Tärnlund, editor, *Proc. Second Int'l Conf. on Logic Programming*, pages 127–139, 1984.

Formal Semantics of
Temporal Epistemic Reflection[+]

Wiebe van der Hoek[a,b]
John-Jules Meyer[a,b]
Jan Treur[b]

[a] Utrecht University, Department of Computer Science
P.O. Box 80.089, 3508 TB Utrecht, The Netherlands.
Email: {wiebe,jj}@cs.ruu.nl

[b] Free University Amsterdam, Department of Mathematics and Computer Science
De Boelelaan 1081a, 1081 HV Amsterdam, The Netherlands.
Email: treur@cs.vu.nl.

Abstract
In this paper we show how a formal semantics can be given to reasoning processes in meta-level architectures that reason about (object level) knowledge states and changes of them. Especially the attention is focused on the upward and downward reflections in these architectures. Temporalized epistemic logic is used to specify meta-level reasoning processes and the outcomes of these.

1 Introduction

Meta-level architectures often are used either to model dynamic control of the object level inferences, or to extend the inference relation of the object level. In [Tre92] we introduced formal semantics for meta-level architectures of the first kind based on temporal models. It may be considered quite natural that for such a dynamic type of reasoning system the temporal element of the reasoning should be made explicit in the formal semantics. For the use of meta-level architectures to extend the object level inference relation the situation looks different. In principle one may work out formal semantics in terms of (the logic behind) this extended, non-classical inference relation; e.g., as in the literature for nonmonotonic logics. However, much discussion is possible about this case. Some papers argue that also in the case of a non-monotonic logic the semantics have to make the inherent temporal element explicit; approaches are described in, e.g., [Gab82], [ET93]. In the current paper we adopt this line.

In principle a downward reflection that extends the inference relation of the object level theory disturbs the (classical) object level semantics: facts (assumptions) are added that are not logically entailed by the available object level knowledge. Adding a temporal dimension (in the spirit of [FG92]) enables one to obtain formal semantics

[+] This work is partially supported by ESPRIT III Basic Research Action 6156 (DRUMS II).

of *downward reflection* in a dynamic sense: as a *transition* from the current object level theory to a next one (where the reflected assumption has been added).

In [MH93a] a metaphor of a meta-level architecture was exploited to define a nonmonotonic logic, called Epistemic Default Logic. It was shown how *upward reflection* can be formalized by a *nonmonotonic entailment relation based on epistemic states* (according to [HM84]), and the meta-level reasoning process by a (monotonic) epistemic logic. Compared to a meta-level architecture, what was still missing was a formalization of the step where the conclusions of the meta-level actually were used to change the object level, i.e., where formulas φ are added to the object level knowledge, in order to be able to reason further with them at the object level. This should be achieved by the downward reflection step. In the current paper we introduce a formalization of this downward reflection step in the reasoning pattern as well. We formalize the semantics of the architecture by means of entailment on the basis of temporalized Kripke models. Thus, a formalization is obtained of the reasoning pattern as a whole, consisting of a process of generating possible default assumptions by meta-level reasoning and actually assuming them by downward reflection (a similar pattern as generated by the so-called BMS-architecture introduced in [TT91]).

The temporal epistemic meta-level architecture described here is a powerful tool to reason with dynamic assumptions: it enables one to introduce and retract additional assumptions during the reasoning, on the basis of explicit epistemic information on the current knowledge state and meta-knowledge to determine adequate additional assumptions. This covers a whole class of reasoning patterns of meta-level architectures; we call them *temporal epistemic meta-level architectures* (TEMA). In such an architecture upward reflection is restricted to the transfer to the meta-level of epistemic information: information about what currently is known and what is not known at the object level, and downward reflection is restricted to introducing additional information (assumptions) to the object level, based on the conclusions derived at the meta-level. In this architecture a number of interesting reasoning patterns can be formalized in temporal semantics in a quite intuitive and transparent manner. In [Tre90], [Tre91] various applications of the architecture are shown, including: hypothetical reasoning, the method of indirect proof, reasoning about knowledge, reasoning about actions integrated with executing them.

The formalization of downward reflection was inspired by [Tre92], where it is pointed out how temporal models can provide an adequate semantics of meta-level architectures for dynamic control, and [ET93] where similar ideas have been worked out to obtain a linear time temporal semantics for default logic. The general idea is that conclusions derived at the meta-level essentially are statements about the state of the object level reasoning at the next moment of time. Thus, downward reflection is a shift of time in a (reasoning) process that is described by temporal logic.

To get the idea we will use the following informal description of an example reasoning pattern, involving diagnostics (of a car that cannot start) by elimination of hypotheses (for a more extensive description, see [Tre90]). Later on (in Section 7.3) we will come back to this example to show how it can be formalized by our approach. Suppose at the object level we have the following (causal) knowledge about a car:

- *if the battery is empty*
    - *then the lights cannot burn*
- *if the battery is empty, or the sparking-plugs are tuned up badly*
    - *then the car does not start*
- *the car does not start*

At the meta-level we control a diagnostic reasoning process to find out whether it can be excluded that an empty battery is the cause of the problems (the only hypothesis that we fully consider in this example). To this end we have the following (simplified) meta-knowledge that enables to propose hypotheses for elimination, and to reject them if indeed they turn out to fail:

- *if it is known that the car does not start*
    - *and it is not known whether the hypothesis "the battery is empty" holds*
    - *then "the battery is empty" is an adequate hypothesis to focus on*
- *if it is known that the car does not start*
    - *and it is known that the hypothesis "the battery is empty" does not hold*
    - *and it is not known whether the hypothesis*
        - *"the sparking-plugs are tuned up badly" holds*
    - *then "the sparking-plugs are tuned up badly" is an adequate hypothesis to focus on*
- *if we have focused on a hypothesis X*
    - *and assuming X we have derived that the observable Y should be the case*
    - *and it has been observed that Y is not the case,*
    - *then our hypothesis X should be rejected*
- *it has been observed that the lights can burn*

Using these two knowledge bases in a temporal epistemic meta-level architecture we can perform the following dynamic reasoning pattern:

1. Draw all conclusions that are possible at the object-level
2. Reflect upwards that at the object level it is not known whether the battery is empty
3. Draw the conclusion at the meta-level that 'battery is empty' is an adequate hypothesis to focus on
4. Reflect this hypothesis downward: introduce it at the object level as additional information (assumption)
5. Draw the conclusion at the object level that the lights cannot burn
6. Reflect upwards the information that at the object level it has been found that the lights cannot burn
7. Use the observation at the meta-level that the lights can burn, and notice the contradictory situation
8. At the meta-level draw the conclusion that the focus hypothesis 'battery is empty' should be rejected
9. Reflect downwards that 'battery is empty' is considered to be not true.

This example shows some of the possibilities of temporal epistemic reflections. Notice especially the manner in which we treat epistemic states and additional information in a dynamic manner. This implies that during the reasoning the (states of) object level and meta-level both change in (reasoning) time; temporal logic is exploited to describe these changes in a logical manner.

Our approach provides a formalization of upward reflection by a nonmonotonic entailment relation based on epistemic states and downward reflection by an entailment relation based on temporal models. As in the literature reflections usually either are kept rather restricted (and static), or are described in a procedural or syntactic manner (e.g., by means of generalized inference rules; see [GTG93]), the novelty of our approach is that it introduces a semantic formalization for reflections involving epistemic and temporal aspects.

In this paper in Section 2 the basic notions of epistemic logic are presented. In Section 3 we do the same for temporal logic and the notion of temporalization of a logic system. In Section 4 we show how upward reflection can be formalized based on epistemic states. In Section 5 we introduce the logic used to formalize the meta-level reasoning: S5P*. In Section 6 we define a labelled branching time temporal formalization of downward reflection. In Section 7 we show how the overall formalization can be obtained from its parts and how it works for the running example reasoning pattern. In Section 8 we draw some conclusions. In this version, proofs are omitted; they can be found in the full paper [HMT94b].

2 Basic notions and properties of Epistemic Logic

2.1 Epistemic Logic

2.1. DEFINITION (epistemic formulas). Let P be a set of propositional constants (atoms); $P = \{p_k \mid k \in I\}$, where I is either a finite or countably infinite set. The set FORM of *epistemic formulas* φ, ψ,... is the smallest set containing P, closed under the classical propositional connectives and the epistemic operator K, where $K\varphi$ means that φ is known. An *objective formula* is a formula without any occurrences of the modal operator K. For $\Gamma \subseteq$ FORM, we denote by $\text{Prop}(\Gamma)$ the set of objective formulas in Γ.

2.2. DEFINITION (S5-Kripke models). An S5-model is a structure $\mathbb{M} = \langle M, \pi \rangle$ where M is a non-empty set, π is a truth assignment function of type $M \rightarrow (P \rightarrow \{t, f\})$ such that for all $m_1, m_2 \in M$: $\pi(m_1) = \pi(m_2) \Rightarrow m_1 = m_2$. The class of S5-models is denoted by $\text{Mod}(S5)$.

The set of states in an S5-model represents a collection of alternative states that are considered (equally) possible on the basis of (lack of) knowledge. We shall use S5-models as representations of the knowledge of agents.

2.3 REMARK For any $m \in M$, the function $\pi(m): p \rightarrow \pi(m)(p)$ is a valuation. Since we have that in an S5-model it holds that $\pi(m_1) = \pi(m_2) \Leftrightarrow m_1 = m_2$, we may identify states m with their valuations $\pi(m)$, and write $m \equiv \pi(m)$ for $m \in M$. So, without loss of generality, we may consider S5-models of the form $\mathbb{M} = \langle M, \pi \rangle$ with $M \subseteq \mathbb{W}$.

2.4. DEFINITION (submodels and union of S5-models). We define a subset relation on S5-models by: $M_1 \subseteq M_2$ iff $M_1 \subseteq M_2$. Moreover, if $M_1 = \langle M_1, \pi_1 \rangle$ and $M_2 = \langle M_2, \pi_2 \rangle$ are two S5-models, their union is defined as: $M_1 \cup M_2 = \langle M, \pi \rangle$, where $M = M_1 \cup M_2$ and $\pi(m) = \pi_i(m)$ if $m \in M_i$, $i = 1, 2$.

2.5. DEFINITION (interpretation of epistemic formulas). Given $M = \langle M, \pi \rangle$, we define the relation $(M, m) \vDash \varphi$ by induction on the structure of the epistemic formula φ:

$$(M, m) \vDash p \quad \Leftrightarrow \quad \pi(m)(p) = t \text{ for } p \in P$$
$$(M, m) \vDash \varphi \wedge \psi \quad \Leftrightarrow \quad (M, m) \vDash \varphi \text{ and } (M, m) \vDash \psi$$
$$(M, m) \vDash \neg\varphi \quad \Leftrightarrow \quad (M, m) \nvDash \varphi$$
$$(M, m) \vDash K\varphi \quad \Leftrightarrow \quad (M, m') \vDash \varphi \text{ for all } m' \in M$$

2.6. DEFINITION (validity).

i φ is *valid in* $M = \langle M, \pi \rangle$, denoted $M \vDash \varphi$, if for all $m \in M$: $(M, m) \vDash \varphi$.

ii φ is *valid*, notation $Mod(S5) \vDash \varphi$, if $M \vDash \varphi$ for all S5-models M.

Validity w.r.t. S5-models can be axiomatized by the system S5:

2.7. DEFINITION (system S5). The logic S5 consists of the following:

Axioms:

(A1)   All propositional tautologies

(A2) $(K\varphi \wedge K(\varphi \rightarrow \psi)) \rightarrow K\psi$ *Knowledge is closed under logical consequence.*

(A3) $K\varphi \rightarrow \varphi$ *Known facts are true.*

(A4) $K\varphi \rightarrow KK\varphi$ *An agent knows that he knows something.*

(A5) $\neg K\varphi \rightarrow K\neg K\varphi$ *An agent knows that he does not know something.*

Derivation rules:

(R1) $\vdash \varphi, \vdash \varphi \rightarrow \psi \Rightarrow \vdash \psi$ *Modus Ponens*

(R2) $\vdash \varphi \Rightarrow \vdash K\varphi$ *Necessitation*

That φ is a theorem derived by the system S5 is denoted by $S5 \vdash \varphi$.

2.8. THEOREM (Soundness and completeness of S5). $S5 \vdash \varphi \Leftrightarrow Mod(S5) \vDash \varphi$

2.2 Epistemic States and Stable Sets

In this paper we simply define an epistemic state as an S5-model:

2.9. DEFINITION. An *epistemic state* is an S5-model $M = \langle M, \pi \rangle$. The set M is the set of epistemic alternatives allowed by the epistemic state M.

2.10. DEFINITION. Let M be an S5 model. Then $K(M)$ is the set of facts known in M: $K(M) = \{\varphi \mid M \vDash K\varphi\}$. We call $K(M)$ the *theory of* M or *knowledge in* M.

We mention here that the knowledge in M are exactly the validities in M (Cf. [MH94]), i.e. we have $K(M) = \{\varphi \mid M \vDash K\varphi\} = \{\varphi \mid M \vDash \varphi\}$.

2.11. LEMMA. For any S5 models M_1 and M_2:

i   If $M_1 \subseteq M_2$ then $\mathrm{Prop}(K(M_2)) \subseteq \mathrm{Prop}(K(M_1))$.

ii   If the set of atoms **P** is finite, then also
$\mathrm{Prop}(K(M_2)) \subseteq \mathrm{Prop}(K(M_1)) \Rightarrow M_1 \subseteq M_2$.

2.12. PROPOSITION (Moore [Moo85]).

i   The theory $\Sigma = K(M)$ of an epistemic state M is a so-called *stable set*, i.e.,
satisfies the following properties:

      (St 1)   all instances of propositional tautologies are elements of Σ;

      (St 2)   if $\varphi \in \Sigma$ and $\varphi \rightarrow \psi \in \Sigma$ then $\psi \in \Sigma$;

      (St 3) $\varphi \in \Sigma \Leftrightarrow K\varphi \in \Sigma$

      (St 4) $\varphi \notin \Sigma \Leftrightarrow \neg K\varphi \in \Sigma$

      (St 5) Σ is propositionally consistent.

ii   Every stable set Σ of epistemic formulas determines an S5-Kripke model M_Σ
for which it holds that $\Sigma = K(M_\Sigma)$. Moreover, if **P** is a finite set, then M_Σ is
the unique S5-Kripke model with this property.

2.13. PROPOSITION. A stable set is uniquely determined by its objective formulas.

3   Basic notions and properties of Temporal Logic

We start (following [FG92]) by defining the temporalized models associated to any
class of models and apply it to the classes of models as previously discussed. In
contrast to the reference as mentioned we use labelled flows of time. We use one fixed
set L of labels, viz. $L = 2^I$, the powerset of some index set I. However, in most
definitions we do not use this fact, but only refer to (elements τ of) L.

3.1 Flows of time

3.1. DEFINITION (discrete labelled flow of time).

Suppose L is a set of labels. A *(discrete) labelled flow of time* (or *lft*) , labelled by L
is a pair $T = (T, (<_\tau)_{\tau \in L})$ consisting of a non-empty set T of time points and a
collection of binary relations $<_\tau$ on T. Here for s, t in T and τ in L the expression
$s <_\tau t$ denotes that t is a (immediate) *successor* of s with respect to an arc labelled by
τ. Sometimes it is convenient to leave the indices out of consideration and use just
the binary relation s < t denoting that $s <_\tau t$ for some τ (for some label τ they are
connected). Thus we have that $< = \cup_\tau <_\tau$. We also use the (non-reflexive) transitive
closure « of this binary relation: $« = <^+$.

We will make additional assumptions on the flow of time; for instance that it
describes a discrete tree structure, with one root and where time branches in the
direction of the future.

3.2. DEFINITION (labelled time tree)

An lft $T = (T, (<_\tau)_{\tau \in L})$ is called a labelled time tree (ltt) if the following conditions
are satisfied (recall that $< = \cup_\tau <_\tau$):

i   the graph $\langle T,< \rangle$ is a directed rooted tree.

ii   Successor existence: Time points have at least one <-successor.

iii Label-deterministic: For every label τ there is at most one τ-successor.

3.3. DEFINITION (branch and path)

a   A *branch* in an lft $\mathbf{T}$ is an lft $\mathbf{B} = (T', (<'_\tau)_{\tau \in L})$ with (i) $T' \subseteq T$, (ii) $s <'_\tau t \Rightarrow$
$s <_\tau t$, (iii) every $s \in T'$ has at most one $<'$-successor $t \in T'$, (iv) for all $s, t \in T'$:
$s <_\tau t \Rightarrow s <'_\tau t$, and (v) every element of T that is in between elements of T' is
itself in T': for all $s' \in T', t \in T, u' \in T' : s' \ll t \ll u' \Rightarrow t \in T'$.

b   A branch in an ltt $\mathbf{T} = (T, (<_\tau)_{\tau \in L})$ is *maximal* if contains the root r of T.

c   A *path* is a finite sequence of successors: $s_0, \ldots, s_n$ such that: $s_i < s_{i+1}$ for all 0
$\le i \le n-1$. We call s_0 the starting point and s_n the end point of the path.

3.4. PROPOSITION Any branch of an lft $\mathbf{T}$ is an ltt.

3.5. DEFINITION (time stamps).
Given an ltt $(T, (<_\tau)_{\tau \in L})$. A mapping $|\cdot| : T \to N$ is
called a *time stamp mapping* if for the root r it holds that $|r| = 0$, and for all time
points s, t it holds $s < t \Rightarrow |t| = |s| + 1$.

3.2 Temporal models

We first define our temporal formulas:

3.6. DEFINITION (temporal formulas).

Given a logic $\mathbf{L}$, *temporal formulas over* (the language of) $\mathbf{L}$ are defined as follows:

i   if φ is a formula of $\mathbf{L}$ then $C\varphi$ is a temporal formula (also called a *temporal atom*)

ii   if φ and ψ are temporal formulas, then so are:

$$\neg\varphi, \varphi \wedge \psi, \varphi \to \psi, X_{\exists,\tau}\varphi, X_\exists\varphi, X_{\forall,\tau}\varphi, X_\forall\varphi, F_\exists\varphi, F_\forall\varphi, G_\exists\varphi, G_\forall\varphi.$$

Note how the C-operator acts as a kind of 'separator' between the basic language for $\mathbf{L}$
and the actual temporal formulas: from the temporal language point of view, formulas
of the form $C\varphi$ may be conceived as a kind of 'atoms'; in the truth-definition,
occurrences of the C enforce a 'shift' in the evaluation of formulas, taking us from a
temporal model $\underline{\mathbb{M}}$ to some of it snapshots $\underline{\mathbb{M}}_t$, which are in their turn models for $\mathbf{L}$:

3.7. DEFINITION (temporal models)

a   Let MOD be a class of models, and $\mathbf{T} = (T, (<_\tau)_{\tau \in L})$ a labelled flow of time. A
temporal MOD-*model over* $\mathbf{T}$ is a mapping $\underline{\mathbb{M}}: T \to$ MOD. For $t \in T$ we
sometimes denote $\underline{\mathbb{M}}(t)$ (the snapshot at time point t) by $\underline{\mathbb{M}}_t$.

b   If we apply a) to the classes of models ModSet(**PC**) or Mod(**S5**) we call these
temporalized models *temporal valuation-set-models* (abbreviated *temporal V-*
models) and *temporal S5-models over* $\mathbf{T}$, respectively. Similarly for the class of
S5P*-models that will be introduced in Section 5.

c   Given an lft $\mathbf{T}$, the temporal formulas are interpreted on MOD-models as follows:

   i   Conjunction and implication are defined as usual; moreover

        $\underline{\mathbb{M}}, s \vDash \neg\varphi$ iff not $\underline{\mathbb{M}}, s \vDash \varphi$;

   ii   The temporal operators are interpreted as follows:

1) $C\varphi$ means that in the current state φ is true, i.e.
   $\underline{M}, s \vDash C\varphi$ iff $\underline{M}_s \vDash \varphi$

2) $X_{\exists,\tau}\varphi$ means that φ is true in some τ-successor state i.e.,
   $\underline{M}, s \vDash X_{\exists,\tau}\varphi$ iff there exists a time point t with $s <_\tau t$ such that $\underline{M}, t \vDash \varphi$

3) $X_\exists\varphi$ means that there is a τ with some τ-successor in which φ is true.
   $\underline{M}, s \vDash X_\exists\varphi$ iff there exists a time point t with $s < t$ such that $\underline{M}, t \vDash \varphi$

4) $X_{\forall,\tau}\varphi$, meaning that φ is true in all τ-successor states, i.e.,
   $\underline{M}, s \vDash X_{\forall,\tau}\varphi$ iff for all time points t with $s <_\tau t$ it holds $\underline{M}, t \vDash \varphi$

5) $X_\forall\varphi$ means that φ is true in all immediate successors:
   $\underline{M}, s \vDash X_\forall\varphi$ iff for all time points t with $s < t$ it holds $\underline{M}, t \vDash \varphi$

6) $F_\exists\varphi$ means that φ is true in some future state, i.e.,
   $\underline{M}, s \vDash F_\exists\varphi$ iff there exists a time point t with $s \ll t$ such that $\underline{M}, t \vDash \varphi$

7) $F_\forall\varphi$, means that for all future paths there is a time point where φ is true:
   $\underline{M}, s \vDash F_\forall\varphi$ iff for all branches $\underline{B}$ starting in s there is a t in $\underline{B}$ with $\underline{M}, t \vDash \varphi$

8) $G_\exists\varphi$ means that φ is true along some future path, i.e.,
   $\underline{M}, s \vDash G_\exists\varphi$ iff there exists a branch $\underline{B}$ starting in s,
   with $\underline{M}, t \vDash \varphi$ for all t in $\underline{B}$.

9) $G_\forall\varphi$, means that φ is true all future states i.e.
   $\underline{M}, s \vDash G_\forall\varphi$ iff for all time points t with $s \ll t$ it holds $\underline{M}, t \vDash \varphi$.

4 Formalizing Upward Reflection Using Epistemic States

In order to let the meta-level manipulate the information that is (explicitly or implicitly) encoded at the object-level, somehow it has to be reflected upward what is known at the object-level, and what is not. The former, i.e. to reflect upward what is known, is straightforward: if an objective formula φ is true at the object-level, we simply reflect this as $K\varphi$ being true. More interestingly, we also want to reflect upward those facts that are (currently) *not* known at the object level. Moreover, we somehow want to implement the idea that the facts that are true at the object level is *all that is known* at the current time point.

The converse relation of $\subseteq$ on Kripke models (Cf. Definition 2.4), will play an important role in the sequel. $M_1 \supseteq M_2$ means that the model M_2, viewed as a representation of the knowledge of an agent, involves a *refinement* of the knowledge associated with model M_1. This has to be understood as follows: in the model M_2 less (or the same) states are considered possible by the agent as compared by the model M_1. So, in the former case the agent has less doubts about the true nature of the world. It will turn out that our definitions below will work in such a way that this means that with respect to model M_2 the agent has at least the knowledge associated with model M_1, and possibly more. So in a transition of M_1 to M_2 we may say that knowledge is gained by the agent. Thus the relation '$\supseteq$' acts as an knowledge ordering on the set of S5-models.

We already noted in Section 2, that the set of states in an S5-model represents the states that are considered possible on the basis of (lack of) knowledge). We also mentioned, that the validities of such a model exactly determine what was called an epistemic state. On the basis of such epistemic states, Halpern & Moses define an entailment relation $\mathrel{\vdash\mkern-7mu\sim}$ with which one can infer what is known, and, more importantly, what is unknown in such epistemic states.

4.1. DEFINITION. Given a set $M \subseteq \mathbf{W}$, we define the *associated S5-model* $\Phi(M)$, given by $\Phi(M) = \langle M, \pi \rangle$ with $\pi: M \times \mathbf{P} \to \{\mathbf{t}, \mathbf{f}\}$ such that $\pi: (m, p) \to m(p)$.

4.2. DEFINITION. Given some objective formula φ, we define M_φ as the set of valuations satisfying φ, i.e., $M_\varphi = \{m \in \mathbf{W} \mid m \vDash \varphi\}$. We denote the epistemic state $\Phi(M_\varphi)$ associated with M_φ by $\mathbb{M}_\varphi$.

4.3. PROPOSITION. $M_\varphi = \cup \{M \mid M \vDash \varphi\} = \cup \{M \mid M \vDash K\varphi\}$.

Thus, in order to get M_φ, we can consider all S5-models of φ and take their union to obtain one 'big' S5-model. We denote the mapping $\varphi \to M_\varphi$ by $\mu: \mu(\varphi) = M_\varphi$.

4.4. DEFINITION (Nonmonotonic epistemic entailment).
For $\varphi \in \mathrm{Prop}(\mathrm{FORM})$, and $\psi \in \mathrm{FORM}$: $\varphi \mathrel{\vdash\mkern-7mu\sim} \psi \Leftrightarrow \psi \in K(\mathbb{M}_\varphi)$.

Informally, this means that ψ is entailed by φ, if ψ is contained in the theory (knowledge) of the "largest S5-model" $\mathbb{M}_\varphi$ of φ. Halpern & Moses showed in [HM84] that this "largest model" need not always be a model of φ itself if we allow φ to contain epistemic operators. However, in our case where we only use objective formulas φ, $\mathbb{M}_\varphi$ is always the largest model for φ. Moreover, Halpern & Moses have shown that in this case the theory $K(\mathbb{M}_\varphi)$ of this largest model is a stable set that contains φ and such that for all stable sets Σ containing φ it holds that $\mathrm{Prop}(K(\mathbb{M}_\varphi)) \subseteq \mathrm{Prop}(\Sigma)$, thus $K(\mathbb{M}_\varphi)$ is the "propositionally least" stable set that contains φ. So $\mathrel{\vdash\mkern-7mu\sim}$ can also be viewed as a *preferential entailment* relation in the sense of Shoham [Sho88], where, in our paper, the preferred models of φ are the largest ones, viz. $\mathbb{M}_\varphi$, where the least objective knowledge is available.

We denote the mapping $\varphi \to K(\mathbb{M}_\varphi)$ by $\kappa: \kappa(\varphi) = K(\mathbb{M}_\varphi)$, the stable set associated with knowing only φ. Alternatively viewed, $\kappa(\varphi)$ is the $\mathrel{\vdash\mkern-7mu\sim}$-closure of φ. Note that since $\kappa(\varphi) = K(\mathbb{M}_\varphi)$ is a stable set, it is also propositionally closed. We now give a few examples to show how the entailment $\mathrel{\vdash\mkern-7mu\sim}$ works: Let p and q be two distinct primitive propositions. Then:

$$p \mathrel{\vdash\mkern-7mu\sim} K(p \vee q) \qquad\qquad p \mathrel{\vdash\mkern-7mu\sim} \neg Kq$$
$$p \mathrel{\vdash\mkern-7mu\sim} Kp \wedge \neg Kq \qquad\qquad p \mathrel{\vdash\mkern-7mu\sim} \neg K\neg q$$
$$p \wedge q \mathrel{\vdash\mkern-7mu\sim} K(p \wedge q) \wedge Kp \wedge Kq \qquad\qquad p \mathrel{\vdash\mkern-7mu\sim} \neg K(p \wedge q)$$
$$p \vee q \mathrel{\vdash\mkern-7mu\sim} K(p \vee q) \qquad\qquad p \vee q \mathrel{\vdash\mkern-7mu\sim} \neg Kp \wedge \neg Kq$$

Obviously, the entailment relation $\mathrel{\vdash\mkern-7mu\sim}$ is nonmonotonic; for instance, we have $p \mathrel{\vdash\mkern-7mu\sim} \neg Kq$, while *not* $p \wedge q \mathrel{\vdash\mkern-7mu\sim} \neg Kq$; it even holds that $p \wedge q \mathrel{\vdash\mkern-7mu\sim} Kq$.

5 The Meta-level: The Epistemic Preference Logic S5P*

The "upward reflection" entailment relation $\vdash$ enables us to derive information about what is known and what is not known. In this section we show how we can use this information to perform meta-level reasoning. To this end we extend our language with operators that indicate that something is a *possible assumption* to be introduced at the object level and thus has a different epistemic status than a *certain fact*. In this way the proverbial "make an assumption" is not made directly in the logic, but a somewhat more cautious approach is taken. The "assuming" itself is part of the downward reflection, to be discussed in Section 6.

Let I be a finite set of indexes. The logic **S5P** (introduced in [MH91] and developed further in [MH92, MH93a,b]) is an extension of the epistemic logic **S5** by means of special modal operators P_i denoting *possible assumption* (w.r.t. situation or frame of mind i), for $i \in I$, and also generalisations P_τ, for $\tau \subseteq I$. Informally, $P_i\varphi$ is read as "φ is a possible assumption (within frame of reference i)". As we shall see below, a frame of reference (or mind) refers to a preferred subset of the whole set S of epistemic alternatives. This operator is very close to the PA-operator of [TT91] and the D-operator of [Doh91]. The generalisation $P_\tau\varphi$ is then read as a possible assumption with respect to the (intersection of the) frames of reference occurring in τ. Also, we have an operator K to denote what is *known* and an operator B to describe what is true (*believed*) *under the hypothesis (under focus)*.

The logic **S5P*** is an extension of **S5P** by allowing an arbitrary set A of additional symbols that denote primitive meta-level propositions; for the moment, one may think of them as a way to express that certain propositions have been observed, or that a given hypothesis is 'in focus'. From a logical point of view, such assertions are just atoms, whose truth is governed by some 'meta truth assignment function'. It may well be, that upon closer examination, there are some logical laws steering the truth of such atoms, but in this stage we will not investigate the inner structure of the given meta-level propositions.

Formally, S5P*-formulas are interpreted on Kripke-structures (called S5P*-models) of the form

$$\mathbb{M} = \langle U, M, \pi, \{M_i \mid i \in I\}, MV \rangle,$$

where:
- U is a collection of states (universe), and $M \subseteq U$ is non-empty (the current focus).
- $\pi: U \times P \to \{t, f\}$ is a truth assignment to the primitive propositions per world
- $M_i \subseteq M$ ($i \in I$) are sets ('frames') of preferred worlds
- MV: $A \to \{t, f\}$ is a valuation for the additional primitive meta-level propositions.

When writing $\mathbb{M}_1 \subseteq \mathbb{M}_2$ we mean that the set of states of $\mathbb{M}_1$ is a subset of those of $\mathbb{M}_2$. Again we identify states s and their truth assignments $\pi(s)$. We let Mod(**S5P***) denote the collection of Kripke-structures of the above form. Given an S5P*-model $\mathbb{M}$ = $\langle U, M, \pi, \{M_i \mid i \in I\}, MV \rangle$, we call the S5-model $\mathbb{M}_M = \langle M, \pi|M \rangle$ the *focused S5-reduct* of $\mathbb{M}$ and $\mathbb{M}_U = \langle U, \pi|U \rangle$ the *universal* or *general S5-reduct* of $\mathbb{M}$.

5.1. DEFINITION (interpretation of S5P*-formulas).

Given a model $M = \langle U, M, \pi, \{M_i \mid i \in I\}, MV \rangle$, we give the following truth definition. Let φ be a formula, $\alpha \in A$ and $m \in U$. The cases in which $\varphi \in P$, $\varphi = (\varphi_1 \wedge \varphi_2)$ or $\varphi = \neg\psi$ are dealt with as in Definition 2.5, for the other cases are as follows:

$(M, m) \vDash K\varphi$ iff for all $m' \in U$, $(M, m') \vDash \varphi$;
$(M, m) \vDash B\varphi$ iff for all $m' \in M$, $(M, m') \vDash \varphi$;
$(M, m) \vDash P_i\varphi$ iff for all $m' \in M_i$, $(M, m') \vDash \varphi$;
$(M, m) \vDash P_\tau\varphi$ iff for all $m' \in M_\tau$, $(M, m') \vDash \varphi$, where $M_\tau = \bigcap_{i \in \tau} M_i$ and $\tau \subseteq I$.
$(M, m) \vDash \alpha$ iff $MV(\alpha) = t$, for α in A.

We see that the clauses state that $P_i\varphi$ is true if φ is a possible assumption w.r.t. subframe M_i, whereas the latter says that $P_\tau\varphi$ is true if φ is a possible assumption w.r.t. the intersection of the subframes M_i, $i \in \tau$. This intersection is denoted by M_τ. We assume that, for $\tau = \varnothing$, $M_\varnothing = \bigcap_{i \in \varnothing} M_i = M$. So in this special case we get that the P_τ modality coincides with the belief operator B. Validity and satisfiability is defined analogously as before.

It is possible to axiomatize (the theory of) **S5P*** by adjusting the axiom system **S5P** of [MH93b]; we will not go into full details here but give some main principles:

5.2. DEFINITION (system **S5P***). In the following, i ranges over I, and τ over subsets of I. Moreover, $\square$, $\square_1$ and $\square_2$ are variables over $\{K, B, P_i, P_\tau \mid i \in I, \tau \subseteq I\}$.

(B1)   All propositional tautologies;

(B2) $(\square\varphi \wedge \square(\varphi \rightarrow \psi)) \rightarrow \square\psi$;
(B3) $\square\varphi \rightarrow \square\square\varphi$;
(B4) $\neg\square\varphi \rightarrow \square\neg\square\varphi$;

(B5) $K\varphi \rightarrow \varphi$;
(B6) $K\varphi \rightarrow \square\varphi$;

(B7) $\neg\square_1\bot \rightarrow (\square_1\square_2\varphi \leftrightarrow \square_2\varphi)$;

(B8) $P_i\varphi \leftrightarrow P_{\{i\}}\varphi$
(B9) $P_\tau\varphi \rightarrow P_{\tau'}\varphi$ $\tau \subseteq \tau'$
(B10) $P_\varnothing\varphi \leftrightarrow B\varphi$
(B11) $\neg B\bot$

(R1)   Modus Ponens
(R2)   Necessitation for K: $\vdash \varphi \Rightarrow \vdash K\varphi$.

We call the resulting system **S5P***. In the sequel we will write $\Gamma \vdash_{\text{S5P*}} \varphi$ or $\varphi \in \text{Th}_{\text{S5P*}}(\Gamma)$ to indicate that φ is an **S5P***-consequence of Γ. We mean this in the more liberal sense: it is possible to derive φ from the assertions in Γ by means of the axioms and rules of the system **S5P***, *including the necessitation rule*. (So, in effect we consider the assertions in Γ as additional axioms: $\Gamma \vdash_{\text{S5P*}} \varphi$ iff $\vdash_{\text{S5P*} \cup \Gamma} \varphi$, cf. [MH93b])

5.3. THEOREM $\Gamma \vdash_{\text{S5P*}} \varphi \Leftrightarrow$ (for all $M \in \text{Mod(\textbf{S5P*})}$: $M \vDash \Gamma \Rightarrow M \vDash \varphi$)
PROOF. Combine the arguments given in [MH92, MH93b].

6 A Temporal Formalization of Downward Reflection

In the previous sections it has been described how the upward reflection can be formalized by a (nonmonotonic) inference based on epistemic states, and the meta-level process by a (monotonic) epistemic logic. In the current section we will introduce a formalization of the downward reflection step in the reasoning pattern. The meta-level reasoning can be viewed as the part of the reasoning pattern where it is determined what the possibilities are for additional assumptions to be made, based on which information is available at the object level and which is not. The outcome at the meta-level concerns conclusions of the form Pφ, where φ is an object-level formula. What is missing still is the step where the assumptions are actually made, i.e., where such formulas φ are added to the object level knowledge, in order to be able to reason further with them at the object level. This is what should be achieved by the downward reflection step. Thus the reasoning pattern as a whole consists of a process of generating possible assumptions and actually assuming them.

By these downward reflections at the object level a hypothetical world description is created (as a refinement or revision of the previous one). This means that in principle not all knowledge available at the object level can be derived already from the object level theory OT: downward reflection creates an essential modification of the object level theory. Therefore it is excluded to model downward reflection according to reflection rules as sometimes can be found in the literature, e.g., "If at the meta-level it is provable that Provable(φ) then at the object level φ is provable" (Cf. [Wey80]):

$$\frac{\text{MT} \vdash \text{Provable}(\varphi)}{\text{OT} \vdash \varphi}$$

A reflection rule like this can only be used in a correct manner if the meta-theory about provability gives a sincere axiomatization of the object level proof system, and in that case by downward reflection nothing can be added to the object level that was not already derivable from the object level theory. Since we essentially extend or modify the object level theory, such an approach cannot serve our purposes here.

In fact, a line of reasoning at the object level is modelled by inferences from subsequently chosen theories instead of inferences from one fixed theory. In principle a downward reflection realizes a shift or transition from one theory to another. In [GTG93] such a shift between theories is formalized by using an explicit parameter referring to the specific theory (called 'context' in their terms) that is concerned, and by specifying relations between theories. In their case downward reflection rules ('bridge rules' in their terms) may have the form:

$$\frac{MT \vdash Provable(OT', \varphi)}{OT' \vdash \varphi}$$

or, in their notation

$$\frac{< Th("\varphi", "OT'"), MT >}{< \varphi, OT'>}$$

Here, the second element of the pair denotes the context in which the formula that is the first element holds. At the meta-level, knowledge is available to derive conclusions about provability relations concerning a variety of object level theories OT. So, if at the object level from a (current) theory OT some conclusions have been derived, and these conclusions have been transformed to the meta-level, then the meta-level may derive conclusions about provability from another object level theory OT'. Subsequently one can continue the object level reasoning from this new object level theory OT'. The shift from OT to OT' is introduced by use of the above reflection rule.

In the approach as adopted here we give a temporal interpretation to these shifts between theories. This can be accomplished by formalizing downward reflection by temporal logic (as in [Tre92]). In a simplified case, where no branching is taken into account, the temporal axiom $(CP\varphi \to X\varphi)$ can be used to formalize downward reflection, for every objective formula φ.

In the general case we want to take into account branching and the role to be played by an index τ in $P_\tau\varphi$. We will use this index τ to label branches in the set of time points. By combining S5P* with the temporal logic obtained in this manner we obtain a formalization of the whole reasoning pattern.

During the reasoning process we modify the information we have at the object level, and accordingly change the focus set M of possible worlds. We can formulate this property as follows:

6.1. DEFINITION. A temporal S5P*-model *obeys downward reflection* if the following holds for any s and τ :
the cluster M_τ in $\underline{M}_s$ is non-empty $\Leftrightarrow$ there is a t with $s <_\tau t$ and for all such t the focus set of states M of $\underline{M}_t$ equals M_τ

Now we are ready to zoom in into the models we like to consider here, the temporal epistemic meta-level architecture (TEMA-) models.

6.2. DEFINITION (TEMA- models)
A TEMA-model $\underline{M}$ is a temporal S5P*-model over an lft T such that:
i T is a labelled time tree;
ii   For every time point s, there is exactly one t with $s <_\varnothing t$;
iii $\underline{M}$ obeys downward reflection.
iv $\underline{M}$ is conservative: if $s < t$ then $(\underline{M}_s)_U \supseteq (\underline{M}_t)_U$

Notice that conservatism refers to the universe U. Sometimes also M is shrinking during the reasoning process: in case of an accumulation of assumptions that are never retracted (e.g., see [HMT93]), but in general M may vary arbitrarily within U.

6.3. THEOREM.

TEMA-models have the following validities:

T0   All the operators of $\{X_{\forall,\tau}, X_\forall, F_\forall, G_\forall\}$ satisfy the K-axiom (C too) and generalisation;

T1 $\vdash_{S5P^*} \varphi \Rightarrow \vDash_{TEMA} C\varphi$ (introduction of C)

T2 $\neg X_\forall \bot$ (successor existence)

T3 $X_{\exists,\tau}\varphi \leftrightarrow X_{\forall,\tau}\varphi$ (label-deterministic)

T4 $X_{\forall,\tau}\varphi \leftrightarrow \neg X_{\exists,\tau}\neg\varphi$ (duality)

T5 $X_\forall\varphi \leftrightarrow \neg X_\exists \neg\varphi$ (duality)

T6 $X_\forall\varphi \leftrightarrow \wedge_{\tau \subseteq I} X_{\forall,\tau}\varphi$ ($<$ is union of $<_\tau$)

T7 $X_\exists\varphi \leftrightarrow \vee_{\tau \subseteq I} X_{\exists,\tau}\varphi$ (dual of T6)

T8 $C(\neg P_\tau \bot \wedge P_\tau\varphi) \leftrightarrow X_{\exists,\tau} CK\varphi$, if φ is objective   (allowing downward reflection)

T9 $G_\forall\varphi \to X_\forall\varphi$ ($< \subseteq \ll$)

T10 $G_\forall\varphi \to X_\forall G_\forall\varphi$ (since $\ll$ is transitive closure of $<$)

T11 $G_\forall(\varphi \to X_\forall\varphi) \to (X_\forall\varphi \to G_\forall\varphi)$ (induction)

T12 $(C\varphi \to X_\forall C\varphi) \wedge (CK\varphi \to X_\forall CK\varphi)$, if φ is objective   (conservativity)

T13 $CK\varphi \to G_\forall CK\varphi$ (from conservativity and induction)

6.4. REMARK. The Theorem above says that the formulas T1 - T13 are at least *sound*; yet we have not been concerned by designing a *complete* logic for TEMA-models.

7   Overall Formalization

In this section we will show how the different parts of the reasoning pattern as described in previous sections can be combined.

7.1 EMA-theories and EMA-entailment

In the language of S5P* we can express meta-knowledge. By combining the formal apparatus of S5P* with Halpern & Moses' nonmonotonic epistemic entailment we obtain a framework in which we can perform static epistemic reasoning: reasoning about the current epistemic state without reflecting the conclusions downwards. We call this framework Epistemic Meta-level Architecture (EMA).

7.1. DEFINITION (EMA-theory). An *EMA-theory* Θ is a pair (W, Δ), where W is a finite, consistent set of objective (i.e. non-modal) formulas describing facts about the world, and Δ is a finite set of S5P*-formulas. The sets W and Δ are to be considered as sets of axioms; we may apply necessitation to them.

7.2. DEFINITION (EMA-entailment). Given an EMA-theory $\Theta = (W, \Delta)$, we define the nonmonotonic inference relation $\vdash_\Theta$ as follows. Let W^* be the conjunction of the formulas in W, and let φ be an objective formula. Then we define the *EMA-entailment relation* $\vdash_\Theta$ w.r.t. Θ as follows: $\varphi \vdash_\Theta \psi \Leftrightarrow_{def} \kappa(\varphi \wedge W^*) \cup \Delta \vdash_{S5P*} \psi$.

7.3. EXAMPLE (*Hypothesis elimination*).
We come back to our example in the introduction. Let the EMA-theory $\Theta = (W, \Delta)$ be defined by

$$W = \{e \rightarrow \neg b\}$$
$$\Delta = \{K\neg s \rightarrow ((K\neg e \wedge \neg K\neg e) \rightarrow P_1 e),$$
$$\text{initially\_}\neg s, \text{ initially\_z } \rightarrow Kz\},$$

where e means "empty battery" and b "lights can burn". Moreover, we have used primitive meta-level propositions "`initially_z`" to indicate that the information "z" is present initially, that is, before the reasoning process starts. Now we have the following:

$$W^* \vdash \neg Ke \wedge \neg K\neg e \quad \text{and} \quad \Delta \cup \{\neg Ke \wedge \neg K\neg e\} \vdash_{S5P*} P_1 e$$

Therefore $\vdash_\Theta P_1 e$

In [HMT94a] we show how this temporal framework also covers default reasoning. Also, we think our framework of temporalizing **S5P*** can easily be adapted to give an account of counterfactual reasoning (as treated in [MH93c]) too.

7.2 TEMA-models and TEMA-entailment

7.4. DEFINITION Let $\Theta = (W, \Delta)$ be an EMA-theory. Then we define a *TEMA-model of* Θ as a TEMA-model $\underline{M}^\Theta$ such that:

i (basis: the root) $\underline{M}^\Theta_r$ is an S5P*-model such that
   (a) the universal S5-reduct of $\underline{M}^\Theta_r$ is the S5-model M_{W^*},
   (b) $\underline{M}^\Theta_r$ satisfies the meta-knowledge, i.e., $\underline{M}^\Theta_r \vDash \Delta$.
ii (induction step) Suppose that we are given an S5P*-model at snapshot $\underline{M}^\Theta_s$. Then we have that for a(n S5P*-) model $\underline{M}^\Theta_t$ with $s <_\tau t$, it holds that:
   (a) $(\underline{M}^\Theta_t)_U$ is the S5-model M_τ as it appeared as a cluster in $\underline{M}^\Theta_s$, and
   (b) $\underline{M}^\Theta_t$ satisfies the meta-knowledge again, i.e., $\underline{M}^\Theta_t \vDash \Delta$.

In general, there are multiple TEMA-models of an EMA-theory Θ. Note that clause ii(a) reflects the downward reflection operation with respect to the P_τ-assumptions.

Even in less trivial reasoning processes we may be interested in some kind of final outcome (a conclusion set). In our case the universe U always shrinks, but not always the focus set M does so. We can view U as a core of derived facts that after all survives during the reasoning pattern; this is reflected in the following definition:

7.5. DEFINITION (limit model).
Suppose $\underline{M}$ is a conservative temporal V-model. The intersection of the models $\underline{M}(s)$ for all s in a given branch $\underline{B} = (T, (<_\tau)_{\tau \in L})$ of the lft T is called the *limit model* of the branch, denoted $\lim_{\underline{B}} \underline{M}$. The set of limit models for all branches is called the *set*

of limit models of $\underline{M}$. These definitions straightforwardly extend to temporal S5- and S5P-models, by identifying M with its set of states, U.

We might formulate definitions of entailment of objective formulae related to any model, or any model based on the standard tree. But it may well happen that there are branches in such models, for instance labelled by the empty set only, that contain no additional information as compared to the background knowledge. It is not always realistic to base entailment on such informationally poor branches in a model. Thus we define:

7.6. DEFINITION (informationally maximal)
We define for branches $\underline{B}_1$ and $\underline{B}_2$ with the same set of time stamps of a TEMA-model $\underline{M}$ that $\underline{B}_2$ is *informationally larger* than $\underline{B}_1$, denoted $\underline{B}_1 \leq \underline{B}_2$, if for all $i \in N$ and s, t with $|s| = |t| = i$ it holds $\underline{B}_2(s) \subseteq \underline{B}_1(t)$.
We call $\underline{B}_1$ *informationally maximal* if it is itself the only branch of $\underline{M}$ that is informationally larger.

7.7. DEFINITION (regular model)
A TEMA-model $\underline{M}$ is called regular if all branches are informationally maximal. The submodel based on all time points t included in at least one informationally maximal branch is called the *regular core* of $\underline{M}$, denoted by reg($\underline{M}$).

7.8. DEFINITION
i   for $k \in N$ we define $\underline{M}^{(k)} = \cup_{t \in reg(\underline{M}), |t| = k} \underline{M}(t)$
ii $\underline{M}^{\omega} = \cap_{k \in N} \underline{M}^{(k)}$.

7.9. THEOREM (From [HMT94a]). Let M be a TEMA-model. Then:
i   For $k \leq k'$ we have $\underline{M}^{(k')} \subseteq \underline{M}^{(k)}$
ii $\underline{M}^{\omega} = \cap_{\underline{B} \text{ branch of } reg(\underline{M})} \lim_{\underline{B}} \underline{M}$.

Since (regular) TEMA models describe a reasoning process over time, it seems natural to have notions of entailment that exploit the conservativity of such models, expressing the monotonic growth of knowledge of objective formulas. The latter restriction is important, since it may be the case that initially an atom p is not known, yielding $\neg Kp$ and hence $K\neg Kp$, while at some later point p has been learnt, giving $\neg K\neg Kp$, expressing that some knowledge (i.e., knowledge about ignorance!) has been lost. So, in the sequel we will be interested in formulas of the form $CK\varphi$, where φ is an objective formula. We will call such formulas *currently known objective formulas* (cko's) and use α and β as variables over them.

In the literature on non-monotonic reasoning, one usually distinguishes so called *sceptical* (true in *all* obtained models) and *credulous* (true in *some* of them) notions of entailment. Due to the fact that we have an (infinite) branching time structure, we have a great variety of combining these notions, although for the formulas that we are interested in, the current objective formulas, various of such notions do collapse. This observation is made explicit in the following theorem.

7.10. THEOREM Let α and β be cko-formulas. Then, on TEMA models, we have the following equivalences:

i $G_\exists \alpha \equiv \alpha$ iii $G_\forall \alpha \equiv X_\forall \alpha$

ii $\alpha \equiv (\alpha \wedge G_\forall \alpha)$ iv $F_\exists \alpha \equiv F_\exists(G_\forall \alpha \wedge \alpha)$

7.11. DEFINITION (sceptical entailment) Let $\underline{M}$ be a TEMA-model with root r and α an cko-formula. We define the *sceptical entailment relation* by:

$$\underline{M} \models_{scep} \alpha \qquad \text{iff} \qquad \underline{M},r \models F_\forall \alpha$$

Due to our remark that we made before definition 7.6, it makes most sense to use this notion of sceptical entailment for (sub) models of type reg($\underline{M}$).

7.12. PROPOSITION Let $\underline{M}$ be a TEMA-model with root r and α a cko-formula. The following are equivalent:

i   reg($\underline{M}$) $\models_{scep} \alpha$

ii $\lim_{\underline{B}} \underline{M} \models \alpha$ for every maximal branch $\underline{B}$ of the regular core of $\underline{M}$.

For our definition of credulous entailment we can be less restrictive; we do not need to bother about informationally maximal branches. Especially, too little information in one branch can always be overcome by another, informationally larger branch.

7.13. DEFINITION (credulous entailment) Let $\underline{M}$ be a TEMA-model and α a cko-formula. We define: $\underline{M} \models_{cred} \alpha$ iff $\underline{M},r \models F_\exists \alpha$

7.14. PROPOSITION
Let $\underline{M}$ be a TEMA-model and α a cko-formula. The following are equivalent:

i $\underline{M} \models_{cred} \alpha$

ii $\lim_{\underline{B}} \underline{M} \models \alpha$ for some maximal branch $\underline{B}$

We finally can give the definitions of sceptical and credulous entailment.

7.15. DEFINITION (entailment from an EMA-theory)
Let $\Theta = (W, \Delta)$ be an EMA-theory and φ an objective formula.

$\Theta \models_{scep} \varphi$ iff   for all models $\underline{M}$ of Θ it holds reg($\underline{M}$) $\models_{scep} \varphi$

$\Theta \models_{cred} \varphi$ iff   for all models $\underline{M}$ of Θ it holds $\underline{M} \models_{cred} \varphi$

7.16. PROPOSITION
Let $\Theta = (W, \Delta)$ be an EMA-theory and φ an objective formula. Then it holds

$$\Theta \models_{scep} \varphi \Rightarrow \Theta \models_{cred} \varphi$$

7.3 The example reasoning pattern

We return to our running example and show how it obtains its natural semantics. To cover the whole reasoning pattern, we take $W = \{(e \vee p) \to \neg s, e \to \neg b\}$ and Δ as:

$\Delta = \{K\neg s \to ((\neg Ke \wedge \neg K\neg e) \to P_1 e),$ *if $\neg s$ is known,*

and we do not know whether $\neg e$ holds, we may choose e as an hypothesis

$(K\neg s \wedge K\neg e) \to ((\neg Kp \wedge \neg K\neg p) \to P_2 p),$ *we investigate p if we know $\neg s$ and $\neg e$*

```
observed_b, initially_¬s,
initially_z → Kz,
possible_hyp_e, possible_hyp_p,
(possible_hyp_z ∧ Pᵢz) → X∃,ᵢ focus_z,
(focus_z ∧ observed_w ∧ B¬w) → bad_hyp_z,
(bad-hyp_z ∧ focus_z) → G∀K¬z
```

$z \in \{e, p\}, i \in \{1, 2\}$

$z \in \{e, p\}, w \in \{b\}$

$z \in \{e, p\}\}$

We denote worlds by 4-tuples $(w, x, y, z) \in \{0,1\}^4$, to be interpreted as the truth-values of the tuple (b, e, p, s). We construct a model that is linear.

* Initial snapshot

We obtain the initial model $\underline{M}^{(0)} = \langle U^{(0)}, M^{(0)}, \pi^{(0)}, M_1^{(0)}, M_2^{(0)}, MV^{(0)} \rangle$. The universe $U^{(0)}$ is based on all models of W; we take $M^{(0)} = U^{(0)}$. Furthermore, we have seen in Example 7.3, how we can derive $P_1 e$ from (W, Δ), defining the cluster $M_1^{(0)}$. In this model $M_2^{(0)} = \varnothing$, (Cf. our remarks at the end of this section); $MV^{(0)}$ is to be understood as the minimal valuation on the atoms of A so that Δ is satisfied;

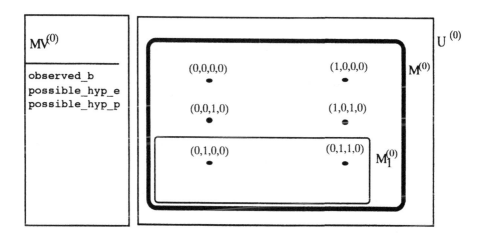

* Next snapshot

Using T8, $C(\neg P_1 \bot \wedge P_1 e) \leftrightarrow X_{\exists,1} CKe$, we obtain a downward reflection into the next model $\underline{M}^{(1)} = \langle U^{(1)}, M^{(1)}, \pi^{(1)}, M_1^{(1)}, M_2^{(1)}, MV^{(1)} \rangle$, where $MV^{(1)}$ is like

$MV^{(0)}$, but under $MV^{(1)}$, now also `focus_e` is true. Moreover, $U^{(1)} = U^{(0)}$, and the new cluster $M^{(1)}$ equals the old $M_1^{(0)}$:

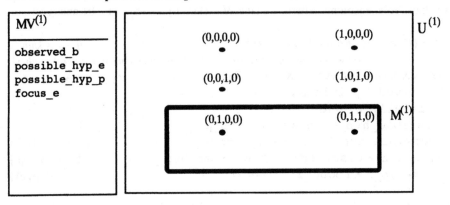

* Further snapshots

Now, the cluster $M^{(1)}$ is the one which is in the current focus: at its object-level theory we can derive that $\neg b$, since in $\underline{M}^{(1)}$ we have $\underline{M}^{(1)} \vDash B\neg b$. We use the meta-knowledge (`focus_z` $\wedge$ `observed_w` $\wedge B\neg w) \rightarrow$ `bad_hyp_z` with e for z and b for w. Using Δ again this yields $G_\forall K\neg e$, so that for all $\underline{M}^{(2)}$ with $\underline{M}^{(1)} < \underline{M}^{(2)}$ it holds that $\underline{M}^{(2)} \vDash \neg e$; semantically this amounts to saying that the universe $U^{(2)}$ of all future models $\underline{M}^{(2)}$ is at most the set $U^{(1)} \setminus M^{(1)}$, and one may proceed and investigate the hypothesis of bad plugs (p) for the failure of the non-starting car.

Some remarks are in order here. Firstly, we have only analyzed some *intended* model for the theory $\Theta = (W, \Delta)$; a model that carries the assumption that the theory (W, Δ) is *all* that we know; this justified for example that we took $M_2^{(1)}$ to be empty. Secondly, since the atoms in A are rather application-dependent, we decided to add all assumptions about them to Δ; however, for classes of applications, one might consider to add some of those properties as axioms to S5P*. Finally, in this particular example we did not exploit the fact that we have a *branching* time model; one may change the example in such a way that the two possible hypotheses p and e are so to speak investigated *simultaneously*. In our example there may be specific reasons to investigate one hypothesis before the other: because one of them (i.e., e) is more easy to refute, for example; this preference is also explicitly modelled in Δ.

8 Conclusions

In [MH93a,b] an Epistemic Default Logic (EDL) was introduced inspired by the notion of meta-level architecture that also was the basis for the BMS-approach introduced in [TT91]. In EDL drawing a default conclusion has no other semantics than that of adding a modal formula to the meta-level. No downward reflection takes place to be able to reason with the default conclusions at the object level (by means of which assumptions actually can be made). In [TT91] downward reflection to actually make assumptions takes place, but no logical formalization was given: it was defined only in a procedural manner.

In principle a downward reflection that enlarges the object level theory disturbs the object level semantics: facts are added that are not logically entailed by the available object level knowledge. Adding a temporal dimension (in the spirit of [FG92]) enables one to obtain formal semantics of downward reflection in a dynamic sense: as a transition from the current object level theory to a next one (where the reflected assumption has been added). This view, also underlying the work presented in [ET93] and [Tre92], turns out to be very fruitful. A number of notions can be formalized in temporal semantics in a quite intuitive and transparent manner.

The temporal epistemic meta-level architecture described here is a powerful tool to reason with dynamic assumptions: it enables one to introduce and retract additional assumptions during the reasoning, on the basis of explicit meta-knowledge. In [Tre90], [Tre91] various applications of this architecture are shown. In the current paper we formalized the semantics of a temporal epistemic meta-level architecture by means of entailment on the basis of temporalized Kripke models.

Acknowledgements

Discussions about the subject with Joeri Engelfriet have played a stimulating role for the development of the material presented here.

REFERENCES

[Doh91]   P. Doherty, NM3 - A Three-Valued Cumulative Non-Monotonic Formalism, in: *Logics in AI*, (J. van Eijck, ed.), LNCS 478, Springer, Berlin, 1991, pp. 196-211.

[ET93]   J. Engelfriet, J. Treur, A Temporal Model Theory for Default Logic, in: M. Clarke, R. Kruse, S. Moral (eds.), Proc. ECSQARU '93, Springer Verlag, 1993, pp. 91-96.

[FG92]   M. Finger, D. Gabbay, Adding a temporal dimension to a logic system, Journal of Logic, Language and Computation 1 (1992), pp. 203-233.

[Gab82]   D.M. Gabbay, Intuitionistic basis for non-monotonic logic, In: G. Goos, J. Hartmanis (eds.), 6th Conference on Automated Deduction, LNCS 138, Springer Verlag, 1982, pp.260-273

[GTG93]   E. Giunchiglia, P. Traverso, F. Giunchiglia, Multi-Context Systems as a Specification Framework for Complex Reasoning Systems, in: J. Treur, Th. Wetter (eds.), Formal Specification of Complex Reasoning Systems, Ellis Horwood, 1993, pp. 45-72

[HM84]   J.Y. Halpern & Y.O. Moses, Towards a Theory of Knowledge and Ignorance, *Proc. Workshop on Non-Monotonic Reasoning*, 1984, pp. 125-143.

[HMT94a]   W. van der Hoek, J.-J. Ch. Meyer & J. Treur, Temporalizing Epistemic Default Logic. Techn. Report, Free University Amsterdam, 1994

[HMT94b]   W. van der Hoek, J.-J. Ch. Meyer & J. Treur, Formal Semantics of Temporal Epistemic Reflection. Techn. Report, Free University Amsterdam, 1994.

[MH91]   J.-J.Ch. Meyer & W. van der Hoek, Non-Monotonic Reasoning by Monotonic Means, in: J. van Eijck (ed.), *Logics in AI (Proc. JELIA '90)*, LNCS 478, Springer, 1991, pp. 399-411.

[MH92]   J.-J.Ch. Meyer & W. van der Hoek, A Modal Logic for Nonmonotonic Reasoning, in: W. van der Hoek, J.-J. Ch. Meyer, Y.H. Tan & C. Witteveen (eds.), *Non-Monotonic Reasoning and Partial Semantics*, Ellis Horwood, Chichester, 1992, pp. 37-77.

[MH93a]   J.-J. Ch. Meyer & W. van der Hoek, An Epistemic Logic for Defeasible Reasoning using a Meta-Level Architecture Metaphor. Techn. Report IR-329, Free University, Amsterdam, 1993.

[MH93b]   J.-J. Ch. Meyer & W. van der Hoek, A Default Logic Based on Epistemic States. in: M. Clarke, R. Kruse, S. Moral (eds.), Proc. ECSQARU '93, Springer Verlag, 1993, pp. 265-273. Full version to appear in Fundamenta Informatica.

[MH93c]   J.-J. Ch. Meyer & W. van der Hoek, Counterfactual reasoning by (means of) defaults in: *Annals of Mathematics and Artificial Intelligence* 9 (III-IV), (1993), pp. 345-360.

[MH94]   J.-J. Ch. Meyer & W. van der Hoek, *Epistemic Logic for AI and Computer Science*, forthcoming.

[Moo85]   R.C. Moore, Semantical Considerations on Nonmonotonic Logic, *Artificial Intelligence* 25, 1985, pp. 75-94.

[Sho88]   Y. Shoham, *Reasoning about Change*, MIT Press, Cambridge, 1988.

[TT91]   Y.-H. Tan & J. Treur, A Bi-Modular Approach to Non-Monotonic Reasoning, in: Proc. WOCFAI'91 (M. DeGlas & D. Gabbay, eds.), Paris, 1991, pp. 461-475.

[Tre90]   J. Treur, Modelling nonclassical reasoning patterns by interacting reasoning modules, Report IR-236, Free University Amsterdam, 1990.

[Tre91]   J. Treur, Interaction types and chemistry of generic task models, in: B. Gaines, M. Linster (eds), Proc. EKAW'91, GMD Studien 211, 1992, Bonn.

[Tre92]   J. Treur, Towards dynamic and temporal semantics of meta-level architectures, Report IR-321, Free University Amsterdam, 1992. Also: Proc. META'94, 1994 (this volume).

[Wey80]   R.W. Weyhrauch, Prolegomena to a Theory of Mechanized Formal Reasoning, Artificial Intelligence J. 13 (1980), pp. 133-170.

Temporal Semantics of Meta-Level Architectures for Dynamic Control of Reasoning[+]

Jan Treur

Free University Amsterdam, Department of Mathematics and Computer Science
De Boelelaan 1081a, 1081 HV Amsterdam, The Netherlands
Email: treur@cs.vu.nl

Abstract

Meta-level architectures for dynamic control of reasoning processes are quite powerful. In the literature many applications in reasoning systems modelling complex tasks are described, usually in a procedural manner. In this paper we present a declarative framework based on temporal (partial) logic that enables one to describe the dynamics of reasoning behaviour by temporal models. Using these models the semantics of the behaviour of the whole (meta-level) reasoning system can be described by a set of (intended) temporal models.

1 Introduction

In the literature on meta-level architectures and reflection two separate streams can be distinguished: a logical stream (e.g., [3], [9], [25]) and a procedural stream (e.g., [5], [6]). Unfortunately there is a serious gap between the two streams. In the logical stream one restricts oneself often to *static reflections*; i.e., of facts the truth of which does not change during the reasoning: e.g., **provable**(A) (with **A** an object-level formula). In the procedural stream usually facts are reflected the truth of which changes during the whole reasoning pattern; e.g. control statements like **current_goal**(A), with **A** an object-level formula, that are sometimes true and sometimes false during the reasoning. If applications to dynamic control of complex reasoning tasks are concerned these *dynamic reflections* are much more powerful (for applications see, e.g.: [6], [5], or [4], [8], [18], [19], [21], [23]). However, a logical basis for this is still lacking. The current paper provides a logical foundation (based on temporal logic) of meta-level architectures for dynamic control. Our logical framework enables one to study these dynamic meta-level architectures by logical means. It can be viewed as a contribution to bridge the gap between the logical stream and the procedural stream.

A meta-level architecture consists of two interacting components that reason at different levels: the object-level component and the meta-level component. The interactions between the components are: upward reflection (information transfer from object-level component to meta-level component) and downward reflection

---

[+] This work is partially supported by ESPRIT III Basic Research Action 6156 (DRUMS II).

(information transfer from meta-level component to object-level component). In a meta-level architecture each of the reasoning processes in one of the components can be assigned its own *local semantics* (local in view of the whole system) that can be formally described according to well-known approaches to static (declarative) and dynamic (procedural) semantics (for instance as known from logic programming). In this local semantics a *static view* (the contents; declarative) and a *dynamic view* (the control; procedural) can be distinguished, and these views can be treated (to a certain extent) as orthogonal (e.g., see [15]).

The crucial point of a meta-level architecture for dynamic control is that the semantics of the meta-level component relates in some manner to (the control of) the reasoning process of the object-level component. To obtain an *overall semantics* for the whole system, the crux is to formally describe the precise *semantic connection* between the two components. The semantic connection is used in a bidirectional manner. Firstly, meta-level reasoning *is about* (or refers to) process aspects of the reasoning of the object-level component and uses information about that (via upward reflection). Secondly, the results of its reasoning may *affect* the control of this object-level reasoning process by changing its control settings (downward reflection). Therefore meta-level architectures enable one to represent control knowledge in the system in an explicit declarative manner.

A formal semantic connection between the two components should relate some (formal) description of procedural (inference process control) aspects of the object-level component to the formal declarative description of the meta-level component. Therefore for the overall semantics of this type of reasoning system a global distinction between a static view and a dynamic view essentially cannot be made (as an extension of the distinction that can be made within each of the components). The views are not orthogonal in this case: to a certain extent they are defined in terms of each other. In particular, it is impossible to provide independent declarative semantics for such systems without taking into account the dynamics (of the object-level component). For the overall architecture a formal semantical description is needed that systematically integrates both views. The lack of such overall semantics for complex reasoning systems (with meta-level reasoning capabilities) was one of the major open problems that were identified during the ECAI'92 workshop on Formal Specification Methods for Complex Reasoning Systems (see [24]).

In this paper we develop a formal framework where partial models are used to explicitly represent (current) information states (see also [12]). This enables us to represent inference processes within each of the components as transitions between partial models, and a trace of a reasoning process as a partial temporal model. First, in Section 2 some basic notions from Temporal Partial Logic are introduced. Next, in Section 3 we give a formalization of a static view and a dynamic view on the object-level reasoning component. In Section 4 a temporal interpretation of meta-level reasoning is introduced and in Section 5 formal semantics for reasoning patterns of meta-level architectures for dynamic control are presented based on partial temporal models that formalize the overall reasoning traces. Finally, in Section 6 an example is presented. In the full report, under an additional finiteness condition a temporal theory is constructed that in some sense is an axiomatisation of the meta-level architecture: its minimal models are just the overall reasoning traces.

2 Basic notions of Temporal Partial Logic

In this section we will introduce the formal notions of partial logic and temporal partial logic that are needed later on.

Definition 2.1 (signature and propositional formula)
a) A *signature* Σ is an ordered sequence of (propositional) atom names, or a sequence of sort, constant, function and predicate symbols in (many-sorted) predicate logic. By $\Sigma_1 \subseteq \Sigma_2$ we denote that Σ_1 is a *subsignature* of Σ_2. The *disjoint union* of two signatures Σ_1 and Σ_2 is denoted by $\Sigma_1 \oplus \Sigma_2$. A *mapping of signatures* $\alpha : \Sigma_1 \to \Sigma_2$ is a mapping from the set of symbols of Σ_1 into the set of symbols of Σ_2 such that sorts are mapped to sorts, constants to constants, predicates to predicates, functions to functions, and the arities and argument-sort relations are respected.
b) The set $At(\Sigma)$ is the *set of (ground) atoms* based on Σ. By a (ground) *formula* of signature Σ we mean a proposition built from (ground) atoms using the connectives $\wedge$, $\to$, $\neg$. We will call these formulae *propositional formulae*, in contrast to the temporal formulae introduced later on. Below we will assume all formulas ground (closed and without quantifiers). For a finite set F of formulae, $con(F)$ denotes the conjunction of the elements of F; in case F is the empty set then by definition $con(F)$ is **true**.
By $Lit(\Sigma)$ we denote the *set of ground literals* of signature Σ.

As discussed in [12], partial models can be used to represent information states in a reasoning system; therefore we define:

Definition 2.2 (partial models as information states)
a) An *information state* or *partial model* M for the signature Σ is an assignment of a truth value from $\{0, 1, u\}$ to each of the atoms of Σ, i.e. $M: At(\Sigma) \to \{0, 1, u\}$. An atom a is *true* in M if 1 is assigned to it, and *false* if 0 is assigned; else it is called *undefined* or *unknown*. A literal L is called *true* in M, denoted by $M \models^+ L$ (resp. *false* in M, denoted by $M \models^- L$) we mean $M(L) = 1$ (resp. $M(L) = 0$) if L is an atom and $M(a) = 0$ (resp. $M(a) = 1$) if $L = \neg a$ with $a \in At(\Sigma)$. By $Lit(M)$ we denote the *set of literals* which are true in M.
We call a partial model M *complete* if no $M(a)$ equals u for any $a \in At(\Sigma)$.
b) The *set of all information states* for Σ is denoted by $IS(\Sigma)$. If $\Sigma_1 \subseteq \Sigma_2$ then this induces an embedding of $IS(\Sigma_1)$ into $IS(\Sigma_2)$; we will identify $IS(\Sigma_1)$ with its image under this embedding: $IS(\Sigma_1) \subseteq IS(\Sigma_2)$. Furthermore, $IS(\Sigma_1 \oplus \Sigma_2)$ can (and will) be identified with the Cartesian product $IS(\Sigma_1) \times IS(\Sigma_2)$.
c) We call N a *refinement* of M, denoted by $M \leq N$, if for all atoms $a \in At(\Sigma)$ it holds: $M(a) \leq N(a)$ where the partial order on truth values is defined by
$$u \leq 0, u \leq 1, u \leq u, 0 \leq 0, 1 \leq 1.$$
d) If K is a set of formulae of signature Σ, a complete model M of signature Σ is called *a model of* K if all formulae of K are true in M. An information state is *consistent* with K if it can be refined to a complete model that is a model of K. By $IS_K(\Sigma)$ we denote the set of all information states for Σ that are consistent with K.

e) If **M** is a partial model for the signature Σ and $S \subseteq At(\Sigma)$, then by **M|S** we denote the *restriction* or *reduct* of **M** to S, defined by

$$\mathbf{M|S(a)} \quad = \quad \mathbf{M(a)} \quad \text{if } \mathbf{a} \in S$$
$$\mathbf{u} \qquad \text{otherwise (i.e., if } \mathbf{a} \in At(\Sigma)\backslash S)$$

If $S = At(\Sigma')$ for some subsignature Σ' of Σ, then we denote **M|S** by **M|Σ'**.

Notice that for partial models **M, N** for Σ it holds $\mathbf{M} \leq \mathbf{N}$ if and only if $\mathbf{M} \vDash^+ \mathbf{L} \Rightarrow \mathbf{N} \vDash^+ \mathbf{L}$ for all literals $\mathbf{L} \in Lit(\Sigma)$. We base the interpretation of propositional formulae on the Strong Kleene truth tables for the logical connectives (see also Definition 2.6 below); more details and possibilities of partial semantics can be found in [2], [13].

Definition 2.3 (labeled flow of time)

Let L be a set of labels.

a) A *(discrete) labeled flow of time* , labeled by L is a pair $\mathbb{T} = (T, (<_i)_i \in L)$ consisting of a nonempty set **T** of time points and a collection of binary relations $<_i$ on **T**. Here for **s, t** in **T** and $i \in L$ the expression $\mathbf{s} <_i \mathbf{t}$ denotes that **t** is a (immediate) *successor* of **s** with respect to an arc labeled by **i**. Sometimes we use just the binary relation $\mathbf{s} < \mathbf{t}$ denoting that $\mathbf{s} <_i \mathbf{t}$ for some **i** (for some label **i** they are connected). Thus $<$ is defined as $\cup_i <_i$. We will assume that this relation $<$ is irreflexive, antisymmetric and antitransitive.

We also use the (irreflexive) transitive closure $\ll$ of this binary relation, defined as $<^+$.

b) We call $\mathbb{T}$ *linear* if $\ll$ is a linear ordering and *rooted* with root **r** if **r** is a (unique) least element: for all **t** it holds $\mathbf{r} = \mathbf{t}$ or $\mathbf{r} \ll \mathbf{t}$. We say $\mathbb{T}$ satisfies *successor existence* if every time point has at least one successor: for all $\mathbf{s} \in \mathbf{T}$ there exists a $\mathbf{t} \in \mathbf{T}$ such that $\mathbf{s} < \mathbf{t}$.

Definition 2.4 (partial temporal model)

Let Σ be a signature.

a) A *labeled (linear time) partial temporal model* of signature Σ with labeled flow of time $\mathbb{T}$ is a mapping

$$\mathbf{M: T} \to \mathbf{IS(\Sigma)}$$

For any fixed time point **t** the partial model **M(t)** is also denoted by $\mathbf{M_t}$; the model **M** can also be denoted by $(\mathbf{M_t})_{t \in T}$. If **a** is an atom, and **t** is a time point in $\mathbb{T}$, and $\mathbf{M_t(a)} = 1$, then we say in this model **M** *at time point* **t** *the atom* **a** *is true*. Similarly we say that *at time point* **t** *the atom* **a** is *false*, respectively *unknown*, if $\mathbf{M_t(a)} = 0$, respectively $\mathbf{M_t(a)} = \mathbf{u}$.

b) The *refinement relation* $\leq$ between partial temporal models is defined as: $\mathbf{M} \leq \mathbf{N}$ if **M** and **N** have the same flow of time and for all time points **t** and atoms **a** it holds $\mathbf{M_t(a)} \leq \mathbf{N_t(a)}$.

c) **M** is called *conservative* if for all $\mathbf{s, t} \in \mathbb{T}$ with $\mathbf{s} < \mathbf{t}$ it holds $\mathbf{M_s} \leq \mathbf{M_t}$.

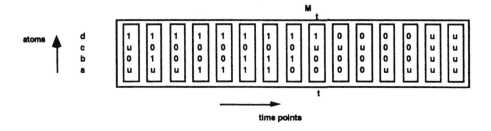

Fig 1 Example of a partial temporal model

From now on we will assume that all used labeled flows of time are linear, rooted and satisfy successor existence. This is equivalent to $\mathbb{T}$ being order-isomorphic to the natural numbers $\mathbb{N}$. Therefore in the rest of the paper we will use $\mathbb{N}$ as our flow of time.

We introduce three temporal operators, **X, P** and **C**, referring to the *next* information state, *past* information states and the *current* information state, respectively. Intuitively, the temporal formula $X\alpha$ is true at time t means that viewed from time point t, the formula α is true in the next information state. We use labeled next operators to be able to distinguish different types of steps. The temporal formula $P\alpha$ is true at time t means that α is true in some past information state. Furthermore we will need an operator that expresses the fact that *currently* α is true (in the *current* information state); this will be the operator **C**. Definition 2.5 makes this formal. Notice that sometimes we will denote the application of the temporal operators like $F(\alpha)$; if no confusion is expected, for shortness we write $F\alpha$. We will not need nested operators in this paper, although it would certainly be possible to use them.

Definition 2.5   (semantics of the temporal operators)
Let a propositional formula α, a labeled partial temporal model **M**, a label $i \in L$ and a time point $t \in \mathbb{N}$ be given. Then:

a) $(M, t) \vDash^+ X_i\alpha$ $\Leftrightarrow$ $\exists s \in \mathbb{N} \; [\, t <_i s \; \& \; (M, s) \vDash^+ \alpha \,]$

       $(M, t) \vDash^- X_i\alpha$ $\Leftrightarrow$ $(M, t) \nvDash^+ X_i\alpha$

b) $(M, t) \vDash^+ C\alpha$ $\Leftrightarrow$ $(M, t) \vDash^+ \alpha$

       $(M, t) \vDash^- C\alpha$ $\Leftrightarrow$ $(M, t) \nvDash^+ C\alpha$

c) $(M, t) \vDash^+ P\alpha$ $\Leftrightarrow$ $\exists s \in \mathbb{N} \; [\, s \ll t \; \& \; (M, s) \vDash^+ \alpha \,]$

       $(M, t) \vDash^- P\alpha$ $\Leftrightarrow$ $(M, t) \nvDash^+ P\alpha$

Now we can make new formulae using conjunctions, negations and implications of these temporal operators. From now on the word (temporal) formula will be used to denote a formula possibly containing any of the new operators, unless stated otherwise. As we do not need nesting of temporal operators, for convenience we will only consider non-nested formulae.

Definition 2.6 (temporal formulae and their interpretation)

Let Σ be a signature, let M be a labeled partial temporal model for Σ, and $t \in \mathbb{N}$ a time point.

a) A *temporal atom* of signature Σ is a formula $O\alpha$ where O is one of the temporal operators in Definition 2.5 and α a propositional formula of signature Σ.

A *temporal formula* of signature Σ is a formula built from temporal atoms of signature Σ, using the logical connectives $\neg, \wedge, \rightarrow$.

b) Any propositional atom $p \in At(\Sigma)$ is interpreted according to:

$$(M, t) \vDash^+ p \quad \Leftrightarrow \quad M(t, p) = 1$$
$$(M, t) \vDash^- p \quad \Leftrightarrow \quad M(t, p) = 0$$

For the interpretation of a temporal atom, see Definition 2.5.

c) For any two temporal or propositional formulae φ and ψ:

(i) $(M, t) \vDash^+ \varphi \wedge \psi \quad \Leftrightarrow \quad (M, t) \vDash^+ \varphi$ and $(M, t) \vDash^+ \psi$

  $(M, t) \vDash^- \varphi \wedge \psi \quad \Leftrightarrow \quad (M, t) \vDash^- \varphi$ or $(M, t) \vDash^- \psi$

(ii) $(M, t) \vDash^+ \varphi \rightarrow \psi \quad \Leftrightarrow \quad (M, t) \vDash^- \varphi$ or $(M, t) \vDash^+ \psi$

  $(M, t) \vDash^- \varphi \rightarrow \psi \quad \Leftrightarrow \quad (M, t) \vDash^+ \varphi$ and $(M, t) \vDash^- \psi$

(iii) $(M, t) \vDash^+ \neg \varphi \quad \Leftrightarrow \quad (M, t) \vDash^- \varphi$

  $(M, t) \vDash^- \neg \varphi \quad \Leftrightarrow \quad (M, t) \vDash^+ \varphi$

d) For any temporal or propositional formula φ:

$(M, t) \nvDash^+ \varphi \quad \Leftrightarrow \quad (M, t) \vDash^+ \varphi$ does not hold

$(M, t) \nvDash^- \varphi \quad \Leftrightarrow \quad (M, t) \vDash^- \varphi$ does not hold

$(M, t) \vDash^u \varphi \quad \Leftrightarrow \quad (M, t) \nvDash^+ \varphi$ and $(M, t) \nvDash^- \varphi$

e) For a partial model M and a set of formulae K, by $M \vDash^+ K$ we mean $M \vDash^+ \varphi$ for all $\varphi \in K$. By $M \vDash^+ \varphi$ we mean $(M, t) \vDash^+ \varphi$ for all $t \in \mathbb{N}$ and by M is a *temporal model of* K, denoted $M \vDash^+ K$, we mean $M \vDash^+ \varphi$ for all $\varphi \in K$, where K is a set of temporal or propositional formulae. A model M of K is called a *minimal model* of K if for any model M' of K with $M' \leq M$ it holds $M' = M$.

The temporal approach provides declarative semantics for systems that behave dynamically, essentially since time has been put into the domain of consideration in an explicit manner: one reasons both on world states and the time points on which they occur. This means that non-conservative changes in truth values of a statement **b** referring to a changing world state are accounted for by considering the statement in fact as two (or more) statements: one (**t**, **b**) referring to one time point **t**, and another one (**s**, **b**) referring to another time point **s**. The truth values of these two statements do not change; e.g., it will always remain true that at time point **t** the statement **b** holds. Thus a dynamic system is described in a declarative manner. Its set of intended models can be constructed in the temporal sense described above. One specific behaviour of the system corresponds to one of these temporal models. We will work out this general idea for the case of a meta-level architecture. More details on temporal logic can be found in [1], [10].

3 Static and dynamic view on the object-level reasoning

In this section we use the notion of a *partial model* to formalize the information state of the object-level reasoning component at a certain moment. A transition of one information state to another one can be formally described by a *mapping between the partial models* specifying the information states. In a reasoning process such a transition is induced by a reasoning step where a knowledge unit **K** (e.g., a set of rules or a single rule) is used to derive some additional conclusions. The *dynamic interpretation* of such a knowledge unit **K** can be defined as the mapping in the set of all relevant partial models induced by **K**.

Note that information states are defined in terms of literals. This implies that in principle only literal conclusions count in inferences. Therefore we can take advantage of the fact that inference relations, restricted to literal conclusions, that are sound with respect to the classical Tarski semantics are also sound with respect to the partial Strong Kleene semantics (and vice versa), as has been established in [17] (cf. Theorem 2.3, p. 464). In the sequel by $\vdash$ we will denote any sound inference relation that is not necessarily complete (e.g, one of: natural deduction, chaining, full resolution, SLD- resolution, unit resolution, etc.).

3.1 The static view on the object-level reasoning

In this subsection we define the underlying language, logical theory and inference relation of the object-level component. Moreover we define the notions of deductive and semantic closure.

Definition 3.1 (static view on the object-level component)
The *static view on the object-level reasoning component* is a tuple
$$\langle\ \Sigma_0,\ OT,\ \vdash\ \rangle$$
with

Σ_0	a signature, called the object-signature
OT	a set of propositional ground formulae expressed in terms of the object-signature
$\vdash$	a classical inference relation (assumed sound but not necessarily complete)

Notice that a literal formula is true in a partial model **M** if and only if according to the classical semantics the formula is true in every complete refinement of **M**.

Definition 3.2 (deductive and semantic closure)
Let **K** be a set of propositional formulae of signature Σ and $\vdash$ a (sound) inference relation or (semantic) entailment relation.

a) For $M \in IS_K(\Sigma)$ we define the partial model $cl_K^{\vdash}(M)$ by

$$cl_K^{\vdash}(M) \models^+ L \ \Leftrightarrow\ K \cup Lit(M) \ \vdash\ L$$

for any literal L. This model is called the *closure* of M under K (with respect to $\vdash\!\sim$). We call M *closed* under K (with respect to $\vdash\!\sim$) if $M = cl_K^{\vdash}(M)$, or, equivalently, if

$$M \vDash^+ L \iff K \cup \text{Lit}(M) \vdash\!\sim L$$

b) If $\vdash\!\sim$ is an inference relation $\vdash$ we denote $cl_K^{\vdash}(M)$ by $dc_K^{\vdash}(M)$ and call it the *deductive closure of* M under K (with respect to $\vdash$). We call M *deductively closed* under K if

$$K \cup \text{Lit}(M) \vdash L \iff M \vDash^+ L$$

i.e., if it is its own deductive closure under K.

c) For the classical semantic consequence relation $\vDash$ (based on complete models) we denote $cl_K^{\vDash}(M)$ by $sc_K(M)$ and call it the *semantic closure of* M. We call M *semantically closed* under K if

$$K \cup \text{Lit}(M) \vDash L \iff M \vDash^+ L$$

i.e., if it is its own semantic closure under K.

Definition 3.3   (conservation, monotonicity, idempotency)
Let K be a set of propositional formulae of signature Σ.
We call the mapping $\alpha : IS_K(\Sigma) \to IS_K(\Sigma)$:

  (i) *conservative*   if $M \leq \alpha(M)$ for all $M \in IS_K(\Sigma)$

  (ii) *monotonic*   if $\alpha(M) \leq \alpha(N)$ for all $M, N \in IS_K(\Sigma)$
                                                    with $M \leq N$

  (iii) *idempotent*   if $\alpha(\alpha(M)) = \alpha(M)$ for all $M \in IS_K(\Sigma)$

Proposition 3.4
Let K be a set of propositional formulae of signature Σ and $\vdash\!\sim$ a (sound) inference relation or the semantic consequence relation.

Then the mapping $cl_K^{\vdash}: IS_K(\Sigma) \to IS_K(\Sigma)$ is conservative, monotonic and idempotent.

Moreover, for any $M \in IS_K(\Sigma)$ and any model N of K that is a complete refinement of M it holds $cl_K^{\vdash}(M) \leq N$. In particular this holds for the semantic closure mapping.

3.2   Object level reasoning traces and controlled inference functions

In Subsection 3.1 we have assumed that the deduction is exhaustive with respect to the specific set K; this is not a realistic assumption. In practice often only some of the inferences that are possible are applied, depending on additional control information. However, the full deductive closure always gives an upper bound: if control is involved leading to non-exhaustive reasoning, the actual outcome is a model M' with

$$M \leq M' \leq dc_K^{\vdash}(M) \leq sc_K(M)$$

In this paper we will assume that controlled inference is deterministic, depending on an assignment of values to some set of control parameters. In that case controlled inference can be described as follows.

Definition 3.5 (controlled inference function)

Suppose K is a set of formulae of signature Σ and $\vdash$ is a (sound) inference relation or the (semantic) entailment relation. The mapping $\alpha: IS_K(\Sigma) \to IS_K(\Sigma)$ is called a *controlled inference function* for K based on $\vdash$ if it is conservative and monotonic and for all $M \in IS_K(\Sigma)$ it holds

$$\alpha(M) \leq cl_K^{\vdash}(M).$$

Notice that we do not require that a controlled inference function is idempotent. If reasoning is not exhaustive idempotency is often lost. Controlled inference functions can be viewed as functions α_K^N where instead of a general entailment relation $\vdash$ a variant is used that is parameterized by certain control information N. Two examples of control parameters and the corresponding inference functions are:

- information about which atoms are the *goals* for the reasoning

In this case the control information N expresses that the conclusions should be restricted to what already is available and the set of atoms G; i.e.,

$$\alpha_K^N(M)(a) = \quad M(a) \qquad\qquad \text{if } M(a) \neq u$$
$$cl_K^{\vdash}(M)|G(a) \qquad \text{otherwise}$$

- information about the *selection of elements of the knowledge base* to be used

Here the control information N expresses that only formulae of a subset K' of the theory K can be used in the reasoning, i.e.,

$$\alpha_K^N(M) = cl_{K'}^{\vdash}(M) \leq cl_K^{\vdash}(M)$$

Notice that these examples of control apply not only to the case of an inference relation but also to the semantic consequence relation. In this sense control can be defined in a semantic (inference relation independent) manner.

In a meta-level architecture the control information N is determined by the meta-level reasoning. What is needed to formalize a meta-level architecture is a formalization of this control information on the right level of abstraction; i.e., in such a manner that it can be subject of a (meta-level) inference process. We will come back to this point in Section 4.

Definition 3.6 (object-reasoning trace)

Let $\langle \Sigma_0, OT, \vdash \rangle$ be a static view on the object-level reasoning component, where $\vdash$ is a sound inference relation. A partial temporal model $(M_t)_{t \in N}$ is called an *(object-reasoning) trace* for $\langle \Sigma_0, OT, \vdash \rangle$ if for all $s, t \in N$ with $s < t$ it holds

$$M_s \leq M_t \leq dc_{OT}^{\vdash}(M_s).$$

Theorem 3.7   (approximation of an intended model: soundness)
Let $\langle \Sigma_0, OT, \vdash \rangle$ be a static view on the object-level reasoning component and N
a model of OT (the intended model).
a) If $(M_t)_{t \in N}$ is a trace for $\langle \Sigma_0, OT, \vdash \rangle$ with root r and $M_r \leq N$ then for all
$t \in N$ it holds $M_t \leq N$, i.e.:
$$M_r \leq \ ... \leq M_t \leq \ ... \leq N$$
b) Let for any $t \in \mathbb{N}$ a controlled inference function $\alpha_t : IS_{OT}(\Sigma_0) \to IS_{OT}(\Sigma_0)$ for
OT be given.
Then for any starting point $M_0 \in IS_K(\Sigma)$ a trace $(M_t)_{t \in \mathbb{N}}$ for $\langle \Sigma_0, OT, \vdash \rangle$ can be
defined by:
$$M_{t+1} = \alpha_t(M_t) \qquad \text{for all } t \in \mathbb{N}$$

Given the formal framework as set up here, the proof of this theorem is not difficult.
The above results show a direct connection between the semantics on the basis of
partial models (as used here) and the classical Tarski semantics. In our terms this
connection can be stated as follows. Reasoning of the object-level component is
always on one specific, intended (complete) model that is a (Tarski) model of the
knowledge base. An information state is a partial description of this intended model: a
partial model with the intended model as one of its complete refinements. During
reasoning this partial description is (step-wise) refined, but (in sound reasoning
processes) always remains within the intended model. In our model reasoning can be
viewed as constructing a partial model, approximating the intended complete model
better and better by refinement steps. This even holds if additional observations are
allowed, based on the intended model (this point is left out of the current paper). Since
at any moment in time the intended complete model is not known, in principle we
have to take into account all complete refinements of the current information state that
are models of the knowledge base. Thus the approach discussed here relates *static
semantics* and *dynamic semantics* to each other in one formal framework.

3.3   Control information and dynamic view on object-level reasoning

In this section we will introduce a formalization of control aspects of the object-level
reasoning. The intended model of the object-level component is (a formal
representation of) a specific world situation. As the meta-level component reasons
about the reasoning process of the object-level component, the intended model of this
is a formal description of (relevant aspects of) the inference process of the object-level
component. Considered from the viewpoint of the meta-level component, the object-
level component can be considered as some exotic world situation with as a crucial
characteristic that it is dynamic: each time the meta-level component starts a new
reasoning session, its associated world situation may have changed. Note that we
assume that object-level and meta-level reasoning processes are alternating: during the
meta-level reasoning the object-level component is not reasoning, so changes of the
object-level state occur only between the reasoning sessions of the meta-level
component.

This observation leads us to introduce a *control signature* that defines at an abstract level a number of descriptors that can be used to characterize the control and process states of the object-level reasoning: a lexicon in terms of which all relevant control information can be expressed. A truth assignment to the ground atoms of such a meta-signature is called a *control-information state*. Such a control-information state can serve as a (partial) model for the meta-level component. The question of what are the semantics of the meta-level component is equivalent to the question of what is described by the control-information state related to an object-level component. We illustrate this idea by some examples (for a more specific example, see Section 6):

- the fact that the object-level statement **h** is (currently) considered a *goal* for the reasoning process; e.g., expressed by the (ground) control-atom **goal(h)** where **h** is the name of an atom in the object-level language;
- a *selection* or *priority* of object-level knowledge elements to be used; e.g., expressed by the (ground) control-atom **rule_priority(r)**, where **r** is the name of a rule in the object-level knowledge base, or **goal_priority(h, 0.9)**, with **h** as above;
- the degree of *exhaustiveness* of the reasoning; e.g., expressed by **exhaustiveness(any)**, meaning that it is enough to determine only one of the current goals (the one with highest possible priority).

A control-information state formalizes at a high level of abstraction the parameter N in a controlled inference function as introduced earlier in Section 3. We assume that the control-information state specifies all information relevant to the control of the (future) reasoning behaviour; i.e., the object-information state and the control-information state together determine in a deterministic manner the behaviour of the object-level reasoning component during its next activation. Of course it depends on the specific inference procedure that is used which control aspects can be influenced and which aspects cannot.

In principle for execution we would expect that all atoms of the control signature have a truth value assigned to it (i.e. the control-information state is a complete model). However, we allow partial control-information states as well.

Definition 3.8 (**dynamic view on the object-level component**)
A *dynamic view* related to the static view $\langle \Sigma_0, OT, \vdash \rangle$ on an object-level component is a tuple $\langle \Sigma_c, \mu_{OT}^{\vdash}, \nu_{OT}^{\vdash} \rangle$ with Σ_c a signature called *control signature* and $\mu_{OT}^{\vdash}, \nu_{OT}^{\vdash}$ mappings

$$\mu_{OT}^{\vdash} \quad : IS(\Sigma_0) \times IS(\Sigma_c) \rightarrow IS(\Sigma_0)$$
$$\nu_{OT}^{\vdash} \quad : IS(\Sigma_0) \times IS(\Sigma_c) \rightarrow IS(\Sigma_c)$$

We call $\mu_{OT}^{\vdash}$ the *(controlled) inference function* for the object-level, and $\nu_{OT}^{\vdash}$ the *process state update function*. For any $N \in IS(\Sigma_c)$ the mapping
$$\mu_{OT}^{N} : IS(\Sigma_0) \rightarrow IS(\Sigma_0)$$
is defined by
$$\mu_{OT}^{N} : M \mapsto \mu_{OT}^{\vdash}(M, N)$$

We assume that for any $N \in IS(\Sigma_c)$ this $\mu_{OT}{}^N$ is a controlled inference function, i.e., it is conservative and monotonic and satisfies

$$\mu_{OT}{}^N(M) \leq dc_{OT}{}^{\vdash}(M) \quad \text{for all} \quad M \in IS(\Sigma_0).$$

When no confusion is expected, we will leave out the subscript and superscript of $\mu_{OT}{}^{\vdash}$ and $\upsilon_{OT}{}^{\vdash}$ and write shortly μ and υ.

In a control signature sometimes reference will be made to (names of) elements of the language based on the object-level signature. On the other hand, also control-atoms are possible that do not refer to specific object-level language elements (e.g., exhaustiveness). We do not prescribe in a generic manner if and how reference is made to object-level language elements. In examples this will always be determined in a more specific manner.

The process state update function expresses what the process brings about with respect to the process state descriptors. Examples: an object-atom was unknown, but becomes known during the reasoning; an object-atom that was a goal has failed to be found.

The functions μ and υ for partial control-information states can be defined from the values of the functions for complete control-information states as follows:

$$\mu(M, N) \quad = \quad gci \; \{\mu(M, N') \mid N \leq N' \; \& \; N' \text{ complete } \}$$
$$\upsilon(M, N) \quad = \quad gci \; \{\upsilon(M, N') \mid N \leq N' \; \& \; N' \text{ complete } \}$$

where the greatest common information state $gci(S)$ of a set S of information states is defined by

$$
\begin{array}{llll}
gci(S)(a) & = & 1 & \text{if for all } M \in S \text{ it holds } M(a) = 1 \\
 & & 0 & \text{if for all } M \in S \text{ it holds } M(a) = 0 \\
 & & u & \text{otherwise}
\end{array}
$$

A (combined) information state is a pair $\langle M, N \rangle$ where M is an object-information state and N a control-information state. A (combined) trace is a sequence of (combined) information states. An object-level *execution step* on the basis of a combined information state $\langle M_t, N_t \rangle$ provides the *successor* information state defined by:

$$\langle M_{t+1}, N_{t+1} \rangle = \langle \mu(M_t, N_t), \upsilon(M_t, N_t) \rangle$$

A combined reasoning trace can be obtained by alternating object-level execution steps and interaction steps between the two levels to obtain new control-information states N. We will work this out in more detail in Subsection 4.2.

4 Temporal interpretation of the meta-level reasoning

Locally, at each of the two reasoning levels, the system behaves conservative and monotonic, but the whole cycle implies non-conservative changes of information states: the actions induced by the upward and downward reflections are not conservative. To describe this we can label the information states with an explicit time parameter (e.g., expressed by natural numbers). The non-conservatism can be covered by our declarative formal model, assuming each new (object-meta) cycle is labeled with the next (successor) time label. This approach implies that the meta-level reasoning component has semantics that relates states of the object-level reasoning component at time t to states of this component at time $t + 1$. In this manner, statements like

"if the atom a *is unknown, then the atom* b *is proposed as a goal"*

after downward reflection can be interpreted in a temporal manner:

"If at time t *the atom* a *is unknown*   *(in the object-level reasoning component)*
then at time t+1 *the atom* b *is a goal "*   *(for the object-level reasoning process)*

Assuming that the meta-level's proposals are always accepted (this assumption is sometimes called causal connection), downward reflection is just a shift in time, replacing the goals at the object-level by new goals (the ones proposed by the meta-level). This can be expressed as follows

$$\neg \: \mathbf{known(a)} \quad \rightarrow \quad \mathbf{proposed\_goal(b)} \qquad \text{(meta-knowledge)}$$

$$\mathbf{C(proposed\_goal(b))} \quad \rightarrow \quad \mathbf{X(goal(b))} \qquad \text{(downward reflection)}$$

where $\mathbf{C}$ means "holds in the current state" and $\mathbf{X}$ "holds in the next state".

Within the *meta-information* involved in the meta-reasoning we distinguish two special types: a) information on relevant aspects of the *current* (control-)state of the object-level reasoning process (possibly also including facts inherited from the past), and b) information on proposals for control parameters that are meant to guide the object-level reasoning process in the near *future*. Therefore we assume that in the meta-signature a copy of the control signature of the object-level component is included as a subsignature that refers to the current state. Moreover, we assume that a second copy of this control signature is included referring to the proposed truth values for the next state of the object-level reasoning process. For example, if $\mathbf{goal(h)}$ is an atom of the control signature, then there are copies $\mathbf{current\_goal(h)}$ and $\mathbf{proposed\_goal(h)}$ in the set of atoms for the meta-signature. A syntactic function transforming a meta-atom into a current variant and a proposed variant of it can simply be defined by two (injective) mappings $\mathbf{c}$ and $\mathbf{p}$ of predicates, leaving the arguments the same e.g.,

$$\mathbf{c(goal)} \: = \: \mathbf{current\_goal}, \quad \mathbf{p(goal)} \: = \: \mathbf{proposed\_goal}.$$

We assume that the reasoning of the meta-level itself has no sophisticated control: for simplicity we assume that it concerns taking deductive closures with respect to the inference relation used at the meta-level. Under this assumption a dynamic view on the meta-level component is completely determined by a static view.

Definition 4.1 (static and dynamic view on the meta-level component)

a) The signature Σ_m is called a *meta-signature* related to Σ_c if there are two injective mappings $c : \Sigma_c \rightarrow \Sigma_m$ and $p : \Sigma_c \rightarrow \Sigma_m$. In this case the subsignatures $c(\Sigma_c)$ and $p(\Sigma_c)$ are denoted by $\Sigma_m{}^c$ and $\Sigma_m{}^p$; they are referring to the *current state control-information* of the object-level and the *proposed state control-information* for the object-level.

b) The *static view on the meta-level component* is a tuple

$$\langle \Sigma_m, MT, \vdash_m \rangle$$

with

Σ_m a signature, called the meta-signature related to Σ_c

MT a set of propositional ground formulae expressed in terms of the meta-signature

$\vdash_m$ a classical inference relation (assumed sound but not necessarily complete)

c) The *inference function of the meta-level* $\mu_{MT}{}^\vdash m$ (or shortly μ^*)

$$\mu_{MT}{}^\vdash m : IS_{MT}(\Sigma_m{}^c) \rightarrow IS_{MT}(\Sigma_m)$$

is defined by the exhaustive inference function based on $\vdash$; i.e., by the transition function

$$\mu_{MT}{}^\vdash m : N \mapsto dc_{MT}{}^\vdash m(N)$$

This function μ^* defines the *dynamic view on the meta-level component*, related to the static view

$$\langle \Sigma_m, MT, \vdash_m \rangle.$$

Note that we essentially use propositional logic to describe the meta-language; if needed a propositional signature can be defined based on the set of all ground atoms expressible in a given (many-sorted) predicate logic signature. In fact it does not matter how language elements at the meta-level are denoted, but how their semantics is defined (in terms of the controlled inference function).

5 Temporal models of overall reasoning patterns

After having introduced the required concepts in the previous sections, in this section it will turn out to be easy to compose them to semantics for the dynamics of a meta-level architecture. The information states of the meta-level component of the reasoning system will have a direct impact on the control-information state of the object-component, and vice versa. These connections will be defined formally in this section. Notice that in our approach the object-level component and the meta-level component do not reason at the same time, but are alternating.

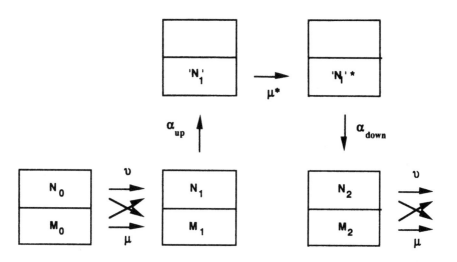

Fig 2 Transitions of information states in a meta-level architecture

In fact the following four types of actions take place (see Fig. 2). For a formal description, see Definitions 5.1 and 5.2 below.

- *object-level reasoning*
The reasoning of the object-level component can be described by the functions μ and υ as defined in Definition 3.8.

- *upward reflection*
Information from the control-information state of the object-level component is transformed (by a transformation function α_{up} defined in Definition 5.1 below) to the next information state of the meta-level component. This will provide input information for the subsequent reasoning of the meta-level component (see Definition 4.1).

- *meta-level reasoning*
The reasoning of the meta-level component can be described by the inference function μ^* as defined in Definition 4.1.

- downward reflection

Information from the of the meta-level component is transformed (by the mapping α_{down}; see Definition 5.1 below) to the next control-information state to be used in the control of the object-level component. This will affect the reasoning behaviour during the subsequent object-level reasoning.

Definition 5.1 (meta-level architecture for dynamic control)

a) A *meta-level architecture for dynamic control* is described by a tuple

$$\mathbf{MLC} = \langle\, \langle\, \Sigma_0, \mathrm{OT}, \vdash_0 \,\rangle; \langle\, \Sigma_c, \mu, \upsilon \,\rangle; \langle\, \Sigma_m, \mathrm{MT}, \vdash_m \,\rangle \;; \langle\, c, p \,\rangle\,\rangle$$

where

$$\langle\, \Sigma_0, \mathrm{OT}, \vdash_0 \,\rangle$$
$$\langle\, \Sigma_c, \mu, \upsilon \,\rangle$$

are a static and a dynamic view on the object-level component,

$$\langle\, \Sigma_m, \mathrm{MT}, \vdash_m \,\rangle$$

is a static view on the meta-level component, where Σ_m is related to the control signature Σ_c by the injective functions $c : \Sigma_c \to \Sigma_m$ and $p : \Sigma_c \to \Sigma_m$ and $\vdash_m$ is an inference relation. Moreover, $\mathbf{MT}$ is a meta-knowledge base satisfying

$$\mathrm{IS}_{MT}(\Sigma_m{}^c) = \mathrm{IS}(\Sigma_m{}^c)$$

i.e., no information state in $\mathrm{IS}(\Sigma_m{}^c)$ is inconsistent with $\mathbf{MT}$.

Based on $\mathbf{MLC}$ we can define the function μ^* according to Definition 4.1.

b) Let $\mathbf{MLC}$ be as in a). The *upward reflection function* is the mapping

$$\alpha_{up}: \mathrm{IS}(\Sigma_c) \to \mathrm{IS}(\Sigma_m)$$

defined for $N \in \mathrm{IS}(\Sigma_c)$ and $b \in \mathrm{At}(\Sigma_m)$ by

$$\alpha_{up}(N)(b) \quad = \quad N(a) \qquad \text{if } b = c(a) \text{ for some } a \in \mathrm{At}(\Sigma_c)$$
$$u \qquad \text{otherwise}$$

The *(left) inverse upward reflection function* β is the mapping

$$\beta: \mathrm{IS}(\Sigma_m) \to \mathrm{IS}(\Sigma_c)$$

defined for $N \in \mathrm{IS}(\Sigma_m)$ and $a \in \mathrm{At}(\Sigma_c)$ by

$$\beta(N)(a) \quad = \quad N(c(a))$$

The *downward reflection function* is the mapping

$$\alpha_{down}: \mathrm{IS}(\Sigma_m) \to \mathrm{IS}(\Sigma_c)$$

defined for $N \in \mathrm{IS}(\Sigma_m)$ and $a \in \mathrm{At}(\Sigma_c)$ by

$$\alpha_{down}(N)(a) \quad = \quad 1 \qquad \text{if } N(p(a)) = 1$$
$$0 \qquad \text{otherwise}$$

The *time shift function* is the mapping

$$\sigma: \mathrm{IS}(\Sigma_m) \to \mathrm{IS}(\Sigma_m)$$

defined by

$$\sigma(N)(b) \quad = \quad N(p(a)) \qquad \text{if } b = c(a) \text{ for some } a \in \mathrm{At}(\Sigma_c)$$
$$\text{and } N(P(a)) \neq u$$
$$0 \qquad \text{if } b = c(a) \text{ for some } a \in \mathrm{At}(\Sigma_c)$$
$$\text{and } N(P(a)) = u$$
$$u \qquad \text{otherwise}$$

Reasoning activities are modifying object-information states in a conservative manner (making refinements). Notice, however, that execution of upward and downward reflection may induce non-conservative changes. Notice that we force the new control state resulting from downward reflection to be two-valued. This is to avoid nondeterministic phenomena and to allow that the meta-level only provides the relevant (partial) information on control. In the rest of the paper **MLC** will denote a tuple as defined in Definition 5.1.

The following relations hold for the functions defined above:
$$\beta\alpha_{up} = id, \quad \alpha_{down} = \beta\sigma, \quad \sigma = \alpha_{up}\alpha_{down}.$$

Definition 5.2 **(overall semantics based on traces)**
a) An *overall trace* for the meta-level architecture **MLC** is a labeled linear partial temporal model
$$(M_t \oplus N_t)_{t \in \mathbb{N}}$$
(also denoted by $M \oplus N$) over $IS(\Sigma_o \oplus \Sigma_m)$ with set of labels $L = \{re, sh\}$ (re for a reasoning step, sh for a time shift step) satisfying the following conditions for all $s, t \in \mathbb{N}$:
(i) If $s <_{re} t$

$$\begin{aligned} M_t &= \mu(M_s, \beta(N_s)) \\ N_t &= \mu^*(\alpha_{up}(\upsilon(M_s, \beta(N_s)))) \end{aligned}$$

(ii) If $s <_{sh} t$

$$\begin{aligned} M_t &= M_s \\ N_t &= \mu^*(\sigma(N_s)) \end{aligned}$$

b) The *(intended) semantics* of the meta-level architecture **MLC** is the set of traces as defined in a), denoted by **Traces(MLC)**.
c) A trace is called *alternating* if for all $r, s, t \in \mathbb{N}$ and $i, j \in L$ with $r <_i s <_j t$ it holds $i \neq j$.

Although usually we are most interested in alternating traces, there may be cases were we are interested in other traces as well: e.g. if we allow multiple activations of the object-level without intervenience of the meta-level.

Temporal models can be defined using our framework by traces of information states. These traces are constructed during reasoning. At each moment of time only a partial (in time) fragment of such a trace-model has been constructed. The set of completed traces can be viewed as the set of intended overall models of the meta-level architecture. The meta-level architecture as a whole approximates an intended model in a conservative manner by subsequently adding elements to the trace according to time steps. This view will be made more precise in the following theorem.

Theorem 5.3 **(approximation of a trace)**
Let **MLC** be a meta-level architecture for dynamic control.
a) The set of alternating traces of **MLC** is parameterized by the initial states, together with the label (from $\{re, sh\}$) of the initial transition.
b) Let $M \oplus N$ be a trace for **MLC**. Define for any time point t

$$M^{(t)}{}_s(a) = \begin{array}{ll} M_s(a) & \text{if } s \ll t \text{ or } s = t \\ u & \text{otherwise} \end{array}$$

$$N^{(t)}{}_s(a) \quad = \quad \begin{array}{ll} N_s(a) & \text{if } s \ll t \text{ or } s = t \\ u & \text{otherwise} \end{array}$$

Then for all time points t it holds

$$M^{(0)} \oplus N^{(0)} \leq \ldots \leq M^{(t)} \oplus N^{(t)} \leq M^{(t+1)} \oplus N^{(t+1)} \leq \ldots \leq M \oplus N$$

Notice that in this section we do not (yet) add temporal elements to the languages of the reasoning components themselves, but we attribute temporal semantics to the whole system by interpreting the object reasoning process and the downward reflection in a temporal manner. This means that within each of the components (locally) we retain our original (non-temporal) semantics. The temporal semantics only serves as a foundation for the composition principle to define an overall semantics composed from the local semantics of each of the components at the two levels.

6 An example reasoning pattern

To illustrate the concepts introduced here we will give a trace of a meta-level architecture for reasoning with dynamic hypotheses that are used as goals. The meta-level reasoning performs hypothesis selection whereas the object-level reasoning performs testing of hypotheses by trying to derive them from observation information (in a goal-directed manner). Control is needed to direct the object-level reasoning to the goal that is to be posed: the selected hypothesis. The meta-level contains declarative knowledge on which hypothesis to select under which circumstance (state of the object-level reasoning process). The downward reflection transforms this information about the selected hypothesis to control-information in the form of a goal set for the object-level reasoning; this enables the system to effectuate control. The upward reflection provides the information for the meta-level component on the current state of what is already known and what is not yet known in the object-level reasoning process. The knowledge in this example system is not realistic, but it enables one to get an impression of the reasoning pattern.

A. *Static view on the object-level reasoning component*

Object-signature (propositional):

$$\Sigma_0 = \langle s_1, s_2, s_3, h_1, h_2 \rangle$$

Object theory (knowledge base of the object-level component) OT:

$$\begin{array}{ll} s_2 \wedge s_3 & \rightarrow h_1 \\ \neg s_3 \wedge s_1 & \rightarrow h_2 \\ \neg s_3 & \rightarrow \neg h_1 \end{array}$$

Inference relation: $\vdash_{ch}$ (chaining)

B. *Dynamic view on the object-level reasoning component*

This reasoning component is used in a goal-directed fashion with *chaining* as *inference relation*. We will not involve the possibility to acquire additional information from the outside of the system.

Control-signature :
$$\langle \ \text{true\_s}_1, \ \text{false\_s}_1, \ \text{true\_s}_2, \ \text{false\_s}_2, \ \text{true\_s}_3, \ \text{false\_s}_3,$$
$$\text{known\_h}_1, \ \text{known\_h}_2, \ \text{goal\_h}_1, \text{goal\_h}_2 \ \rangle$$

Inference function :

The dependency of the *inference function* μ on the control-information state is concentrated in information expressed by goal-statements **goal_h$_i$** (meaning that **h$_i$** is a goal for the object-level reasoning process). As a formal definition we can take
$$\mu(M, \ N)(a) \ = \quad \text{dc}_{\text{OT}}{}^{\vdash}\text{ch}(M)(a) \quad \text{if } \ a = h_i \text{ and } N(\text{goal\_h}_i) = 1$$
$$M(a) \qquad\qquad \text{otherwise}$$

Process state update function

The *process state update function* υ is defined as follows:
- The statement **known_h$_i$** gets truth value 1 in the control-information state if in the object-information state **h$_i$** has truth value 1 or 0; it gets truth value 0 otherwise (i.e., if **h$_i$** has truth value **u** in the object-information state).
- The statement **true_s$_i$** has truth value 1 in the control-information state if in the object-information state **s$_i$** has truth value 1; it gets truth value 0 otherwise (i.e., if **s$_i$** has truth value **u** or 0 in the object-information state).
- The statement **false_s$_i$** has truth value 1 in the control-information state if in the object-information state **s$_i$** has truth value 0; it gets truth value 0 otherwise (i.e., if **s$_i$** has truth value **u** or 1 in the object-information state).
- The other truth values remain unchanged.

C. *Meta-level reasoning component*

The *meta-signature* is taken as the disjoint union of two copies of the control signature above: **c_at** (currently **at**), and **p_at** (proposed **at**), where **at** is an atom of the control signature.

Knowledge base of the meta-level component (**MT**):

$$\text{c\_true\_s}_2 \ \wedge \ \neg \ \text{c\_known\_h}_1 \ \rightarrow \ \text{p\_goal\_h}_1$$
$$\text{c\_false\_s}_3 \ \wedge \ \neg \ \text{c\_known\_h}_2 \ \rightarrow \ \text{p\_goal\_h}_2$$

The meta-level wil use *chaining* as its *inference relation*.
The inference function $\mu*$ is the deductive closure function under **MT**.

Trace of an example session

In Fig. 3 a session with initial state $\langle s_1, s_2, s_3 \rangle : \langle 1, 1, 0 \rangle$ is described. Here for convenience partial models are denoted by the list of atomic statements and negations of atomic statements that are true. For the (combined) information states (named p_i) of the object-level component both the object-information states and the control-information states are depicted (separated by a colon ;). For the meta-level component only the object-information states are depicted (named t_i). For shortness only some relevant (literal) facts are written in the information states.

object-level component   *meta-level component*

p_0 : $[s_1, s_2, \neg s_3]$; $[\neg known\_h_1, \neg known\_h_2]$

p_1 : $[s_1, s_2, \neg s_3]$; $[true\_s_2, \neg known\_h_1, \neg known\_h_2]$

   t_0: $[c\_true\_s_2, \neg c\_known\_h_1, \neg c\_known\_h_2]$

   t_1: $[c\_true\_s_2, \neg c\_known\_h_1, \neg c\_known\_h_2, p\_goal\_h_1]$

p_2 : $[s_1, s_2, \neg s_3]$; $[true\_s_2, \neg known\_h_1, \neg known\_h_2, goal\_h_1]$

p_3 : $[s_1, s_2, \neg s_3, \neg h_1]$; $[true\_s_2, false\_s_3, known\_h_1, \neg known\_h_2]$

   t_2: $[c\_true\_s_2, c\_false\_s_3, c\_known\_h_1, \neg c\_known\_h_2]$

   t_3: $[c\_true\_s_2, c\_false\_s_3, c\_known\_h_1, \neg c\_known\_h_2, p\_goal\_h_2]$

p_4 : $[s_1, s_2, \neg s_3, \neg h_1]$; $[true\_s_2, false\_s_3, known\_h_1, \neg known\_h_2, goal\_h_2]$

p_5 : $[s_1, s_2, \neg s_3, \neg h_1, h_2]$; $[true\_s_1, true\_s_2, false\_s_3, known\_h_1, known\_h_2]$

Fig 3 Trace of an example session

This example shows that it does not matter how language elements at the meta-level are denoted, but how their semantics is defined. Using a propositional language at the meta-level is possible, but the more concise syntactical notation of predicate logic has practical advantages. Therefore, often a predicate logic language is used at the meta-level. For the semantics of the whole reasoning pattern this makes no essential difference.

7 Conclusions

The semantic framework as discussed provides integration of static and dynamic aspects in two different forms. On the one hand we connect the partial semantics as used to describe information states to the standard Tarski semantics: a partial model corresponds to the set of all of its complete refinements. On the other hand the integration between static and dynamic aspects takes place by introducing the notion of an explicit (declarative) control-information state in the object-level component.

Our logical framework has been partly inspired by Weyhrauch's view on the role of partial models (or simulation structures) in meta-level architectures ([25], [9]). What is different in our case is that the partial models may be dynamic. Furthermore, similarities can be found to the approach called dynamic interpretation of natural language (e.g., see [7], [10], [11]). In this approach the dynamic interpretation of a sentence in natural language is defined as an operator that transforms the current information state into a new one where the content of the sentence is included.

With respect to dynamics the type of meta-level architecture covered here is less restricted than sometimes studied in logical approaches, where meta-level predicates are meant to express only static properties of the object-level, e.g., provability, et cetera. We believe that the semantic model as discussed here can help to bridge the gap between the (more restricted) logic-based approaches and (less restricted) procedural approaches to meta-level architectures.

It is not difficult to use our framework to model meta-level reasoning that looks ahead more than one step. One can transfer a part of the information at the meta-level over the time shift and thus connect the reasoning in different activations of the meta-level. The results presented here can also be extended easily to the case of higher meta-levels where also the control of the meta-level is guided in a dynamic manner (in this case a refinement of the time scale can be made).

Our logical framework has been implemented and applied in a number of practical applications, often in projects in cooperation with companies (e.g., [4], [8]), and has also been used to model an approach to defeasible reasoning with explicit control (see [18], [19]). The type of meta-level architecture as discussed can be designed and formally specified using our formal specification framework DESIRE (framework for DEsign and Specification of Interacting REasoning modules; see [14]). In DESIRE various types of reasoning components are covered; e.g., goals can be used to guide the reasoning, and various measures of exhaustiveness can be used (for a more extensive survey, see [14]). By means of DESIRE complex reasoning systems can be designed and specified according to what we call a *compositional architecture:* an architecture composed from a number of formally specified reasoning components using formal and standard composition principles (see [14]).

In our current practical applications (using DESIRE) of the framework as described the control-information states are two-valued, i.e., no **u**'s occur as truth values. In other words: at each moment all state descriptors defined by the control signature have a determined truth value. On the one hand this corresponds to the intuition that, since the meta-level reasoning is about the states of the object-level reasoning process, input information about this can be acquired from the system itself: there is essentially no incompleteness of incoming information. On the other hand the two-valuedness of the control-information states is related to the fact that we require that the control of the object-level reasoning is completely determined by the truth values of the control-atoms, and vice versa. Therefore, if we require complete deterministic specification of the behaviour of the reasoning system, all control-atoms should have determined truth values (and vice versa). If we would allow non-deterministic control (e.g., by only specifying some, but not all aspects of the control), the control-information states may be viewed as essentially three-valued.

Since for the notion of a compositional architecture and the framework DESIRE, an essential use is made of the notion of a meta-level architecture, the (formal) semantics of the static and dynamic semantics of DESIRE depend on these semantics of meta-level architectures. In literature not much work is reported on such foundations. As this paper contributes an approach to the semantics of meta-level architectures, this can be used to work out a semantics for DESIRE. This work is planned in the near future.

Acknowledgements

Guszti Eiben, Joeri Engelfriet and Pieter van Langen have read and commented upon earlier drafts of this paper. This, and the reviewers comments have led to a number of improvements in the text.

References

1. J.F.A.K. van Benthem, The logic of time: a model-theoretic investigation into the varieties of temporal ontology and temporal discource, Reidel, Dordrecht, 1983.

2. S. Blamey, Partial Logic, in: D. Gabbay and F. Guenthner (eds.), Handbook of Philosophical Logic, Vol. III, 1-70, Reidel, Dordrecht, 1986.

3. K. Bowen and R. Kowalski, Amalgamating language and meta-language in logic programming. In: K. Clark, S. Tarnlund (eds.), Logic programming. Academic Press, 1982.

4. H.A. Brumsen, J.H.M. Pannekeet and J. Treur, A compositional knowledge-based architecture modelling process aspects of design tasks, Proc. 12th Int. Conf. on AI, Expert systems and Natural Language, Avignon'92 (Vol. 1), 1992, pp. 283-294.

5. W.J. Clancey and C. Bock, Representing control knowledge as abstract tasks and metarules, in: Bolc, Coombs (eds.), Expert System Applications, 1988.

6. R. Davis, Metarules: reasoning about control, Artificial Intelligence 15 (1980), pp. 179-222.

7. T. Fernando, Transition systems and dynamic semantics, Proc. JELIA'92 Workshop on Logic and AI, Berlin, 1992.

8. P.A. Geelen and W. Kowalczyk, A knowledge-based system for the routing of international blank payment orders, Proc. 12th Int. Conf. on AI, Expert systems and Natural Language, Avignon-92 (Vol. 2), 1992, pp. 669-677.

9. E. Giunchiglia, P. Traverso and F. Giunchiglia, Multi-context Systems as a Specification framework for Complex Reasoning Systems, In: [24], 1993, pp. 45-72.

10. R. Goldblatt, Logics of Time and Computation. CSLI Lecture Notes, Vol. 7. 1987, Center for the Study of Language and Information.

11. J.A.W. Kamp, A theory of truth and semantic representation, In: Formal methods in the study of language. Mathematical Centre Tracts 135, Amsterdam, 1981.

12. P.H.G. van Langen and J. Treur, Representing world situations and information states by many-sorted partial models, Report PE8904, University of Amsterdam, Department of Mathematics and Computer Science, 1989.

13. T. Langholm, Partiality, Truth and Persistance, CSLI Lecture Notes No. 15, Stanford University, Stanford, 1988.

14. I.A. van Langevelde, A.W. Philipsen, J. Treur, Formal specification of compositional architectures, In: B. Neumann (ed.), Proc. 10th European Conference on Artificial Intelligence, ECAI'92, Wiley and Sons, 1992, pp. 272-276.

15. J.W. Lloyd, Foundations of logic programming, Springer Verlag, 1984.

16. P. Maes, D. Nardi (eds), Meta-level architectures and reflection, Elsevier Science Publishers, 1988.

17. Y.H. Tan and J. Treur, A bi-modular approach to nonmonotonic reasoning, In: De Glas, M., Gabbay, D. (eds.), Proc. World Congress on Fundamentals of Artificial Intelligence, WOCFAI'91, 1991, pp. 461-476.

18. Y.H. Tan and J. Treur, Constructive default logic and the control of defeasible reasoning, In: B. Neumann (ed.), Proc. 10th European Conference on Artificial Intelligence, ECAI'92, Wiley and Sons, 1992, pp. 299-303.

19. Y.H. Tan and J. Treur, Constructive default logic in a meta-level architecture, in: A Yonezawa, B.C. Smith (eds.), Proc. International Workshop on new Models in Software Architecture (IMSA) 1992, Reflection and Meta-level Architectures, 1992, pp. 184-189.

20. J. Treur, Completeness and definability in diagnostic expert systems, Proc. European Conference on Artificial Intelligence, ECAI'88, München, 1988, pp. 619-624.

21. J. Treur, On the use of reflection principles in modelling complex reasoning, International Journal of Intelligent Systems 6 (1991), pp. 277-294.

22. J. Treur, Declarative functionality descriptions of interactive reasoning modules, In: H. Boley, M.M. Richter (eds.), Processing Declarative Knowledge, Proc. of the International Workshop PDK'91, Lecture Notes in Artificial Intelligence, vol. 567, Springer Verlag, 1991, pp. 221-236.

23. J. Treur, P. Veerkamp, Explicit representation of design process knowledge, in: J.S. Gero (ed.), Artificial Intelligence in Design '92, Proc. AID'92, Kluwer Academic Publishers, 1992, pp. 677-696.

24. J. Treur and Th. Wetter (eds.), Formal Specification of Complex Reasoning Systems, Ellis Horwood, 1993, pp 282.

25. R.W. Weyhrauch, Prolegomena to a theory of mechanized formal reasoning, Artificial Intelligence 13 (1980), pp. 133-170.

Gödel as a Meta-Language
for Composing Logic Programs

Antonio Brogi and Simone Contiero

Dipartimento di Informatica, Università di Pisa
Corso Italia 40, 56125 Pisa - Italy

Abstract. Increasing attention is being paid to Gödel, a new declarative programming language aimed at diminishing the gap between theory and practice of programming with logic. An intriguing question is whether or not existing logic programs can be suitably re-used in Gödel. We investigate the possibility of employing Gödel as a meta-language for re-using and composing existing definite programs. Two alternative implementations of a set of meta-level operations for composing definite programs are presented. The first implementation consists of an extended vanilla meta-interpreter using the non-ground representation of object level programs. The second implementation exploits the meta-programming facilities offered by Gödel, which support the construction of meta-interpreters using the ground representation of object level programs.

1 Introduction

The idea of programming with logic [15] found its first realisation in the Prolog language, which has been widely considered as *the* logic programming language for the last twenty years. The development of logic programming, however, has shown the deficiencies of Prolog as a declarative language and the consequent gap between theory and practice of programming with logic [2]. Several efforts have been devoted to design new programming languages aimed at being the declarative successors of Prolog (e.g., [9, 22]). The Gödel language [14] is starting to emerge as a new declarative general-purpose programming language in the family of logic programming languages. Gödel is a strongly typed language, has a module system and places considerable emphasis on meta-programming. These features of Gödel, in particular the emphasis on meta-programming, suggest the intriguing question whether or not existing logic programs can be suitably re-used in Gödel.

The ultimate objective of this work is to investigate the adequacy of Gödel as a meta-language for re-using and composing existing definite programs. In this perspective, we consider a simple extension of logic programming which consists of introducing a set of meta-level operations for composing definite programs. These operations, originally presented in [4, 19], form an algebra of logic programs with interesting properties for reasoning about programs and program compositions. From a programming perspective, the operations enhance the expressive power of logic programming by supporting a wealth of programming techniques, ranging from software engineering to artificial intelligence applications [5, 6, 7]. In this paper the implementation in Gödel of such a set of program composition operations is discussed.

Meta-level operations over object level programs can be naturally implemented by means of meta-programming techniques. More precisely, as shown in [4, 6], several composition operations over logic programs can be implemented by extending the well known *vanilla* meta-interpreter [24]. The actual realisation of extended vanilla meta-interpreters in Gödel presents various implementation choices that lead to different solutions. One of these choices is the representation of object level constructs, which is one of the crucial issues in meta-programming. In logic programming, object level expressions are usually represented by terms at the meta-level, and the critical issue is how to represent object level variables at the meta-level. Two basic alternative representations of object level variables are employed, as ground terms and as variables (or, more generally, non-ground terms). The first is called *ground* representation and the second *non-ground* representation [13]. The ground representation is very versatile and adequate for many applications of meta-programming, such as program transformation and compiler writing. Though the non-ground representation is less versatile, it has been widely adopted in the practice of meta-programming with logic, e.g. for the construction of several expert systems [24]. Indeed, in the absence of suitable support for managing the complexity of meta-programming with the ground representation, programmers have been attracted by the simplicity and efficiency of the non-ground representation, which are due to the fact that the non-ground representation directly exploits the basic unification mechanism of the meta-language. A thorough discussion of the two representations can be found for instance in [8, 13, 16, 17, 20, 23].

Two alternative implementations in Gödel of program composition operations are presented here. Both implementations are based on an extended vanilla meta-interpreter, and they differ in the choice of the representation of object level programs.

The first implementation adopts the non-ground representation of object level programs. This choice offers a simple and concise way of extending the vanilla meta-interpreter to deal with program composition operations. The implementation is equipped with suitable support for the non-ground representation, which is not directly supported by the Gödel system. This support frees the user from the need of explicitly providing the non-ground representation of object programs.

The second implementation adopts the ground representation of object level programs, for which Gödel provides considerable support. The Gödel approach to meta-programming is strongly based on abstract data types. For instance, the system module Programs offers a large number of operations on an abstract data type that is the type of terms representing object level programs. The abstract data type view supports a declarative high-level style of meta-programming as the user has to be concerned neither with the internal representation of the data type nor with the implementation of the associated operations. The ground representation offered by Gödel, however, relies on a naming policy that does contrast with the naming policy employed in the context of logic program composition. We show how this problem can be tackled by suitably extending Gödel's support for generating the ground representation of object programs.

The two implementations are then compared in order to highlight the merits of the alternative representations and, most important, the adequacy of Gödel as a meta-language for composing logic programs. The analysis of the implementations also outlines some possible extensions of Gödel which may improve the flexibility of the language.

The plan of the paper follows. A suite of meta-level operations for composing logic programs is introduced in Section 2. Two alternative implementations of these operations in Gödel are described in Sections 3 and 4. Finally in Section 5 some conclusions are drawn.

2 Program Composition Operations

This section is devoted to briefly introduce a set of composition operations that form an algebra of logic programs, originally defined in [4, 19].

Four basic operations for composing definite logic programs are introduced: Union (denoted by $\cup$), intersection ($\cap$), encapsulation ($*$), and import ($\triangleleft$). The operations are defined in a semantics-driven style, following the intuition that if the meaning of a program P is denoted by the corresponding *immediate consequence operator* $T(P)$ then such a meaning is a homomorphism for several interesting operations on programs. Notice that the standard least Herbrand model semantics of logic programming is not appropriate to model compositions of programs in that it does not enjoy the compositionality requirement. In fact, the least Herbrand model of a program cannot be obtained, in general, from the least Herbrand models of its clauses. Each program P is therefore denoted by the corresponding $T(P)$.

Recall that, for a logic program P, the immediate consequence operator $T(P)$ is a continuous mapping over Herbrand interpretations defined as follows [25]. For any Herbrand interpretation I:

$$A \in T(P)(I) \Longleftrightarrow (\exists \bar{B} : A \leftarrow \bar{B} \in ground(P) \wedge \bar{B} \subseteq I)$$

where $\bar{B}$ is a (possibly empty) conjunction of atoms and $ground(P)$ denotes the ground version of program P. The powers of $T(P)$ are defined as usual [1]:

$$T^0(P)(I) = I$$
$$T^{n+1}(P)(I) = T(P)(T^n(P)(I))$$
$$T^\omega(P)(I) = \bigcup_{n<\omega} T^n(P)(I).$$

The semantics of program compositions can be given in a compositional way by extending the definition of T with respect to the first argument. For any Herbrand interpretation I:

$$T(P \cup Q)(I) = T(P)(I) \cup T(Q)(I)$$
$$T(P \cap Q)(I) = T(P)(I) \cap T(Q)(I)$$
$$T(P^*)(I) = T^\omega(P)(\emptyset)$$
$$T(P \triangleleft Q)(I) = T(P)(I \cup T^\omega(Q)(\emptyset))$$

The above definition generalises the notion of immediate consequence operator from programs to compositions of programs. The operations of union and intersection of programs directly relate to their set-theoretic equivalent. For any interpretation I, the set of immediate consequences of the union (resp. intersection) of two programs is the set-theoretic union (resp. intersection) of the sets of immediate consequences of the separate programs. For any interpretation I, the formulae that may be derived in an encapsulated program P^* are all the formulae that may be derived from P in an arbitrary (finite) number of steps. Finally, for any interpretation I, the set of immediate consequences of the import of two programs $P \lhd Q$ is the set of formulae that may be derived in the importing program P in a single deduction step from I and from the set of formulae that may be derived in the imported program Q in an arbitrary (finite) number of steps.

The operations $\cup$, $\cap$, $*$ and $\lhd$ satisfy a number of algebraic properties such as associativity, commutativity and distributivity. The resulting algebra [4] extends the algebra presented in [19] and provides a formal basis for proving properties of program compositions. For instance, syntactically different program compositions may be compared and simplified by means of the properties of program composition operations.

Program composition operations can be equivalently defined by characterising the operational behaviour of program compositions. Such a characterisation can be expressed by directly extending the standard notation of SLD refutation [1, 18, 25] to deal with program expressions. The standard SLD refutation relation may be defined by means of inference rules of the form

$$\frac{Premise}{Conclusion}$$

asserting that *Conclusion* holds whenever *Premise* holds. We write $P \vdash G$ if there exists a refutation for a goal G in a program P.

$$\frac{}{P \vdash Empty} \tag{1}$$

$$\frac{P \vdash G_1 \,\wedge\, P \vdash G_2}{P \vdash (G_1, G_2)} \tag{2}$$

$$\frac{P \vdash (A \leftarrow G) \,\wedge\, P \vdash G}{P \vdash A} \tag{3}$$

Rule (1) states that the empty goal, denoted by *Empty*, is solved in any program P. Rule (2) deals with conjunctive goals. It states that a conjunction (G_1, G_2) is solved in a program P if G_1 is solved in P and G_2 is solved in P. Finally, rule (3) deals with atomic goal reduction. To solve an atomic goal A, choose a clause from program P and recursively solve the body of the clause in P. Notice that, for the sake of simplicity, substitutions are omitted in that we are interested here in characterising only the (ground) success set of a program.

Program clauses are represented by means of the following rule:

$$\frac{P \text{ is a plain program } \wedge\ A \leftarrow G \in ground(P)}{P \vdash (A \leftarrow G)} \tag{4}$$

The derivation relation $\vdash$ can be generalised to the case of program compositions in a simple way. Namely, each composition operation is modelled by adding new inference rules to rules (1)—(4).

$$\frac{P \vdash (A \leftarrow G)}{P \cup Q \vdash (A \leftarrow G)} \tag{5}$$

$$\frac{Q \vdash (A \leftarrow G)}{P \cup Q \vdash (A \leftarrow G)} \tag{6}$$

$$\frac{P \vdash (A \leftarrow G_1) \ \wedge \ Q \vdash (A \leftarrow G_2)}{P \cap Q \vdash (A \leftarrow G_1, G_2)} \tag{7}$$

$$\frac{P \vdash A}{P^* \vdash (A \leftarrow Empty)} \tag{8}$$

$$\frac{P \vdash (A \leftarrow G_1, G_2) \ \wedge \ Q \vdash G_2}{P \triangleleft Q \vdash (A \leftarrow G_1)} \tag{9}$$

Rules (5) and (6) state that a clause $A \leftarrow G$ belongs to the program expression $P \cup Q$ if it belongs either to P or to Q. Rule (7) states that a clause $A \leftarrow G$ belongs to $P \cap Q$ if there is a clause $A \leftarrow G_1$ in P and a clause $A \leftarrow G_2$ in Q such that $G = (G_1, G_2)$. Rule (8) states that the program expression P^* contains a unit clause $A \leftarrow Empty$ for each atom A that is provable in P. Finally, rule (9) deals with the import operation. It states that the clauses in $P \triangleleft Q$ are obtained from the clauses in P by dropping the calls to Q, provided that they are provable in Q.

The extended derivation relation $\vdash$ defined by rules (1)—(9) characterises the operational behaviour of arbitrary composition of programs. It is worth noting that the operational and the $T(P)$-based definitions are equivalent. As shown in [4], for any program expression P:

$$A \in T^\omega(P)(\emptyset) \iff P \vdash A.$$

The use of the composition operations $\cup$, $\cap$, $*$ and $\triangleleft$ for programming finds natural application in several domains, ranging over expert systems, hypothetical and hierarchical reasoning, knowledge assimilation and modularisation. The description of such applications is outside the scope of this paper and is reported in [4, 5, 6, 7].

3 Non-Ground Representation of Object Programs

We now present a first implementation in Gödel of the set of program composition operations introduced in the previous Section. The implementation consists of an extended vanilla meta-interpreter using the non-ground representation of object level programs. The definition of the vanilla meta-interpreter in Gödel is illustrated in Subsection 3.1. The extended meta-interpreter is presented in Subsection 3.2, and the associated support for the non-ground representation of object definite programs is described in Subsection 3.3.

3.1 Vanilla Meta-Interpreter

The standard vanilla meta-interpreter [24] using the non-ground representation of object programs can be written in Gödel as illustrated in Chapter 10 of [14]. We consider here a more general form of the vanilla meta-interpreter, where the Solve predicate has an extra argument to explicitly denote the name of the object program to be interpreted. The module `Vanilla` below contains the definition of this more general form of the vanilla meta-interpreter in Gödel.

```
MODULE      Vanilla.
IMPORT      Object_Program.
PREDICATE   Solve : Program_Name * OFormula.
DELAY       Solve(x,y)  UNTIL GROUND(x) & NONVAR(y).

Solve(x, Empty).
Solve(x, y And z) <- Solve(x, y) & Solve(x, z).
Solve(x, y) <- Statement(x, y If z) & Solve(x, z).
```

According to Gödel's syntax, the `PREDICATE` declaration declares `Solve` to be a binary predicate whose first argument has type `Program_Name` and whose second argument has type `OFormula`. The type `Program_Name` is used for the type of meta-level terms representing the name (viz. a constant) of the object level program. The type `OFormula` is used for the type of meta-level terms representing object level formulae. The connectives `&` and `<-` are represented by the functions `And` and `If`, respectively. The `DELAY` declaration is a control declaration stating that calls to `Solve` will delay until first argument (i.e. the program name) is ground and the second argument is not a variable. Statements in the object program to be interpreted are represented in the imported module `Object_Program` using the predicate `Statement` and the constant `Empty`.

The module `Object_Program` (imported by `Vanilla`) contains the meta-level representation of the program to be interpreted. For instance, consider the program consisting of the module `M` below, which defines the relations `Arc` and `Path` over a graph.

```
MODULE      M.
BASE        Node.
CONSTANT    Bristol, London, Pisa : Node.
PREDICATE   Arc, Path : Node * Node.

Path(x, y) <- Arc(x, y).
Path(x, y) <- Arc(x, z) & Path(z, y).
Arc(Bristol, London).
Arc(London, Pisa).
```

The meta-level representation of `M` is reported below. Object level symbols are represented by themselves, including object level variables which are represented by meta-level variables.

```
EXPORT      Object_Program.
BASE        Program_Name, OFormula, Node.
CONSTANT    Empty : OFormula;
            M : Program_Name;
            Bristol, London, Pisa : Node.
FUNCTION    And : xFy(110) : OFormula * OFormula -> OFormula;
            If : xFy(100) : OFormula * OFormula -> OFormula;
```

```
          Arc, Path : Node * Node -> OFormula.
PREDICATE  Statement : Program_Name * OFormula.

LOCAL      Object_Program.
Statement(M, Path(x,y) If Arc(x,y)).
Statement(M, Path(x,y) If Arc(x,z) And Path(z,y)).
Statement(M, Arc(Bristol,London) If Empty).
Statement(M, Arc(London,Pisa) If Empty).
```

According to Gödel's syntax, the module Object_Program consists of an EXPORT and a LOCAL part. The EXPORT part contains the declarations of types, constants, functions and predicates that are exported by the module. The BASE declaration declares the types used in the object program (viz. Node), as well as the types Program_Name and OFormula, which are also used in the importing module Vanilla. The CONSTANT declaration declares the constant Empty, the name of the object program to be interpreted, and the constants occurring in the object program. The FUNCTION declaration declares the function symbols used in both modules (If, And), as well as the object level predicates (e.g. Arc), which are represented by functions at the meta-level. Notice that the module conditions of Gödel require that types, constants and functions used in both modules (such as OFormula, Empty and If) must be declared in the imported module Object_Program. Finally, the LOCAL part of module Object_Program contains the clauses defining the predicate Statement, which is used to represent the object program to be interpreted by the vanilla meta-interpreter.

It is worth making a couple of remarks here. First, the lack of parametric modules in Gödel does not allow a flexible use of the Solve meta-interpreter since the name of the module containing the object program to be interpreted must be fixed in the module Vanilla. The availability of parametric modules would increase the flexibility of Gödel and, in particular, the possibility of parameterising a module w.r.t. the modules to be imported would allow a more flexible use of the meta-interpreter. Second, Gödel does not provide any special support for meta-programming with the non-ground representation. This means that, though it is easy to write a vanilla meta-interpreter in Gödel, the non-ground representation of the object programs must be given *explicitly* by the programmer [14].

3.2   Extended Vanilla Meta-Interpreter

The Solve meta-interpreter can be extended in a simple and concise way in order to implement meta-level operations for composing logic programs. Following [4, 6], each program composition operation is represented at the meta-level by a function symbol: ∪ by Union, ∩ by Intersection, * by Encapsulate, and ◁ by Import. The idea is to use the first argument of Solve for representing arbitrary compositions of object programs, such as P Union Q, rather than just a single object program. The meaning of each function symbol denoting a program composition operation can be defined by extending the definition of the vanilla meta-interpreter. Intuitively, this corresponds to turning the inference rules given in Section 2 into meta-level axioms.

The following module Extended_Vanilla contains the definition of the Solve meta-interpreter suitably extended to deal with programs composition operations.

```
EXPORT      Extended_Vanilla.
IMPORT      Object_Programs.
PREDICATE   Solve : Program_Expression * OFormula.
DELAY       Solve(x,y)  UNTIL GROUND(x) & NONVAR(y).


LOCAL       Extended_Vanilla.
PREDICATE   Clause : Program_Expression * OFormula.
Solve(x, Empty).
Solve(x, y And z) <-
     Solve(x, y) &
     Solve(x, z).
Solve(x, y) <-
     Clause(x, y If z) &
     Solve(x, z).
Clause(x Union y, z If w) <-
     Clause(x, z If w).
Clause(x Union y, z If w) <-
     Clause(y, z If w).
Clause(x Intersection y, z If (w1 And w2)) <-
     Clause(x, z If w1) &
     Clause(y, z If w2).
Clause(Encapsulate(x), y If Empty) <-
     Solve(x, y).
Clause(x Import y, w If z) <-
     Clause(x, w If u) &
     Partition(u, z, v) &
     Solve(y, v).
Clause(x, y If z) <-
     Statement(x, y If z).
...
```

The EXPORT part contains the declaration of the predicate Solve, whose first argument now has type Program_Expression. This is the type of a meta-level term representing a program expression, that is a term constructed via the functions Union, Intersection, Encapsulation and Import starting from a set of program names.

The definition of Solve in the LOCAL part of Extended_Vanilla extends the definition of Solve given in the module Vanilla. The only differences are the type of the first argument (which is now Program_Expression) and the substitution of the predicate Statement with a new predicate Clause. The latter is introduced for the meta-level representation of compositions of object programs. Intuitively speaking, the definition of Clause extends the definition of Statement by induction on the structure of program expressions. For instance, the definition of Clause in the case of Union states that a clause z <- w belongs to the meta-level representation of a program composition x ∪ y if it belongs either to the meta-level representation of x or to the meta-level representation of y. The definition of Clause for Intersection states that a clause z <- (u & v) belongs to the meta-level representation of the composition x ∩ y if z1 <- w1 belongs to the meta-level representation of x, z2 <- w2 belongs to the meta-level representation of y, z1 and z2 unify via a mgu ϑ, and $z = (z1)\vartheta$, and $(u$ & $v)$ $= (w1$ & $w2)\vartheta$. Notice that the adoption of the non-ground representation of object programs allows this statement to exploit the basic unification mechanism. The meta-level representation of an encapsulated program expression Encapsulate(x) consists of assertions of the form Clause(Encapsulate(x), y If Empty) for each y provable in x. The Import operation is defined as follows. The statements in a composition x Import y are obtained from the statements of x by possibly dropping part of their body if this is provable in the imported y and possibly instantiating the remaining part of the body. Finally the last definition of Clause resorts to the predicate Statement (used to represent the single object programs to be composed), which is defined in the imported module Object_Programs.

Basic modularity principles suggest that the Solve vanilla meta-interpreter and the representation of the object programs should be arranged into separate modules. Such a separation makes it easier to use the meta-interpreter with different collections of object programs. The structuring of the module Extended_Vanilla partly supports such a possibility in the sense that the meta-interpreter and the representation of the object programs are arranged into two separate modules:

Extended_Vanilla
↓
Object_Programs

Notice that such a separation requires the employment of two predicates (Clause and Statement) for the meta-level representation of object program compositions. This is due to the module conditions of Gödel that do not allow to spread the definition of a predicate over different modules.

We also implemented a more modular solution that establishes a one-to-one correspondence between object level programs and Gödel modules containing their meta-level representation, as illustrated by the following figure:

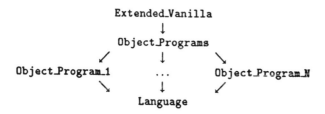

Roughly speaking, each object program P_i is represented by means of a predicate Statement_i defined in a Gödel module Object_Program_i. In this second implementation, Object_Programs simply plays the role of a bridge between the extended vanilla meta-interpreter and the representation of the object programs, and contains a clause

```
Statement(x, y If z) <- Statement_i(x, y If z).
```

for each imported module Object_Program_i. Finally, types and symbols used in both modules are declared in the module Language at the bottom of the hierarchy, as required by Gödel's module conditions.

3.3 Support for the Non-Ground Representation

As we pointed out, Gödel provides considerable support for meta-programming with the ground representation, while it does not provide any special support for the non-ground representation. This means that, though it is easy to write a vanilla meta-interpreter in Gödel, the non-ground representation of the object programs must be given *explicitly* by the programmer. When using Gödel as a meta-language for composing object definite programs, the problem is how to automatically generate a Gödel module containing the non-ground representation of a (collection of) given object definite program(s).

In order to free programmers from the need of explicitly providing the non-ground representation of their object definite programs, we developed suitable support for the generation of such a representation. The main issue to be faced is concerned with types, since Gödel is a strongly typed language while definite programs do not contain any type declaration. One solution might be to try to infer as much type information as possible from the object program, by resorting to program analysis techniques (e.g. [3]). Rather than trying to infer (incomplete) type information from untyped logic programs, we simply employed a single type OTerm to represent *any* object level term.

The module Non_Ground_IO supports the generation of the non-ground representation of a (collection of) object definite program(s) to be imported by the Extended_Vanilla module. Notice that we actually implemented two variants of the support corresponding to the two module structures for the extended vanilla presented in the previous Subsection. Since these supports have a similar structure we shall discuss only the support for the first solution.

```
EXPORT      Non_Ground_IO.
IMPORT      FlocksIO.
PREDICATE   Represent : List(String) * String.
DELAY       Represent(x,y) UNTIL GROUND(x) & GROUND(y).
```

Non_Ground_IO imports FlocksIO, containing the abstract data type Flock that has revealed to be very convenient for the parsing of the files containing the object logic programs. The predicate Represent can be used to generate the non-ground representation of a collection of object definite programs. The first argument of Represent is a list of strings that denotes the names of the files containing the definite programs to be represented at the meta-level. The second argument of Represent is a string denoting the name of the Gödel module in which the non-ground representation of the object programs will be written. Notice that if the object level programs are intended to be interpreted by the vanilla meta-program of module Extended_Vanilla then the second argument of Represent is necessarily the string "Object_Programs", since Gödel does not provide parametric modules at the moment. The possibility of specifying the name of the target Gödel module is however offered by the support also in the perspective of the availability of parametric modules in the near future [12].

To illustrate the use of the support, suppose that the user wants to query program expressions obtained by composing a collection of given object definite programs. First, the non-ground representation of the object programs is generated by means of the module Non_Ground_IO that creates the Gödel module Object_Programs.

```
[Non_Ground_IO] <- Represent(["P","Q","R"], "Object_Programs").
```

Then, the user loads the module Extended_Vanilla (which imports Object_Programs) and queries it by means of meta-level goals such as

```
[Extended_Vanilla] <- Solve(P Union (Q Import R), G(x)).
```

4 Ground Representation of Object Programs

We now present a second implementation of the set of program composition operations introduced in Section 2. This second implementation exploits the meta-programming facilities offered by Gödel for the ground representation of object level programs. As in the previous Section, the vanilla meta-interpreter is illustrated first. Then we show the extended meta-interpreter and the associated support for the generation of the ground representation of object logic programs.

4.1 Vanilla Meta-Interpreter

Gödel provides considerable support for meta-programming with the ground representation, in which object level expressions are represented by ground terms at the meta-level. For instance, the vanilla meta-interpreter discussed in Subsection 3.1 (module Vanilla) can be written in Gödel using the ground representation of object level programs [10], as illustrated in the module Ground_Vanilla below.

```
EXPORT     Ground_Vanilla.
IMPORT     Programs.
PREDICATE Demo : Program * Formula * TermSubst.

LOCAL      Ground_Vanilla.
Demo(program, goal, answer) <-
     StandardiseFormula(goal, 0, var_index, new_goal) &
     EmptyTermSubst(empty_subst) &
     Solve(program, new_goal, var_index, _, empty_subst, subst) &
     RestrictSubstToFormula(new_goal, subst, answer).

PREDICATE
Solve : Program * Formula * Integer * Integer * TermSubst * TermSubst.
Solve(program, goal, v, v, subst, subst) <-
     EmptyFormula(goal).
Solve(program, goal, v_in, v_out, subst_in, subst_out) <-
     And(left, right, goal) &
     Solve(program, left, v_in, new_v, subst_in, new_subst) &
     Solve(program, right, new_v, v_out, new_subst, subst_out).
Solve(program, goal, v_in, v_out, sub_in, sub_out) <-
     Atom(goal) &
     MyStatementMatchAtom(program, goal, stment) &
     Resolve(goal, stment, v_in, new_v, sub_in, new_sub, new_goal) &
     Solve(program, new_goal, new_v, v_out, new_sub, sub_out).
```

The module Ground_Vanilla imports the system module Programs, which contains a variety of predicates for handling the ground representation of Gödel programs. The

PREDICATE declaration declares Demo to be a ternary predicate with arguments of type Program, Formula and TermSubst, which represent the ground representation of a program, of a formula and of a term substitution, respectively.

The LOCAL part of Ground_Vanilla contains the definition of the predicate Demo. The Demo predicate first calls the predicates StandardiseFormula and EmptyTermSubst to initialise the variables in the object goal and to get the representation of the empty substitution. Demo then calls Solve that performs the interpretation of the object program and returns the representation of the computed substitution. Finally such a substitution is restricted to the initial object goal.

The definition of Solve closely resembles the definition of Solve with the non-ground representation of object programs (module Vanilla in Subsection 3.1). To illustrate the reading of this meta-interpreter, let us focus on the last statement in the definition of Solve, namely the case of atomic goal reduction. Given an atomic goal, the predicate MyStatementMatchAtom selects a statement in the program whose predicate (proposition) in the head is the same as the predicate (proposition) in the goal. The predicate Resolve performs the clause reduction step: Given a goal, a selected statement and a substitution returns the new goal and a new substitution obtained by composing the initial one with the substitution computed during the reduction step.

One of the annoying aspects of meta-programming with the ground representation is that unification and substitutions must be handled explicitly. Gödel offers several system modules, such as Programs, which provide a large set of operations for working with the ground representation. It is worth noting that the abstract data type view of modules hides most of the complexity of these operations and supports a declarative style of meta-programming.

4.2 Extended Vanilla Meta-Interpreter

We now extend the vanilla meta-interpreter using the ground representation in order to implement meta-level operations for composing definite programs. The idea is to follow the same approach presented in Section 3. As in the non-ground case, program composition operations are represented by function symbols whose meaning is defined by meta-level axioms. The key difference is that in the ground case object programs are represented by ground terms rather than referred to by constant names as in the non-ground case. The ground representation of a single object program can be generated by means of the system predicate ProgramCompile, which given the name of a program returns the term containing its ground representation.

There is, however, a main problem to face when extending the ground vanilla meta-interpreter with the ground representation to deal with program expressions. Indeed Gödel meta-programming facilities strictly rely on the internal ground representation of Gödel object programs, which coherently mirrors the naming policy of the language. Each symbol is internally represented by a *flat name* that is a quadruple containing name, category and arity of the symbol, as well as the name of the module in which the symbol is declared. Therefore symbols with same name, category and arity that are declared in different modules are distinguished. This does contrast with the naming policy adopted in the logic program composition setting, where a predicate definition may be spread over different modules (e.g. see [4, 6, 21]).

A first solution to this obstacle is to exploit the meta-programming support offered by Gödel to generate the ground representation of object programs, and to meta-program the unification mechanism of flat names in order to identify symbols with same name, category and arity declared in different modules. The implementation of this solution presented two major drawbacks. First, it is necessary to introduce an extra representation of substitutions and therefore to construct from scratch a corresponding support. Second, the introduction of a further interpretation layer for handling the unification mechanism heavily affects the performance of the system.

An alternative approach consists of extending the support for the generation of the ground representation of object programs in order to avoid the undesired distinction among flat names. This way, the extended vanilla meta-interpreter is let free from the problem of flat names, which is completely solved during the pre-processing phase. Before presenting the extended support for this second solution (Subsection 4.3), let us show the corresponding extended vanilla meta-interpreter.

```
EXPORT     Extended_Ground_Vanilla.
IMPORT     Program_Expressions.
PREDICATE  Demo: Program_Expression * Formula * TermSubst.

LOCAL      Extended_Ground_Vanilla.
Demo(pexp, goal, answer) <-
      StandardiseFormula(goal, 0, var_index, new_goal) &
      EmptyTermSubst(empty_subst) &
      Solve(pexp, new_goal, var_index, _, empty_subst, subst) &
      RestrictSubstToFormula(new_goal, subst, answer).

PREDICATE Solve :
Program_Expression * Formula * Integer * Integer * TermSubst * TermSubst.
...
Solve(pexp, goal, v_in, v_out, sub_in, sub_out) <-
      Atom(goal) &
      MyStatementsMatchAtom(pexp, goal, v_in, v1, stment) &
      Resolve(goal, stment, v1, new_v, sub_in, new_sub, new_goal) &
      Solve(pexp, new_goal, new_v, v_out, new_sub, sub_out).

PREDICATE MyStatementsMatchAtom :
Program_Expression * Formula * Integer * Integer * Formula.

MyStatementsMatchAtom(p Union q, goal, v_in, v_out, statement) <-
      MyStatementsMatchAtom(p, goal, v_in, v_out, statement).
MyStatementsMatchAtom(p Union q, goal, v_in, v_out, statement) <-
      MyStatementsMatchAtom(q, goal, v_in, v_out, statement).
...
MyStatementsMatchAtom(Enc(p), goal, v_in, v_out, statement) <-
      EmptyTermSubst(empty_subst) &
      Solve(p, goal, v_in, v_out, empty_subst, subst_out) &
      BuildStatement(goal, subst_out, statement).
...
MyStatementsMatchAtom(Prog(_,p), goal, v, v1, statement) <-
```

```
MyStatementMatchAtom(p, goal, statement1) &
StandardiseFormula(statement1,v,v1,statement).
```

Extended_Ground_Vanilla imports the module Program_Expressions, which contains the declaration of the type Program_Expression with the associated function symbols. The only change in the definition of the Demo predicate is the type of its first argument. The definition of Solve extends the definition of Solve given in the module Ground_Vanilla much in the same way as in the non-ground case of Section 3. Intuitively, the predicate MyStatementsMatchAtom extends MyStatementMatchAtom in the same way as Clause extends Statement in the vanilla meta-interpreter of Subsection 3.2. It is worth noting that in this extended ground meta-interpreter subsitutions must be explicitly handled. For instance, in the case of an encapsulated program expression Enc(p), the representation of the empty substitution as well as of the answer computed by Solve must be explicitly handled in order to build a statement belonging to the ground representation of Enc(p).

4.3 Support for the Ground Representation

In order to employ the module Extended_Ground_Vanilla for querying an arbitrary composition of object logic programs, the meta-interpreter must be provided with the ground representation of such a program expression. We have therefore equipped the module Extended_Ground_Vanilla with a suitable support capable of generating the ground representation of a composition of object definite programs.

Given a program expression E over definite logic programs, the support generates the Gödel term denoting the ground representation of E. Such a term is obtained by composing the ground representations of the single object programs forming the program expression. The ground representation of a definite program P is obtained by first transforming P into an equivalent Gödel program P_G, and then applying ProgramCompile to P_G. According to Subsection 3.3, each logic program is transformed into an untyped Gödel module by employing a unique type Void.

The support is in charge of solving the problem of flat names, so that symbols with same name, category and arity, though occurring in different programs, are identified in their ground representation. This is done by assigning the same name to all the modules and by arranging them into separate directories. The structure of this support is more complex than the non-ground one, since it must combine Gödel modules and utilities, and Unix commands. Notice, however, that the complexity of the ground support is transparent to the user, which only has to invoke a Unix shell called ProgramsCompile. Indeed, in order to compile a collection of object programs, the user simply calls ProgramsCompile with the names (of the files containing) the object programs. For instance, the compilation of the object programs P, Q and R is obtained by invoking:

```
csh ProgramsCompile P Q R
```

Finally, in order to allow the user to query the extended vanilla meta-interpreter, we developed a user interface module Test_Ground_Vanilla reported below. The predicate Go allows the user to specify both the program expression and the goal of the intended query as strings. First the ground representation of the program expression is built via the predicate StringToProgramExpression. Then the string identifying the goal is converted into a formula w.r.t. the name of any of the object programs (which have been given the same name M during the pre-processing phase). Finally, the answer possibly computed by Demo is converted back from the internal Gödel representation of substitutions into a string.

```
MODULE      Test_Ground_Vanilla.
IMPORT      Extended_Ground_Vanilla, Program_ExpressionsIO, Answers.
PREDICATE   Go : String * String * String.

Go(pe_string, goal_string, answer_string) <-
      StringToProgramExpression(pe_string, program_expression) &
      ProgramInProgramExpression(program_expression, program) &
      StringToProgramFormula(program, "M", goal_string, [goal]) &
      Demo(program_expression, goal, answer) &
      RestrictSubstToFormula(goal, answer, computed_answer) &
      AnswerString(program, "M", computed_answer, answer_string).
```

---

5 Discussion

The ultimate objective of our work was to investigate the adequacy of Gödel as a meta-language for re-using and composing existing definite programs. First of all, we would like to draw from our own experience a few considerations on Gödel as a programming language. Gödel offers a number of features to which logic programmers may not be used to. For instance Gödel is a strongly typed language that forces programmers to assign a type to each symbol in a program (except for variables). After feeling initially reluctant to employ types, we found that Gödel's type discipline greatly assists the correct development of programs. The module system is another feature of Gödel that, along with the associated abstract data type view, is essential for the incremental and disciplined development of software. Moreover some Gödel system modules, such as Units and Flocks, have revealed to be very convenient for analysing and handling terms and programs.

Another relevant feature of Gödel is the emphasis placed on meta-programming. Gödel provides ample support for meta-programming with the ground representation. Several system modules offer operations for generating and handling the representation of programs. We found Gödel's abstract data type view of modules particularly powerful in the case of meta-programming. Indeed, system modules such as Programs and Syntax support a declarative style of meta-programming by hiding the complexity of the ground representation. In this respect, however, it is worth mentioning that one of the deficiencies of the the current implementation of Gödel is the tracer. For instance terms with types used in the ground representation, such as Formula, are not visible to the user. As a consequence the tracer is often of little help when analysing the behaviour of meta-programs.

We developed two alternative implementations in Gödel of a suite of meta-level operations for composing definite programs. The two implementations are based on an extended vanilla meta-interpreter and they differ each other for the chosen representation of object level programs. The first implementation, described in Section 3, employs the non-ground representation of object programs and is quite simple and efficient.

The second implementation described in Section 4 heavily exploits Gödel's support for meta-programming. The choice of addressing the naming problem during the pre-processing phase notably simplifies the structure of the extended meta-interpreter with the ground representation, which closely resembles its non-ground correspondent.

Even though the non-ground implementation offers much better performances than the ground one, the latter is definitely more promising than the former. Indeed the ground implementation can greatly benefit from the application of tools that are currently under development for Gödel and that rely on the ground representation. For instance, a partial evaluator called SAGE for Gödel programs is currently under experimentation [10, 11]. The extended vanilla meta-interpreter using the ground representation was designed according to the requirements of SAGE. We expect that the partial evaluation of the meta-interpreter will sensibly reduce the efficiency gap between the two implementations. Another advantage of the implementation with the ground representation is its extensibility to deal with composition of object Gödel programs. Actually the presented implementation of program composition operations already deals with the composition of object Gödel programs in that definite programs are translated into corresponding Gödel programs. Such programs, however, do not contain control declarations or type information. An interesting research direction is to investigate how to extend the implementation of program composition operations to deal with arbitrary Gödel programs. This may allow to push the program composition approach to extend the Gödel language itself.

Finally, we would like to mention some other possible extensions and improvements that might be encorporated in Gödel. The availability of parametric modules [12], for instance, would notably increase the flexibility of the language. In the case of the non-ground implementation described in Section 3, it would allow the parameterisation of the extended vanilla meta-interpreter w.r.t. the object programs, thus enhancing its actual usability. Even more importantly, the possibility of parameterising the language of a Gödel module would allow to overcome one of the most severe limits of the language. Namely, Gödel meta-programming support does not allow to meta-program extensions of the language itself. This is due to the fact that the ProgramCompile operation generating the ground representation of a program requires the latter to be a (pure) Gödel program. A more flexible form of ProgramCompile could also offer the possibility of specifying different names for the program to be compiled and for the file containing it.

Acknowledgements

We would like to thank J.W. Lloyd for the many suggestions and encouragement. Thanks also to A. Bowers, D. Pedreschi and to the referees for their valuable comments. This work was partly supported by Esprit BRA 6810 *Compulog 2* and by Progetto Finalizzato Sistemi Informatici e Calcolo Parallelo of C.N.R. under grant n. 92.01564.PF69.

References

1. K. R. Apt. Logic programming. In J. van Leeuwen, editor, *Handbook of Theoretical Computer Science*, pages 493–574. Elsevier, 1990. Vol. B.

2. K. R. Apt. Declarative programming in Prolog. In D. Miller, editor, *Proc. International Symposium on Logic Programming*, pages 11–35. MIT Press, 1993.

3. R. Barbuti and R. Giacobazzi. A bottom-up polymorphic type inference in logic programming. *Science of Computer Programming*, 19(3):281–313, 1992.

4. A. Brogi. *Program Construction in Computational Logic*. PhD thesis, University of Pisa, March 1993.

5. A. Brogi, P. Mancarella, D. Pedreschi, and F. Turini. Hierarchies through Basic Metalevel Operators. In M. Bruynooghe, editor, *Proceedings of the Second Workshop on Meta-programming in Logic*, pages 381–396, 1990.

6. A. Brogi, P. Mancarella, D. Pedreschi, and F. Turini. Modular Logic Programming. *ACM Transactions on Programming Languages and Systems*, 16(3), 1994.

7. A. Brogi and F. Turini. Metalogic for Knowledge Representation. In J.A. Allen, R. Fikes, and E. Sandewall, editors, *Principles of Knowledge Representation and Reasoning: Proceedings of the Second International Conference*, pages 100–106. Morgan Kaufmann, 1990.

8. A. Brogi and F. Turini. Semantics of meta-logic in an algebra of programs. In S. Abramski, editor, *Proceedings Ninth Annual IEEE Symposium on Logic in Computer Science*. IEEE Society Press, 1994.

9. K.L Clark, F.G. McCabe, and S. Gregory. IC-Prolog language features. In K.L. Clark and S.A. Tarnlund, editors, *Logic Programming*, pages 253–266. Academic Press, 1982.

10. C.A. Gurr. A guide to specialising Gödel programs with the partial evaluator sage. Technical report, University of Edinburgh, 1994.

11. C.A. Gurr. *A self-applicable partial evaluator for the logic programming language Gödel*. PhD thesis, University of Bristol, 1994.

12. P.M. Hill. A parameterised module system for constructing typed logic programs. In R. Bajcsy, editor, *Proceedings IJCAI'93*, pages 874–880. Morgan Kaufmann, 1993.

13. P.M. Hill and J.W. Lloyd. Analysis of metaprograms. In H.D. Abramson and M.H. Rogers, editors, *Metaprogramming in Logic Programming*, pages 23–52. The MIT Press, 1989.

14. P.M. Hill and J.W. Lloyd. *The Gödel Programming Language*. The MIT Press, 1994.

15. R.A. Kowalski. Predicate logic as a programming language. In *IFIP 74*, pages 569–574, 1974.

16. R.A. Kowalski. Problems and Promises of Computational Logic. In J.W. Lloyd, editor, *Computational Logic, Symposium Proceedings*, pages 1–36. Springer-Verlag, 1990.

17. G. Levi and D. Ramundo. A Formalization of Metaprogramming for Real. In D.S. Warren, editor, *Proceedings Tenth International Conference on Logic Programming*, pages 354–373. The MIT Press, 1993.

18. J.W. Lloyd. *Foundations of logic programming*. Springer-Verlag, second edition, 1987.

19. P. Mancarella and D. Pedreschi. An algebra of logic programs. In R. A. Kowalski and K. A. Bowen, editors, *Proceedings Fifth International Conference on Logic Programming*, pages 1006–1023. The MIT Press, 1988.

20. B. Martens and D. De Schreye. Why untyped non-ground meta-programming is not (much of) a problem. Technical Report CW159, Departement Computerwetenschappen, K.U.Leuven, Belgium, December 1992. Revised November 1993, Abridged version to appear in The Journal of Logic Programming.

21. L. Monteiro and A. Porto. Contextual logic programming. In G. Levi and M. Martelli, editors, *Proceedings Sixth International Conference on Logic Programming*, pages 284–302. The MIT Press, 1989.

22. L. Naish. Negation and quantifiers in NU-Prolog. In E. Shapiro, editor, *Proceedings Third International Conference on Logic Programming*, pages 624–634. Springer-Verlag, 1987.

23. D. De Schreye and B. Martens. A Sensible Least Herbrand Semantics for Untyped Vanilla Meta-Programming and its Extension to a Limited Form of Amalgamation. In A. Pettorossi, editor, *Proceedings of the Third Workshop on Meta-programming in Logic*, pages 127–141, 1992.

24. L. Sterling and E. Shapiro. *The Art of Prolog*. The MIT Press, 1986.

25. M. H. van Emden and R. A. Kowalski. The semantics of predicate logic as a programming language. *Journal of the ACM*, 23(4):733–742, 1976.

A Module System for Meta-Programming

P.M. Hill*

School of Computer Studies, University of Leeds, Leeds LS2 9JT
hill@scs.leeds.ac.uk

Abstract. The need for modules in the development of large programs is well known while meta-programming is widely regarded as a simple yet powerful methodology for knowledge representation and reasoning. Thus if we wish to reason about large knowledge bases, it is desirable that meta-programs should be designed to reason about modular programs. This paper describes a module system which allows the modules to be parametrised over the language symbols exported by the module and shows that this provides a natural environment for meta-programming where both the meta and object programs enjoy the same parametrised module system.

1 Introduction

A meta-program, regardless of the nature of the programming language, is a program whose data denotes another (object) program. The object program does not necessarily have to be written in the same programming language as the meta-program although this is often the case and most theoretical work on meta-programming in logic programming (as in this paper) makes this assumption. There are many applications for meta-programming: compilers, interpreters, program analysers, program transformers, and so on. Furthermore, because a logic program is often used in artificial intelligence and defines some knowledge, a meta-program is often viewed as a meta-reasoner for reasoning about this knowledge. Meta-programming for logic was first studied by Bowen and Kowalski [2] and continues to be an important area of research. Barklund [1] has recently written an overview of this area together with a survey of the literature. In [6], Hill and Gallagher have contributed an in-depth survey of the area focusing on those aspects of meta-programming particularly relevant to logic programming.

The key to the semantics of a meta-program with respect to an object program is the way the object program is represented in the meta-program. Such a representation is often called a *naming relation*. Both the syntax and the semantics of this relation has to be defined. Normally, the naming relation is given for each symbol of the object language. Then simple rules of construction can be used to define the representation of the constructed terms and formulas. However, there are two basic representations for the object variables: either as

* This work is supported by SERC Grant GR/H/79862

ground terms or as variables (or, more generally, non-ground terms). The first is called the *ground representation* and the second, the *non-ground representation*. Clearly, the ground representation is the most versatile since the object expressions may then be regarded as an arbitrary string of characters. The main factor in favour of the non-ground representation is the fact that the semantics of the object program can easily be represented in the meta-program [8]. The meta-programming in this paper is presented using the non-ground representation although the underlying ideas are also applicable to meta-programming using a ground representation.

Logic programming has been used extensively for representing and reasoning about knowledge bases. For large knowledge bases, we require a means of segmenting the program so that small component parts of the knowledge base can be developed. These can then be used to build larger components, and so on, until the program is completed. These components are called *modules*. Modules have been researched from a number of points of view including software engineering, object-oriented programming, and theory construction. The module system discussed in this paper is designed with the software engineering application in mind, in particular with regard to the use of modules for program construction. There are a number of requirements for such a module system.

1. There must be a means of combining modules. This is normally achieved by allowing one module to *import* another.
2. Part of a module should be protected from unintended use by other modules. This is called *encapsulation*. Usually a module is divided into two parts. One part defines a language that can be used by an importing module. The other part extends this language with symbols only required locally.
3. It should be possible to develop a module independently of other modules that it does not import. Thus the import relation is normally restricted to defining a partial order on the modules in a program. The order of compilation of the modules must then respect this ordering.
4. A module should be usable in as many contexts as possible. A module providing an abstract data type such as a stack or a definition of an abstract relation such as transitivity needs to be *re-usable* and not tied to a specific application.

The programming language Gödel [9] has a simple module system that supports importation, encapsulation, and separate compilation as well as allowing for modules defining abstract data types. The Gödel module system does not support re-usable modules defining abstract relations such as transitivity. This means that many generic relations have to be redefined in each module in which they are used. Many meta-programs such as interpreters are generic so that a meta-programming module will have to be modified for each object program it is to reason about. Moreover, when the object program is also modular, the problem is more acute. In order to realise the first three criteria of importation, encapsulation, and separate compilation, Gödel has a number of module conditions (see Section 3). These force a modular logic program to be represented in the meta-program as a single module. Thus, the encapsulation and separate compilation of the object module are lost in the representation. In [7], we described a parametrised module system that extended the Gödel module system and provided better facilities for defining abstract relations. In this paper, we focus on the application of this system for meta-programming using the non-ground representation.[2]

Important issues for meta-programming in logic programming are *amalgamation* and *self-applicability*. For these concepts, it is necessary to assume that there is one system for writing both the object program and the meta-program. If the meta-program can be applied to a representation of itself, then it is said to be *self-applicable*. If the language of the meta-program is an extension of the object language and each statement in the object program is a logical consequence of the meta-program, then we say that the meta-program and object program are *amalgamated*. For both self-application and amalgamation, the meta-program and object program must be written in the same (logic) programming language and hence be constructed by means of the same module system. Thus it is desirable that a programming language be able to represent its own module system. We will show in this paper, that the parametrised module system described here (and defined more formally in [7]) satisfies this requirement.

Since the ideas in this paper are based on the Gödel programming language, we give below a simple Gödel program for readers not familiar with the language. This program defines a family and the relations between them[3]. Further examples illustrating the ideas of this paper will be modified forms of this program.

```
MODULE    TheJonesFamilyTree.
BASE      Person.
CONSTANT  Mary,Pat,Sid,Tim: Person.

PREDICATE Mother,Father: Person * Person.
Father(Sid,Pat).
Father(Tim,Mary).
Mother(Pat,Mary).

PREDICATE Par: Person * Person.
Par(x,y) <- Mother(x,y) \/ Father(x,y).

PREDICATE Anc: Person * Person.
Anc(x,y) <- Par(x,z) & Anc(z,y).
Anc(x,y) <- Par(x,y).
```

---

[2] The language Gödel does not provide any support for the non-ground representation for meta-programming. However, this does not prevent this style of meta-programming in Gödel. For example, the meta-program Vanilla in [9, pages 145–147] illustrates an interpreter using this representation.

[3] Par, Rels, Anc are short for *Parent*, *Relations*, *Ancestor*, respectively.

In the above program, there are a number of *declarations* and *statements*. The statements are formulas in the language defined by the language declarations. These declarations begin with a key word that indicates the category[4] and arity of the symbol being declared. The type in a constant declaration or on the right of the -> in a function declaration is the *range type* of the declaration. Constant and function declarations define the top-level constructors of their range types. Thus **Person** is declared to be a base type; the constant declaration for **Mary**, **Pat**, etc., defines **Person**. **Mother** and **Father** are declared to be predicates with arguments of type **Person**. Every symbol used in a program (other than the logical symbols and variables) must be declared in the program. The module system in Gödel will be explained later in Section 3.

The paper is organised as follows. In the next section, we present a non-ground representation and illustrate it using the above example with a simple meta-interpreter. The Gödel module system and its extension to include parametrised modules is given in Section 3. We combine these ideas in Section 4 and show how a modular program can be represented and interpreted by a meta-interpreter using the same module system. The final section highlights the basic advantages of the parametrised module system for meta-programming and discusses the relation of this work with other relevant published work.

2   The Non-Ground Representation

The non-ground representation requires the variables in the object program to be represented by non-ground terms in the meta-program. This paper assumes, as in Prolog, that object variables are represented as variables in the meta-program. The representation for the non-logical symbols is summarised as follows.

Object Symbols	Meta Symbols
Constant	Constant
Function of arity n	Function of arity n
Proposition	Constant
Predicate of arity n	Function of arity n

It is often assumed that distinct symbols (including the variables) in the object language are represented by symbols in the meta-language distinct from the object language symbols. For example, **Tim** and **Pat** in **TheJonesFamilyTree** can be represented by constants, say **OTim** and **OPat**. The predicates such as **Father** and **Mother** in **TheJonesFamilyTree** can be similarly represented by functions, say **OFather** and **OMother** of the same arity. However, the name of the object symbol can also be used as the name of the symbol that represents it. This trivial naming relation, which is the one adopted in Prolog, does not in itself cause any amalgamation of the object and meta-languages and may be regarded as just a syntactic convenience for the programmer. Note however

---

[4] The categories are: type constructor, function, or predicate. A base, constant, or proposition is regarded as a constructor, function, or predicate, respectively, of arity 0.

that, in the case of the constants, and functions, a symbol can represent itself, thereby facilitating the strong amalgamation of the meta and object levels and many levels of meta-reasoning.

We can construct terms that represent terms and atoms of the object language in the obvious way. However, to represent formulas, a representation of the logical connectives is required. Our representation is as follows.

Object Symbols	Meta Symbols
Binary connective	Function of arity 2
Unary connective	Function of arity 1

Continuing the above representation, suppose the connectives &, \/, and ˜ are represented by functions AND/2, OR/2, and NOT/1, respectively. Then the formula Par(x,z) & Anc(z,y) will be represented by the term Par(x,z) AND Anc(z,y) and Mother(x,y) \/ Father(x,y) by Mother(x,y) OR Father(x,y) .

We assume here that a logic program[5] is a set of clauses of the form h <- b or h where h is an atom, called the *head*, and b is any formula constructed using atoms and the connectives &, \/, and ˜, called the *body*. It is clearly necessary, that if the meta-program is to reason about the object program there must be a way of identifying which formulas represent the statements. In Prolog, each statement in the program is represented as a fact (that is, a statement with the empty body). This has the advantage that the variables in the fact are automatically standardised apart each time the fact is used. We adopt this representation here. In particular, we assume that the language of the meta-program contains the distinguished predicate Clause/2[6] and each clause in the object program of the form

h <- b.

is represented as a fact

Clause(h,b).

and each atomic clause

h.

is represented as a fact

Clause(h,Empty).

Hence, with this representation, the object program is represented in the meta-program by the definition of Clause/2. The representation of the family database and its use by the well known 'vanilla' interpreter (see [13]) is given below.

Before continuing with the discussion of the representation of modules, we examine (informally) the semantics of this program. An issue that has had much attention is whether the facts defining Clause accurately represent the object program. The problem is a consequence of the fact that, in Prolog, the language is not explicitly defined but assumed to be determined by the symbols used in the program and goal. Thus, the variables in the object program range over the terms in the language of the object program while the variables in the definition

---

[5] See [10] for the underlying theoretical basis of logic programming assumed in this paper.

[6] We use the predicate name Clause/2 although statements such as h <- $b \vee c$ are not clauses. Such a statement is equivalent to two clauses h <- b and h <- c.

of `Clause` range not only over the terms representing terms in the object program but also over the terms representing the formulas of the object program.

```
MODULE    Vanilla.
BASE      ObjectForm.
CONSTANT  Empty : ObjectForm.
FUNCTION  AND : xFy(110) : ObjectForm * ObjectForm -> ObjectForm;
          OR : xFy(110) : ObjectForm * ObjectForm -> ObjectForm;
          NOT : Fy(120) : ObjectForm -> ObjectForm.

PREDICATE Solve : ObjectForm.
Solve(Empty).
Solve(x AND y) <- Solve(x) & Solve(y).
Solve(x OR y) <- Solve(x) \/ Solve(y).
Solve(NOT x) <- ~ Solve(x).
Solve(x) <- Clause(x, y) & Solve(y).

BASE      Person.
CONSTANT  Mary,Pat,Sid,Tim : Person.
FUNCTION  Mother, Father : Person * Person -> ObjectForm.
FUNCTION  Par, Anc : Person * Person -> ObjectForm.

PREDICATE Clause: ObjectForm * ObjectForm.
Clause(Father(Sid,Pat), Empty).
Clause(Father(Tim,Mary), Empty).
Clause(Mother(Pat,Mary, Empty)).
Clause(Par(x,y), Mother(x,y) OR Father(x,y)).
Clause(Anc(x,y), Par(x,y)).
Clause(Anc(x,y), Par(x,z) AND Anc(z,y)).
```

Thus, in the family database example, the terms in the object language are the constants Mary, Pat, Sid, Tim, while in an untyped meta-program containing a representation of just the clauses in this program, the terms would not only include these constants but also Father(Tim,Mary), Mother(Mary,Tim), Father(Tim,Mother(Mary,Tim)) and so on. Thus in the statement

```
Anc(x,y) <- Par(x,z) & Anc(z,y).
```

in the object program, x and y are universally quantified in the domain of people, while in the fact

```
Clause(Anc(x,y), Par(x,z) And Anc(z,y)).
```

in the (untyped) meta-program, x and y are universally quantified in a domain that contains a denotation of all terms and formulas of the object program.

There is a simple solution, that is, assume that the intended interpretation of the meta-program is typed. The types distinguish between the terms that represent terms in the object program and terms that represent the formulas. This approach has been developed by Hill and Lloyd in [8][7]. An alternative solution is to assume an untyped interpretation but restrict the object program so that the semantics of this representation is preserved. This approach has been explored by De Schreye and Martens in [11] using the concept of *language independence*. (Informally, a program is language independent when the Herbrand model is not affected by the addition of new constant and function symbols.)

It can be seen that the problem is avoided in the examples here by assuming that, as in Gödel, the programming language is typed, so that all the symbols and their types are explicitly declared.

3 Modules

In this section, the Gödel module system is explained by means of an example. (We provide here an informal description of the Gödel module system. A complete definition is given in [9].) In the program below, the family database program in Section 1 is divided into two modules, one specific to the particular family where the mother and father relations are defined, and the other defining the parent and ancestor relations.

Each module is in two parts called *export* and *local*. A part begins with a module declaration stating whether it is the export or local part and the name of the module[8]. The export part contains language declarations for symbols that can be used in this module and also in other modules that import it. Thus the type **Person** can be used in either part of **TheJonesRels** as well as in **TheJonesFamily**. Symbols declared in the local part of a module are only available for use within this part. Hence, since **TheJonesRels** declares **Par** in the local part, **Par** cannot be used outside this module. Statements are only allowed in the local part of a module. These define the predicates declared in either part of the module. **TheJonesRels** also has a module declaration that begins with the key word **IMPORT**. This makes all the symbols declared in the export part of **TheJonesFamily** available for use in **TheJonesRels**. Note that to avoid unnecessary overloading, a symbol name (for a given arity and category) must have at most one declaration in a module.

---

[7] Note that this does not mean that the program has to be explicitly typed, just that the intended interpretation is typed so that all the clauses in the program and goals for the program respect the intended typing. With simple many-sorted typing as required for meta-programming, no type checking in the execution of a correctly typed goal and program is needed.

[8] The **Module** declaration, as used in the example in Section 1, indicates it is the *local* part and that there is no *export* part to this module.

```
EXPORT    TheJonesFamily.
BASE      Person.
CONSTANT  Mary,Pat,Sid,Tim: Person.
PREDICATE Mother,Father: Person * Person.

LOCAL     TheJonesFamily.
Father(Sid,Pat).
Father(Tim,Mary).
Mother(Pat,Mary).
```

```
EXPORT    TheJonesRels.
IMPORT    TheJonesFamily.
PREDICATE Anc: Person * Person.

LOCAL     TheJonesRels.
PREDICATE Par: Person * Person.
Par(x,y) <- Mother(x,y) \/ Father(x,y).
Anc(x,y) <- Par(x,z) & Anc(z,y).
Anc(x,y) <- Par(x,y).
```

A *Gödel program* for a module m (called the *main module*) is the smallest set of modules that includes m and is closed wrt the modules named in the import declarations. The program must satisfy the following three conditions.

1. The module names can be partially ordered so that if m' occurs in an import declaration in a module named m then $m' < m$.
2. Every symbol appearing in (the export part of) a module, must be declared in or imported into (the export part of) the module.
3. Each constructor or predicate with a non-empty definition in a module must be declared in that module.

These conditions enable independent compilation and protect procedures defined in one module from being modified by another. The set of modules {TheJonesFamily, TheJonesRels} form a Gödel program.

The Vanilla module in Section 2 includes the representation of the object program. It is clearly desirable to split this module into two modules, one containing the vanilla interpreter and the other the family database. The interpreter part should be generic and hence, independent of the object programs that it can interpret. However, the three Gödel module conditions make this impossible. In particular, the declarations for ObjectForm, Empty, AND, OR, and NOT have to be included in the module containing the representation of the object program. Furthermore, if the third module condition was deleted so that the representation of the object program was in a separate module, then the interpreter module would have to import this representation by means of an import declaration, and so be specialised for the particular object program.

We now describe a parametrised module system. This is a straightforward extension of the Gödel module system. We explain the system using the same family database as before.

```
EXPORT      TheJones.
%
IMPORT      Rels(Jones,Ma,Pa).
BASE        Jones.
PREDICATE Ma,Pa: Jones * Jones.
CONSTANT   Mary,Pat,Sid,Tim: Jones.

LOCAL       TheJones.
Father(Sid,Pat).
Father(Tim,Mary).
Mother(Pat,Mary).
```

```
EXPORT      Rels(Person,Mother,Father).
BASE        Person.
PREDICATE Mother,Father: Person * Person.
%
PREDICATE Anc: Person * Person.

LOCAL       Rels(Person,Mother,Father).
IMPORT      Trans(Person, Par).
PREDICATE Par: Person * Person.
Par(x,y) <- Mother(x,y) \/ Father(x,y).
Anc(x,y) <-  Tr(x,y).
```

```
EXPORT      Trans(Point,Connect).
BASE        Point.
PREDICATE Connect: Point * Point.
%
PREDICATE Tr: Point * Point.

LOCAL       Trans(Point, Connect).
Tr(x,y) <- Connect(x,y).
Tr(x,y) <- Connect(x,z) & Tr(z,y).
```

In this example, there are three declared modules, identified by **TheJones**, **Rels**, and `Trans`. The `TheJones` module, which is not parametrised, is the main module which imports an instance of the `Rels` module. This is parametrised with respect to the base `Person` and predicates `Mother` and `Father`. Note that it is the `TheJones` module that imports the `Rels` module which defines the relations `Par` and `Anc` whereas, in the previous example based on the Gödel module system, the importation was in the opposite direction. The module name that follows the key words `EXPORT` and `LOCAL` consists of an *identifier* with 0 or more symbols as arguments. The set of declarations for these symbols (which must be in the export part of the module) is called the *signature* of the module. For example, `Rels(Person,Mother,Father)` is a module name with identifier `Rels` and signature

```
BASE       Person.
PREDICATE Mother,Father: Person * Person.
```

Symbols that are declared in a module but are not in the signature are said to be *completely specified* by the module. For example, the base `Jones` is completely specified in `TheJones`.

The written module is the *initial* module. *Instances* of these modules can be obtained by substituting new symbols for symbols occurring in the module name. In the example program, the following instance of the initial `Trans` module is imported into `Rels(Person,Mother,Father)`

```
EXPORT     Trans(Person,Par).
BASE       Person.
PREDICATE Par: Person * Person.
%
PREDICATE Tr: Person * Person.


LOCAL      Trans(Person,Par).
Tr(x,y) <- Par(x,y).
Tr(x,y) <- Par(x,z) & Tr(z,y).
```

The substituted symbols used to obtain the instance module must be distinct from symbols that were completely specified by its initial module. For example, the module name `Trans(Person,Tr)` is not a legal module name for importation since `Tr` is completely specified in `Trans(Person,Connect)`. Therefore, if a symbol is completely specified in a module, it will be completely specified in every module that is an instance of it. In the example, as the predicate `Tr` is completely specified by `Trans(Point,Connect)`, it is also completely specified by `Trans(Person,Par)`. This rule ensures that the definitions of constructors and predicates that are completely specified cannot be modified by modules that import them.

We define a *modular program* in a similar way to the definition of a Gödel program above. Informally, it the smallest set of initial modules with a distinguished module called the main module and closed with respect to the identifiers in the import declarations. It must satisfy similar module conditions to those given for the Gödel module system.

1* The identifiers in the module names can be partially ordered so that if I' is an identifier in an import declaration in a module with identifier I then $I' < I$.

2* Every symbol name appearing in (the export part of) a module m, must either be declared in (the export part of) m or be completely specified by the export part of a module that is imported into (the export part of) m.

3* Each constructor or predicate name declared in or imported into a module and completely specified by an imported module n may only have a non-empty definition in n or in imported modules that are also imported into n.

A modular program for the `TheJones` consists of the following set of modules

{`TheJones`, `Rels(Person,Mother,Father)`, `Trans(Point,Connect)`}

The semantics of a modular program are given by renaming the symbols in the different instances of the generic modules to ensure that overloaded names are standardised apart and then combining the renamed modules to form a program consisting of a single flattened module. The semantics of the modular program are then the semantics of the combined module. Details of the flattening procedure for this parametrised module system can be found in [7].

4 The Representation of Modules

We now present the `Vanilla` interpreter using parametrised modules.

```
EXPORT    Vanilla(ObjectForm,Clause).
BASE      ObjectForm.
PREDICATE Clause : ObjectForm * ObjectForm.
%
IMPORT    Representation(ObjectForm).
PREDICATE Solve : ObjectForm.

LOCAL     Vanilla(ObjectForm,Clause).
Solve(Empty).
Solve(x AND y) <- Solve(x) & Solve(y).
Solve(x OR y) <- Solve(x) \/ Solve(y).
Solve(NOT x) <- ~ Solve(x).
Solve(x) <- Clause(x, y) & Solve(y).
```

The interpreter assumes a particular representation of the object program which is first defined in another module called `Representation(Formula)` and then imported into the `Vanilla` module.

```
EXPORT     Representation(Formula).
BASE       Formula.
%
CONSTANT   Empty : Formula.
FUNCTION   AND : xFy(110) : Formula * Formula -> Formula;
           OR : xFy(110) : Formula * Formula -> Formula;
           NOT : Fy(120) : Formula -> Formula.
```

The **Vanilla** interpreter can be used for any program provided it uses the representation declared in **Representation(Formula)** module. Each module of the object program can now be represented as a single module in the meta-program. The following is such a representation of the **TheJones** program

{TheJones, Rels(Person,Mother,Father), Trans(Point,Connect)}.

```
EXPORT     OTheJones(ObjectForm,Clause).
BASE       ObjectForm.
PREDICATE  Clause: ObjectForm * ObjectForm.
%
IMPORT     Representation(ObjectForm).
IMPORT     ORels(Jones,Ma,Pa,ObjectForm,Clause).
BASE       Jones.
FUNCTION   Ma,Pa: Jones * Jones -> ObjectForm.
CONSTANT   Mary,Pat,Sid,Tim: Jones.

LOCAL      OTheJones(ObjectForm,Clause).
Clause(Father(Sid,Pat), Empty).
Clause(Father(Tim,Mary), Empty).
Clause(Mother(Pat,Mary), Empty).
```

```
EXPORT     ORels(Person,Mother,Father,ObjectForm,Clause).
BASE       Person.
FUNCTION   Mother,Father: Person * Person -> ObjectForm.
BASE       ObjectForm.
PREDICATE  Clause: ObjectForm * ObjectForm.
%
IMPORT     Representation(ObjectForm).
FUNCTION   Anc: Person * Person -> ObjectForm.
```

```
LOCAL     ORels(Person,Mother,Father,ObjectForm,Clause).
IMPORT    OTrans(Person, Par,ObjectForm,Clause).
FUNCTION  Par: Person * Person -> ObjectForm.
Clause(Par(x,y), Mother(x,y) OR Father(x,y)).
Clause(Anc(x,y), Tr(x,y)).
```

```
EXPORT    OTrans(Point,Connect,ObjectForm,Clause).
BASE      Point.
FUNCTION  Connect: Point * Point -> ObjectForm.
BASE      ObjectForm.
PREDICATE Clause: ObjectForm * ObjectForm.
%
IMPORT    Representation(ObjectForm).
FUNCTION  Tr: Point * Point -> ObjectForm.

LOCAL     OTrans(Point,Connect,ObjectForm,Clause).
Clause(Tr(x,y), Connect(x,y)).
Clause(Tr(x,y), Connect(x,z) AND Tr(z,y)).
```

Note that this representation maintains the module structure of the underlying object program. We have added two arguments ObjectForm and Clause to each module name in the object program. Language declarations for ObjectForm and Clause and an import declaration for the Representation(ObjectForm) module have been added to each module. Note also, that because Clause occurs in the signature of each of these modules, the definition of Clause consists of the union of the sets of facts in these modules that contain the predicate Clause. This representation transformation, illustrated in the above example, is defined for an object module independent of its use in an object program. Thus the representation of a modular program consisting of a set of modules $\{M_1, \ldots, M_n\}$ is the set of modules $\{M_1', \ldots, M_n'\}$, where M_i' represents M_i, for $i \in \{1, \ldots, n\}$.

This representation is also independent of any particular meta-program. To use Vanilla for this representation of the object program, another module is required that imports both the representation of the main module of the object program and the Vanilla module. Such a module is given below.

```
EXPORT    InterpretTheJones.
IMPORT    Vanilla(ObjectForm, Clause).
IMPORT    OTheJones(ObjectForm, Clause).
BASE      ObjectForm.
PREDICATE Clause: ObjectForm * ObjectForm.
```

A goal for the program with main module InterpretTheJones is

```
<- Solve(Anc(Sid, x)).
```

5 Discussion

The parametrised module system described here is ideal for meta-programming for at least two reasons. First, good software engineering practice requires modules to be independent program components. This means that the meta-program should be a generic module, independent of any specific object program that it has to manipulate. The second concerns the need to improve the efficiency of meta-programs. By including the object program in the meta-program, meta-programming predicates defining the semantics of the object program can be implemented directly using the object program. In this form of meta-programming, the programming languages used for the object and meta-programs have to be the same so that a representation must be defined for modular programs. By preserving the module structure of the object program in its representation, it is straightforward to keep the locality of symbol names, thereby preventing accidental interference between overloaded symbol names.

We have assumed that the module system is a fundamental requirement for a programming language and will therefore be provided as one of the main features of the language. Meta-programming techniques have also been used to provide a module system on top of a logic programming language that has no built-in module system. However, a built-in module system has many advantages. One of these is that it can support separate compilation for the individual modules. Moreover, at run-time, only the flattened form of the program need actually be executed while, with the meta-programming method, there is the overhead of the extra level of interpretation. One disadvantage of the parametrised modules is that separate compilation of the modules appears to be more difficult. Work on the implementation of parametrised modules is required to determine the best way to resolve this problem.

The meta-programming approach was first proposed in [2]. Here it was shown how, by representing programs as terms, predicates defining the inference procedure can interpret the programs in a constrained way, thereby limiting particular reasoning strategies to different parts of the object program. This idea has been further developed in a number of papers by Brogi, Mancarella, Pedreschi, Turini and Contiero. In particular, the papers [3] and [4] consider the combining of modules and meta-programming from a perspective opposite to that considered in this paper. They show how using a simple extension of interpreters for logic programs (using both non-ground and ground representations), operators for combining export and local parts of modules as well as importing one module into another module can be defined. Our module system does have certain advantages over the one proposed in these papers. In particular, there is no apparent way in which the definition of a predicate exported by a module can be protected from alteration by an importing module. In a programming language such as Gödel, such lack of protection would inhibit efficient implementation of the numeric modules and cause unpredictable results in the presence of negation in the bodies of statements. Further work needs to be done to see if the parametrised module system described here can be implemented using techniques similar to those described in these papers.

We have based our approach in this paper on the module and type systems of the programming language Gödel. The language 'LOG was developed by Cervesato and Rossi [5] with similar aims as Gödel, at least as far as providing declarative facilities for meta-programming. Each expression has basically two representations. One as a constant naming the expression and the other as composite object representing the structure of the expression. A particular feature of this language is the ability to define parts of object programs as *inner* programs for a meta-program. This provides a limited means of defining and interpreting programs as modules.

References

1. J. Barklund. Metaprogramming in logic. Technical Report UPMAIL Technical Report 80, Department of Computer Science, University of Uppsala, Sweden, 1994. to be published in Encyclopedia of Computer Science and Technology, A. Kent and J.G. Williams (eds.), Marcell Dekker, New York, 1994/5.
2. K.A. Bowen and R.A. Kowalski. Amalgamating language and metalanguage in logic programming. In K.L. Clark and S.-Å. Tärnlund, editors, *Logic Programming*, pages 153–172. Academic Press, 1982.
3. A. Brogi and S. Contiero. Gödel as a meta-language for composing logic programs. In F. Turini, editor, *Meta-Programming in Logic, Proceedings of the 4th International Workshop*, Pisa, Italy. Springer-Verlag, 1994.
4. A. Brogi, P. Mancarella, D. Pedreschi, and F. Turini. Meta for modularising logic programming. In A. Pettorossi, editor, *Proceedings of the Third Workshop on Meta-programming in Logic*, pages 105–119. Springer-Verlag, 1992.
5. I. Cervesato and G. F. Rossi. Logic meta-programming facilities in 'LOG. In A. Petterossi, editor, *Proceedings of the Third Workshop on Meta-programming in Logic*, pages 148–161. Springer-Verlag, 1992.
6. P. M. Hill and J. Gallagher. Meta-programming in logic programming. Technical Report 94.22, School of Computer Studies, University of Leeds, 1994. To be published in *Handbook of Logic in Artificial Intelligence and Logic Programming, Vol. 5*, Oxford Science Publications, Oxford University Press.
7. P.M. Hill. A parameterised module system for constructing typed logic programs. In R. Bajcsy, editor, *Proceedings of 13th International Joint Conference on Artificial Intelligence*, Chambéry, France, pages 874–880. Morgan-Kaufmann, 1993.
8. P.M. Hill and J.W. Lloyd. Analysis of meta-programs. In H.D. Abramson and M.H. Rogers, editors, *Meta-Programming in Logic Programming*, pages 23–52. MIT Press, 1989. Proceedings of the Meta88 Workshop, June 1988.
9. P.M. Hill and J.W. Lloyd. *The Gödel Programming Language*. MIT Press, 1994.
10. J.W. Lloyd. *Foundations of Logic Programming*. Springer-Verlag, 2nd edn., 1987.
11. B. Martens and D. De Schreye. A perfect Herbrand semantics for untyped vanilla meta-programming. In K. Apt, editor, *Proceedings of the Joint International Conference on Logic Programming*, Washington, USA, pages 511–525, 1992.
12. B. Martens and D. De Schreye. Why untyped non-ground meta-programming is not (much of) a problem. Technical Report CW 159, Department of Computer Science, Katholieke Universiteit Leuven, 1992.
13. L. Sterling and E. Shapiro. *The Art of Prolog*. MIT Press, 1986.

Building Proofs in Context

Giuseppe Attardi and Maria Simi *

Dipartimento di Informatica
Università di Pisa
Corso Italia 40
I-56125 Pisa, Italy
net: {attardi, simi}@di.unipi.it

Abstract. When reasoning with implicitly defined contexts or theories, a general notion of proof in context is more appropriate than classical uses of reflection rules. Proofs in a multicontext framework can still be carried out by switching to a context, reasoning within it, and exporting the result. Context switching however does not correspond to reflection or reification but involves changing the level of nesting of theory within another theory. We introduce a generalised rule for proof in context and a convenient notation to express nesting of contexts, which allows us to carry out reasoning in and across contexts in a safe and natural way.

1 Introduction

A general notion of relativised truth can be useful for reasoning in and about different theories in a formal setting. For example to reason about the reasoning of different agents, to model temporal evolution of knowledge, to split a large knowledge base into manageable chunks or microtheories that can be related to each other by means of transfer rules or lifting axioms.

There are several approaches to the formalization of a notion of relativised truth: by means of a predicate expressing "provability" like for example $PR(T, P)$ in [22] and $demo(T, P)$ in [7], or with a notion of truth in context like for example $ist(c, p)$ in [11, 14, 15, 8] and p^c in [20], or with a notion of entailment from a set of assumptions like $in(P, vp)$ [3, 21, 5].

Most of these are syntactic approaches where theories are modeled as collections of *reified statements* or *statement names* in First Order Predicate Calculus. The object theory is extended with a meta-theory consisting of statements about statements. The relation between general validity and truth relativised to a subtheory is usually expressed by means of a pair of reflection/reification rules. These are classically formulated as follows [13, 22]:

$$\frac{T \vdash P}{pr \vdash demo(T, P)} \qquad (Reification1)$$

$$\frac{pr \vdash demo(T, P)}{T \vdash P} \qquad (Reflection1)$$

which say that if formula P is derivable from the set of statements T, then $demo(T, P)$ is derivable in the meta-theory from theory pr and vice versa, where pr is a theory containing a suitable axiomatisation of the demo predicate.

Unfortunately carrying out proofs dealing with multiple theories is not simple. When reasoning about reasoning, one often needs to carry out some proof steps within a different theory from the current one and then to lift the conclusions back into the original theory. The deductive rules required to carry out these steps involve either *reflection* principles or some other notion of *proof in context*.

Standard formulations of the reflection rules assume explicit knowledge of the theory one reasons about. In many interesting application however it is not possible to explicitly state once for all the assumptions of a theory; this is often the case for theories representing agents and is always the case for infinite theories, for theories which refer to each other or reflective theories, as those required for expressing common knowledge.

---

* This work has been done while the authors were visiting the International Computer Science Institute, Berkeley CA.

Moreover dynamic extension of theories, provided by lifting axioms, requires the ability to define theories implicitly.

In this paper we discuss different approaches to contextual reasoning according to whether implicit and mutually referential theories are allowed. We will argue that implicit contexts are necessary for most significant applications. When reasoning with implicit contexts however, reflection rules are not the right way to tranfer facts from one context to another.

Context switching in a natural deduction proof is better seen as nesting or unnesting of contexts justified by suitable rules for proof in context. A notation is introduced to write more readable proofs where context switching is interpreted according to this semantics.

The three approaches discussed in the paper use three slightly different notations for relativised truth, whose correspondence is shown below:

$demo(T, P)$ Kowalski
$ist(T, P)$ McCarthy
$in(P, T)$ Attardi and Simi

2 Approaches to proofs in context

2.1 Explicit vs. implicit theories

When reasoning with contexts or theories, an important distinction is whether we are dealing with explicit theories or with theories which are only implicitly defined.

In the first case we assume that a theory can be explicitly and completely characterised by a finite set of statements representing the relevant assumptions of the theory. If names are used for theories they should be considered as linguistic shortcuts for the set of statements. This is an adequate model in many practical situations; for example one can reason about a subset of an agent's beliefs *as if* it was complete, without having to know all of them.

In many interesting applications theories cannot be defined explicitly as a finite set of statements and the ability to characterize theories implicitly is required. The language must then allow for constants, or more in general terms, denoting theories or contexts.

For example, one may want to characterize theories by means of assertions, like McCarthy's *lifting rules* [McCarthy 93], stating that whenever a formula satisfying some condition holds in a theory vp_1 then a related formula holds in another theory vp_2. As a special case, subsumption between theories can be expressed as follows in our notation:

$$\forall x \,.\, in(x, vp_1) \Rightarrow in(x, vp_2)$$

Incidentally, the possibility of quantification over statements inside the in predicate, not provided by modal logics, appears essential here; hence our preference for a syntactic treatment of in, our notion of relativised truth.

Another example could be the evolution of state of affairs, as in:

$$in('Clear(A) \wedge Clear(B)', sit_1) \Rightarrow in('On(A, B)', sit(puton(A, B), sit_1))$$

This allows for compact statements of problems and leaves to the logic machinery the burden of incrementally specifying theories when the rules for their construction are known.

Implicit theories are also required for expressing self referential statements. Here are a few examples, where we use in to express *belief*:

John believes that he has a false belief
$in('\exists x \,.\, in(x, vp(John)) \wedge False(x)', vp(John))$

Agent a believes that whatever he and agent b believe is true, while b does not believe so
$in('\forall x \,.\, in(x, vp(a)) \vee in(x, vp(b)) \Rightarrow True(x)', vp(a))$
$in('\neg \forall x \,.\, in(x, vp(a)) \vee in(x, vp(b)) \Rightarrow True(x)', vp(b))$

Agent a and agent b have *common knowledge* (or belief) that A

$$\text{in}('A', CK) \qquad\qquad (CK\text{-}1)$$
$$\forall x \,.\, \text{in}(x, CK) \Rightarrow \text{in}(x, vp(a)) \wedge \text{in}(x, vp(b)) \qquad\qquad (CK\text{-}2)$$
$$\forall x \,.\, \text{in}(x, CK) \Rightarrow \text{in}(\text{in}(x, vp(a)) \wedge \text{in}(x, vp(b)), CK) \qquad\qquad (CK\text{-}3)$$

Note that for common knowledge it is not enough that both agents know A but it is also required that they know that they know A, that they know that they know that they know A, ... and so on. Such infinite nesting calls for a recursive definition like $CK\text{-}3$.

The approach of implicit contexts is carried to its extreme consequences in the proposal of McCarthy, where contexts are primitive objects denoted by symbols in the language, and are never explicitly characterised.

2.2 Reflective vs. layered theories

Syntactic approaches to provability differ in the degree of connection between object-theory and meta-theory, ranging from a semantic connection of completely separate theories as in [12], to the reflection principles of FOL [22] between two still distinct theories, to user defined bridge rule connecting theories with distinct languages, to the proposal of a single amalgamated theory encompassing object and meta-level [7] or, more in general, reflective theories.

Nevertheless, a satisfactory first order theory of relativised truth is not easy to develop since one must face delicate issues of semantics and must avoid the pitfalls of paradoxes arising from self referential statements, which trickle in by diagonalization [Montague 63].

A simple way out of paradoxes is to keep the object language separate from the metalanguage [18, 22] and when nested beliefs are involved, to build a hierarchy of languages, each one being a meta-language for the previous [12]. Self reference is not allowed and the construction of paradoxical statements is blocked.

However this forbids also non paradoxical self referential statements like those mentioned in the previous section. This lack of expressiveness may be considered a major drawback [19] since self referential statements about truth, beliefs or knowledge arise naturally in common sense reasoning. The complex machinery required by the layered approach also does not seem convenient for implementation within reasoning programs nor natural as a formalisation of common sense.

When self or mutually referential theories are allowed, they must be accounted for in the semantics of the logic. One way to do so is to use non well founded sets [1] as denotation for theories and rely on Barwise solution lemma to ensure that solutions to the recursive equations exist.

A different approach is the one pursued in the theory of viewpoints [5], where viewpoints denote recursive set of statements and the interpretation of in statements is done in a layer by layer fashion so as to properly account for paradoxical self referential statements.

In a reflective framework reflection rules are to be carefully formulated to avoid falling into an inconsistent theory because of the results of [17] [2]. A possibility is to be conservative and use reflection rules, such as the ones in [7], which do not add any new theorem with respect to the amalgamated theory without reflection rules. But useful non conservative formulations, still preserving consistency, are possible, as the one we propose for the theory of viewpoints.

2.3 Reflection rules vs. nesting/unnesting

When reasoning with multiple theories one needs rules for context switching, i.e. for moving from one theory to another, performing some deduction there and then transfering elsewhere some of the derived consequences.

For theories which are specially related by being one the meta-theory for the other, reflection and reification may be used for this purpose.

All standard formulations of the reflection rule postulate that only theorems can be object of reflection, in fact the premise of the rule is something like

$$pr \vdash \mathsf{demo}(T, P)$$

where pr is just an axiomatization of provability. In other words, $\mathsf{demo}(T, P)$ cannot be just asserted as part of the statement of the problem or assumed during a hypothetical line of reasoning.

---

[2] Note that this could happen even in a layered framework if the layering collapses due to strong bridge rules

When dealing with implicit theories, the following is a very useful pattern of reasoning. One could assert or assume:

$$\text{demo}(T, P)$$

and given that $P \vdash Q$, one would like to conclude that

$$\text{demo}(T, Q)$$

Standard reflection would not be applicable for such kind of reasoning since an asserted or assumed demo statement is not a logical theorem. It is possible to strengthen reflection to make it applicable in such cases, as proposed for instance in [10] or [6], provided the conclusion still depends on the assumptions made to derive the antecedent.

Reification should be restricted accordingly to prevent unsound consequences. For example the antecedent $T \vdash Q$ of reification should not be the consequence of an assumption or something which is just stated but rather must correspond to a real ability to prove. Differently one could use reification to conclude $\text{demo}(T, S)$ in any theory from the fact that S is true (or is asserted to be derivable) in theory T. These restrictions would again prevent the application of reification for concluding $\text{demo}(T, Q)$.

For a slightly different reason we believe that reification is also used improperly in the solution to the three wise men puzzle presented in [13] in the framework of the amalgamated logic of Bowen and Kowalski [7]. The pattern of reasoning used there corresponds to constructing a new object level theory consisting of those statements P such as

$$\text{demo}(T, P)$$

holds. This theory should not be confused with T, since it is rather an image of T from the perspective of the current reasoner; let us call it T'. Once the conclusion Q is reached in T' reification is used, improperly, to conclude $\text{demo}(T, Q)$.

These considerations suggest that another kind of context switching is more appropriate when reasoning with implicit contexts which are dynamically nested during deduction: this kind of contextual reasoning can be made explicit, by representing somehow the structure of nesting.

For example, in their formalization of contexts [8] Buvač and Mason propose rules for context switching which correspond to this idea, and introduce indexes made of sequences of context names to represent nested contexts. The rule they present is bidirectional and reads as follows:

$$\frac{\vdash_{\overline{k}*k_1} \phi}{\vdash_{\overline{k}} \text{ist}(k_1, \phi)} \qquad (CS)$$

The index $\overline{k}$ represents a sequence of contexts and the rule expresses that a statement about the truth of ϕ in a series of nested contexts $\overline{k}$ can be turned into the fact ϕ holding in the series of contexts $\overline{k}*k_1$. Keeping track of the level of nesting is crucial for the correctness of the rule.

Following this idea we will formally justify our notion proof in context, introduce a notation for nesting/unnesting of contexts and present a solution to the three wise men which does not make use of reflection rules, but still exploits a very general mechanism for context switching.

3 Proof theory

In order to discuss the problems and subtle issues hinted in the previous sections, we introduce a formal deductive system for proofs in contexts developed in connection with the theory of viewpoints [5].

The theory of viewpoints is a reflective first order theory with explicit or implicit viewpoints (viewpoint constants and functions). A complete semantic account of viewpoints is presented in [5].

Viewpoints are sets of reified statements and the expression $\text{in}('P', vp)$ means that a statement P is contextually entailed by the set of assumptions represented by vp.

More precisely, given a term t_1 denoting a statement and a viewpoint expression t_2 denoting a set of statements, $in(t_1, t_2)$ is true at a model $\mathcal{M}$ iff the statement denoted by t_1 is true in any model of the statement denoted by t_2 which is "coherent" with $\mathcal{M}$ in the interpretation of viewpoint constants and functions.

The proof theory for viewpoints can be conveniently presented in the style of natural deduction.

3.1 Inference rules for classical natural deduction

As customary, the notation $\Gamma \vdash P$ indicates the pending assumptions in rules where some of the assumptions are discharged, like in the cases of implication introduction and negation introduction. When the pending assumptions are.the same in the antecedent and consequent of a rule they are omitted.

The rules for natural deduction are quite standard. For example:

$$\frac{P, Q}{P \wedge Q} \quad (\wedge\ I) \qquad\qquad\qquad \frac{P \wedge Q}{P, Q} \quad (\wedge\ E)$$

are the rules for conjunction introduction and elimination, respectively, and

$$\frac{\Gamma \cup \{P\} \vdash Q}{\Gamma \vdash (P \Rightarrow Q)} \quad (\Rightarrow\ I) \qquad\qquad \frac{P, P \Rightarrow Q}{Q} \quad (\Rightarrow\ E)$$

are the rules for implication introduction and elimination. The full set of classical rules used is presented in the appendix.

3.2 Metalevel axioms and inference rules

The behaviour of in is characterised by the following axioms and inference rules, which allow classical reasoning to be performed inside any viewpoint.

The first axiom asserts that all the statements which constitute a viewpoint hold in the viewpoint itself, while the second establishes a principle which could be called *positive introspection*, if we chose an epistemic interpretation for in. The third axiom states monotonicity of viewpoints.

$$in('P', \{\ldots, 'P', \ldots\}) \tag{Axiom1}$$

$$in('P', vp) \Rightarrow in('in('P', vp)', vp) \tag{Axiom2}$$

$$in('P', vp) \Rightarrow in('P', vp \cup \{'Q'\}) \tag{Axiom3}$$

Moreover we have a meta-inference rule for each classical natural deduction inference rule. For example:

$$\frac{in('P', vp), in('Q', vp)}{in('P \wedge Q', vp)} \tag{Meta $\wedge$ I}$$

$$\frac{in('P \wedge Q', vp)}{in('P', vp), in('Q', vp)} \tag{Meta $\wedge$ E}$$

$$\frac{in('Q', vp \cup \{'P'\})}{in('P \Rightarrow Q', vp)} \tag{Meta $\Rightarrow$ I}$$

$$\frac{in('P', vp), in('P \Rightarrow Q', vp)}{in('Q', vp)} \tag{Meta $\Rightarrow$ E}$$

The full set of meta-inference rules is presented in the appendix.

3.3 Reflection rules

The following are the reflection and reification rules for the theory of viewpoints: they are more powerful than those of [7], but still safe from paradoxes as discussed in [4].

$$\frac{vp_1 \vdash in('P', vp_2)}{vp_1 \cup vp_2 \vdash P} \qquad\qquad\qquad (Reflection)$$

$$\frac{vp \vdash_C P}{\vdash in('P', vp)} \qquad\qquad\qquad (Reification)$$

The notation $\vdash_C$ stands for "classically derivable" or "derivable without using the reflection rules".

We have argued elsewhere for the usefulness of the strong version of reflection [4], [5]. As a consequence we have:

Theorem 1. $in('P', \{'Q'\}) \Rightarrow (Q \Rightarrow P)$

Reification is a derived inference rule; in fact any proof at the object level can be completely mirrored at the metalevel using the meta-level inference rules. In fact also the stronger

$$vp \vdash_C P \quad iff \quad \vdash_C in('P', vp)$$

holds.

This can be proved by induction on the length of the proof, with the base case being provided by *Axiom*1, or derived as a consequence of theorem 3 below.

3.4 Proof in context

Proof in context is a powerful mechanism for reasoning across multiple contexts. In this section we present the results which provide the formal justification for this technique.

First we note that logical theorems can be used in proofs within any context, and then that classical deductions can be carried out within any viewpoint.

The first result is a consequence of reification and *Axiom*1:

Theorem 2. $in('P', vp)$, *for any logical theorem P and viewpoint vp.*

Theorem 3 Proof in context. $\qquad \dfrac{\Gamma \vdash_C P}{in('\Gamma', vp) \vdash_C in('P', vp)}$

where the Γ is a finite set of sentences and the consequent should be read as the conjunction of the formulae in Γ.

Proof. The proof is by induction on the length of the derivation and by showing that each step can be mirrored at the metalevel using the corresponding meta-inference rule. The only interesting case is the one for implication introduction. Suppose that this rule was used in the last step of the derivation to prove $Q \Rightarrow R$. We have by induction hypothesis:

$$\Gamma \cup \{Q\} \vdash_C R$$
$$in('\Gamma \wedge Q', vp) \vdash_C in('R', vp)$$

Therefore:

$in('\Gamma', vp)$	$(premise)$
$in('Q', vp \cup \{'Q'\})$	$(Axiom1)$
$in('\Gamma \wedge Q', vp \cup \{'Q'\})$	$(Meta \wedge I)$
$in('R', vp \cup \{'Q'\})$	$(induction)$
$in('Q \Rightarrow R', vp)$	$(Meta \Rightarrow I)$

The following inference rules can be established from theorem 3:

$$\frac{\text{in}('P', vp), P \vdash_C Q}{\text{in}('Q', vp)} \qquad\qquad\qquad (\textit{Proof in context})$$

which generalises to:

$$\frac{\{x \mid \text{in}(x, vp)\} \vdash_C P}{\text{in}('P', vp)} \qquad\qquad\qquad (\textit{Generalised proof in context})$$

The antecedent of the rule corresponds to the condition that in order to exploit a proof carried out in another context one must know at least that the premises of the proof are in that context.

Notice that the consequent of theorem 3 is again a classical derivation, therefore the theorem can be applied repeatedly, to carry out a proof at any level of nesting within viewpoints. Therefore, if $P \vdash_C Q$ then:

$$\text{in}('P', vp_1) \vdash_C \text{in}('Q', vp_1)$$
$$\text{in}('\text{in}('P', vp_1)', vp_2) \vdash_C \text{in}('\text{in}('Q', vp_1)', vp_2)$$
$$...$$

Similarly the rules of proof in context can be extended to deal with arbitrary level of nesting.

3.5 Entering and leaving contexts

Another useful mechanism to build proofs in context is the ability to switch contexts and perform natural deduction proofs within viewpoints. The safest way to interpret context switching in the framework of natural deduction proofs with pending assumptions and implicit contexts is simply to go one level deeper or shallower in nesting, or in other words *unnesting* and *nesting*.

This means for instance that in order to prove a statement of the form

$$\text{in}('P', vp_1)$$

one may pretend to move inside vp_1, and perform a proof using available facts of the form $\text{in}('...', vp_1)$. If the formula P is itself of the form $\text{in}('Q', vp_2)$ one will have to go one level deeper to prove Q by using this time just facts of the form $\text{in}('\text{in}('...', vp_2)', vp_1)$.

Later we will provide safe rules for importing and exporting facts in a context.

4 A proof method and notation

Our proofs will become more readable and intuitive with the aid of a graphical notation, which emphasises the boundaries and nesting of contexts. The notation we introduce is an extension of the box notation introduced by Kalish and Montague [16].

4.1 Rules for classical natural deduction

We show here some examples of proof schemas for classical natural deduction.

The following schema corresponds to the rule of $\Rightarrow I$ and should be read as: "if assuming P you succeed in proving Q, then you have proved $P \Rightarrow Q$".

$$\boxed{\begin{array}{l} P \quad (assum.) \\ ... \\ Q \end{array}}$$

$$P \Rightarrow Q$$

Similarly, the schema corresponding to the inference rule of $\neg I$ is the following:

$$\neg P$$

The box notation is useful to visualise the scope of the assumptions made during a natural deduction proof. In performing a proof within a box one can use facts proved or assumed in the same box or in enclosing boxes. Facts cannot be exported from within a box to an enclosing or unrelated box.

4.2 Rules for proofs in context

For proofs in context we introduce a different kind of box, with a double border, to suggest boundaries which are more difficult to traverse. The double box represents a viewpoint, i.e. a theory, whose assumptions, if known, are listed in the heading of the box. If the assumptions are not known the name of the viewpoint is shown. The only two rules for bringing facts in and out of a double box are the rules corresponding to unnesting and nesting.

Importing a fact in a viewpoint:

$$(unnesting)$$

Exporting a fact from a viewpoint:

$$(nesting)$$

The only way to import a fact P in a double box vp is to have a statement $in('P', vp)$ in the environment immediately outside the box. Symmetrically you can obtain $in('P', vp)$ in the environment immediately outside a double box vp if P appears in a line immediately inside the double box (not inside a further single or double box within the double box). Note that to import a fact in nested double boxes an appropriate number of crossing double lines must be justified.

According to *Axiom* 1, the assumptions of a viewpoint, if known, can also be used inside the viewpoint:

and, in the case of explicit viewpoints, in can be introduced as follows:

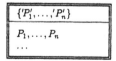

$$in('P', \{'P_1', \ldots, 'P_n'\})$$

Theorem 3 justifies the possibility of carrying on regular natural deduction proofs within a double box. For example the following deduction schema is valid:

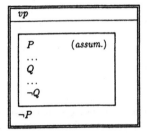

$$\text{in}('\neg P', vp)$$

This is just a combination of the schemas introduced above for classical negation introduction and *nesting*. Notice that opening a single box within a double box to make an assumption corresponds to adding the assumption to those of the viewpoint in the box. In practice, it should be considered as an alternative notation for:

$$\boxed{\begin{array}{l} vp \cup \{'P'\} \\ \hline \\ \cdots \\ Q \\ \cdots \\ \neg Q \end{array}}$$

$$\text{in}('\neg P', vp)$$

This schema provides us a mean to carry out proofs by contradiction, which naturally occur in the solution of the three wise men puzzle.

5 Examples

To illustrate the method just described we will use two examples taken from the recent literature where proofs are composed of subproofs in different contexts.

The first example was used by John McCarthy [15] to illustrate the power of "lifting axioms", which allow extrapolating facts from one theory to another and transforming them at the same time into a different format. Even though no formal proof theory was provided, the example was meant to suggest the kind of proofs one would like to be able to perform. In this case, things are complicated by the fact that the proof is carried out in a natural deduction setting, so there are pending assumption when switching from one context to another.

The second one is a solution to the classical three wise men puzzle, which as been tackled in so many different ways in the knowledge representation literature. In [6] we discussed for example the approach taken by Kowalski and Kim [13] where a rule of *reification* is used, we believe improperly, to lift a conclusion reached in a common knowledge theory $wise_0$ to the theory of the third wise man. Our solution is meant to show that a natural representation of the kind of knowledge and reasoning involved in the puzzle requires the ability to express common knowledge through recursive definition (hence implicit contexts) and a mechanism for nesting and unnesting.

5.1 Lifting rules and natural deduction

In performing proofs involving multiple theories, one would like to be able to move easily from one theory to another, reason within a theory with traditional means, for instance by natural deduction, and then to carry outside some of the consequences obtained.

One must be careful however, not to leave behind in an innermost context essential assumptions and not to extrapolate to an unrelated context.

The example presented in [15] is useful to illustrate these issues.

A fixed theory, called *AboveTheory*, is used to represent the basic facts about the blocks world which do not depend on situations. One would like to make these facts and their consequences available, in the appropriate form, in another theory c where situations are accounted for. The correspondence between these theories is established by axioms written in viewpoint c.

An outer context, c_0, is also needed for lifting facts deduced in *AboveTheory* to c. Using our notation, the statement of the problem can be expressed as in Figure 1.

To simplify the notation, we have dropped the quotation marks used to represent meta-level statements.

---

c_0

(1) $\forall p \; in(p, AboveTheory) \Rightarrow in(in(p, AboveTheory), c)$

> *AboveTheory*
>
> (2) $\forall x, y \; on(x, y) \Rightarrow above(x, y)$
> (3) $\forall x, y, z \; above(x, y) \wedge above(y, z) \Rightarrow above(x, z)$

> c
>
> (4) $\forall x, y, s \; on(x, y, s) \Leftrightarrow in(on(x, y), c(s))$
> (5) $\forall x, y, s \; above(x, y, s) \Leftrightarrow in(above(x, y), c(s))$
> (6) $\forall p, s \; in(p, AboveTheory) \Rightarrow in(p, c(s))$

Figure 1. Statement of the lifting problem

The lifting axiom (1) was missing in the sketch of proof presented by McCarthy [15] but it is necessary in order to lift

$$in(\forall x, y \; on(x, y) \Rightarrow above(x, y), AboveTheory)$$

from c_0 to c where it can be exploited by axiom (6). Without this additional assumption step (10) below could not be accounted for by any sound rule, producing a case of improper lifting.

The full proof appears in Figure 2.

5.2 The three wise men

The statement of this well known puzzle is the following [13].

> *A king, wishing to determine which of his three wise men is the wisest, puts a white spot on each of their foreheads, and tells them that at least one of the spots is white. The king arranges the wise men in a circle so that they can see and hear each other (but cannot see their own spots) and asks each wise man in turn what is the colour of his spot. The first two say that they don't know, and the third says that his spot is white.*

Several solutions to the three wise men puzzle have appeared in the literature, some of which quite reasonable. Our aim here is to illustrate the adequacy of a general method for reasoning about nested beliefs by presenting a solution which is both a sound and, we believe, intuitive. In this spirit we will limit ourselves to the core aspects of the puzzle, without dealing with the derivation of negative knowledge or the evolution of time as for example in [2, 9].

In our solution common knowledge is grouped in a single theory and lifting axioms are provided for each agent to access it. The advantages are a more compact statement of the problem which does not rely on "ad hoc" initialization of theories.

A common approach to the representation of nested beliefs is to introduce explicitly a number of different theories according to the different views that an agent has of other agents. In the three wise men puzzle we

```
┌─────────────────────────────────────────────────────────────────────────┐
│ c₀                                                                        │
│                                                                           │
│  (7) in(on(A, B, S₀), c)                                    (assumption)  │
│                                                                           │
│  ┌──────────────────────────────────────────────────────────────────┐    │
│  │ c                                                                  │    │
│  │                                                                    │    │
│  │  (8) on(A, B, S₀)                              (unnesting, 7)      │    │
│  │  (9) in(on(A, B), c(S₀))                           (8 and 4)       │    │
│  │  (10) in(∀x, y on(x, y) ⇒ above(x, y), AboveTheory)                │    │
│  │                                    (2, nesting, 1, unnesting)      │    │
│  │  (11) in(∀x, y on(x, y) ⇒ above(x, y), c(S₀))                      │    │
│  │                               (proof in context, 6 and 10)         │    │
│  │                                                                    │    │
│  │  ┌─────────────────────────────────────────────────────────┐      │    │
│  │  │ c(S₀)                                                     │      │    │
│  │  │                                                           │      │    │
│  │  │  (12) on(A, B)                        (unnesting, 9)      │      │    │
│  │  │  (13) ∀x, y on(x, y) ⇒ above(x, y)    (unnesting, 11)     │      │    │
│  │  │  (14) above(A, B)            (proof in context, 12 and 13)│      │    │
│  │  └─────────────────────────────────────────────────────────┘      │    │
│  │                                                                    │    │
│  │  (15) in(above(A, B), c(S₀))                    (nesting, 14)      │    │
│  │  (16) in(above(A, B), c(S₀)) ⇒ above(A, B, S₀)  (instance of 5)    │    │
│  │  (17) above(A, B, S₀)         (proof in context, 15 and 16)        │    │
│  └──────────────────────────────────────────────────────────────────┘    │
│                                                                           │
│  (18) in(above(A, B, S₀), c)                                              │
└─────────────────────────────────────────────────────────────────────────┘
```

Figure 2. Proof of the lifting problem

would have the theory that $wise_3$ has about $wise_2$, the theory that $wise_3$ has about the theory that $wise_2$ has about $wise_1$, ... and so on [12]. The construction of tower of theories, each one being "meta" for the one below, is what justifies the use of reflection and reification principles to transfer information between them (for a recent solution along these lines see for example [9]).

It seems to us quite unnatural to be forced to conceive from the beginning an appropriate number of theories according to the number of agents and the nesting level of the reasoning which is required: in this simple puzzle, which requires a nesting level of three, one should theoretically conceive of 27 different theories (even without considering the evolution of time).

On the fly construction of theories on the other hand requires some additional mechanism such as dealing with sequences of contexts as in [8].

Our solution is not radically different but, we believe, more natural. The nesting of viewpoints implicitly takes care of the different perspectives.

Finally our solution does not rely on axioms like *confidence* [2, 13] or *wiseness* [3] which are used in other solutions to make a wise man believe any conclusion of another wise man that he is aware of.

The following viewpoints are used.

$wise_1$: viewpoint of the first wise man
$wise_2$: viewpoint of the second wise man
$wise_3$: viewpoint of the third wise man
CK: viewpoint including the common knowledge

The predicate $white_i$ means the color of the spot of wise man i is white. The common knowledge viewpoint is shown in Figure 3.

Two axioms, external to the CK and wise men viewpoints are needed for the wise men to obtain the common knowledge.

(1) $\forall x \ in(x, CK) \Rightarrow$
     $in(x, wise_1) \wedge in(x, wise_2) \wedge in(x, wise_3)$
(2) $\forall x \ in(x, CK) \Rightarrow$
     $in(in(x, wise_1) \wedge in(x, wise_2) \wedge in(x, wise_3), CK)$

CK

(3) $white_1 \lor white_2 \lor white_3$	(at least one spot is white)
(4) $white_1 \Rightarrow in(white_1, wise_2) \land in(white_1, wise_3)$	($wise_2$ and $wise_3$ see $wise_1$'s spot)
(4') $\neg white_1 \Rightarrow in(\neg white_1, wise_2) \land in(\neg white_1, wise_3)$	
(5) $white_2 \Rightarrow in(white_2, wise_1) \land in(white_2, wise_3)$	($wise_1$ and $wise_3$ see $wise_2$'s spot)
(5') $\neg white_2 \Rightarrow in(\neg white_2, wise_1) \land in(\neg white_2, wise_3)$	
(6) $white_3 \Rightarrow in(white_3, wise_1) \land in(white_3, wise_2)$	($wise_1$ and $wise_2$ see $wise_3$'s spot)
(6') $\neg white_3 \Rightarrow in(\neg white_3, wise_1) \land in(\neg white_3, wise_2)$	
(7) $\neg in(white_1, wise_1)$	(asserted by first man)
(8) $\neg in(white_2, wise_2)$	(asserted by second man)

Figure 3. Statement of the problem in the three wise men puzzle

Axioms (1) and (2) provide a proper account of common knowledge, allowing to derive the commonly known facts in any viewpoint, no matter how nested. In particular axiom (2) is used to achieve the appropriate level of nesting in CK, axiom (1) to lift from the CK viewpoint to any other viewpoint. The details of the derivation of common knowledge are omitted from the proof.

We can formally account, as shown in Figure 4, for the reasoning of the third wise man after the first and second one have spoken. The third wise man is in fact able to prove that his spot is white.

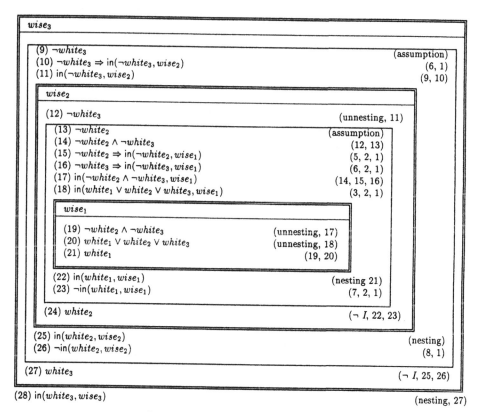

Figure 4. Proof of the three wise men puzzle

6 Conclusions

We have discussed several approaches to the realization of formal systems for contextual reasoning. Important issues in this respect are whether explicit or implicit theories are allowed and whether theories are stratified or one unique reflexive theory is allowed.

We described a set of inference rules for proof in context based on the theory of viewpoints and a notation for their application which expands on the box notation introduced by Kalish and Montague for natural deduction.

We suggested that when dealing with partially specified theories or contexts a generalised notion of proof in context is more appropriate than reflection rules and gave an account of what "entering" and "leaving" a context should be in the setting of natural deduction proofs.

Acknowledgments We wish to thank the International Computer Science Institute in Berkeley, for providing the support and the right atmosphere to get this work done and Saša Buvač for interesting and useful discussions which helped us to better understand McCarthy's notion of context.

References

1. P. Aczel (1988). *Non-well-founded sets*, CSLI lecture notes, **12**, Center for the Study of Language and Information, Stanford, California.
2. L. C. Aiello, d. Nardi, M. Schaerf (1988). Reasoning about Knowledge and Ignorance, in *International Conference of 5th generation Computer System*, pages 618-627, Tokyo.
3. G. Attardi and M. Simi (1984). Metalanguage and reasoning across viewpoints, in *ECAI84: Advances in Artificial Intelligence*, T. O'Shea (ed.), Elsevier Science Publishers, Amsterdam.
4. G. Attardi and M. Simi (1991). Reflections about reflection, in Allen, J. A., Fikes, R., and Sandewall, E. (eds.) *Principles of Knowledge Representation and Reasoning: Proceedings of the Second International Conference.* Morgan Kaufmann, San Mateo, California.
5. G. Attardi and M. Simi (1993). A formalisation of viewpoints, TR-93-062, International Computer Science Institute, Berkeley.
6. G. Attardi and M. Simi (1994). Proofs in context, in Doyle, J. and Torasso, P. (eds.) *Principles of Knowledge Representation and Reasoning: Proceedings of the Fourth International Conference.* Morgan Kaufmann, San Mateo, California.
7. K.A. Bowen and R.A. Kowalski (1982). Amalgamating language and metalanguage in logic programming, in *Logic Programming*, K. Clark and S. Tarnlund (eds.), Academic Press, 153-172.
8. S. Buvač and I.A. Mason (1993). Propositional Logic in Context, *Proc. of the Eleventh AAAI Conference*, Washington DC, 412-419.
9. A. Cimatti and L. Serafini (1994). Multi Agent Reasoning with Belief Contexts: the Approach and a Case Study", proceedings of ECAI-94, Workshop on Agent Theories, Architectures, and Languages.
10. F. Giunchiglia, L. Serafini, Multilanguage hierarchical logics (or: how we can do without modal logics), *Artificial Intelligence*, 65:29-70, 1994.
11. R.V. Guha (1991). Contexts: a formalization and some applications, MCC Tech. Rep. ACT-CYC-42391.
12. K. Konolige (1982). A first order formalization of knowledge and action for a multiagent planning system, *Machine Intelligence* 10.
13. R. Kowalski and Kim J.S. (1991). A metalogic programming approach to multi-agent knowledge and belief, in Vladimir Lifschitz (ed.), *Artificial Intelligence and the Mathematical Theory of Computation: Papers in Honor of John McCarthy*, Academic Press, 1991, Academic Press, 231-246.
14. J. McCarthy, Generality in Artificial Intelligence, *Communications of the ACM*, **30**(12), 1987, 1030-1035.
15. J. McCarthy (1993). Notes on Formalizing Context, *Proceedings of the Thirteenth International Joint Conference on Artificial Intelligence*, Chambery.
16. D. Kalish and R. Montague (1964). *Logic: techniques of formal reasoning*, New York, Harcourt, Brace & World.
17. R. Montague (1963). Syntactical treatment of modalities, with corollaries on reflexion principles and finite axiomatizability, *Acta Philosoph. Fennica*, **16**, 153-167.
18. R. C. Moore (1977). Reasoning about knowledge and action, *Proc. of IJCAI77*, Cambridge, MA, 223-227.
19. D. Perlis (1985). Languages with self-reference I: foundations, *Artificial Intelligence*, 25:301-322.
20. Y. Shoham (1991). Varieties of contexts, in Vladimir Lifschitz (ed.), *Artificial Intelligence and the Mathematical Theory of Computation: Papers in Honor of John McCarthy*, Academic Press, 393-407.
21. M. Simi (1991). Viewpoints subsume belief, truth and situations, in *Trends in Artificial Intelligence, Proc. of 2nd Congress of the Italian Association for Artificial Intelligence*, Ardizzone, Gaglio, Sorbello (Eds), *Lecture Notes in Artificial Intelligence* 549, Springer Verlag, 38-47.
22. R.W. Weyhrauch (1980). Prolegomena to a theory of mechanized formal reasoning, *Artificial Intelligence*, **13**(1,2):133-170.

A APPENDIX

A.1 Inference rules for classical natural deduction

$$\frac{P,Q}{P \wedge Q} \qquad\qquad (\wedge\ I)$$

$$\frac{P \wedge Q}{P,Q} \qquad\qquad (\wedge\ E)$$

$$\frac{\Gamma \cup \{P\} \vdash Q}{\Gamma \vdash (P \Rightarrow Q)} \qquad\qquad (\Rightarrow\ I)$$

$$\frac{P, P \Rightarrow Q}{Q} \qquad\qquad (\Rightarrow\ E)$$

$$\frac{P}{P \vee Q, Q \vee P} \qquad\qquad (\vee\ I)$$

$$\frac{P \vee Q, P \Rightarrow R, Q \Rightarrow R}{R} \qquad\qquad (\vee\ E)$$

$$\frac{P \Rightarrow Q, P \Rightarrow \neg Q}{\neg P} \qquad\qquad (\neg\ I)$$

$$\frac{\neg\neg P}{P} \qquad\qquad (\neg\ E)$$

$$\frac{P[y/x]}{\forall y\ .\ P} \qquad \textit{where y is a new variable} \qquad\qquad (\forall\ I)$$

where the notation $P[t/x]$ stands for P with all the free occurrences of variable x substituted by t.

$$\frac{\forall x\ .\ P}{P[t/x]} \qquad \textit{where t does not contain variables occurring in P} \qquad\qquad (\forall\ E)$$

$$\frac{P[t/x]}{\exists x\ .\ P} \qquad\qquad (\exists\ I)$$

$$\frac{\exists x\ .\ P}{P[y/x]} \qquad \textit{where y is a new variable} \qquad\qquad (\exists\ E)$$

A.2 Metalevel axioms and inference rules

$$\text{in}('P', \{\ldots, 'P', \ldots\}) \qquad\qquad (Axiom1)$$

$$\text{in}('P', vp) \Rightarrow \text{in}('\text{in}('P', vp)', vp) \qquad\qquad (Ax2)$$

$$\frac{\text{in}('P', vp), \text{in}('Q', vp)}{\text{in}('P \wedge Q', vp)} \qquad\qquad (Meta\ \wedge\ I)$$

$$\frac{\text{in}('P \wedge Q', vp)}{\text{in}('P', vp), \text{in}('Q', vp)} \qquad\qquad (Meta\ \wedge\ E)$$

$$\frac{\text{in}('Q', vp \cup \{'P'\})}{\text{in}('P \Rightarrow Q', vp)} \qquad\qquad (Meta\ \Rightarrow\ I)$$

$$\frac{\mathsf{in}('P',vp),\mathsf{in}('P \Rightarrow Q',vp)}{\mathsf{in}('Q',vp)} \qquad\qquad (Meta \Rightarrow E)$$

$$\frac{\mathsf{in}('P',vp)}{\mathsf{in}('P \vee Q',vp),\mathsf{in}('Q \vee P',vp)} \qquad\qquad (Meta \vee I)$$

$$\frac{\mathsf{in}('P \vee Q',vp),\mathsf{in}('P \Rightarrow R',vp),\mathsf{in}('Q \Rightarrow R',vp)}{\mathsf{in}('R',vp)} \qquad\qquad (Meta \vee E)$$

$$\frac{\mathsf{in}('P',vp),\mathsf{in}('P \Rightarrow Q \wedge \neg Q',vp)}{\mathsf{in}('\neg P',vp)} \qquad\qquad (Meta \neg I)$$

$$\frac{\mathsf{in}('\neg\neg P',vp)}{\mathsf{in}('P',vp)} \qquad\qquad (Meta \neg E)$$

$$\frac{\mathsf{in}('P[y/x]',vp)}{\mathsf{in}('\forall x . P',vp)} \qquad\qquad (Meta \,\forall\, I)$$

$$\frac{\mathsf{in}('\forall x . P',vp)}{\mathsf{in}('P[t/x]',vp)} \qquad\qquad (Meta \,\forall\, E)$$

$$\frac{\mathsf{in}('P[t/x]',vp)}{\mathsf{in}('\exists x . P',vp)} \qquad\qquad (Meta \,\exists\, I)$$

$$\frac{\mathsf{in}('\exists x . P',vp)}{\mathsf{in}('P[y/x]',vp)} \qquad\qquad (Meta \,\exists\, E)$$

Introspective Metatheoretic Reasoning[*]

Fausto Giunchiglia[1,2] Alessandro Cimatti[1]

[1] IRST, 38050 Povo, Trento, Italy
[2] University of Trento, Via Inama 5, 38100, Trento, Italy

Abstract. This paper describes a reasoning system, called GETFOL, able to introspect (the code implementing) its own deductive machinery, to reason deductively about it in a declarative metatheory and to produce new executable code which can then be pushed back into the underlying implementation. In this paper we discuss the general architecture of GETFOL and the problems related to its implementation.

1 The Goal

The work partially described in this paper tackles the problem of investigating criteria and techniques for the development of real introspective/reflective systems. We describe a system, called GETFOL[3] [Giu92], able to introspect its own code, to reason deductively about it in a declarative metatheory and, as a result, to produce new executable code which can then be pushed back into the underlying implementation. The behaviour of this system is schematized in three steps (see figure 1). We call *Lifting* the first step, projecting computation into deduction. With lifting, the source code of the system is processed, and a formal metatheory describing it is automatically generated. Then, during the *Reasoning* step, the theorem proving capabilities of the system are used to reason about the lifted theory. This allows to deduce theorems which can be interpreted as specifications of new system functionalities. We call *Flattening* the last step, dual of lifting, projecting deduction into computation. By flattening, new pieces of executable code are automatically generated from statements proved with the reasoning step, and added to the system code. In this way the system may be extended, adding new inference procedures, and modified, substituting old versions of inference procedures with new, possibly optimized, ones.

Figure 2 shows the conceptual dependencies among the entities involved in the process of figure 1. The starting point is the object theory OT. We give a computational account of OT, i.e. we mechanize it, by writing code. By devising MT, we give a declarative characterization of OT. The lifting and flattening procedures act as a bridge between the code implementing OT and its metatheory MT. We call the process described in figure 1, *introspective metatheoretic* reasoning (IMR). *Introspective* because the system is able to formalize and reason

---

[*] This work has been done at IRST as part of the MAIA project.

[3] GETFOL is a reimplementation/extension of the FOL system [Wey80, GW91]. GETFOL has, with minor variations, all the functionalities of FOL plus extensions, some of which described here, to allow for metatheoretic theorem proving.

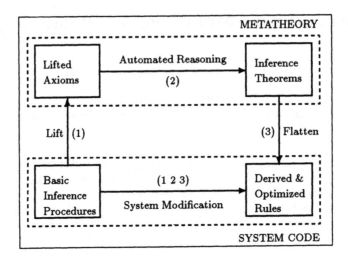

Fig. 1. The lifting-reasoning-flattening cycle.

about (parts of) its own code. *Metatheoretic* because we require that MT be a *metatheory* of the theory OT mechanized by the system code. Notice that reasoning in a formal metatheory gives the right perspective for describing inference mechanisms. In general one would like to prove that a new inference procedure is derived from, admissible or consistent with the already existing ones. This is the kind of results that metatheoretic reasoning provides (see for instance [BM81, GS88, Pau89]). However MT is different from any other metatheory defined in the past. Not only does it describe the object level logic, but it also takes into account the computations implementing it. Its theorems can be interpreted in terms of both the object level logic and its mechanization. To point out this fact, we say that MT is a *metatheory of a mechanized object theory.*

Some observations. The work closest in spirit to ours is Smith's [Smi83]. In Smith's reflective 3-Lisp, the flow of computation can be *inspected* by reifying the status of the interpreter in explicit data structures, and *modified* by suitably updating these data structures. In both his and our approaches, what Smith calls an *embedded account* of the system, i.e. a (partial) description of the system in the system itself, can be automatically generated and the system's behaviour can be modified. The basic difference is in the way the system is modified. In 3-Lisp these changes can not be declaratively reasoned about but only performed procedurally by computation.

On the other hand, our architecture is based on a sharp distinction between *computation* and *deduction*. There are many reasons underlying this choice. Epistemologically, we believe that deduction and computation are two fundamentally different phenomena. Technically, using logic gives a semantics which allows us to make and prove correctness statements. Implementationally, we can use and integrate different techniques for solving the different aspects of the problem. For instance theorem proving techniques can be used for the automation of the reasoning step, while techniques for the synthesis, optimization and verification

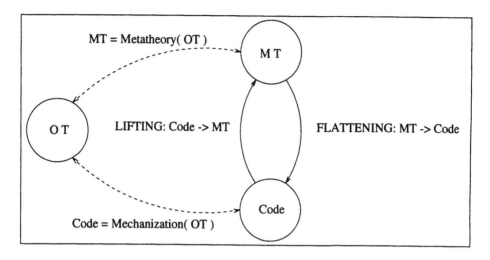

Fig. 2. The conceptual schema of the system

of the correctness of programs can be used for the implementation of the lifting and flattening processes. Furthermore, from a cognitive science viewpoint, formal deduction seems more suited than computation for the modeling of reasoning (but see [JL83]).

Though distinguished, computation and deduction are related. This is a well known fact in the mathematical logic literature. It is in fact possible to express deduction in terms of computation (e.g. in the λ-calculus) and viceversa deduction (e.g. in Peano arithmetics) may represent computation. A formal definition of the lifting and flattening processes, together with a proof of their completeness and correctness, could be seen as a re-statement of some of the results in the mathematical logic literature. However, these similarities are more superficial than it might look at a first sight. The results in the mathematical logic literature allow us to connect in a general way (to the extent that it is possible) deduction with computation in idealized programming languages (e.g. functional languages). In our work we have connected deduction in a formal metatheory (see section 2), particularly suited for representing search strategies [GT92], and the code of a real running system, GETFOL, with more than one MB of source code. Establishing this connection has required facing a lot of problems which are irrelevant from the point of view of mathematical logic and which are due to the fact that we are dealing with a real system. These problems and our proposed solutions are informally (and partially) described in the following of this paper. Thus, in section 2 we describe the structure of the metatheory MT, and various ways of reasoning about and extending inference mechanisms. In section 3 we discuss the problem of writing code which can be lifted. In section 4 we focus on the problems related to lifting and flattening the code of a real system, showing how MT can be automatically generated. Finally, in section 5 we draw some conclusions.

2 MT: a metatheory of a mechanized object theory

$$\frac{A \wedge B}{A} \wedge\mathbf{E} \quad \left| \quad \frac{A(y)}{\forall x.A(x)} \; \forall\mathbf{I} \quad \right| \quad \frac{\forall x.A(x)}{A(t)} \; \forall\mathbf{E}$$

Fig. 3. Some inference rules

The metatheory MT described in this section is a minor extension of the metatheory defined and formally studied in [GT91, GT92, GT94]. In this section we summarize the ideas described in those papers to the extent that it is needed for the goals of this paper.

We suppose that OT uses a first order classical sequent logic. By *sequent* we mean a pair (Γ, A), where A is a formula and Γ a set of formulas. The inference rules are a sequent presentation of Prawitz' natural deduction calculus [Pra65]. They allow introduction and elimination in the post-sequent A. Some of the rules are shown in figure 3. They are one of the two conjunction elimination rules, and the rules for introduction and elimination of the universal quantifier. MT contains all the "standard" information about OT. That is, its language must have names for all the objects of OT (constants, variables, sequents, ...). It must have axioms that say, for instance, that x is a variable (i.e. $Var("x")$, where "x" is the name of x), that a sequent s is a conjunction (i.e. $Conj("s")$) and so on. This construction is routinary and it is not reported here. More interestingly, each (piece of code implementing an) inference rule of OT is represented in MT by a function symbol. The (subroutines implementing the) inference rules listed above are represented in MT by the function symbols *alli*, *alle* and *ande*. Their behaviour is described by the axioms of figure 4. T represents provability for (the mechanization of) sequents. $\mathcal{A}_{ande}$ states that, if (the subroutine for) and elimination is applied to any (data structure representing a) provable sequent x whose formula is a conjunction, the (data structure representing the) result $ande(x)$ is also a provable sequent. The other axioms have a similar interpretation. The predicates Var, $Term$ and $Forall$ represent the (code computing the) properties of being an individual variable, a term and a sequent with a universally quantified formula. $NoFree$ represents the (code computing the) restriction on the generalized variable in the rule of forall introduction.

$$
\begin{aligned}
&(\mathcal{A}_{ande}) : \forall x.(T(x) \wedge Conj(x) \supset T(ande(x))) \\
&(\mathcal{A}_{alli}) : \forall x \; y \; z.(T(x) \wedge Var(y) \wedge Var(z) \wedge NoFree(z,x) \supset T(alli(x,y,z))) \\
&(\mathcal{A}_{alle}) : \forall x \; y.(T(x) \wedge Term(y) \wedge Forall(x) \supset T(alle(x,y)))
\end{aligned}
$$

Fig. 4. Axiomatization of primitive inference rules

The axioms in figure 4 allow only for reasoning about correct proofs, i.e. compositions of applicable inference steps. However, while trying to prove a theorem, rarely the user (or the system itself) has a detailed proof schema in mind, and inference rules may be tried without knowing whether they are applicable or not. In the code of GETFOL possible failures are taken into account by suitable subroutines, called *tactics*. Tactics call the basic inference rules when applicability conditions are satisfied, otherwise return a particular data structure representing failure. In order to implement the cycle of figure 1, tactics, and therefore failure, must be represented in MT. Failure is represented by the individual constant *fail*. The basic concept of being a failure is expressed by the predicate *Fail* defined as

$$\forall x.Fail(x) \leftrightarrow x = fail$$

The axiom

$$(\mathcal{A}_{nottfail}) : \forall x.\neg(Fail(x) \wedge T(x))$$

states that T and *Fail* are disjoint. Being a proof step (either successful or failing) is expressed by the predicate *Tac*, defined by the axiom

$$(\mathcal{A}_{Tac}) : \forall x.Tac(x) \leftrightarrow (Fail(x) \vee T(x))$$

However, this is not enough. The metalevel representation of a tactic must have a structural similarity (very close to a one-to-one mapping) with the tactic itself. This makes the lifting, theorem proving and flattening of new tactics easier and more natural to understand and to perform. Tactics are basically programs and, as such, make extensive use of conditional constructs (and their derivatives). We have therefore extended the first order language with conditional term constructors (i.e. **trmif** *wff* **then** *term* **else** *term*), whose meaning is given by the inference rules for elimination and introduction of figure 5 (the elimination rule for A is not listed, being very similar to that for $\neg A$). The resulting theory is a conservative extension of first order logic. Figure 6 shows how the tactics implementing the inference rules of figure 3 are represented in MT.

$$\frac{\begin{matrix}[A] & [\neg A] \\ \vdots & \vdots \\ B(t_1) & B(t_2)\end{matrix}}{B(\textbf{trmif } A \textbf{ then } t_1 \textbf{ else } t_2)}\; trmifI \qquad \frac{\neg A \quad B(\textbf{trmif } A \textbf{ then } t_1 \textbf{ else } t_2)}{B(t_2)}\; trmifE_{\neg}$$

Fig. 5. Rules for conditional terms

$$\begin{aligned}
(\mathcal{A}_{andetac}) &: \forall x.andetac(x) = \textbf{trmif } T(x) \wedge Conj(x) \\
&\qquad \textbf{then } ande(x) \\
&\qquad \textbf{else } fail \\
(\mathcal{A}_{allitac}) &: \forall x\ y\ z.allitac(x,y,z) = \textbf{trmif } T(x) \wedge Var(y) \wedge Var(z) \wedge NoFree(z,x) \\
&\qquad \textbf{then } alli(x,y,z) \\
&\qquad \textbf{else } fail \\
(\mathcal{A}_{alletac}) &: \forall x\ y.alletac(x,y) = \textbf{trmif } T(x) \wedge Term(y) \wedge Forall(x) \\
&\qquad \textbf{then } alle(x,y) \\
&\qquad \textbf{else } fail
\end{aligned}$$

Fig. 6. Definition of primitive tactics

Notice that tactics implement the total version of inference rules by returning an explicit failure. Failures are used for this particular purpose. Therefore, it may not be the case that inference rules return a failure when the applicability conditions are not satisfied. This fact is formalized in MT by

$$(\mathcal{A}_{allinofail}) : \forall x \; y \; z . \neg alli(x, y, z) = fail$$

and allows to keep reasoning about correct inference and general inference completely separated.

$$
\begin{array}{l}
\forall x \; y . alleandeallitac(x, y) = \textbf{trmif} \; T(x) \wedge Forall(x) \wedge Var(y) \wedge \\
\qquad\qquad\qquad\qquad Conj(alle(x, y)) \wedge NoFree(y, ande(alle(x, y))) \\
\qquad\qquad \textbf{then} \; alli(ande(alle(x, y)), y, y) \\
\qquad\qquad \textbf{else} \; fail
\end{array}
$$

Fig. 7. The definition of *alleandeallitac*

The ultimate goal of reasoning in MT about an inference procedure is to prove the corresponding *admissibility statement*, and then to flatten it down as system code. For inference rules, such statements have the form of axioms of figure 4, i.e.

$$\forall \underline{x} . Applicability(\underline{x}) \supset T(rule(\underline{x}))$$

The admissibility statements for the corresponding tactics have the form

$$\forall \underline{x} . Tac(ruletac(\underline{x}))$$

In MT the admissibility statements for an inference rule and for the corresponding tactic can be derived from each other.

In MT and extensions of MT it is possible to synthesize/optimize tactics. In the following we will try to give the flavour of how this can be done via two examples. As an example of possible reasoning in MT, let us consider the class of *finite compositions* of inference rules. Arbitrary compositions of function symbols can be proved to satisfy *Tac* in MT. Consider the rule which is a composition of a forall elimination, a conjunction elimination and finally a forall introduction. From the admissibility statement for *allitac*, we can prove in MT $\forall x \; y . Tac(allitac(andetac(alletac(x, y)), y, y)))$, stating that the composition is a correct inference rule. From the point of view of the logic provably equal terms are completely indistinguishable. However, the composition of tactics as above generates very redundant and inefficient code. An equivalent term defining the very same inference steps and corresponding to more *optimized* code, can be deduced by composing the axioms of figure 4:

$$
\begin{array}{l}
\forall x \; y . T(x) \wedge Forall(x) \wedge Var(y) \wedge Conj(alle(x, y)) \wedge NoFree(y, ande(alle(x, y))) \supset \\
\qquad\qquad T(alli(ande(alle(x, y)), y, y))
\end{array}
$$

The corresponding tactic is defined as:

$$
\begin{array}{l}
\forall x \; y . alleandeallitac(x, y) = \textbf{trmif} \; T(x) \wedge Forall(x) \wedge Var(y) \wedge \\
\qquad\qquad\qquad\qquad Conj(alle(x, y)) \wedge NoFree(y, ande(alle(x, y))) \\
\qquad\qquad \textbf{then} \; alli(ande(alle(x, y)), y, y) \\
\qquad\qquad \textbf{else} \; fail
\end{array}
$$

alleandeallitac is optimized with respect to the composition of primitive tactics, as the applicability is expressed in a simplified way. For instance, the theoremhood tests performed by *andetac* and *allitac* are eliminated as they are implied by the other applicability conditions. Notice that this framework allows for incremental reasoning about inference procedures: first a basic version of inference rule can be considered, (e.g. by composing tactics) and then it can be more and more optimized by deducing further properties.

$$\forall\ x.uniclosetac(x) = \textbf{trmif } T(x)$$
$$\textbf{then let } (var\ .\ get\text{-}free(x))$$
$$\textbf{in trmif } Var(var)$$
$$\textbf{then } uniclosetac(allitac(x, var, var))$$
$$\textbf{else } x$$
$$\textbf{else } fail$$

Fig. 8. Definition of *uniclosetac*

In (extensions of) MT it is also possible to reason about *admissible* inference procedures, i.e. rules which do not enlarge the provability relation of OT, but which can not be expressed as a finite composition of primitive inference steps. Being able to reason about this kind of inference rules gives the system the capability of extending itself with non trivial functionalities: typical examples are a tautology decider or a normalization procedure. (Our treatment of admissible rules is similar to Boyer and Moore's work on metafunctions [BM81].) As a simple example, consider the inference rule performing the universal closure of an arbitrary sequent. This rule succeeds provided that none of the free variables of the formula occurs free in the dependencies. This rule can be represented in MT by the function symbol *uniclosetac*, defined by the formula of figure 8. The definition exploits another conservative extension of first order language, the environment term constructor, i.e. **let** (*var . term*) **in** *term* (which can be defined similarly to what done for conditional terms). The corresponding admissibility statement is the formula $\forall x.Tac(uniclosetac(x))$. Its proof requires reasoning by induction with the axiom schema of figure 9. The intuition underlying the proof is to perform an induction on the structure of well formed formulas (wffs). At every step the selected free variable is proved to be quantified by the application of *allitac*.

$$\mathcal{A}_{ind} : \forall x\ y.(Wff(x) \land Wff(y) \supset (\Phi(x) \land \Phi(y) \supset \Phi(mkor(x,y)))) \land \ldots \land$$
$$\forall x\ y.(Var(x) \land Wff(y) \supset (\Phi(y) \supset \Phi(mkforall(x,y)))) \land \ldots \supset$$
$$\forall x.(Wff(x) \supset \Phi(x))$$

Fig. 9. The structural induction axiom schema for wffs

3 Writing liftable code

The goal is to produce code which, once lifted, will generate MT (see figure 2). However, this is not a trivial consequence of the fact that the code im-

```
(DEFLAM alliprf (X Y Z)
 (IF (AND (THEOREM X) (VAR Y) (VAR Z) (NOFREE Z X))
  (proof-add-theorem (alli X Y Z))
  (print-error-message)))
```

Fig. 10. Unliftable code for forall introduction

plements the deductive machinery of OT. Writing liftable code is not a simple operation of implementation. It is not enough to satisfy the usual software engineering requirements (e.g. bug-free, understandable). The code must preserve a form of structural similarity with the entity being represented, in this case OT. Every data structure, every step of computation, the system structure and the abstraction levels are determined in terms of the concepts that are intended to be relevant. To emphasize this point we say that we do *representation theory using programs as representational tools*. We call this way of writing code, *mechanization*, and distinguish it from simple implementation.

Consider the code implementing the simple inference rule of forall introduction, shown in figure 10 (the programming language is HGKM [GC89], a SCHEME-like language with first order semantics, used as the implementation language of GETFOL). THEOREM evaluates to TRUE if the argument is (a data structure representing) a provable sequent, VAR evaluates to true if the argument is a variable, alli builds a sequent and proof-add-theorem adds it to the proof. Its reading is very natural: if the inference rule is applicable then build the (data structure representing) the resulting sequent and store it in the current proof, otherwise report an error message. Everyone would agree that this is well written code. But the correspondence with the description in MT of ∀I, is only partially preserved. For instance, whilst *allitac* in MT is a function from sequents to sequents, alliprf has a side effect on the system (either adds a sequent to the proof or prints an error message) and does not return a value.

The work on mechanizing GETFOL has required different rewritings of the code, during which we have devised a general schema for the development of liftable code (see figure 11). We call *state* the information stored in the system. The state is of different kinds, and is justified by different reasons. For instance, the axioms of the object theory are stored because of their logical relevance. A counter for the automatic generation of different names for skolem functions is stored for user interface purposes. In decision procedures, global variables are used to avoid explicit argument passing and optimize performances. We call *logical state* (LS) the part of the state containing logical information, *physical state* (PS) the remaining state. This distinction separates the information which is relevant for lifting (i.e. LS) from the one which must be suitably "hidden" by identifying in the code a liftable abstraction level. The system code is then separated in operations which are functional on the state, the *computation machinery* (CM), and operations which change the state, the *update machinery* (UM). Notice that this classification is completely general and independent of

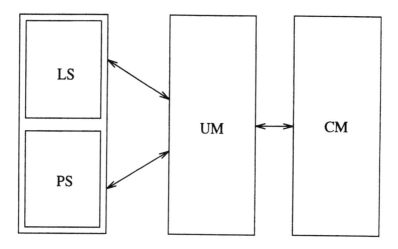

Fig. 11. The structure of the system code.

the application.

The implementation of forall introduction according to this schema is given in figure 12. `allitac` and `alli` are CM primitives. `allitac` implements the forall introduction tactic, while `alli` implements $\forall I$ assuming that the arguments satisfy the applicability conditions. `alli` and `allitac` are functions over a fixed state, as their execution does not add a theorem to the current proof nor does it print an error message. `proof-add-theorem` is the UM primitive "hiding" part of LS, i.e. the current proof, to which it adds the given theorem. `print-error-message` "hides" part of the physical state, i.e. the output channel, on which it prints an error message. These operations are called by the UM primitive `alliprf`, according to the result of `allitac`. Compare the two implementations of $\forall I$ in figure 10 and in figure 12. The computations described in the two cases are very similar, undistinguishable from many points of view. However, the abstraction levels in the "mechanized" solution are sharply identified, the functions are separated from the actions on the state. Even more important, notice the correspondence between the subroutines `alli` and `allitac` with the axiomatization in MT of the rule of forall introduction (figure 6).

4 Lifting and flattening code

As already pointed out in section 1, the relation between computation and deduction has been widely studied in mathematical logic. In computer science, programming languages are given formal account to allow formal reasoning for program synthesis, optimization and verification. Independently of the features of the language being described, i.e. functional [BM79] or imperative [MW87a, MW87b], all these approaches are based on a *uniform* mapping. In most cases, the computing subroutines are immersed into the logic with a mapping which is basically one-to-one. Our approach takes a different perspective. Lifting and

```
;;; UPDATE MACHINERY
(DEFLAM alliprf (X Y Z)
 (maybe-proof-add-theorem (allitac X Y Z)))

(DEFLAM maybe-proof-add-theorem (X)
 (IF (FAIL? X)
  (print-error-message X) ;;; UPDATE THE PHYSICAL STATE
  (proof-add-theorem X))) ;;; UPDATE THE LOGICAL  STATE

;;; COMPUTATION MACHINERY
(DEFLAM allitac (X Y Z)
 (IF (AND (THEOREM X) (VAR Y) (VAR Z) (NOFREE Z X))
  (alli X Y Z)
  fail))
```

Fig. 12. Liftable code for forall introduction

flattening *are not uniform*: different parts of the code are treated in different ways.

In this paper, for lack of space, we do not consider flattening. We rely on the intuition that the flattening step of the schema in figure 1 is the inverse of the lifting. Given a definitional axiom, it generates the corresponding one-to-one CM definition. For instance, flattening the definition of *uniclose* given in figure 8 gives the CM function **uniclosetac** (figure 13). Statements about the system state, once proved in MT, are flattened onto UM primitives. For instance, the admissibility statement $\forall x.Tac(uniclosetac(x))$ is flattened in the UM primitive **unicloseprf**.

Lifting is defined according to the classification of the code discussed in previous section. Basically, for each of LS, CM and UM there is a corresponding lifting procedure generating an appropriate subset of axioms of MT. In this paper we do not consider lifting of LS. The intuition is that LS contains information related to the objects of OT (e.g. the language, the axioms), and lifting LS gives their metatheoretic description, e.g. the formula $Var("x")$ for the (data structure implementing the) variable x of OT. We focus here only on the harder cases of lifting CM and lifting UM.

4.1   Lifting CM

The code in CM is a collection of functions defined using the functional subpart of HGKM. Therefore lifting CM can exploit the usual techniques for reasoning about functional programs: function definitions are immersed via a one-to-one mapping into the logic of MT and become definitional axioms. For instance, lifting the function definition of **allitac** (figure 12) gives the axiom $\mathcal{A}_{allitac}$ (figure 6).

```
(DEFLAM unicloseprf (x)
 (maybe-proof-add-theorem (uniclosetac x)))

(DEFLAM uniclosetac (seq)
 (IF (NOT (THEOREM seq)) fail
  (LET ((v (get-free seq)))
   (IF (VAR v)
    (uniclosetac (allitac seq v v))
    v)))))
```

Fig. 13. Flattening *uniclose*

In principle, it would be possible to lift all the function definitions of the system going down to the basic primitives, e.g. CAR, CDR, CONS. This is for instance what Boyer and Moore do; higher levels of functional abstraction are added incrementally during the theorem proving activity [BM79]. Their "bottom-up" approach is motivated by the fact that their goal is to prove the termination and the correctness (with respect to a certain specification) of user defined functions. However, we are interested in developing systems which reason selectively about portions of their underlying implementation code. The idea is to keep MT as partial as possible still maintaining all the needed information. This in order to make the process of self-extension focused and feasible in practice. Therefore, we take a slightly different approach. First of all, we want reasoning to be *local*: for instance, we do not want to consider the system deciders when reasoning about inference rules. Furthermore, reasoning should be *at the right level of abstraction*. For instance, in order to reason about tactics, we are not interested in the internal structure of inference rules, and we take alli to be a primitive object, i.e. as if it were a black box. We call *abstract machine* the collection of all and only the primitive objects involved in our reasoning (e.g. alli). Notice that the choice of what code we lift, and at which level of abstraction we describe it, strongly depends on the goals: for instance, taking inference rules as primitive objects is particularly suited for synthesizing new tactics.

The hard problem in lifting CM is to lift the suitable formalization for the abstract machine. Indeed, this can not be described with the one-to-one lifting of its subroutine definitions: this would imply reasoning at a different (lower) level of abstraction. The properties of an abstract machine must be lifted without analyzing its internal structure. General criteria for axiomatizing abstract machines have been defined. A first issue is the characterization in MT of partial functions (e.g. alli). A partial function is associated with a total version (e.g. allitac) which returns the special data structure fail when the value is not defined. In order to avoid confusion, partial functions never return fail. This general property has been exploited to define a lifting mechanism returning the corresponding axiom: in the case of alli and allitac, the resulting axiom is $\mathcal{A}_{allinofail}$ (see page 6). A second issue is the lifting of code imple-

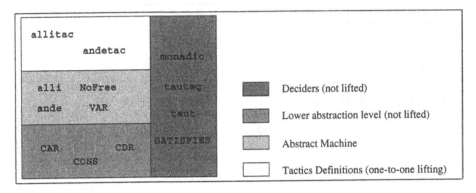

Fig. 14. Lifting the structure of CM

menting inductively defined data structures. Consider the code in figure 15. It is a (simplified) version of the implementation of wffs. In this code it is possible to identify constructors (**mkor**), recognizers (**DISJ**) and selectors (**lfor, rtor**). Starting from such considerations, we can lift the axioms defining the mutual relations of constructors and selectors, e.g.

$$\mathcal{A}_{lformkor} : \forall x\ y.lfor(mkor(x,y)) = x$$

The concrete implementation of this code is not shown, as lifting is independent of it. The recursively defined predicate **WFF** "glues together" the various objects into an inductively defined data type. From each of its clauses, we can lift the type information for constructors, e.g.

$$\mathcal{A}_{typemkforall} : \forall x\ y.Var(x) \wedge Wff(y) \supset Wff(mkforall(x,y))$$

In the case of inductively defined data, it is also possible to lift a principle of structural induction. For **WFF** the result is the axiom $\mathcal{A}_{ind}$ used in the admissibility proof of *uniclose* (see figure 9).

4.2 Lifting UM

Lifting the theory of inference is somehow analogous to lifting the theory of wffs. Indeed, the set of provable sequents of OT can be seen analogously to wffs, i.e. an inductively defined set, whose constructors are the inference rules. (One minor difference is that for wffs we only consider the partial functions, (e.g. the computation of **lfor** may return a random value or even raise an exception if the argument is not a **DISJ**); in the case of inference we consider both (partial) inference rules and the corresponding total versions with **fail**. Under this interpretation, axiom $\mathcal{A}_{ande}$ is the dual of the typing axiom $\mathcal{A}_{typemkforall}$.) A possible solution could be therefore to lift $\mathcal{A}_{typemkforall}$ from CM, as done for wffs, given a recursive definition of the provability predicate (dual of **WFF**). To do this, however, we would also need the recognizers for different inference steps (dual of **UNIQUANT**). The problem is that theoremhood is actually a property which is not decidable,

```
(DEFLAM mkor (wff1 wff2) ...)
(DEFLAM lfor (wff) ...)
(DEFLAM rtor (wff) ...)
(DEFLAM DISJ (wff) ...)
...
(DEFLAM mkforall (var wff) ...)
(DEFLAM bvarof (wff) ...)
(DEFLAM matrix (wff) ...)
(DEFLAM UNIQUANT (wff) ...)
...
(DEFLAM WFF (wff)
 (IF (DISJ wff) (AND (WFF (lfor wff)) (WFF (rtor wff)))
  ...
  (IF (UNIQUANT wff) (AND (WFF (matrix wff)) (VAR (bvarof wff)))
   ...
   FALSE)))
```

Fig. 15. The code implementing the data type WFF

and we would like to avoid using not recursive recognizers. One (partial) solution could be to axiomatize proofhood, rather than theoremhood. This is somehow similar to the (type-theoretical/algebraic) approach of [BC91, CAB+86]. In this approach inference rules, rather than being functions on sequents, are functions on whole proofs. One problem is that, potentially, objects have to be recomputed from scratch any time they are introduced in the system. This is acceptable for wffs, whose data structures are indeed generated at parsing time. For sequents, though, building the corresponding proof may become a very hard task.

Lifting UM is based on a different idea. In GETFOL, the state is used to save partial computations: this allows us to reuse objects which have already been computed. Sequents, once proved, are asserted as theorems in the part of LS implementing the current proof (by proof-add-theorem). Verifying theoremhood (cfr. THEOREM in figure 12) amounts to searching in LS. Therefore the current proof in LS can be seen as an approximation of the non-recursive set of all the provable sequents. Lifting UM is based on the idea that, if an UM primitive adds the result of a computation to the state approximating a certain set, then the result also belongs to the set, i.e. the approximation is sound. In the particular case of inference, alliprf adds the result of allitac to the state approximating the set of (successful and failing) inference steps (this state is hidden by maybe-proof-add-theorem). Being Tac the description in MT of this set involved, by lifting alliprf we get the admissibility statement for $allitac$, i.e. $\forall x\ y\ z.Tac(allitac(x, y, z))$. As shown in section 2, for our purposes this is equivalent to the axiom $\mathcal{A}_{alli}$. Following this approach, it is possible to lift from the UM all the necessary axioms. In particular, to lift the induction principle

$$\forall \underline{x}.\Phi(ruletac_1(\underline{x})) \wedge \ldots \wedge \forall \underline{x}.\Phi(ruletac_n(\underline{x})) \supset (\forall x.Tac(x) \supset \Phi(x))$$

it is sufficient to notice that the *ruletac$_i$* are the only inference rules in the system.

5 Conclusions

In this paper we have presented the architecture and the issues related to the development of an *introspective metatheoretic* reasoning system, able to *lift* its own code into a declarative metatheory, *reason* about it and *flatten* down the resulting theorems into new system code. The theoretical foundations of this work (e.g. the formal definition of lifting and flattening, the proof of their full symmetry and correctness) are discussed in the companion paper [GC92].

References

[BC91]   D. Basin and R. Constable. Metalogical Frameworks. In *Proceedings of the Second Workshop on Logical Frameworks*, Edinburgh, Scotland, 1991. To Appear as a chapter in a Cambridge University Press book.

[BM79]   R.S. Boyer and J.S. Moore. *A Computational Logic*. Academic Press, 1979. ACM monograph series.

[BM81]   R.S. Boyer and J.S. Moore. Metafunctions: proving them correct and using them efficiently as new proof procedures. In R.S. Boyer and J.S. Moore, editors, *The correctness problem in computer science*, pages 103–184. Academic Press, 1981.

[CAB⁺86]   R.L. Constable, S.F. Allen, H.M. Bromley, et al. *Implementing Mathematics with the NuPRL Proof Development System*. Prentice Hall, 1986.

[GC89]   F. Giunchiglia and A. Cimatti. HGKM Manual - a revised version. Technical Report 8906-22, IRST, Trento, Italy, 1989.

[GC92]   F. Giunchiglia and A. Cimatti. Introspective Metatheoretic Theorem Proving. Technical Report 9211-22, IRST, Trento, Italy, 1992. Upgraded and extended version forthcoming.

[Giu92]   F. Giunchiglia. The GETFOL Manual - GETFOL version 1. Technical Report 92-0010, DIST - University of Genova, Genoa, Italy, 1992.

[GS88]   F. Giunchiglia and A. Smaill. Reflection in constructive and non-constructive automated reasoning. In H. Abramson and M. H. Rogers, editors, *Proc. of META-88, Workshop on Metaprogramming in Logic*, pages 123–145. MIT Press, 1988. Also IRST-Technical Report 8902-04 and DAI Research Paper 375, University of Edinburgh.

[GT91]   F. Giunchiglia and P. Traverso. Reflective reasoning with and between a declarative metatheory and the implementation code. In *Proc. of the 12th International Joint Conference on Artificial Intelligence*, pages 111–117, Sydney, 1991. Also IRST-Technical Report 9012-03, IRST, Trento, Italy.

[GT92]   F. Giunchiglia and P. Traverso. A Metatheory of a Mechanized Object Theory. Technical Report 9211-24, IRST, Trento, Italy, 1992. Submitted for publication to: Journal of Artificial Intelligence.

[GT94]   F. Giunchiglia and P. Traverso. Program Tactics and Logic Tactics. In *Proceedings 5th Intnl. Conference on Logic Programming and Automated Reasoning (LPAR'94)*, Kiev, Ukraine, July 16-21, 1994. Also IRST-Technical

Report 9301-01, IRST, Trento, Italy. Presented at the Third International Symposium on Artificial Intelligence and Mathematics, Fort Lauderdale, Florida, January 1994.

[GW91] F. Giunchiglia and R.W. Weyhrauch. FOL User Manual - FOL version 2. Manual 9109-08, IRST, Trento, Italy, 1991. Also DIST Technical Report 91-0006, DIST, University of Genova.

[JL83] P. N. Johnson-Laird. *Mental Models*. Cambridge University Press, 1983.

[MW87a] Z. Manna and R. Waldinger. The Deductive Synthesis of Imperative LISP Programs. In *Sixth National Conference on Artificial Intelligence*. AAAI, 1987.

[MW87b] Z. Manna and R. Waldinger. How to clear a block: A Theory of Plans. Technical Report STAN-CS-87-1141, Department of Computer Science, Stanford University, 1987.

[Pau89] L. Paulson. The Foundation of a Generic Theorem Prover. *Journal of Automated Reasoning*, 5:363–396, 1989.

[Pra65] D. Prawitz. *Natural Deduction - A proof theoretical study*. Almquist and Wiksell, Stockholm, 1965.

[Smi83] B.C. Smith. Reflection and Semantics in LISP. In *Proc. 11th ACM POPL*, pages 23–35, 1983.

[Wey80] R.W. Weyhrauch. Prolegomena to a Theory of Mechanized Formal Reasoning. *Artificial Intelligence*, 13(1):133–176, 1980.

Abstract Debugging of Logic Programs

Marco Comini[1], Giorgio Levi[1] and Giuliana Vitiello[2]

[1] Dipartimento di Informatica, Università di Pisa, Corso Italia 40, 56125 Pisa, Italy
[2] Dipartimento di Informatica ed Applicazioni, Università di Salerno, Baronissi (Salerno), Italy

1 Introduction

Abstract debugging of logic programs is a novel combination of three known techniques, i.e. *algorithmic (declarative) debugging* [25, 15, 22], the *s-semantics approach* to the definition of program denotations modeling various observable behaviors [13, 14, 16, 4, 3], and *abstract interpretation* [10, 11, 12]. A similar approach was developed in [5] for imperative languages.

The debugging problem can formally be defined as follows. Let P be a program, $[P]_\alpha$ be the behavior of P w.r.t. the observable property α, and $\mathcal{I}_\alpha$ be the specification of the *intended* behavior of P w.r.t. α. The debugging consists of comparing $[P]_\alpha$ and $\mathcal{I}_\alpha$ and determining the "errors" and the program components which are sources of errors, when $[P]_\alpha \neq \mathcal{I}_\alpha$.

The above formulation is parametric w.r.t. the *observable* α, which is considered in the specification $\mathcal{I}_\alpha$ and in the actual behavior $[P]_\alpha$.

Declarative debugging is concerned with model-theoretic properties rather than with the operational behavior. The specification is therefore the intended declarative semantics of the program, which is the least Herbrand model in [25, 22], and the set of atomic logical consequences in [15].

Abstract debugging is a generalization of declarative debugging, where we consider operational properties. An observable is any property which can be extracted from a goal computation, i.e. *observables are abstractions of SLD-trees*. Examples of useful observables are *computed answers, finite failures* and *call patterns* (i.e. procedure calls selected in an *SLD*-derivation). Other examples come from program analysis, e.g. *depth(l)-answers* (i.e. answers containing terms whose depth is $\leq l$), *types, modes* and *ground dependencies*. As we will discuss later, the relation among the observables can naturally be understood in terms of abstract interpretation. A similar approach to the semantics of logic programs can be found in [18].

Here are some motivations for abstract debugging.

- The most natural abstract debugging for positive logic programs is debugging w.r.t. computed answers, which leads to a more precise analysis, since declarative debugging is related to correct answers only.
- Debugging w.r.t. finite failures allows us to verify another relevant behavior, which has also a logical interpretation.
- Less abstract observables, such as call patterns, can be useful to verify the control and data flow between different procedures, as we usually do in interactive debugging. For example, the intended behavior that we specify

might be a set of assertions of the form "the execution of the procedure call $p(t_1, \ldots, t_n)$ leads to the procedure call $q(s_1, \ldots, s_m)$".

- Debugging w.r.t. depth(l)-answers makes debugging w.r.t. computed answers effective, since both $\mathcal{I}_\alpha$ and $[P]_\alpha$ are finite.
- Debugging w.r.t. types, modes and ground dependencies allows us to verify partial program properties. For example, if $\mathcal{I}_\alpha$ specifies the intended program behavior w.r.t. types, abstract debugging boils down to type checking.

In declarative debugging, the specification is usually assumed to be given by means of an *oracle*. This approach is feasible even in abstract debugging. However, since our method can handle abstractions, we can easily come out with finite observable behaviors and specify them in an extensional way.

Our theory of abstract debugging is built on an algebraic semantic framework for positive logic programs [8], based on the formalization of observables as abstractions. A complete description of the framework is outside the scope of this paper. In Section 2 we summarize the main properties of the framework. In Section 3 we state the debugging problem in the general case, where we have to compare the behaviors for all the possible goals. However, there exists an important class of observables with strong semantic properties, which allows us to consider the behaviors for most general atomic goals only. These observables are called *s-observables* and are discussed in Section 4. The debugging problem and the diagnosis algorithms for s-observables are considered in Section 5. Both correctness and completeness can be verified by diagnosis algorithms, which only need to consider the behavior for finitely many goals. We also show that the existing declarative debugging methods can be reconstructed as instances of abstract debugging w.r.t. s-observables. Finally section 6 discusses a weaker class of observables (i-observables), which are meant to model the abstraction involved in program analysis and we show that only the completeness diagnosis algorithm is applicable.

2 Observables

Following [8], we consider Constraint Logic Programs (*CLP*) [21] with the PRO-LOG (leftmost) selection rule. We assume the reader to be familiar with the notions of SLD-resolution and SLD-tree (see [23, 1]).

We represent, for notational convenience, SLD-trees as sets of nodes. Let T be an SLD-tree rooted at the goal G. A *node* in T is a 4-tuple $\langle G, c, \tilde{b}, k \rangle$, where $c \,\Box\, \tilde{b}$ is the goal associated to the node (c is the accumulated constraint and $\tilde{b}$ is a conjunction of atoms), and k is the sequence of (renamed apart) clauses used in the derivation of $c \,\Box\, \tilde{b}$ from G.

We define a partial order $\leq$ on nodes. $\langle G, c, \tilde{b}, k \rangle \leq \langle G, c', \tilde{b}', k' \rangle$ if and only if $k' = k :: k''$ for some k'' A set of nodes A is *well formed* if $\rho \in A \Rightarrow \forall \rho' \leq \rho, \rho' \in A$.

An SLD-tree is represented as a well formed set of nodes. $\mathcal{R}$ is the domain of all the well formed sets of nodes (modulo variance), partially ordered by set inclusion. We can now define the *behavior* $[P]$ of a program P as the set of all the nodes of SLD-trees of G, for any goal G.

Observables are abstractions of *SLD*-trees. More precisely, an *observable* is a function α from $\mathcal{R}$ to an abstract domain $\mathcal{D}$, which preserves the partial orders. α is an abstraction function according to abstract interpretation, i.e. there exists a function γ (concretization) from $\mathcal{D}$ to $\mathcal{R}$, such that (α, γ) is a Galois connection.

The *abstract behavior* $[P]_\alpha$ w.r.t. the observable α is simply defined as $\alpha([P])$. α induces an *observational equivalence* $=_\alpha$ on programs. Namely $p_1 =_\alpha p_2$ iff $[P_1]_\alpha = [P_2]_\alpha$, i.e. if P_1 and P_2 cannot be distinguished by looking at their abstract behaviors. Observational equivalences can be used to define a partial order $\leq$ on observables. Namely $\alpha' \leq \alpha$ (α is stronger than α') if $p_1 =_\alpha p_2$ implies $p_1 =_{\alpha'} p_2$.

Let $\sigma \in \mathcal{R}$ be a well formed set of nodes. The following functions are examples of observables.

- (computed answer constraints)

$$\xi(\sigma) = \{ \langle G, c \rangle \mid \langle G, c, true, k \rangle \in \sigma \}.$$

- (correct answer constraints)

$$\psi(\sigma) = \{ \langle G, c \rangle \mid \langle G, c', true, k \rangle \in \sigma, c \leq c' \}.$$

- (solutions of answer constraints)

$$\phi(\sigma) = \{ \langle G, \vartheta \rangle \mid \langle G, c, true, k \rangle \in \sigma, \vartheta \text{ is a solution of } c \}.$$

- (answers with depth)

$$\chi(\sigma) = \{ \langle G, c, n \rangle \mid \langle G, c, true, k \rangle \in \sigma, n \text{ is the length of } k \}.$$

- (*l*-answers with depth)

$$\varXi(\sigma) = \{ \langle G, c, n \rangle \mid \langle G, c, true, k \rangle \in \sigma,$$
$$n \text{ is the length of } k,$$
$$n \leq l \}.$$

- (depth(*l*)-answers)

$$\zeta(\sigma) = \{ \langle G, c \rangle \mid \langle G, c, true, k \rangle \in \sigma, \text{the depth of all terms in } c \text{ is } \leq l \}.$$

- (call patterns)

$$\eta(\sigma) = \{ \langle G, c, a \rangle \mid \langle G, c, (a, \tilde{b}), k \rangle \in \sigma \}.$$

- (finite failures)

$$\theta(\sigma) = \{ G \mid \text{the set of nodes in } \sigma \text{ containing } G \text{ as first element is}$$
$$\text{finite, non-empty and does not contain nodes of the}$$
$$\text{form } \langle g, c, true, k \rangle \}.$$

- (ground dependencies in answer constraints)
The ground dependencies among the variables occurring in the goal are represented as propositional formulas (see the domain *Prop* [9]). We assume the

answer constraints to be equalities of the form $X = t$. Abstract constraints are given by the function ν defined as follows:

$$\nu(c_1 \wedge c_2) = \nu(c_1) \wedge \nu(c_2),$$

$$\nu(X = t) = X \longleftrightarrow \bigwedge X_i \text{ s.t. } X_i \in Vars(t).$$

The observable is then

$$\pi(\sigma) = \{ \, \langle G, \nu(c) \rangle \mid \langle G, c, true, k \rangle \in \sigma \, \}.$$

3 Abstract debugging

Let P be a program, α be an observable and $\mathcal{I}_\alpha$ be the specification of the intended behavior of P w.r.t. α. The following definitions are straightforward extensions of the definitions given in the case of declarative debugging [25, 15, 22].

- P is *partially correct* w.r.t. $\mathcal{I}_\alpha$, if $[P]_\alpha \subseteq \mathcal{I}_\alpha$.
- P is *complete* w.r.t. $\mathcal{I}_\alpha$, if $\mathcal{I}_\alpha \subseteq [P]_\alpha$.
- P is *totally correct* w.r.t. $\mathcal{I}_\alpha$, if $[P]_\alpha = \mathcal{I}_\alpha$.

Incorrectness and incompleteness symptoms are elements (of the abstract domain $\mathcal{D}$) on which the specification and the program disagree. According to the above definitions, in order to detect the symptoms we need to consider the abstract behaviors for all the goals. However there exists a large class of observables (*s-observables*), for which we can consider only the behaviors for most general atomic goals. As we will show in Section 5, abstract debugging can be reformulated in much simpler terms for *s*-observables.

4 *s*-observables

s-observables are observables for which we can define a *denotation*, which generalizes the properties of the *s*-semantics [3]. The program denotation is defined by collecting the behaviors for *most general atomic goals*, i.e. goals of the form $p(\tilde{X})$ consisting of the application of a predicate symbol to a tuple of distinct variables. This property holds for our basic observable, i.e. *SLD*-trees. We will discuss the properties in the case of *SLD*-trees first and then move to the abstractions.

The *SLD-trees denotation* $\mathcal{O}(P)$ of a program P is the set of all the nodes of *SLD*-trees of most general atomic goals. *SLD*-trees have the following properties, first proved in [16, 17] and later formalized in algebraic terms [8]:

- *AND*-compositionality, i.e. the *SLD*-tree of any (conjunctive) goal can be determined from $\mathcal{O}(P)$.
- Correctness and full abstraction of the denotation $\mathcal{O}(P)$, i.e.

$$[P_1] = [P_2] \iff \mathcal{O}(P_1) = \mathcal{O}(P_2).$$

- Equivalent bottom-up construction of the denotation, i.e. $\mathcal{O}(P)$ can be obtained as least fixpoint of the operator T_P defined in the following. Let I be an element of $\mathcal{R}$.

$$T_P(I) = \{ \quad \langle q(\tilde{X}), c, \tilde{b}, ks \rangle \mid$$

    1. k is the (renamed apart) clause of P
        $q(\tilde{t}) :- c_k \; \square \; q_1(\tilde{t_1}), \ldots, q_m(\tilde{t_m}), a_{m+1}, \ldots, a_n,$
    2. $\langle q_j(\tilde{X_j}), c_j, true, ks_j \rangle,\ 1 \leq j < m,$
        $\langle q_m(\tilde{X_m}), c_m, \tilde{b_m}, ks_m \rangle,$
        are renamed apart elements of I,
    3. $c = \exists(\tilde{X}{=}\tilde{t} \wedge c_k \wedge \tilde{X_1}{=}\tilde{t_1} \wedge c_1 \wedge \cdots \wedge \tilde{X_m}{=}\tilde{t_m} \wedge c_m)_{\tilde{X}, \tilde{b}},$
    4. $\tilde{b} = \tilde{b_m}, a_{m+1},\ \ldots, a_n,$
    5. $ks = [k] :: ks_1 :: \ldots :: ks_m \}.$

The top-down and bottom-up denotations corresponding to an observable α can be defined as follows.

- $\mathcal{O}_\alpha(P) = \alpha(\mathcal{O}(P)),$
- $T_{P,\alpha} = \alpha \circ T_P \circ \gamma$ (where γ is the concretization function).

It is worth noting that in the top-down construction we abstract the *SLD*-trees of most general atomic goals, while, in the bottom-up one, we perform the abstraction at each fixpoint approximation step.

$\mathcal{O}_\alpha(P)$ and $T_{P,\alpha}$ are meaningful only for observables which satisfy suitable conditions formally stated in [8]. Informally, an *s*-observable is an abstraction α of *SLD*-trees, for which we have a lifting (to more general goals) property and which is *AND*-compositional. Moreover, α must be *compatible* with the bottom-up operator T_P, namely $T_{P,\alpha} \circ \alpha = \alpha \circ T_P$.

s-observables have the same properties of *SLD*-trees, namely *AND*-compositionality, correctness (and full abstraction) of the denotation and equivalent bottom-up denotation.

Examples of *s*-observables are (computed and correct) answer constraints, solutions of answer constraints, answers with depth, *l*-answers with depth and call patterns. Some of the specialized bottom-up operators $T_{P,\alpha} = \alpha \circ T_P \circ \gamma$ are the *CLP* versions of existing "immediate consequences operators". In particular,

- $T_{P,\phi}$ is the ground operator defined in [26] (and $\mathcal{O}_\phi(P)$ is the least Herbrand model).
- $T_{P,\psi}$ is the non-ground operator first defined in [6] (and $\mathcal{O}_\psi(P)$ is the least term model).
- $T_{P,\xi}$ is the *s*-semantics operator defined in [13].
- $T_{P,\eta}$ is the call patterns operator defined in [16].

We show two of the operators that will be later used in the examples.

- (computed answer constraints)

$T_{P,\xi}(I) = \{ \quad \langle q(\tilde{X}), c \rangle \mid$
1. there exists a (renamed apart) clause of P
$q(\tilde{t}) :- c_k \,\square\, q_1(\tilde{t_1}), \ldots, q_n(\tilde{t_n}),$
2. $\langle q_j(\tilde{X_j}), c_j \rangle, \ 1 \leq j \leq n,$
are renamed apart elements of I,
3. $c = \exists (\tilde{X} = \tilde{t} \wedge c_k \wedge \tilde{X_1} = \tilde{t_1} \wedge c_1 \wedge \cdots \wedge \tilde{X_n} = \tilde{t_n} \wedge c_n)_{\tilde{X}} \ \}.$

- (l-answers with depth)

$T_{P,\Xi}(I) = \{ \quad \langle q(\tilde{X}), c, m \rangle \mid$
1. there exists a (renamed apart) clause of P
$q(\tilde{t}) :- c_k \,\square\, q_1(\tilde{t_1}), \ldots, q_n(\tilde{t_n}),$
2. $\langle q_j(\tilde{X_j}), c_j, m_j \rangle, \ 1 \leq j \leq n,$
are renamed apart elements of I,
3. $c = \exists (\tilde{X} = \tilde{t} \wedge c_k \wedge \tilde{X_1} = \tilde{t_1} \wedge c_1 \wedge \cdots \wedge \tilde{X_n} = \tilde{t_n} \wedge c_n)_{\tilde{X}},$
4. $m = m_1 + \cdots + m_n + 1, m \leq l \}.$

depth(l)-answers and ground dependencies are not s-observables, since the denotations do not contain enough information to get a precise reconstruction of the behavior of conjunctive goals. However they belong to a weaker class of observables (i-observables), for which we still have reasonable semantic properties (see Section 6). Note that, if we want an s-observable which gives a finite approximation of computed answers, depth(l)-answers cannot be chosen and we have to resort to l-answers with depth.

Finite failures are not even i-observables. For all the observables α in this class, we have to consider a stronger s-observable β ($\alpha \leq \beta$). The corresponding denotation $\mathcal{O}_\beta(P)$ is correct and AND-compositional w.r.t. α, yet it is not fully abstract [8]. In the case of abstract debugging, we are forced to specify the behavior w.r.t. β.

5 Abstract debugging w.r.t. s-observables

Let P be a program and α be an s-observable. Since α is an s-observable, we know that the actual and the intended behaviors of P for all the goals can be uniquely determined from the behaviors for most general goals. Hence the following definitions correspond to the definitions of Section 3. Let $\mathcal{I}_\alpha$ be the specification of the intended behavior of P for most general atomic goals w.r.t. α.

- P is *partially correct* w.r.t. $\mathcal{I}_\alpha$, if $\mathcal{O}_\alpha(P) \subseteq \mathcal{I}_\alpha$.
- P is *complete* w.r.t. $\mathcal{I}_\alpha$, if $\mathcal{I}_\alpha \subseteq \mathcal{O}_\alpha(P)$.
- P is *totally correct* w.r.t. $\mathcal{I}_\alpha$, if $\mathcal{O}_\alpha(P) = \mathcal{I}_\alpha$.

It is worth noting that we can reconstruct the usual definitions of declarative debugging within our more general framework, thus showing that the use of declarative specifications can also be motivated by operational arguments (i.e. they are denotations for AND-compositional observables). In particular,

- the observable ϕ (solutions of answer constraints) gives us the declarative debugging based on the least Herbrand model [25, 22];
- the observable ψ (correct answer constraints) gives us the declarative debugging based on the atomic logical consequences [15].

If P is not totally correct, we are left with the problem of determining the errors, which are based, following [25, 15, 22], on the *symptoms*. Namely,

- an *incorrectness symptom* is an element in $\mathcal{O}_\alpha(P) - \mathcal{I}_\alpha$,
- an *incompleteness symptom* is an element in $\mathcal{I}_\alpha - \mathcal{O}_\alpha(P)$.

Note that our incompleteness symptoms are different from the *insufficiency symptoms* used in [25, 15, 22]. As we will show in the following, our definitions lead to a symmetric diagnosis for incorrectness and incompleteness.

It is straightforward to realize that an element may sometimes be an (incorrectness or incompleteness) symptom, just because of another symptom. The *diagnosis* determines the "basic" symptoms, and, in the case of incorrectness, the relevant clause in the program. This is captured by the definitions of *incorrect clause* and *uncovered element*.

Definition 1. A clause $c \in P$ is *incorrect on* the element σ if

$$\sigma \in \mathcal{O}_\alpha(P), \ \sigma \notin \mathcal{I}_\alpha, \ \sigma \in T_{\{c\},\alpha}(\mathcal{I}_\alpha).$$

An element σ is *uncovered* if

$$\sigma \in \mathcal{I}_\alpha, \ \sigma \notin \mathcal{O}_\alpha(P), \ \sigma \notin T_{P,\alpha}(\mathcal{I}_\alpha).$$

Note that $T_{\{c\},\alpha}$ is the operator associated to the program $\{c\}$, consisting of the clause c only. Note also that $\mathcal{O}_\alpha(P)$ and $\mathcal{I}_\alpha$ play a symmetric role in the diagnosis of incorrect clauses and uncovered elements. The following Theorem is a justification of the diagnosis method.

Theorem 2. *Let P be a program, α be an s-observable and $\mathcal{I}_\alpha$ be the specification of the intended behavior of P for most general atomic goals w.r.t. α.*

Then P is totally correct w.r.t. $\mathcal{I}_\alpha$, if and only if there are no incorrect clauses and uncovered elements according to Definition 1.

The *diagnosis algorithm for incorrect clauses* consists of the following steps:

1. generation of an element $\sigma \in \mathcal{O}_\alpha(P)$;
2. if $\sigma \notin \mathcal{I}_\alpha$ and exists a clause c in P such that $\sigma \in T_{\{c\},\alpha}(\mathcal{I}_\alpha)$, then c is labeled as incorrect.

The generation of the elements in $\mathcal{O}_\alpha(P)$ is performed according to the definition of the denotation. In the top-down case, the algorithm starts with a most general atomic goal and, as soon as an element in the denotation is generated, step 2 is applied. The existence of an equivalent bottom-up denotation allows us to define bottom-up diagnosis algorithms. Step 2 is now applied whenever a new element is generated in the iterative construction of the fixpoint.

The *diagnosis algorithm for uncovered elements* is driven by the elements in $\mathcal{I}_\alpha$.

For each $\sigma \in \mathcal{I}_\alpha$

1. select the most general atomic goal G in σ;
2. compute the set of elements Σ derived from G (top-down);
3. if $\sigma \notin \Sigma$ and $\sigma \notin T_{P,\alpha}(\mathcal{I}_\alpha)$ then σ is an uncovered element.

Note that, both in the specification and in the denotation built by the debugging algorithms, we are concerned with the behaviors of most general atomic goals. Hence there are finitely many goals, i.e. one for each predicate symbol in P. If the abstraction α guarantees that for each most general atomic goal we have finitely many observations, then the specification is finite and our diagnosis is effective. In such a case, as already mentioned, $\mathcal{I}_\alpha$ can be specified in an extensional way and there is no need for the oracle. One example of such an abstraction is the observable Ξ (*l*-answers with depth), that we will consider in the example below.

Example 1. The following program P is a "wrong" version of **append**, that we want to debug w.r.t. the observable Ξ (*l*-answers with depth, $l = 2$).

c_1) app(X, Y, Z) :− Y = [], Z = X □.
c_2) app(X, Y, Z) :− X = [X1|X2], Z = [X1|Z2] □ app(X2, Y, Z2)

The specification of the intended behavior of P is

$$\mathcal{I}_\Xi = \{\langle app(X, Y, Z), (X = [], Y = Z), 1\rangle, \\ \langle app(X, Y, Z), (X = [X1], Z = [X1|Y]), 2\rangle\},$$

while the actual behavior is

$$\mathcal{O}_\Xi(P) = \{\langle app(X, Y, Z), (Y = [], X = Z), 1\rangle, \\ \langle app(X, Y, Z), (Y = [], X = [X1|X2], X = Z), 2\rangle\}.$$

All the elements in $\mathcal{O}_\Xi(P)$ are incorrectness symptoms, while all the elements in $\mathcal{I}_\Xi$ are incompleteness symptoms. In order to detect incorrect clauses and uncovered elements, we have to apply the bottom-up operator $T_{P,\Xi}$ to $\mathcal{I}_\Xi$:

$$T_{\{c_1\},\Xi}(\mathcal{I}_\Xi) = \{\langle app(X, Y, Z), (Y = [], X = Z), 1\rangle\}$$

$$T_{\{c_2\},\Xi}(\mathcal{I}_\Xi) = \{\langle app(X, Y, Z), (X = [X1], Z = [X1|Y]), 2\rangle\}.$$

The diagnosis detects that c_1 is an incorrect clause on $\langle app(X, Y, Z), (Y = [], X = Z), 1\rangle$ and that $\langle app(X, Y, Z), (X = [], Y = Z), 1\rangle$ is an uncovered element.

Even when the diagnosis is effective, our algorithms are forced to compute $\mathcal{O}_\alpha(P)$. This amounts to a fixpoint computation or to the construction of the *SLD*-trees for all the most general atomic goals.

6 Abstract debugging w.r.t. *i*-observables

i-observables are meant to model the abstraction involved in abstract interpretation. According to the approach in [7, 19, 20], the abstraction process is performed by defining a semi-morphism μ from the concrete constraint system to a suitable abstract constraint system. This abstraction has to be combined with the usual abstraction of SLD-trees to generate an i-observable. Namely, an i-observable $\alpha_\mu = \alpha \circ \mu^\sigma$ is the composition of an s-observable α with the function μ^σ, obtained by extending to nodes the constraint semimorphism μ. Note that in general α_μ is not an s-observable.

We follow the so-called "generalized semantics" approach of [19, 20], which is a kind of abstract compilation, where the program P is transformed into an *abstract program* $\mu(P)$, obtained by applying μ to the concrete constraints in P. Once we have the abstract program $\mu(P)$, we get the usual (equivalent) top-down and bottom-up denotations $\mathcal{O}_\alpha(\mu(P))$ and $T_{\mu(P),\alpha} \uparrow \omega$, since α is an s-observable.

The abstract compilation introduces an approximation which is safe in the sense of abstract interpretation, i.e.

$$\mathcal{O}_\alpha(\mu(P)) \supseteq \mathcal{O}_{\alpha_\mu},$$

where $\mathcal{O}_{\alpha_\mu} = \alpha_\mu(\mathcal{O}(P))$ is the abstraction according to α_μ of the concrete denotation $\mathcal{O}(P)$.

In abstract debugging, $\mathcal{I}_{\alpha_\mu}$ is a specification of the intended behavior $\mathcal{O}_{\alpha_\mu}$. Hence we cannot get any information about partial correctness from abstract compilation, since in general the following relation holds (for a complete program):

$$\mathcal{O}_\alpha(\mu(P)) \supseteq \mathcal{I}_{\alpha_\mu}.$$

On the other hand, the definitions given in Section 5 related to completeness (and the corresponding diagnosis algorithm for detecting uncovered elements) are applicable to the case of i-observables as well, as shown by the following example, where we detect uncovered elements w.r.t. a specification related to ground dependencies.

Example 2. We consider the program P of example 1. The overall i-observable ξ_ν is the composition of the computed answer constraints s-observable and of the abstraction on the constraint system ν of Section 2, which gives the ground dependencies.

The abstract program $\nu(P)$ is

$c_1)$ app(X, Y, Z) :– $Y \wedge (X \leftrightarrow Z)\,\square$.
$c_2)$ app(X, Y, Z) :– $(X \leftrightarrow (X1 \wedge X2)) \wedge (Z \leftrightarrow (X1 \wedge Z2))\,\square$ app$(X2, Y, Z2)$

The specification of the intended behavior of P is

$$\mathcal{I}_{\xi_\nu} = \{\langle app(X, Y, Z), X \wedge (Y \leftrightarrow Z)\rangle, \langle app(X, Y, Z), Z \leftrightarrow (X \wedge Y)\rangle\},$$

while the actual behavior of the abstract program is

$$\mathcal{O}_\xi(\nu(P)) = \{\langle app(X, Y, Z), Y \wedge (X \leftrightarrow Z)\rangle\}.$$

All the elements in $\mathcal{I}_{\xi_\nu}$ are incompleteness symptoms. In order to detect uncovered elements, we have to apply the bottom-up operator $T_{\nu(P),\xi}$ to $\mathcal{I}_{\xi_\nu}$.

$$T_{\nu(P),\xi}(\mathcal{I}_{\xi_\nu}) = \{\, \langle app(X, Y, Z), Y \wedge (X \leftrightarrow Z)\rangle, \\ \langle app(X, Y, Z), Z \leftrightarrow (X \wedge Z)\rangle\}$$

The diagnosis detects that $\langle app(X,Y,Z), X \wedge (Y \leftrightarrow Z) \rangle$ is an uncovered element. Note that sometimes disjunctive constraints in $Prop$ are replaced by their greatest lower bound. This solution, if applied to $\mathcal{I}_{\xi_\nu}$ and to $\mathcal{O}_\xi(\nu(P))$, would introduce a further approximation, which, in this example, would not allow the detection of the uncovered element any longer.

7 Conclusions

Our debugging method applies to positive logic programs, under the leftmost selection rule. All our results can easily be extended to *local* selection rules [27].

A more interesting extension is related to the overall PROLOG computation rule, including the depth-first search based on the clause ordering. Abstract debugging can be based on a PROLOG version of the semantic framework in [8], which was recently defined in [24]. The framework generalizes the computed answer semantics for PROLOG given in [2] and introduces a new abstract compilation method, where each constraint is abstracted by a pair of abstract constraints (upper and lower approximations).

Abstract debugging can be viewed as yet another program verification method, where we are concerned with partial correctness properties and the specification is extensional. An interesting problem is that of better understanding the relation among the various techniques for proving program properties, namely debugging, verification and abstract interpretation.

References

1. K. R. Apt. Introduction to Logic Programming. In J. van Leeuwen, editor, *Handbook of Theoretical Computer Science*, volume B: Formal Models and Semantics, pages 495–574. Elsevier, Amsterdam and The MIT Press, Cambridge, 1990.
2. A. Bossi, M. Bugliesi, and M. Fabris. Fixpoint semantics for PROLOG. In D. S. Warren, editor, *Proc. Tenth Int'l Conf. on Logic Programming*, pages 374–389. The MIT Press, Cambridge, Mass., 1993.
3. A. Bossi, M. Gabbrielli, G. Levi, and M. Martelli. The s-semantics approach: Theory and applications. *Journal of Logic Programming*, 19-20:149–197, 1994.
4. A. Bossi, M. Gabbrielli, G. Levi, and M. C. Meo. A Compositional Semantics for Logic Programs. *Theoretical Computer Science*, 122(1-2):3–47, 1994.
5. F. Bourdoncle. Abstract debugging of higher-order imperative languages. In *PLDI'93*, pages 46–55, 1993.
6. K. L. Clark. Predicate logic as a computational formalism. Res. Report DOC 79/59, Imperial College, Dept. of Computing, London, 1979.
7. P. Codognet and G. Filè. Computations, Abstractions and Constraints. In *Proc. Fourth IEEE Int'l Conference on Computer Languages*. IEEE Press, 1992.
8. M. Comini and G. Levi. An algebraic theory of observables. In M. Bruynooghe, editor, *Proc. 1994 Int'l Symposium on Logic Programming*. The MIT Press, Cambridge, Mass., 1994. To appear.
9. A. Cortesi, G. Filè, and W. Winsborough. *Prop* revisited: Propositional Formula as Abstract Domain for Groundness Analysis. In *Proc. Sixth IEEE Symp. on Logic In Computer Science*, pages 322–327. IEEE Computer Society Press, 1991.

10. P. Cousot and R. Cousot. Abstract Interpretation: A Unified Lattice Model for Static Analysis of Programs by Construction or Approximation of Fixpoints. In *Proc. Fourth ACM Symp. Principles of Programming Languages*, pages 238–252, 1977.

11. P. Cousot and R. Cousot. Systematic Design of Program Analysis Frameworks. In *Proc. Sixth ACM Symp. Principles of Programming Languages*, pages 269–282, 1979.

12. P. Cousot and R. Cousot. Abstract Interpretation and Applications to Logic Programs. *Journal of Logic Programming*, 13(2 & 3):103–179, 1992.

13. M. Falaschi, G. Levi, M. Martelli, and C. Palamidessi. Declarative Modeling of the Operational Behavior of Logic Languages. *Theoretical Computer Science*, 69(3):289–318, 1989.

14. M. Falaschi, G. Levi, M. Martelli, and C. Palamidessi. A Model-Theoretic Reconstruction of the Operational Semantics of Logic Programs. *Information and Computation*, 102(1):86–113, 1993.

15. G. Ferrand. Error Diagnosis in Logic Programming, an Adaptation of E. Y. Shapiro's Method. *Journal of Logic Programming*, 4:177–198, 1987.

16. M. Gabbrielli, G. Levi, and M. C. Meo. Observational Equivalences for Logic Programs. In K. Apt, editor, *Proc. Joint Int'l Conf. and Symposium on Logic Programming*, pages 131–145. The MIT Press, Cambridge, Mass., 1992. Extended version to appear in *Information and Computation*.

17. M. Gabbrielli, G. Levi, and M. C. Meo. A resultants semantics for PROLOG. Technical report, Dipartimento di Informatica, Università di Pisa, 1994.

18. R. Giacobazzi. On the Collecting Semantics of Logic Programs. In F. S. de Boer and M. Gabbrielli, editors, *Verification and Analysis of Logic Languages, Proc. of the Post-Conference ICLP Workshop*, pages 159–174, 1994.

19. R. Giacobazzi, S. K. Debray, and G. Levi. A Generalized Semantics for Constraint Logic Programs. In *Proceedings of the International Conference on Fifth Generation Computer Systems 1992*, pages 581–591, 1992.

20. R. Giacobazzi, G. Levi, and S. K. Debray. Joining Abstract and Concrete Computations in Constraint Logic Programming. In M. Nivat, C. Rattray, T. Rus, and G. Scollo, editors, *Algebraic Methodology and Software Technology (AMAST'93), Proceedings of the Third International Conference on Algebraic Methodology and Software Technology*, Workshops in Computing, pages 111–127. Springer-Verlag, Berlin, 1993.

21. J. Jaffar and J. L. Lassez. Constraint Logic Programming. In *Proc. Fourteenth Annual ACM Symp. on Principles of Programming Languages*, pages 111–119. ACM, 1987.

22. J. W. Lloyd. Declarative error diagnosis. *New Generation Computing*, 5(2):133–154, 1987.

23. J. W. Lloyd. *Foundations of Logic Programming*. Springer-Verlag, Berlin, 1987. Second edition.

24. D. Micciancio. Interpretazione astratta di programmi logici con il controllo di PROLOG. Master's thesis, Dipartimento di Informatica, Università di Pisa, 1994. in italian.

25. E. Y. Shapiro. Algorithmic program debugging. In *Proc. Ninth Annual ACM Symp. on Principles of Programming Languages*, pages 412–531. ACM Press, 1982.

26. M. H. van Emden and R. A. Kowalski. The semantics of predicate logic as a programming language. *Journal of the ACM*, 23(4):733–742, 1976.

27. L. Vieille. Recursive query processing: the power of logic. *Theoretical Computer Science*, 69:1–53, 1989.

Authors Index:

Lecture Notes in Computer Science

For information about Vols. 1–808
please contact your bookseller or Springer-Verlag